Lecture Notes in Control and Information Sciences

Edited by M. Thoma and A. Wyner

W0235175

92

Lj. T. Grujić, A. A. Martynyuk, M. Ribbens-Pavella

Large Scale Systems Stability under Structural and Singular Perturbations

Springer-Verlag
Berlin Heidelberg GmbH

Series Editors
M. Thoma · A. Wyner

Advisory Board
L. D. Davisson · A. G. J. MacFarlane · H. Kwakernaak
J. L. Massey · J. Stoer · Ya Z. Tsypkin · A. J. Viterbi

Authors
Ljubomir T. Grujić
Faculty of Mechanical Engineering
P.O. Box 174
27 Marta 80
11001 Belgrade
Yugoslavia

A. A. Martynyuk
Institute of Mathematics
Ukrainian Academy of Sciences
Repin Str. 3
252004 Kiew
USSR

M. Ribbens-Pavella
Unversité De Liège
Institute D'Electricité Montefiore
Circuits Electriques
Sart Tilman, B28
4000 Liège
Belgique

ISBN 978-3-540-18300-6 ISBN 978-3-540-47874-4 (eBook)
DOI 10.1007/978-3-540-47874-4

This work is subject to copyright. All rights are reserved, whether the whole or part of the material is concerned, specifically the rights of translation, reprinting, re-use of illustrations, recitation, broadcasting, reproduction on microfilms or in other ways, and storage in data banks. Duplication of this publication or parts thereof is only permitted under the provisions of the German Copyright Law of September 9, 1965, in its version of June 24, 1985, and a copyright fee must always be paid. Violations fall under the prosecution act of the German Copyright Law.

© Springer-Verlag Berlin Heidelberg 1987
Originally published by Springer-Verlag Berlin Heidelberg New York in 1987

2161/3020-543210

Lecture Notes in Control and Information Sciences

Edited by A. V. Balakrishnan and M. Thoma

For information about Vols. 1–21 please contact your bookseller or Springer-Verlag.

To Aleksandr Mikailovich Liapunov

(1857 - 1918)

PREFACE

This book constitutes an up to date presentation and development of stability theory in the Liapunov sense with various extensions and applications.

Precise definitions of well known and new stability properties are given by the authors who present general results on the Liapunov stability properties of non-stationary systems which are out of the classical stability theory framework.

The study involves the use of time varying sets and is broadened to time varying Lur'e-Postnikov systems and singularly perturbed systems.

A remarkable contribution is proposed by the authors who establish necessary and sufficient conditions, similar to Liapunov's one, for uniform absolute stability of time varying Lur'e-Postnikov systems.

Comparison systems and comparison principle are studied, in general and particular forms, and applied to large scale systems.

In that sense various forms of large-scale systems aggregation are studied and various stability criteria are established under different hypotheses : with invariant structure, with Lur'e-Postnikov form and with singularly perturbed properties. Proposed results are broadened to structural stability analysis aimed at studying stability properties under unknown and unpredictable structural variations. The criteria are developed both in algebraic and frequency domains. They essentially reduce the order and complexity of stability problems.

A number of various aggregation-decomposition forms are also considered for power systems from the large scale systems stand point. Precise definitions are introduced by the authors for various stability domains with application to large-scale systems in general and more specifically to power systems. Stability properties and domains of disturbed power systems are established.

A number of examples and applications presented throughout this book illustrate the various results.

According to the amount and importance of definitions and stability criteria presented I consider that this book initially published in Russian, represents the most complete one on stability theory proposed at this date. It interests all people concerned with stability problems in the largest sense and with security, reliability and robustness.

Professor Pierre BORNE
Lille, France

FOREWORD

Poincare's daring idea to obtain qualitative information on motion directly from the differential equation describing it, i.e. without integration, was realized by Liapunov (1892). With his absolute completeness and irreproachable strictness, Liapunov laid the foundations of a conceptually new approach to the qualitative methods of the theory of differential equations. Nowadays, Liapunov's methods are recognized to be among the most powerful means of stability analysis in exact sciences. These, along with the many extensions further developed, contributed to broaden substantially the classes of problems able of being effectively analyzed by the direct method.

The present book contains an essay of development of the general theory of stability in the sense of Liapunov, elements of the stability theory of comparison systems (systems of ordinary differential equations with monotonous right-hand parts), presentation of the general methods for the analysis of structural stability of large-scale systems, including systems with singular perturbations. The Liapunov functions (scalar, vector and matrix) and his direct method for the stability analysis of the unperturbed motion are used throughout the book. Some of the obtained stability results are applied to the analysis of large-scale electric power systems. The stability of these systems is a very important particular case for which the direct criteria show extremely useful.

The Russian version of this monograph was completed in 1982, the 125th anniversary of Liapunov's birthday. Since, new results of the authors have been added and included in the present version. More specifically Chapter V has been thoroughly revised and completed. Overall, this English version is more than a mere translation of the Russian one.

Our permanent concern has been to write up in a clear, easy to comprehend, way, readable for both engineers who need convenient mathematical machinery for large-scale system stability analysis, and mathematicians who are interested in new problems of the qualitative theory of differential equations.

We have tried to do justice to scientists who the first obtained results in various areas of the large-scale systems stability theory, and to refer to their original papers. It is reflected in the Bigliographies which include more than 400 references. Certainly, even such a list is still incomplete. This can be partly explained by the intensive research efforts and developments in the area, and by the extremely wide domains of its application, beginning with technology and finishing with the problems of populational dynamics. We apologize to all those whose work was not cited or properly described.

ACKNOWLEDGMENTS

Academicians Yu.A. Mitropolsky and Ye.F. Mishchenko, Associate Member of Academy of Sciences of the USSR, V.I. Zubov and Professor Yu.A. Ryabov have got acquainted with the Russian manuscript of the book. Their detailed remarks were extremely valuable. Many conversations of A.A. Martynyuk with Professor A.B. Zhishchenko greatly influenced the presentation of problems connected with the algebraic type of the obtained results.

Collaborators of the Processes Stability Department of the Institute of Mechanics of the Ukrainian Academy of Sciences, I.Yu. Lazareva, Ye.P. Shatilova have contributed much in the course of the technical work on the manuscript. Mrs. M.B. Counet-Lecomte did an outstanding job in typing the final English version. The quality of this camera-ready presentation owes enormously to her expertise.

The authors are cordially thankful to all of them.

Lj.T.G.	A.A.M.	M.R.P.
Belgrade	Kiev	Liège

September 1987.

CONTENTS

Chapter V
LARGE-SCALE POWER SYSTEMS STABILITY

LIST OF BASIC SYMBOLS

All symbols are fully defined at the place where they are first introduced. As a convenience to the reader we have collected some of the more frequently used symbols in several places. The largest collection is the one given below. Additional list for later use can be found in the introduction to Chapter V.

$\mathcal{A}, \mathcal{B}, \mathcal{C}, \ldots$	upper case boldface Script letters denote sets
$\mathcal{A} \setminus \mathcal{B}$	a difference between sets $\mathcal{A}$ and $\mathcal{B}$
$\mathcal{A} \cup \mathcal{B}$, $\mathcal{A} \cap \mathcal{B}$	union, intersection of sets $\mathcal{A}$ and $\mathcal{B}$
$\mathbf{A}, \mathbf{B}, \mathbf{C}, \ldots$	upper case boldface Gothic letters denote matrices with constant or functional entries except $\mathbf{V}$
$\mathbf{a}, \mathbf{b}, \mathbf{c}, \ldots$	lower case boldface Gothic letters denote vectors
$a, b, c, \ldots$	lower case letters denote scalars
$\mathcal{B}_{\Delta}(t_0) = \{\mathbf{x} : \Vert \mathbf{x} \Vert < \Delta(t_0)\}$	the hyperball with the center in the origin and radius equal to $\Delta(t_0)$
$C^{(i,j)}(\mathcal{T}_{\tau} \times \mathcal{N})$	the family of all functions i-times differentiable on $\mathcal{T}_{\tau}$ and j-times differentiable on $\mathcal{N}$
$C(\mathcal{T}_{\tau} \times \mathcal{N})$	the family of all functions continuous on $\mathcal{T}_{\tau} \times \mathcal{N}$
$C^{(i,j)}(\mathcal{T}_{\tau} \times \mathcal{N}, \mathcal{R}^k)$	the family of all functions mapping $\mathcal{T}_{\tau} \times \mathcal{N}$ into $\mathcal{R}^k$ which are in $C^{(i,j)}(\mathcal{T}_{\tau} \times \mathcal{N})$

$D^{+}v(t,\mathbf{x}) = \limsup\left\{\frac{v[t+\theta,\chi(t+\theta;t,\mathbf{x})]-v(t,\mathbf{x})}{\theta} : \theta\to 0^{+}\right\}$	the upper right-hand Dini derivative of v along χ at $(t,\mathbf{x})$
$D_{+}v(t,\mathbf{x}) = \liminf\left\{\frac{v[t+\theta,\chi(t+\theta;t,\mathbf{x})]-v(t,\mathbf{x})}{\theta} : \theta\to 0^{+}\right\}$	the lower right-hand Dini derivative of v along χ at $(t,\mathbf{x})$
$D^{*}v(t,\mathbf{x})$	denotes that both $D^{+}v(t,\mathbf{x})$ and $D_{+}v(t,\mathbf{x})$ can be used
$d(\mathbf{x},A) = \inf[\Vert\mathbf{x}-\mathbf{y}\Vert : \mathbf{y}\in A]$	a distance from $\mathbf{x}$ to A
$d(A,B) = \max\{\sup[d(\mathbf{x},A): \mathbf{x}\in B], \sup[d(\mathbf{x},B): \mathbf{x}\in A]\}$	a distance between A and B
$\mathbf{f}: R\times R^{n}\to R^{n}$	a vector function mapping $R\times R^{n}$ into R^{n}
I_{k}	the $k\times k$ identity matrix
$\mathrm{He}(\cdot)$	the Hermitian part of a matrix $(\cdot)$
$i,j,k,\dots,N$	integers
$j = \sqrt{-1}$	the imaginary unit
$K_{[0,\xi]}$	the class of comparison functions on $[0,\xi]$
N	a time-invariant neighbourhood of the origin of R^{n}, or the set of the first N natural numbers : $N = \{1,2,\dots,N\}$
$N(t)$	a neighbourhood of the origin at $t\in R$
$N_{\tau} = \{(t,\mathbf{x}): t\in T_{\tau}, \mathbf{x}\in N(t)\}$	a neighbourhood of the origin of $R\times R^{n}$ over T_{τ}
$N = \{(t,\mathbf{x}): t\in R, \mathbf{x}\in N(t)\}$	a neighbourhood of the origin of $R\times R^{n}$
$O = \{\mathbf{x}: \mathbf{x}=\mathbf{0}\}$	the singleton containing the origin of R^{n}
R	the set of all real numbers
$R_{+} = [0,+\infty[\subset R$	the set of all non-negative numbers
$\mathring{R}_{+} =]0,+\infty[$	the set of all positive real numbers

R^k	k-th dimensional real vector space
$R\times R^n$	the cartesian product of R and R^n
S	a time-invariant subset of R^n
$S(t)$	a time-varying subset of R^n
S_s	a structural set of a system defining all system structural variations via structural matrices $\mathbf{S}$
$\mathcal{S}_\tau = \{(t,\mathbf{x}): t\in T_\tau, \mathbf{x}\in S(t)\}$	a subset of $T_\tau\times R^n$ associated with $S(t)$
$\mathcal{S} = \{(t,\mathbf{x}): t\in R, \mathbf{x}\in S(t)\}$	a subset of $R\times R^n$ associated with $S(t)$, $\mathcal{S}_\tau=\mathcal{S}$ iff $T_\tau=R$
$\operatorname{sign}\zeta = \zeta\|\zeta\|^{-1}$ iff $\zeta\neq 0$ and $\operatorname{sign} 0 = 0$	the signum type nonlinearity
$\bar{T}_o = [t_o,+\infty] = \{t: t_o\leq t\leq +\infty\}$	the largest time interval beginning with t_o
$T = [-\infty,+\infty] = \{t: -\infty\leq t\leq +\infty\}$	the largest time interval
$T_o^* =]t_o,+\infty] = \{t: t_o< t\leq +\infty\}$	the left semi-open unbounded time interval associated with t_o
$T_o = [t_o,+\infty[= \{t: t_o\leq t< +\infty\}$	the right semi-open unbounded time interval associated with t_o
$T_\tau = [\tau,+\infty[= \{t: \tau\leq t< +\infty\}$, $\tau\in R$	the right semi-open unbounded time interval associated with τ
$T_i\subseteq R$	a time interval of all initial moments t_o under consideration (or, of all admissible t_o)
$t\in R$	a time variable, instant
$t_o\in R$	an initial instant
$V_\zeta(t)$	is the largest connected neighbourhood of $\mathbf{x}=0$ associated with a positive definite v such that $v(t,\mathbf{x})<\zeta$, $\forall \mathbf{x}\in V_\zeta(t)$
$\mathbf{x}(t)$	a state vector of a system at $t\in R$, $\mathbf{x}=(x_1,x_2,\dots,x_n)^T$
$\alpha,\beta,\gamma,\dots$	Greek letters denote scalars unless otherwise specified

$\Delta_M(t_0) = \text{Max}\,\{\Delta : \Delta=\Delta(t_0), \forall \rho>0, \forall x_0 \in B_\Delta, \exists\, \tau(t_0,x_0,\rho) \in]0,+\infty[, \ni \chi(t;t_0,x_0) \in B_\rho, \forall t \in T_\tau\}$	the maximal Δ obeying the definition of attractivity
$\delta_M(t_0,\epsilon) = \text{Max}\,\{\delta : \delta=\delta(t_0,\epsilon) \ni x_0 \in B_\delta(t_0,\epsilon) \Rightarrow \chi(t;t_0,x_0) \in B_\epsilon, \forall t \in T_0\}$	the maximal δ obeying the definition of stability
∂S	the boundary of a set S
$\bar{S}$	the closure of a set S
ϕ	empty set
$\tau_m(t_0,x_0,\rho) = \text{Min}\,\{\tau : \tau=\tau(t_0,x_0,\rho) \ni \chi(t;t_0,x_0) \in B_\rho, \forall t \in T_\tau\}$	the minimal τ satisfying the definition of attractivity
$\lambda_i(\cdot)$	the i-th eigenvalue of a matrix $(\cdot)$
$\Lambda_M(\cdot)$	the maximal eigenvalue of a matrix $(\cdot)$
$\lambda_m(\cdot)$	the minimal eigenvalue of a matrix $(\cdot)$
$\chi(t;t_0,x_0)$	a motion of a system at $t \in R$ iff $x(t_0)=x_0$, $\chi(t_0;t_0,x_0) \equiv x_0$
$\Rightarrow$	"implies"
$\Leftrightarrow$	"iff" ("if and only if")
$\forall$	"for every"
$\exists$	"there exist(s)"
$\nexists$	"there does (do) not exist"
$\ni$	"such that"
$\in$	"belongs to"
$\| \ \|$	the Euclidean norm
[ ]	denotes a closed interval
] [	denotes an open interval
( )	a general interval which can be semi-open, open, or closed.

CHAPTER I

OUTLINE OF THE LIAPUNOV STABILITY THEORY IN GENERAL

I.1. INTRODUCTORY COMMENTS

In the late years of the XIXth century H. Poincaré (1881-1882) started qualitative analysis of non-linear differential equations considering those of the second order. His ideas inspired A.M. Liapunov (1892, p.8) for more general qualitative analysis of the non-linear differential equations which resulted in a genious scientific contribution presented in his famous fundamental dissertation (Liapunov, 1892). Among numerous basic contributions the following are in particular important for what will be considered in this book.

A.M. Liapunov initiated qualitative study of relationships between perturbed (real) motions and an unperturbed (reference, desirable, nominal) motion. Moreover, he considered behaviour of suitable functions along the motions. These functions can represent different types of energy and/or material flow, which illustrates and emphasizes physical importance of the problem. Relying on the concept of continuity of functions, A.M. Liapunov naturally introduced a general definition of stability of the reference motion with respect to some functions requiring $\epsilon-\delta$ closeness of the values of these functions taken along perturbed motions and the reference motion. Although A.M. Liapunov did not explicitely define either attraction or asymptotic stability of the reference motion (with or without respect to some functions) in his dissertation , it has been often accepted [see N.G. Chetaev (1946), J.L. Massera (1949,1956), E.A. Barbashin and N.N. Krasovskii (1952), I.G. Malkin (1954), H.A. Antosiewvicz (1958), T. Yoshizawa (1966) and N. Rouche, P. Habets and M. Laloy (1977)] to associate the notion of asymptotic stability with Liapunov's dissertation. The reason for it is likely following Liapunov's remark on his theorem on stability of the reference motion (Liapunov, 1892, p. 61) :

> "*Remark 2.* If a function V, satisfying the conditions of the theorem, admits simultaneously an infinitely small upper limit, and its derivative is a sign-definite function, then it can be proved that every perturbed motion, sufficiently close to the unperturbed one, will approach it asymptotically."

Altogether, A.M. Liapunov founded a general concept of stability of motions, which has been becoming more and more important for dynamic analysis in mathematics, mechanics, automatic control, systems theory, ecology, biology, economics etc.

Another remarkable Liapunov's contribution is his second method that is also called Liapunov's direct method [see Lasalle and Lefschetz (1961)]. The aim of this method is a stability problem solution via system state differential equation *without use of its solutions*. Information about a stability property is to be deduced directly from the differential equation despite the corresponding stability property definition is phraised in terms of system motions. A.M. Liapunov established the concept of semi-definite and definite functions and used them as a keystone of his second method. Then, A.M. Liapunov transferred stability problems to the problem of behaviour of a corresponding definite function along system motions. In order to solve this problem A.M. Liapunov considered the sign of the total time derivative of the chosen function along system motions. The sign has to be tested at all points belonging to a neighbourhood of the reference motion. The precise requirements on the function and the sign of its total time derivative along system motions were established by A.M. Liapunov for (uniform) stability and (uniform) asymptotic stability of the zero state representing the reference motion.

I.2. ON DEFINITION OF STABILITY PROPERTIES IN LIAPUNOV's SENSE

I.2.1. Liapunov's original definition

A.M. Liapunov started his investigations with the following [A.M. Liapunov (1892), p. 12] :

> "**1.** Let us consider any material system with k degrees of freedom. Let
>
> $$q_1, q_2, \dots, q_k$$
>
> be k independent variables, which we use to determine its position.

We shall assume that quantities taking real values for all real system positions are taken for such variables.

Considering the mentioned variables as functions in time t we shall denote their first time derivatives by

$$q_1', q_2', \dots, q_k' .$$

In every dynamic problem, in which forces are prespecified in certain way, such functions will satisfy some k second order differential equations.

Let any particular solution for such equations be found

$$q_1 = f_1(t) , q_2 = f_2(t) , \dots , q_k = f_k(t) ,$$

in which the quantities q_j are expressed as real functions in t , which at every t give only possible values to them.[1]

To that particular solution will correspond a definite motion of our system. Comparing it in a known sense with others, which are possible under the same forces, we shall call that motion *unperturbed*, and all others, with which it is compared, *perturbed*.

Understanding by t_o a given instant, let us denote the values corresponding to it of quantities q_j, q_j' along any motion with q_{jo}, q_{jo}' .

Let

$$q_{10} = f_1(t_o)+\epsilon_1 , q_{20} = f_2(t_o)+\epsilon_2 , \dots , q_{ko} = f_k(t_o)+\epsilon_k ,$$

$$q_{10}' = f_1'(t_o)+\epsilon_1' , q_{20}' = f_2'(t_o)+\epsilon_2' , \dots , q_{ko}' = f_k'(t_o)+\epsilon_k' ,$$

where ϵ_j, ϵ_j' are real-valued constants.

Prespecifying the constants, which will be called *perturbations*, a perturbed motion is determined. We shall assume that we may prescribe them every number sufficiently small.

By speaking about perturbed motions, *close* to the unperturbed one, we shall comprehend motions, for which the perturbations are numerically sufficiently small.

Let $Q_1, Q_2, \dots, Q_n$ be any given continuous real-valued functions of quantities

$$q_1, q_2, \dots, q_k, q_1', q_2', \dots, q_k' .$$

Along the unperturbed motion they become known functions of t , which will be denoted by $F_1, F_2, \dots, F_n$. Along a perturbed motion they will be functions of quantities

---

[1] It can happen that the quantities q_j by their choice do not take all real values but only those not greater than - and not less than certain bounds.

$$t, \epsilon_1, \epsilon_2, \ldots, \epsilon_k, \epsilon_1', \epsilon_2', \ldots, \epsilon_k' .$$

When all ϵ_j, ϵ_j' are equal to zero, then the quantities

$$Q_1 - F_1 , Q_2 - F_2 , \ldots , Q_n - F_n$$

will be equal to zero for every t. However, if the constants ϵ_j, ϵ_j' are not zero, but all are infinitely small, then a question rises : is it possible to specify such infinitely small bounds on the quantities $Q_s - F_s$ that the latter never become grater than their values ?

A solution of the question, which is the topic of our investigations, depends on both a character of the considered unperturbed motion and a choice of the functions $Q_1, Q_2, \ldots, Q_n$ and the instant t_o. Under a specific choice of the latter, the reply to the question, respectively, will characterize in some sense the unperturbed motion, by determining a feature of the latter, which will be called *stability*, or that contrary to it, will be called *instability*.

We shall be exclusively interested in those cases in which the solution of the considered question does not depend on a choice of the instant t_o, when perturbations are acting. Thus we accept herein the following definition.

Let $L_1, L_2, \ldots, L_n$ *be arbitrary given positive numbers. If for all* L_s *, nevertheless how small they are, can be selected positive numbers* $E_1, E_2, \ldots, E_k, E_1', E_2', \ldots, E_k'$ *so that for all real* ϵ_j, ϵ_j' *, satisfying the conditions*

$$|\epsilon_j| \leq E_j , |\epsilon_j'| \leq E_j' \quad (j = 1, 2, \ldots, k) ,$$ [1]

and for all t *, greater than* t_o *, the inequalities*

$$|Q_1 - F_1| < L_1 , |Q_2 - F_2| < L_2 , \ldots , |Q_n - F_n| < L_n ,$$

are satisfied then the unperturbed motion is stable with respect to the quantities $Q_1, Q_2, \ldots, Q_n$ *; otherwise - unstable* with respect to the same quantities*.*"

[1] In general $|x|$ means the absolute value of a real-, or modulus of a complex quantity x.

The Remark 2 on p. 61 in Liapunov (1892) reproduced on p.2 of our § I.1 has been commonly used as a basis for notions of attraction and asymptotic stability of the reference motion.

I.2.2. Comments on Liapunov's original definition

Comment 1. The inequalities on $|\epsilon_j|$ and $|\epsilon_j'|$ are weak and those on $|Q_j-F_j|$ are strong. This unsymmetry is usually avoided by imposing the same type of inequalities on all $|\epsilon_j|$, $|\epsilon_j'|$ and $|Q_j-F_j|$, which yields stability definitions equivalent to Liapunov's original definition. This equivalence can be easily proved.

Comment 2. Stability of the reference motion was defined by A.M. Liapunov with respect to arbitrary functions Q_j that are continuous in all q_i, q_i'. This has been very thoughtful and physically important because Q_j can represent energy or material flow. In this connection A.M. Liapunov introduced new variables x_i,

$$x_i = Q_i - F_i \ , \quad i=1,2,\dots,n \ ,$$

and accepted the following (Liapunov, 1892, p. 15) :

> "We shall assume that the number n and the functions Q_s, are such, that the order of the system is n and that it is reducible to the normal form
>
> $$\frac{dx_1}{dt} = X_1 \ , \ \frac{dx_2}{dt} = X_2 \ , \dots , \ \frac{dx_n}{dt} = X_n \ , \qquad (1)$$
>
> and everywhere in the sequel we shall consider these last equations, calling them the differential equations of a perturbed motion.
>
> All X_s in the equations (1) are known functions of quantities
>
> $$x_1, x_2, \dots, x_n, t \ ,$$
>
> vanishing for
>
> $$x_1 = x_2 = \dots = x_n = 0 \ . \text{ "}$$

Comment 3. Stability of the reference motion requires arbitrary closeness of the perturbed motions to the reference motion provided their sufficient closeness is assured at the initial instant t_o.

Comment 4. The closeness of the perturbed motions to the reference motion is to be realized over unbounded time interval $T_o^* =]t_o,+\infty]$, i.e. for *all* t *greater* than t_o. This point has been commonly neglected in the literature. Namely, the closeness has been commonly required either on $\bar{T}_o = [t_o,+\infty]$ or on $T_o = [t_o,+\infty[$, i.e. for *all* t *not less* than t_o. This difference can be crucial in cases when system motions are discontinuous at $t = t_o$.

Comment 5. A.M. Liapunov defined stability of the reference motion for cases when it is not influenced by t_o. However, the initial moment can essentially influence stability of the reference motion in cases when system motions are not continuous in t. Besides, t_o can essentially influence the maximal admissible values of all E_j and E_j' even when all system motions are continuous in t.

Comment 6. The stability of the reference motion was defined by A.M. Liapunov with respect to initial perturbations of the general coordinates q_j, q_j', rather than with respect to persistent external disturbances.

Comment 7. The stability definition does not care about the values E_j and E_j' except that they must be positive. Hence, for large values of all L_j, the maximal admissible E_j and E_j' can be so small that they are not useful for engineering needs.

I.2.3. Relationship between the reference motion and the zero solution

Let $2k$ be the order of the system and y_i, $i=1,2,\dots,2k$, be its i-th state variable. Using the basic physical laws (e.g. the law of the energy conservation and the law of the material conservation) we can for a large class of systems get state differential equations in the following scalar form :

$$\frac{dy_i}{dt} = Y_i(t,y_1,\dots,y_{2k}) \ , \quad i=1,2,\dots,2k \ , \tag{1}$$

or in the equivalent vector form

$$\frac{d\mathbf{y}}{dt} = \mathbf{Y}(t,\mathbf{y}) \tag{2}$$

where * $\mathbf{y} = (y_1,y_2,\dots,y_{2k})^T \in \mathbf{R}^{2k}$ and $\mathbf{Y} = (Y_1,Y_2,\dots,Y_{2k})^T$, $\mathbf{Y} : T\times\mathbf{R}^{2k} \to \mathbf{R}^{2k}$. A motion of (2) is denoted by $\eta(t;t_o;\mathbf{y}_o)$, $\eta(t_o;t_o;\mathbf{y}_o) \equiv \mathbf{y}_o$, and the reference motion $\eta_r(t;t_o;\mathbf{y}_{ro})$. From the physical point of view the reference motion should be realizable by the system. From the mathematical point of view this means that the reference motion is a solution of (2),

$$\frac{d}{dt}\,\eta_r(t;t_o;\mathbf{y}_{ro}) \equiv \mathbf{Y}[t,\eta_r(t;t_o;\mathbf{y}_{ro})] \ . \tag{3}$$

Let the Liapunov transformation of coordinates be used,

$$\mathbf{x} = \mathbf{y}-\mathbf{y}_r \ , \tag{4}$$

where $\mathbf{y}_r(t) \equiv \eta_r(t;t_o;\mathbf{y}_{ro})$. Let $\mathbf{f} : T\times\mathbf{R}^{2k} \to \mathbf{R}^{2k}$ be defined by

$$\mathbf{f}(t,\mathbf{x}) = \mathbf{Y}[t,\mathbf{y}_r(t)+\mathbf{x}] - \mathbf{Y}[t,\mathbf{y}_r(t)] \ . \tag{5}$$

It is evident that

$$\mathbf{f}(t,\mathbf{0}) \equiv \mathbf{0} \ . \tag{6}$$

---

* In Liapunov's notation $\mathbf{y} = (q_1,q_2,\dots,q_k,q_1',q_2',\dots,q_k')^T$.

Now (2)-(5) yield

$$\frac{dx}{dt} = f(t,x) . \tag{7}$$

In this way, the behaviour of perturbed motions related to the reference motion (in total coordinates) is represented by the behaviour of the state deviation x with respect to the zero state deviation. The reference motion in the total coordinates y_i is represented by the zero deviation $x=0$ in state deviation coordinates x_i . With this in mind, the following result emphasizes complete generality of both Liapunov's second method and results presented in A.M. Liapunov (1892) for the system (7). Let $Q : R^{2k} \to R^n$, $n=2k$ is admissible but not required.

Theorem 1. *Stability of* $x=0$ *of the system (7) with respect to* $Q=x$ *is necessary and sufficient for stability of the reference motion* η_r *of the system (2) with respect to every vector function* Q *that is continuous in* y .

Proof. Necessity. This part is true because $Q(y) = y$ is continuous in y and evidently stability of $x=0$ with respect to x is implied by stability of η_r with respect to $Q(y) = y$.
Sufficiency. Let $L_i>0$, $i=1,2,\dots,n$,be arbitrarily chosen. Continuity of Q in y implies existence of $\ell_i>0$, $\ell_i = \ell_i(L,y_r)$, $L = (L_1,L_2,\dots,L_n)^T$, $i=1,2,\dots,2k$, such that $|y_i-y_{ri}| < \ell_i$, $\forall i=1,2,\dots,2k$, implies $|Q_i(y)-Q_i(y_r)| < L_i$, $i=1,2,\dots,n$. Stability of $x=0$ of (7) (with respect to x) guarantees existence of $\delta_i>0$, $\delta_i=\delta_i(\boldsymbol{\ell})$, $\boldsymbol{\ell} = (\ell_1,\ell_2,\dots,\ell_{2k})^T$, such that $|x_{io}|<\delta_i$, $i=1,2,\dots,2k$, implies $|x_i(t;t_o;x_{io})| < \ell_i$, $\forall t\ge t_o$, $\forall t_o \in R$, $i=1,2,\dots,2k$, where $\chi(t;t_o;x_o)$, $\chi(t_o;t_o;x_o) \equiv x_o$, is the solution of (7), $\chi = (\chi_1,\chi_2,\dots,\chi_{2k})^T$. Finally, for every $L_i>0$, $i=1,2,\dots,n$, there is $\delta_j^*>0$, $\delta_j^* = \frac{1}{2}\delta_j$, $j=1,2,\dots,n$, such that $|y_{jo}-y_{rjo}| \le \delta_j^*$, $j=1,2,\dots,n$, implies

$$|Q_i[\eta(t;t_o;y_o)] - Q_i[\eta_r(t;t_o;y_{ro})]| < L_i , \quad \forall t\ge t_o , \quad i=1,2,\dots,n . \quad \blacksquare$$

This theorem reduces the problem of the stability of the reference motion of (2) with respect to Q to the stability problem of $x=0$ of (7) with respect to x ; it is stated and proved herein for the first time.

I.2.4. Accepted definitions of stability properties in Liapunov's sense

By the very definition, stationary (time-invariant) systems are those whose motions are not effected by (the choice of) the initial instant $t_o \in R$. However, such property is not characteristic for non-stationary (time-varying) systems. It is therefore natural to consider influence of t_o on stability properties of non-stationary systems, which is motivation for accepting the next definitions based on those by A.M. Liapunov (1892), K.P. Persidskii (1933), N.G. Chetaev (1946), J.L. Massera (1949,1956), E.A. Barbashin and N.N. Krasovskii (1952), I.G. Malkin (1954), H.A. Antosiewicz (1958) [see also N.N. Krasovskii (1959), R.E. Kalman and J.E. Bertram (1960), V.V. Nemytskii and V.V. Stepanov (1960)], V.I. Zubov (1964) [see also W.A. Coppel (1965), T. Yoshizawa (1966), N.P. Bhatia and G.P. Szegö (1967)], B.P. Demidovich (1967), W. Hahn (1967) [see also D.D. Šiljak (1969,1974), E.A. Barbashin (1970)], K.S. Narendra and J.H. Taylor (1973), Lj.T. Grujić (1975,1977) [see also N. Rouche, P. Habets and M. Laloy (1977)]. For the historical reviews of the development of stability definitions until 1975 see T. Yoshizawa (1966) and N. Rouche, P. Habets and M. Laloy (1977).

Definition 1. The state $\mathbf{x}=\mathbf{0}$ of the system (7) is :

(i) *stable with respect to* T_i iff for every $t_o \in T_i$ and every $\epsilon \in \overset{o}{R}_+$ there exists $\delta(t_o,\epsilon) > 0$, such that $\|\mathbf{x}_o\| < \delta(t_o,\epsilon)$ implies

$$\|\chi(t;t_o;\mathbf{x}_o)\| < \epsilon \quad , \ \forall t \in T_o$$

(ii) *uniformly stable with respect to* T_i iff both (i) holds and for every $\epsilon \in \overset{o}{R}_+$ the corresponding maximal δ_M obeying (i) satisfies

$$\inf[\delta_M(t,\epsilon) : t \in T_i] > 0 \ ;$$

(iii) *stable in the whole with respect to* T_i iff both (i) holds and

$$\delta_M(t,\epsilon) \to +\infty \quad \text{as} \quad \epsilon \to +\infty \ , \ \forall t \in T_i \ ;$$

(iv) *uniformly stable in the whole with respect to* T_i iff both (ii) and (iii) hold.

(v) *unstable with respect to* T_i iff there are $t_o \in T_i$, $\epsilon \in]0,+\infty[$ and $\tau \in T_o$, $\tau > t_o$, such that for every $\delta \in]0,+\infty[$ there is $\mathbf{x}_o$, $\|\mathbf{x}_o\| < \delta$, for which

$$\|\chi(\tau;t_o,\mathbf{x}_o)\| \geq \epsilon \ .$$

(vi) The expression "*with respect to* T_i " is omitted from (i)-(v) iff $T_i = R$. ■

These stability properties hold as $t \to +\infty$ but not for $t = +\infty$.

Example 1 (Lj.T. Grujić 1975). Let $x \in R$ and $\frac{dx}{dt} = \frac{1}{1-t}x$. Then,

$$\chi(t;t_o;x_o) = (t-1)^{-1}(t_o-1)\,x_o \quad \text{for} \quad t_o \neq 1 \quad \text{and} \quad t \neq 1 \ .$$

For $t_o = 1$ the motion is not defined and

$$|\chi(t;t_o;x_o)| \to +\infty \quad \text{as} \quad t \to (1-0) \ , \ \forall t_o \in]-\infty,1[\ , \ \forall (x_o \neq 0) \in R \ .$$

Hence,

$$\delta_M(t,\epsilon) = 0 \ , \ \forall\epsilon \in \overset{o}{R}_+ , \ \forall t \in \,]-\infty,1] \ .$$

However,

$$\delta_M(t,\epsilon) = \epsilon \ , \ \forall t \in \,]1,+\infty[\ .$$

The state $x=0$ is uniformly stable in the whole with respect to every $T_i \subseteq \,]-1,+\infty[$, but it is not stable.

Example 2 (Lj.T. Grujić 1975). The first order non-stationary system is defined by

$$\frac{dx}{dt} = \frac{(1 + t\sin t + t^2\cos t)\, x \cdot \exp(-\frac{1}{2}\pi)}{\frac{1}{2}\pi \cdot \exp(-t\sin t) + t \cdot \exp(-\frac{1}{2}\pi)}$$

Solutions are found in the form

$$\chi(t;t_o;x_o) = \frac{\frac{1}{2}\pi + t_o \exp(-\frac{1}{2}\pi + t_o \sin t_o)}{\frac{1}{2}\pi + t\exp(-\frac{1}{2}\pi + t\sin t)}\, x_o \ , \ t_o \neq -\frac{\pi}{2} \ , \ t \neq -\frac{\pi}{2} \ ,$$

so that

$$|\chi(t;t_o;x_o)| \to +\infty \quad \text{as} \quad t \to (-\frac{\pi}{2} - 0) \ , \ \forall t_o \in \,]-\infty,-\frac{\pi}{2}[\ , \ \forall (x_o \neq 0) \in R \ .$$

This result and analysis of $\chi(t;t_o;x_o)$ yield

$$\delta_M(t,\epsilon) = \begin{cases} 0 & t \in \,]-\infty,-\frac{\pi}{2}] \\ \epsilon & t \in \,]-\frac{\pi}{2},0] \\ \epsilon\pi\,[\pi + 2t\exp(-\frac{\pi}{2} + t\sin t)]^{-1} \ , & t \in [0,+\infty[\ . \end{cases}$$

The state $x=0$ is stable in the whole with respect to $]-\frac{\pi}{2},+\infty[$ and uniformly stable in the whole with respect to every bounded $T_i \subset \,]-\frac{\pi}{2},+\infty[$, but it is not stable.

In these examples, the motions χ are not continuous in all $t \in R$.

Proposition 1. *If there is a time-invariant neighbourhood $N \subseteq R^n$ of $x=0$ such that $\chi(t;t_o;x_o)$ is continuous in $(t,t_o,x_o) \in T_o \times R \times N$ then stability of $x=0$ of the system (7) with respect to some non-empty T_i implies its stability.*

This result can be easily proved as well as the following :

Proposition 2. *If $x=0$ of (7) is stable (in the whole) then, respectively, it is uniformly stable (in the whole) with respect to every bounded $T_i \subset R$.*

Example 3 (Lj.T. Grujić 1975). Solutions of the first order non-stationary system

$$\frac{dx}{dt} = -\frac{\beta + 2\gamma t}{\alpha + \beta t + \gamma t^2}\, x \ , \ \alpha > 0 \ , \ \beta^2 < 4\alpha\gamma \ , \ \gamma > 0$$

are given by

$$\chi(t;t_o;x_o) = (\alpha + \beta t_o + \gamma t_o^2)(\alpha + \beta t + \gamma t^2)^{-1} x_o .$$

In this case

$$\delta_M(t,\epsilon) = \frac{(4\alpha\gamma - \beta^2)\epsilon}{8\gamma(\alpha + \beta t + \gamma t^2)} [1 - \text{sign}(t + \frac{\beta}{2\gamma})] + \frac{\epsilon}{2}[1 + \text{sign}(t + \frac{\beta}{2\gamma})]$$

Hence,

$$\inf[\delta_M(t,\epsilon) : t \in R] = 0 , \quad \forall \epsilon \in]0,+\infty[,$$

and

$$\delta_M(t,\epsilon) \to +\infty \quad \text{as} \quad \epsilon \to +\infty , \quad \forall t \in R$$

The state $\mathbf{x}=0$ is stable in the whole but not uniformly.

However, it is uniformly stable in the whole with respect to $T_i = [\zeta,+\infty[$ for any $\zeta \in]-\infty,+\infty[$.

Definition 2. The state $\mathbf{x}=0$ of the system (7) is :

(i) *attractive with respect to* T_i iff for every $t_o \in T_i$ there exists $\Delta(t_o) > 0$ and for every $\zeta > 0$ there exists $\tau(t_o;\mathbf{x}_o,\zeta) \in [0,+\infty[$ such that $\|\mathbf{x}_o\| < \Delta(t_o)$ implies $\|\chi(t;t_o;\mathbf{x}_o)\| < \zeta$, $\forall t \in]t_o + \tau(t_o,\mathbf{x}_o,\zeta),+\infty[$;

(ii) $\mathbf{x}_o$ *-uniformly attractive with respect to* T_i iff both (i) is true and for every $t_o \in T_i$ there exists $\Delta(t_o) > 0$ and for every $\zeta \in]0,+\infty[$ there exists $\tau_u[t_o,\Delta(t_o),\zeta] \in [0,+\infty[$ such that

$$\sup[\tau_m(t_o,\mathbf{x}_o,\zeta) : \mathbf{x}_o \in B_\Delta(t_o)] = \tau_u[t_o,\Delta(t_o),\zeta] ;$$

(iii) t_o *-uniformly attractive with respect to* T_i iff (i) is true, there is $\Delta > 0$ and for every $(\mathbf{x}_o,\zeta) \in B_\Delta \times]0,+\infty[$ there exists $\tau_u(T_i,\mathbf{x}_o,\zeta) \in [0,+\infty[$ such that

$$\sup[\tau_m(t_o;\mathbf{x}_o;\zeta) : t_o \in T_i] = \tau_u(T_i,\mathbf{x}_o,\zeta) ;$$

(iv) *uniformly attractive with respect to* T_i iff both (ii) and (iii) hold, that is, that (i) is true, there exists $\Delta > 0$ and for every $\zeta \in]0,+\infty[$ there is $\tau_u(T_i,\Delta,\zeta) \in [0,+\infty[$ such that

$$\sup[\tau_m(t_o,\mathbf{x}_o,\zeta) : (t_o,\mathbf{x}_o) \in T_i \times B_\Delta] = \tau_u(T_i,\Delta,\zeta) .$$

(v) The properties (i)-(iv) hold *"in the whole"* iff (i) is true for every $\Delta(t_o) \in]0,+\infty[$ and every $t_o \in T_i$.

(vi) The expression *"with respect to* T_i *"* is omitted iff $T_i = R$. ■

Example 4. For the system of Example 1 the following are found :

$$\Delta_M(t) = \begin{cases} 0 , & t \in]-\infty,1[\\ +\infty , & t \in]1,+\infty[\end{cases}$$

$$\tau_m(t,x,\zeta) = \begin{cases} +\infty , & t \in]-\infty,1[\\ \frac{t-1}{\zeta}|x| + 1 , & t \in]1,+\infty[\end{cases} .$$

The state $x=0$ is :

a) attractive in the whole with respect to $T_i =]1,+\infty[$,

b) t_o - uniformly attractive in the whole with respect to any bounded $T_i \subset]1,+\infty[$,

c) x_o - uniformly attractive with respect to $T_i =]1,+\infty[$,

d) uniformly attractive with respect to any bounded $T_i \subset]1,+\infty[$,

e) not attractive.

The next results can be easily verified.

Proposition 3. *If there is a time-invariant neighbourhood $N \subseteq R^n$ of $x=0$ such that $x(t;t_o;x_o)$ is continuous in $(t,t_o,x_o) \in T_o \times R \times N$ then attraction of $x=0$ of the system (7) with respect to some non-empty T_i implies its attraction.*

Proposition 4. *If $x=0$ of (7) is attractive then it is uniformly attractive with respect to every bounded $T_i \subset R$.*

Example 5 (Lj.T. Grujić 1975). We consider the system of Example 3 once again and find :

$$\inf[\Delta_M(t) : t \in R] = +\infty ,$$

$$\tau_m(t,\Delta,\zeta) = \begin{cases} \max[0 , (2\gamma)^{-1}\{[\beta^2 - 4\alpha\gamma + 4\gamma\zeta^{-1} \cdot \Delta(\alpha+\beta t+\gamma t^2)]^{1/2} - \beta\} \\ \qquad \text{for} \quad \Delta \geq (4\alpha\gamma - \beta^2)\,\zeta\,[4\gamma(\alpha+\beta t+\gamma t^2)]^{-1} , \\ 0 , \text{ for} \quad \Delta < (4\alpha\gamma - \beta^2)\,\zeta\,[4\gamma(\alpha+\beta t+\gamma t^2)]^{-1} . \end{cases}$$

Hence,

$$\sup[\tau_m(t,\Delta,\zeta) : t \in R] = +\infty \quad \text{for} \quad \Delta \geq (4\alpha\gamma - \beta^2)\,\zeta\,[4\gamma(\alpha+\beta t+\gamma t^2)]^{-1} .$$

The state $x=0$ is :

a) attractive in the whole,

b) x_o - uniformly attractive in the whole,

c) t_o - uniformly attractive in the whole with respect to any bounded $T_i \subset R$,

d) uniformly attractive in the whole with respect to any bounded $T_i \subset R$,

e) not uniformly attractive.

Definition 3. The state $x=0$ of the system (7) is :

(i) *asymptotically stable with respect to* T_i iff it is both stable with respect to T_i and attractive with respect to T_i ;

(ii) *equi-asymptotically stable with respect to* T_i iff it is both stable with respect to T_i and x_o - uniformly attractive with respect to T_i ;

(iii) *quasi-uniformly asymptotically stable with respect to* T_i iff it is both uniformly stable with respect to T_i and t_o - uniformly attractive with respect to T_i ;

(iv) *uniformly asymptotically stable with respect to* T_i iff it is both uniformly stable with respect to T_i and uniformly attractive with respect to T_i ;

(v) the properties (i)-(iv) hold *"in the whole"* iff both the corresponding stability of $\mathbf{x}=\mathbf{0}$ and the corresponding attraction of $\mathbf{x}=\mathbf{0}$ hold in the whole;

(vi) *exponentially stable with respect to* T_i iff there are $\Delta>0$ and real numbers $\alpha\geq 1$ and $\beta>0$ such that $\|\mathbf{x}_o\|<\Delta$ implies

$$\|\chi(t;t_o;\mathbf{x}_o)\| \leq \alpha\|\mathbf{x}_o\| \exp[-\beta(t-t_o)] \ , \ \forall t\in T_o \ , \ \forall t_o\in T_i \ .$$

This holds *in the whole* iff it is true for $\Delta=+\infty$.

(vii) The expression *"with respect to* T_i *"* is omitted iff $T_i=R$. ∎

Example 6 (Lj.T. Grujić 1975). The second order system is described by

$$\frac{d\mathbf{x}}{dt} = A(t)\,\mathbf{x} \ , \quad A(t) = \frac{1}{1+t^2}\begin{bmatrix} -t \ , & 1 \\ -1 \ , & -t \end{bmatrix} ,$$

and its solutions are found in the form

$$\chi(t;t_o;\mathbf{x}_o) = \frac{1}{1+t^2}\begin{bmatrix} 1+t_o t \ , & t-t_o \\ t_o - t \ , & 1+t_o t \end{bmatrix}\mathbf{x}_o \ .$$

Hence,

$$\delta_M(t,\epsilon) = \frac{\epsilon}{2}[1+(1+t^2)^{-1}\cdot(1-\operatorname{sign} t)+\operatorname{sign} t] \ ,$$

which implies

$$\inf[\delta_M(t,\epsilon): t\in R] = 0 \ , \ \forall\epsilon\in\,]0,+\infty[\ ,$$

and

$$\tau_m(t,\|\mathbf{x}\|,\zeta) = \begin{cases} [(1+t^2)^{1/2}\|\mathbf{x}\|\zeta^{-1}-1]^{1/2} \ , & \text{for } \|\mathbf{x}\|\geq\zeta(1+t^2)^{-1/2} \ , \\ 0 \ , \ \text{for } 0<\|\mathbf{x}\|\leq\zeta(1+t^2)^{-1/2} \ , & \end{cases}$$

which yields

$$\sup[\tau_m(t,\Delta,\zeta): t\in R] = +\infty \quad \text{for} \quad 0<\zeta\leq\Delta(1+t^2)^{1/2} \ , \ \forall\Delta\in\,]0,+\infty[\ .$$

Therefore, the state $\mathbf{x}=\mathbf{0}$ is :

a) asymptotically stable in the whole,

b) equi-asymptotically stable,

c) uniformly asymptotically stable with respect to any bounded $T_i\subset R$,

d) not equi-asymptotically stable in the whole,

e) not uniformly asymptotically stable in the whole with respect to any bounded $T_i\subset R$.

Notice that the system is linear.

The next results are straightforward corollaries to Propositions 1-4.

Proposition 5. *If there is a time-invariant neighbourhood* $N \subset R^n$ *of* $x=0$ *such that* $\chi(t;t_o;x_o)$ *is continuous in* $(t,t_o,x_o) \in T_o \times R \times N$ *then asymptotic stability of* $x=0$ *of the system (7) with respect to some non-empty* T_i *implies its asymptotic stability.*

Proposition 6. *If* $x=0$ *of (7) is asymptotically stable then it is uniformly asymptotically stable with respect to every bounded* $T_i \subset R$.

I.2.5. Equilibrium states

For the sake of clarity we state

Definition 4. State x^* of the system (7) is its *equilibrium state over* T_i iff

$$\chi(t;t_o,x^*) = x^* \ , \ \forall t \in T_o \ , \ \forall t_o \in T_i \ . \tag{8}$$

The expression "*over* T_i " is omitted iff $T_i = R$. ■

Proposition 7. *For* $x^* \in R^n$ *to be an equilibrium state of the system (7) over* T_i *it is necessary and sufficient that both*

(i) *for every* $t_o \in T_i$ *there is the unique solution* $\chi(t;t_o;x^*)$ *of (7), which is defined for all* $t \in T_o$

and

(ii) $f(t,x^*) = 0 \ , \ \forall t \in T_o \ , \ \forall t_o \in T_i$.

Proof. *Necessity.* Necessity of (i) and (ii) for x^* to be an equilibrium state of (7) is evidently implied by (8).
Sufficiency. If x^* satisfies the condition (ii) then $x(t) = x(t;t_o,x^*) = x^*$, $\forall t \in T_o$ and $\forall t_o \in T_i$, obeys

$$\frac{d}{dt} x(t) = 0 = f(t,x^*) = f[t,x(t)] \ , \ \forall t \in T_o \ , \ \forall t_o \in T_i \ .$$

Hence, $\chi(t;t_o;x^*) = x^*$ is a solution of (7) at (t_o,x^*) for all $t_o \in T_i$, which is unique due to the condition (i).
Hence (8) holds. ■

The conditions for existence and uniqueness of the solutions can be found in the books by E.A. Coddington and N. Levinson (1955), P. Hartman (1964), A. Halanay (1966) and L.S. Pontriagin (1970) [see also R. Kalman and J.E. Bertram (1960)].

Proposition 8. *If* $x=0$ *of the system (7) is stable with respect to* T_i *then it is an equilibrium state of the system over* T_i .

Proof. Let $x=0$ of (7) be stable with respect to T_i and $\epsilon > 0$ be arbitrarily small. Then $\|\chi(t;t_o;0)\| < \epsilon$ for all $t \in T_o$ and every $t_o \in T_i$ because $x_o = 0$ and $\|x_o\| = 0 < \delta_M(t_o,\epsilon)$. Let χ_1 and χ_2 be two solutions of (7) through $(t_o, 0)$, $t_o \in T_i$. Then,

$$\|\mathbf{x}_1(t;t_o;0) - \mathbf{x}_2(t;t_o;0)\| \le \|\mathbf{x}_1(t;t_o;0)\| + \|\mathbf{x}_2(t;t_o;0)\| < \epsilon_n \qquad (9)$$

for all $t \in T_o$ and every $t_o \in T_i$ because

$$\|\mathbf{x}_o\| = 0 < \delta_M(t_o, \frac{\epsilon_n}{2}) \ .$$

Let $\epsilon_n \to 0$ as $n \to +\infty$. It now follows from (9) that $\|\mathbf{x}_1(t;t_o;0) - \mathbf{x}_2(t;t_o;0)\|$ is less than ϵ_n no matter how large integer n is taken. Hence,

$$\mathbf{x}_1(t;t_o;0) \equiv \mathbf{x}_2(t;t_o;0)$$

and

$$\|\mathbf{x}_i(t;t_o;0)\| < \epsilon_n \ , \ i=1,2 \ ,$$

for arbitrarily large integer n. It follows that $\mathbf{x}(t;t_o;0) \equiv \mathbf{0}$ is the unique solution of (7) on T_o for all $t_o \in T_i$, which proves that $\mathbf{x}=\mathbf{0}$ is an equilibrium state of (7) over T_i. ■

Let $\mathbf{g} : R^n \to R^n$ define an autonomous system

$$\frac{d\mathbf{x}}{dt} = \mathbf{g}(\mathbf{x}) \ . \qquad (10)$$

Every stability property of $\mathbf{x}=\mathbf{0}$ of (10) is uniform in $t_o \in R$. Besides, Proposition 8 yields

Corollary 1. *If* $\mathbf{x}=\mathbf{0}$ *of the system (10) is its equilibrium state over some non-empty interval* $T_i \subset R$ *then it is an equilibrium state of the system.*

I.3. ON THE LIAPUNOV STABILITY CONDITIONS

I.3.1. Brief outline of Liapunov's original results

A.M. Liapunov (1892, p. 25) defined two essentially different approaches to solving stability problems as follows :

> "All ways, which we can present for solving the question we are interested in, we can divide in two categories.
>
> With one we associate all those, which lead to a direct investigation of a perturbed motion and in the basis of which there is a determination of general and particular solutions of the differential equations (1).
>
> In general the solutions should be searched in the form of infinite series, the simpliest type of which can be considered those from the preceding paragraph. They are series ordered in terms of integer powers of fixed variables. However we shall meet in the sequel series of another character.

The collection of all ways for the stability investigation, which are in this category, we shall call *the first method.*

With another one we associate all those, which are based on principles independent of a determination of any solution of the differential equations of a perturbed motion.

Such one, for example, is the well-known way for an investigation of the equilibrium stability in the case that there is a force function.

All these ways can be reduced to a determination and an investigation of integrals of the equations (1), and in general in the basis of all of them, which we shall meet in the sequel, there will be always a determination of functions of variables $x_1, x_2, \dots, x_n, t$ according to given conditions, which should be satisfied by their total derivatives in t, taken under an assumption that $x_1, x_2, \dots, x_n$ are functions of t satisfying the equations (1).

The collection of all ways of such a category we shall call *the second method.*"

In order to effectively develop the second method A.M. Liapunov introduced the concept of semi-definite and definite functions and the notion of decrescent functions as follows (Liapunov, 1892, p. 56) :

"We shall consider herein real-valued functions of real variables

$$x_1, x_2, \dots, x_n, t \ , \tag{39}$$

obeying conditions of the form

$$t \geq T \ , \ |x_s| \leq H \quad (s = 1, 2, \dots, n) \tag{40}$$

where T and H are constants, the former of which * can be arbitrarily large and the latter may be arbitrarily small (but different than zero) * .

Then we shall consider only functions which are continuous and one-one under the conditions (40) and vanish at

$$x_1 = x_2 = \dots = x_n = 0 \ .$$

Such properties will possess all functions considered by us (even if it were not mentioned). But, besides that, they can possess special features, for definition of which we shall introduce several terms.

Let be considered a function V which is such that under the conditions, if in them T is sufficiently large, and H sufficiently small, it can take, apart from those equal to zero, only values of one arbitrary sign.

Such a function we shall call *signconstant*. When we wish to underline its sign, then we shall say that it is *positive* or *negative function*.

In addition to that, if the function V does not depend on t, and the constant H can be chosen sufficiently small so that, under the conditions (40) the equation $V=0$ can hold only for one set of values of the variables

$$x_1 = x_2 = \dots = x_n = 0 ,$$

then we shall call the function V *signdefinite* one, and wishing to underline its sign - *positive-definite* or *negative-definite*.

We shall use the last notions also with respect to functions depending on t. However, in such a case the function V will be called *signdefinite* only under the condition, if for it it is possible to find such a t-independent positive-definite function W, for which one of two expressions

$$V-W \quad \text{or} \quad -V-W$$

would represent a positive function.

Hence, each of functions

$$x_1^2 + x_2^2 - 2x_1x_2 \cos t \ , \quad t(x_1^2 + x_2^2) - 2x_1x_2 \cos t$$

is signconstant. However, the former is only signconstant, and the latter, if $n=2$, is simultaneously signdefinite.

Every function V, for which the constant H can be chosen so small that for numerical values of that function under the conditions (40) there is an upper bound, will be called *bounded*.

In view of the properties which, under our assumption, possess all functions considered by us, will be such, for example, every function independent of t.

A bounded function can be such that for every positive ϵ, nevertheless how small, there is such non-zero number h, for which for all values of variables, satisfying conditions

$$t \geq T \ , \quad |x_s| \leq h \quad (s = 1,2,\dots,n) \ ,$$

will hold the following :

$$|V| \leq \epsilon \ .$$

This condition will satisfy, for example, every function independent of t. However functions depending on t, even bounded, can violate it. Such a case represents, for example, a function

$$\sin[(x_1 + x_2 + \dots + x_n)\,t] \; .$$

When the function V fulfils the preceding requirement, then we shall speak that it *admits infinitely small upper bound.*

Such is, for example, a function

$$(x_1 + x_2 + \dots + x_n)\,\sin t \; .$$

Let V be a function admitting infinitely small upper bound. Then, if we know that the variables satisfy a condition

$$t \geq T \;,\quad |V| \geq \ell \;,$$

where ℓ is a positive number, hence we conclude that there is another positive number λ , less than which cannot be the greatest quantity among $|x_1|, |x_2|, \dots, |x_n|$. "

In order to examine behaviour of the values of a definite function V along system motions without using the motions themselves A.M. Liapunov (1892, p.58) proposed the following :

"Simultaneously with the function V we shall often consider an expression

$$V' = \frac{\partial V}{\partial x_1} X_1 + \frac{\partial V}{\partial x_2} X_2 + \dots + \frac{\partial V}{\partial x_n} X_n + \frac{\partial V}{\partial t} \;,$$

representing its total time derivative, taken under an assumption that $x_1, x_2, \dots, x_n$ are functions of t, which satisfy differential equations of a perturbed motion.

In such cases we shall always assume that the function V is such that V' as a function of the variables (39) * would be continuous and one-one under the conditions (40) * .

Speaking further about the derivative of the function V, we shall mean that it is the total derivative."

These concepts have been the keystone of the second Liapunov method and for a solution of (uniform) stability of $\mathbf{x}=0$ (Liapunov, 1892, p. 59) :

"*Theorem I. If the differential equations of a perturbed motion are such that it is possible to find a signdefinite function V, the derivative V' of which in view of these equations would be either a signconstant function with the opposite sign to that of V, or identically equal to zero, then the unperturbed motion is stable.*"

In addition to this result A.M. Liapunov (1892) made the "*Remark 2*" reproduced on p.2 of our § I.1 that has become the foundation of the asymptotic stability concept and for a solution of (uniform) asymptotic stability of $\mathbf{x}=0$.

In order to illustrate deepness, generality and importance of Liapunov's results once again, let following his results be cited (Liapunov, 1892, p.75):

> "**Theorem I.** *When the roots* $\kappa_1, \kappa_2, \ldots, \kappa_n$ *of the characteristic equation are such that for a given natural number* m *it is impossible any relationship of the form*
>
> $$m_1\kappa_1 + m_2\kappa_2 + \ldots + m_n\kappa_n = 0 ,$$
>
> *in which all* m_s *are non-negative integers, giving their sum equal to* m*, then it is always possible to find just one whole homogenous function* V *of the power* m *of the quantities* κ_s *satisfying the equation*
>
> $$\sum_{s=1}^{n} (p_{s1}x_1 + p_{s2}x_2 + \ldots + p_{sn}x_n) \frac{\partial V}{\partial x_s} = U \qquad (9)$$
>
> *for arbitrarily given whole homogenous function* U *of the quantities* x_s *of the same power* m."
>
> "**Theorem II.** *When the real parts of all roots* κ_s *are negative and when in the equation (9) there is the function* U *being signdefinite form of any even power* m*, then the form* V *of the power* m *satisfying that equation is also sign definite with the opposite sign to that of* U."

F.R. Gantmakher (1974b) recognized the fundamental potential of these results and deduced from them a theorem (Theorem 3', p.189) which has been commonly referred to as the Liapunov matrix theorem [see S. Barnett and C. Storey (1970)]. This theorem is a fundamental theorem for the stability theory. For its presentation the following is needed.

Definition 5. *A matrix* $\mathbf{H} = (h_{ij}) \in \mathbf{R}^{n\times n}$ *is* :

(i) *positive (negative) semi-definite* iff its quadratic form $V(\mathbf{x}) = \mathbf{x}^T\mathbf{H}\mathbf{x}$ is positive (negative) semi-definite, respectively;

(ii) *positive (negative) definite* iff its quadratic form $V(\mathbf{x}) = \mathbf{x}^T\mathbf{H}\mathbf{x}$ is positive (negative) definite, respectively. ■

Let a k-th order principal minor of the matrix **H** be denoted by

$$\mathbf{H}\begin{bmatrix} i_1 & i_2 & \ldots & i_k \\ i_1 & i_2 & \ldots & i_k \end{bmatrix} = \begin{vmatrix} h_{i_1i_1} & h_{i_1i_2} & \ldots & h_{i_1i_k} \\ h_{i_2i_1} & h_{i_2i_2} & \ldots & h_{i_2i_k} \\ \ldots & \ldots & \ldots & \ldots \\ h_{i_ki_1} & h_{i_ki_2} & \ldots & h_{i_ki_k} \end{vmatrix} ,$$

where

$$i_j \in \{1,2,\ldots,n\} , \; i_j < i_{j+1} , \; j = 1,2,\ldots,k , \; k = 1,2,\ldots,n .$$

The leading principal minor of the k-th order of **H** is

$$H\begin{bmatrix} 1 & 2 & \dots & k \\ 1 & 2 & \dots & k \end{bmatrix} = \begin{vmatrix} h_{11} & h_{12} & \dots & h_{1k} \\ h_{21} & h_{22} & \dots & h_{2k} \\ \dots & \dots & & \dots \\ h_{k1} & h_{k2} & \dots & h_{kk} \end{vmatrix}, \quad k = 1,2,\dots,n .$$

The following criteria are well known [see Gantmacher (1974a)] :

Theorem 2. *Necessary and sufficient for a symmetric* n×n *matrix* **H** *to be :*

(i) *positive semi-definite is that all its principal minors are non-negative*

$$H\begin{bmatrix} i_1 & i_2 & \dots & i_k \\ i_1 & i_2 & \dots & i_k \end{bmatrix} \geq 0 , \quad 1 \leq i_1 < i_2 < \dots < i_k \leq n , \quad k = 1,2,\dots,n ;$$

(ii) *negative semi-definite is that both all its even order principal minors are non-negative and all its odd order principal minors are non-positive*

$$H\begin{bmatrix} i_1 & i_2 & \dots & i_k \\ i_1 & i_2 & \dots & i_k \end{bmatrix} \begin{cases} \geq 0 , & k = 2,4,\dots \\ \leq 0 , & k = 1,3,\dots \end{cases} ;$$

(iii) *positive definite is that all its leading principal minors are positive*

$$H\begin{bmatrix} 1 & 2 & \dots & k \\ 1 & 2 & \dots & k \end{bmatrix} > 0 , \quad k = 1,2,\dots,n ;$$

(iv) *negative definite is that both its first order leading principal minor is negative and all its leading principal minors are alternatively negative and positive*

$$(-1)^k H\begin{bmatrix} 1 & 2 & \dots & k \\ 1 & 2 & \dots & k \end{bmatrix} > 0 , \quad k = 1,2,\dots,n .$$

Notice that a square matrix **A** with all real valued elements is (semi-) definite iff its symmetric part $A_s = \frac{1}{2}(A + A^T)$ is (semi-) definite, and a square matrix **A** with complex valued elements is (semi-) definite iff its Hermitian part $A_H = \frac{1}{2}(A + A^*)$ is (semi-) definite, where A^* is the transpose conjugate matrix of the matrix **A** .

Now, the fundamental theorem of the stability theory - the Liapunov matrix theorem - can be stated as a corollary to the preceding Theorems I and II by Liapunov.

Theorem 3. *In order that real parts of all eigenvalues of a matrix* A, $A \in R^{n\times n}$, *be negative it is necessary and sufficient that for any positive definite symmetric matrix* G, $G \in R^{n\times n}$, *there exists the unique solution* H, $H \in R^{n\times n}$, *of the (Liapunov) matrix equation*

$$A^T H + HA = -G ,$$

which is also positive definite symmetric matrix.

For solving the Liapunov matrix equation, see for example Barnett and Storey (1970).

I.3.2. Brief outline of the classical and novel developments of the Liapunov second method

Following Liapunov (1892), the classical development of his second method consists of a number of stability theorems providing stability conditions are imposed on appropriate *scalar function* V and its total time derivative along system motions over a *time-invariant* neighbourhood of $x=0$. Adequate expositions of the classic development of the Liapunov second method can be found in the books by Yoshizawa (1966) and Rouche, Habets and Laloy (1977).

I.3.2.1. Comparison functions

Comparison functions are used as upper or lower estimates of the function V and its total time derivative. They are usually denoted by ϕ, $\phi : R_+ \to R_+$. The main contributor to the investigation of properties of and use of the comparison functions is Hahn (1967). What follows is mainly based on his definitions and results.

Definition 6. A function ϕ, $\phi : R_+ \to R_+$, *belongs to*

(i) *the class* $K_{[0,\alpha[}$, $0 < \alpha \le +\infty$, iff both it is defined, continuous and strictly increasing on $[0,\alpha[$ and $\phi(0) = 0$;

(ii) *the class* K iff (i) holds for $\alpha = +\infty$, $K = K_{[0,+\infty[}$;

(iii) *the class* KR iff both it belongs to the class K and $\phi(\zeta) \to +\infty$ as $\zeta \to +\infty$;

(iv) *the class* $L_{[0,\alpha[}$ iff both it is defined, continuous and strictly decreasing on $[0,\alpha[$ and $\lim[\phi(\zeta) : \zeta \to +\infty] = 0$;

(v) *the class* L iff (iv) holds for $\alpha = +\infty$, $L = L_{[0,+\infty[}$. ∎

Let ϕ^I denote the inverse function of ϕ, $\phi^I[\phi(\zeta)] \equiv \zeta$.

The next result was established by Hahn (1967).

Proposition 9.

(i) *If* $\phi \in K$ *and* $\psi \in K$ *then* $\phi(\psi) \in K$ *;*

(ii) *If* $\phi \in K$ *and* $\sigma \in L$ *then* $\phi(\sigma) \in L$ *;*

(iii) *If* $\phi \in K_{[0,\alpha[}$ *and* $\phi(\alpha) = \xi$ *then* $\phi^I \in K_{[0,\xi]}$ *;*

(iv) *If* $\phi \in K$ *and* $\lim[\phi(\zeta) : \zeta \to +\infty] = \xi$ *then* ϕ^I *is not defined on* $]\xi, +\infty]$ *;*

(v) *If* $\phi \in K_{[0,\alpha]}$, $\psi \in K_{[0,\alpha]}$ *and* $\phi(\zeta) > \psi(\zeta)$ *on* $[0,\alpha[$ *then* $\phi^I(\zeta) < \psi^I(\zeta)$ *on* $[0,\beta]$ *where* $\beta = \psi(\alpha)$.

Definition 7. A function ϕ, $\phi : R_+ \times R_+ \to R_+$, *belongs to* :

(i) *the class* $KK_{[0;\alpha,\beta[}$ iff both $\phi(o,\zeta) \in K_{[0,\alpha[}$ for every $\zeta \in [0,\beta[$ and $\phi(\zeta,o) \in K_{[0,\beta[}$ for every $\zeta \in [0,\alpha[$;

(ii) *the class* KK iff (i) holds for $\alpha = \beta = +\infty$;

(iii) *the class* $KL_{[0;\alpha,\beta[}$ iff both $\phi(o,\zeta) \in K_{[0,\alpha[}$ for every $\zeta \in [0,\beta[$ and $\phi(\zeta,o) \in L_{[0,\beta[}$ for every $\zeta \in [0,\alpha[$;

(iv) *the class* KL iff (iii) *holds for* $\alpha = \beta = +\infty$. ∎

I.3.2.2. Definite functions and comparison functions

For the sake of preciseness the following definition will be used throughout the book, which is based on the corresponding definition by Liapunov (1892) and Hahn (1967, p.98, Def. 24.3a).

Definition 8. Function $v : R^n \to R$ is :

(i) *positive semi-definite* iff there is a time-invariant neighbourhood N of $x=0$, $N \subseteq R^n$, such that
   a) v is continuous on N : $v(x) \in C(N)$,
   b) v is non-negative on N : $v(x) \geq 0$, $\forall x \in N$,
   c) v vanishes at the origin : $v(0) = 0$;

(ii) *positive semi-definite on a neighbourhood* S *of* $x=0$ iff (i) holds for $N=S$;

(iii) *positive semi-definite in the whole* iff (i) holds for $N=R^n$;

(iv) *negative semi-definite* (*on a neighbourhood* S *of* $x=0$ or *in the whole*) iff $(-v)$ is positive semi-definite (on the neighbourhood S or in the whole, respectively). ∎

Remark 1. It is to be noted that function v defined by $v(x) = 0$ for all $x \in R^n$ is both positive and negative semi-definite. This ambiguity can be avoided by introducing the notion of strictly positive (negative) semi-definite function. A function $v : R^n \to R$ is said to be *strictly positive (negative) semi-definite* iff both it is positive (negative) semi-definite and there is $y \in N$ such that $v(y) > 0$ $(v(y) < 0)$, respectively. Then H is *strictly positive (negative) semi-definite* iff $v(x) = x^T H x$ is strictly positive (negative) semi-definite, respectively. Necessary and sufficient for a symmetric matrix H to be strictly positive semi-definite is obviously that both (i) of Theorem 2 holds and that there is $j \in \{1,2,\dots,n\}$ such that

$$H \begin{bmatrix} i_1 & i_2 & \dots & i_j \\ i_1 & i_2 & \dots & i_j \end{bmatrix} > 0 \ .$$

Necessary and sufficient for a symmetric matrix H to be strictly negative semi-definite is obviously that both (ii) of Theorem 2 holds and that there is odd $j \in \{1,2,\dots,n\}$ such that

$$(-1)^j H \begin{bmatrix} i_1 & i_2 & \dots & i_j \\ i_1 & i_2 & \dots & i_j \end{bmatrix} > 0 \ .$$

Definition 9. Function $v : R^n \to R$ is :

(i) *positive definite* if there is a time-invariant neighbourhood N, $N \subseteq R^n$, of $x = 0$ such that both it is positive semi-definite on N and $v(x) > 0$, $\forall (x \neq 0) \in N$;

(ii) *positive definite on a neighbourhood S of* $x = 0$ iff (i) holds for $N = S$;

(iii) *positive definite in the whole* iff (i) holds for $N = R^n$;

(iv) *negative definite (on a neighbourhood S of* $x = 0$ or *in the whole)* iff $(-v)$ is positive definite (on the neighbourhood S or in the whole, respectively). ■

Hahn (1967) proved

Proposition 10. *Necessary and sufficient for positive definiteness of* v *on a neighbourhood* N *of* $x = 0$ *is existence of comparison functions* $\phi_i \in K_{[0,\alpha[}$, $i=1,2$, *where* $\alpha = \sup\{\|x\| : x \in N\}$, *such that both* $v(x) \in C(N)$ *and* $\phi_1(\|x\|) \le v(x) \le \phi_2(\|x\|)$, $\forall x \in N$.

Definition 10. Function $v : R \times R^n \to R$ is :

(i) *positive semi-definite on* $T_\tau = [\tau, +\infty[$, $\tau \in R$, iff there is a time-invariant connected neighbourhood N of $x = 0$, $N \subseteq R^n$, such that

a) v is continuous in $(t,x) \in T_\tau \times N$: $v(t,x) \in C(T_\tau \times N)$;

b) v is non-negative on N : $v(t,\mathbf{x}) \geq 0$, $\forall(t,\mathbf{x}) \in T_\tau \times N$;

and

c) v vanishes at the origin : $v(t,\mathbf{0}) = 0$, $\forall t \in T_\tau$.

d) Iff the conditions (a)-(c) hold and for every $t \in T_\tau$ there is $\mathbf{y} \in N$ such that $v(t,\mathbf{y}) > 0$, then v is *strictly positive semi-definite on* T_τ ;

(ii) *positive semi-definite on* $T_\tau \times S$ iff (i) holds for $N=S$;

(iii) *positive semi-definite in the whole on* T_τ iff (i) holds for $N=R^n$;

(iv) *negative semi-definite (in the whole) on* T_τ *(on* $T_\tau \times N$*)* iff (-v) is positive semi-definite (in the whole) on T_τ (on $T_\tau \times N$), respectively.

(v) The expression "*on* T_τ " is omitted iff all corresponding requirements hold for every $\tau \in R$. ■

Definition 11. Function $v: R \times R^n \to R$ is :

(i) *positive definite on* T_τ , $\tau \in R$, iff there is a time-invariant connected neighbourhood N of $\mathbf{x}=\mathbf{0}$, $N \subseteq R^n$, such that both it is positive semi-definite on $T_\tau \times N$ and there exists a positive definite function w on N, $w : R^n \to R$, obeying $w(\mathbf{x}) \leq v(t,\mathbf{x})$, $\forall(t,\mathbf{x}) \in T_\tau \times N$;

(ii) *positive definite on* $T_\tau \times S$ iff (i) holds for $N=S$;

(iii) *positive definite in the whole on* T_τ iff (i) holds for $N=R^n$;

(iv) *negative definite (in the whole) on* T_τ *(on* $T_\tau \times N$*)* iff (-v) is positive definite (in the whole) on T_τ (on $T_\tau \times N$) , respectively.

(v) The expression "*on* T_τ " is omitted iff all corresponding requirements hold for every $\tau \in R$. ■

The following result is obtained directly from Proposition 10 and Definition 10.

Proposition 11. *Necessary and sufficient for a function* $v: R \times R^n \to R^n$ *to be positive definite on* $T_\tau \times N$ *when* N *is a time-invariant neighbourhood of* $\mathbf{x}=\mathbf{0}$ *is that*

(i) $v(t,\mathbf{x}) \in C(T_\tau \times N)$;

(ii) $v(t,\mathbf{0}) = 0$, $\forall t \in T_\tau$

and

(iii) *there is* $\phi \in K_{[0,\alpha[}$, *where* $\alpha = \sup\{\|\mathbf{x}\| : \mathbf{x} \in N\}$, *such that* $\phi(\|\mathbf{x}\|) \leq v(t,\mathbf{x})$, $\forall(t,\mathbf{x}) \in T_\tau \times N$.

Definition 12. Set $V_\zeta(t)$ is the largest connected neighbourhood of $\mathbf{x}=0$ at $t\in R$ which can be associated with a function v, $v: R\times R^n \to R$, so that $\mathbf{x}\in V_\zeta(t)$ implies $v(t,\mathbf{x})<\zeta$. ■

Remark 2. In order to understand and appreciate deepness and importance of Liapunov's concept of definite functions let functions v and w be considered, $v,w: R\times R^n \to R$. Let they obey the following on $T_\tau \times N$, where N is a connected neighbourhood of $\mathbf{x}=0$:

(i) v is positive definite on $T_\tau \times N$;
(ii) w is positive semi-definite on $T_\tau \times N$ and $w(t,\mathbf{x})>0$, $\forall(t,\mathbf{x}\neq \mathbf{0}) \in T_\tau\times N$, but it is not positive definite on $T_\tau \times N$.

Let $V_\zeta(t)$ and $W_\zeta(t)$ be associated with v and w in sense of Definition 12. Then, the following is true :

(a) there is $\xi\in\,]0,+\infty[$ such that (Fig. 1) $V_\zeta(t)\subseteq N$, $\forall t\in T_\tau$, $\forall\zeta\in\,]0,\xi]$.

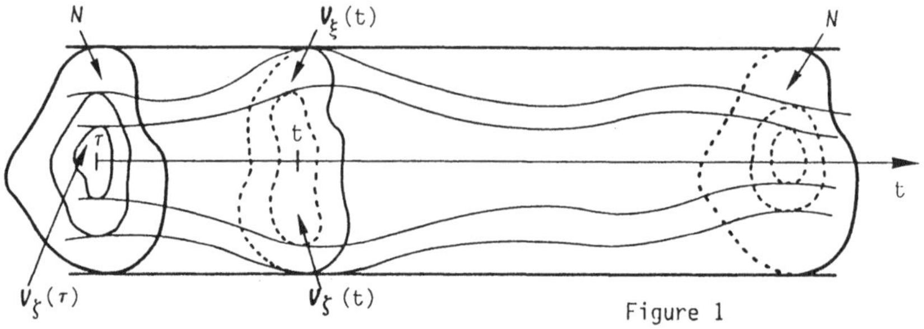

Figure 1

For example, $v(t,\mathbf{x})=(1+t^2)\|\mathbf{x}\|^2$, $v(t,\mathbf{x})=(e^t+e^{-t})\|\mathbf{x}\|^2$ etc;

(b) for any $\zeta\in\,]0,+\infty[$ for which $W_\zeta(\tau)\subseteq N$ there is $t\in T_\tau$, $t>\tau$, such that (Fig. 2) $W_\zeta(t)\setminus N\neq\phi$.

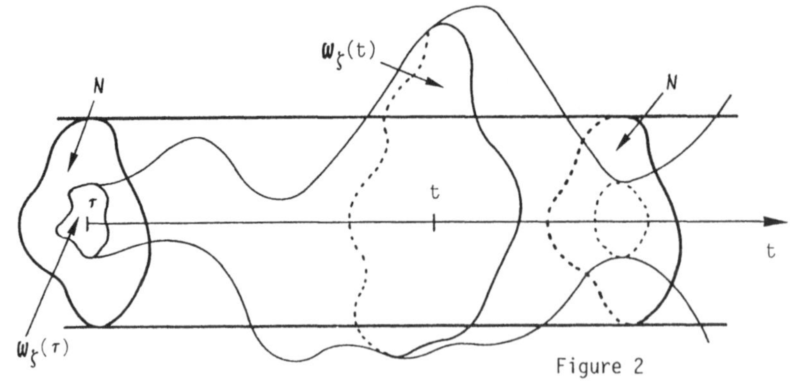

Figure 2

For example, $w(t,\mathbf{x}) = e^{-t}\|\mathbf{x}\|^2$, or $w(t,\mathbf{x}) = (1+t^2)^{-1}\|\mathbf{x}\|^2$.

Definition 13. Function $v : R\times R^n \to R$ is :

(i) *decrescent on* T_τ , $\tau\in R$, iff there is a time-invariant neighbourhood N of $\mathbf{x}=0$ and a positive definite function w on N , $w : R^n \to R$, such that $v(t,\mathbf{x}) \le w(\mathbf{x})$, $\forall(t,\mathbf{x}) \in T_\tau\times N$;

(ii) *decrescent on* $T_\tau\times S$ iff (i) holds for $N=S$;

(iii) *decrescent in the whole on* T_τ iff (i) holds for $N=R^n$.

(iv) The expression "*on* T_τ " is omitted iff all corresponding conditions still hold for every $\tau\in R$. ∎

Proposition 10 and Definition 13 imply

Proposition 12. *Necessary and sufficient for* v *to be decrescent on* $T_\tau\times N$ *when* N *is a time-invariant neighbourhood of* $\mathbf{x}=0$ *is existence of a comparison function* $\phi\in K_{[0,\alpha[}$ *, where* $\alpha = \sup\{\|\mathbf{x}\| : \mathbf{x}\in N\}$ *, such that* $v(t,\mathbf{x}) \le \phi(\|\mathbf{x}\|)$, $\forall(t,\mathbf{x})\in T_\tau\times N$.

Remark 3. In order to clarify meaning and importance of decrescent functions let functions v and w be considered, $v,w : R\times R^n \to R$. Let they obey the following on $T_\tau\times N$:

(i) v is positive definite and decrescent on $T_\tau\times N$;

(ii) w is positive definite but not decrescent on $T_\tau\times N$.

Let $V_\zeta(t)$ and $W_\zeta(t)$ be associated with v and w in the sense of Definition 12. Then, the following holds :

(a) Let $\xi>0$ be the largest number obeying $V_\xi(t)\subseteq N$, $\forall t\in T_\tau$. Then, for every $\zeta\in]0,\xi]$ there is a connected time-invariant neighbourhood N_ζ of $\mathbf{x}=0$, $N_\zeta\subseteq N$, such that $V_\zeta(t)\supseteq N_\zeta$, $\forall t\in T_\tau$ (Fig. 3).

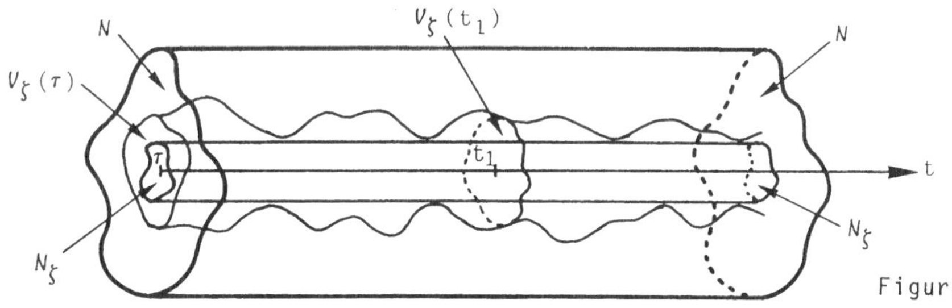

Figure 3.

For example, $v(t,\mathbf{x}) = \dfrac{2+t^2}{1+t+t^2}\|\mathbf{x}\|^2$, $v(t,\mathbf{x}) = (2+\sin t)\|\mathbf{x}\|^2$.

(b) let $\xi > 0$ be the largest number obeying $\mathcal{W}_\xi(t) \subseteq N$, $\forall t \in T_\tau$. Then, for any $\zeta \in]0,\xi]$ and for any connected neighbourhood $N_\epsilon \subseteq N$ (ϵ can be arbitrarily small, $\epsilon > 0$, and denotes the diameter of N_ϵ) there is $t_1(\zeta, N_\epsilon) \in]\tau, +\infty]$ such that $\mathcal{W}_\zeta(t_1)$ is a proper subset of N_ϵ, $\mathcal{W}_\zeta(t_1) \subset N_\epsilon$ (Fig. 4).

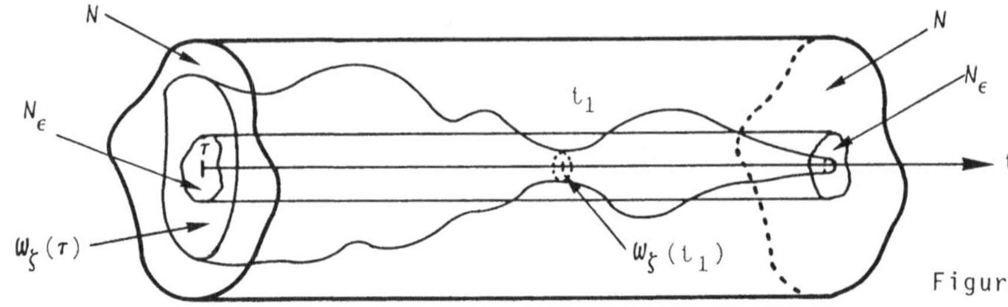

Figure 4

For example, $w(t,\mathbf{x}) = (1+e^t)\|\mathbf{x}\|^2$, $w(t,\mathbf{x}) = (1+t^2)\|\mathbf{x}\|^2$ etc.

Barbashin and Krasovkii (1952, 1954) discovered the concept of radially unbounded functions. They showed necessity of it for asymptotic stability in the whole.

Definition 14. Function $v : R \times R^n \to R$ is :

(i) *radially unbounded on* T_τ, $\tau \in R$, iff $\|\mathbf{x}\| \to +\infty$ implies $v(t,\mathbf{x}) \to +\infty$, $\forall t \in T_\tau$;

(ii) *radially unbounded* iff $\|\mathbf{x}\| \to +\infty$ implies $v(t,\mathbf{x}) \to +\infty$, $\forall t \in T_\tau$, $\forall \tau \in R$. ∎

Following Hahn (1967) the next can be easily verified.

Proposition 13. *Necessary and sufficient for a positive definite in the whole (on* T_τ*) function* v *to be radially unbounded is that there exists* $\phi \in KR$ *obeying, respectively,* $v(t,\mathbf{x}) \geq \phi(\|\mathbf{x}\|)$, $\forall \mathbf{x} \in R^n$, $\forall t \in R$ ($\forall t \in T_\tau$).

For example,

$$v(t,\mathbf{x}) = \frac{(2+t^2)(4+\|\mathbf{x}\|)}{2+\|\mathbf{x}\|^2}\|\mathbf{x}\|$$

is not radially unbounded, but

$$v(t,\mathbf{x}) = \frac{(2+t^2)(4+\|\mathbf{x}\|)}{2+\|\mathbf{x}\|}\|\mathbf{x}\|$$

is radially unbounded. In this case, $\phi(\zeta) = \zeta \in KR$ obeys

$$v(t,\mathbf{x}) \geq \phi(\|\mathbf{x}\|), \quad \forall (t,\mathbf{x}) \in R \times R^n .$$

I.3.2.3. Dini derivative and Eulerian derivative

In this section the notions of upper and lower limit of a function $\psi : R \to R$ are needed [see McSchane (1944)]. In brief [see Demidovich (1967)] they can be explained as follows.

Let t_k be a member of a sequence S_τ^- (S_τ^+) obeying

(i) $t_k \in R$ for every integer k, $t_k < \tau$ ($t_k > \tau$)

and

(ii) $t_k \to \tau^-$ ($t_k \to \tau^+$) as $k \to +\infty$.

Definition 15.

(i) Number $\alpha \in R$ is the *partial limit of the function* ψ *over the sequence* S_τ^- (S_τ^+) iff for every $\epsilon > 0$ there is an integer N such that $k > N$ implies $|\psi(t_k) - \alpha| < \epsilon$;

(ii) the symbol $\alpha = +\infty$ ($\alpha = -\infty$) *is the partial limit of the function* ψ *over the sequence* S_τ^- (S_τ^+) iff for every $\epsilon \in]0, +\infty[$ there is an integer N such that, respectively, $k > N$ implies $\psi(t_k) > \frac{1}{\epsilon}$ ($\psi(t_k) < -\frac{1}{\epsilon}$);

(iii) the greatest (smallest) partial limit of the function ψ over all sequences S_τ^- is its *left upper (lower)* limit at $t = \tau$, respectively, which is denoted by $\limsup[\psi(t) : t \to \tau^-]$, ($\liminf[\psi(t) : t \to \tau^-]$).

(iv) *right upper (lower) limit of* ψ *at* $t = \tau$ is analogously defined when everywhere in (iii) τ^- and S_τ^- are respectively replaced by τ^+ and S_τ^+. ∎

Definition 16. Let $\mathbf{v}$ be a continuous (either scalar or vector) function, $\mathbf{v} : T_\tau \times R^n \to R^s$, $\mathbf{v}(t, \mathbf{x}) \in C(T_\tau \times N)$, and let solutions χ of the system (7) exist and be defined on $T_\tau \times N$. Then, for $(t, \mathbf{x}) \in T_\tau \times N$,

(i) $D^+\mathbf{v}(t,\mathbf{x}) = \limsup \{\frac{\mathbf{v}[t+\theta, \chi(t+\theta; t, \mathbf{x})] - \mathbf{v}(t,\mathbf{x})}{\theta} : \theta \to 0^+\}$

is the *upper right Dini derivative of* $\mathbf{v}$ *along the motion* χ *at* $(t, \mathbf{x})$;

(ii) $D_+\mathbf{v}(t,\mathbf{x}) = \liminf \{\frac{\mathbf{v}[t+\theta; \chi(t+\theta; t, \mathbf{x})] - \mathbf{v}(t,\mathbf{x})}{\theta} : \theta \to 0^+\}$

is the *lower right Dini derivative of* $\mathbf{v}$ *along the motion* χ *at* $(t, \mathbf{x})$;

(iii) $D^-\mathbf{v}(t,\mathbf{x}) = \limsup \{\frac{\mathbf{v}[t+\theta, \chi(t+\theta; t; \mathbf{x})] - \mathbf{v}(t,\mathbf{x})}{\theta} : \theta \to 0^-\}$

is the *upper left Dini derivative of* $\mathbf{v}$ *along the motion* χ *at* $(t, \mathbf{x})$;

(iv) $D_-\mathbf{v}(t,\mathbf{x}) = \liminf \{\frac{\mathbf{v}[t+\theta;\chi(t+\theta;t,\mathbf{x})] - \mathbf{v}(t,\mathbf{x})}{\theta} : \theta \to 0^-\}$

is the *lower left Dini derivative of* $\mathbf{v}$ *along the motion* χ *at* $(t,\mathbf{x})$.

(v) The function $\mathbf{v}$ has *Eulerian derivative* $\dot{\mathbf{v}}$, $\dot{\mathbf{v}}(t,\mathbf{x}) = \frac{d}{dt}\mathbf{v}(t,\mathbf{x})$, at $(t,\mathbf{x})$ along the motion χ iff $D^+\mathbf{v}(t,\mathbf{x}) = D_+\mathbf{v}(t,\mathbf{x}) = D^-\mathbf{v}(t,\mathbf{x}) = D_-\mathbf{v}(t,\mathbf{x}) = D\mathbf{v}(t,\mathbf{x})$ and then $\dot{\mathbf{v}}(t,\mathbf{x}) = D\mathbf{v}(t,\mathbf{x})$. ∎

If v is a scalar function and differentiable at $(t,\mathbf{x})$ then (Liapunov 1892)

$$\dot{v}(t,\mathbf{x}) = \frac{\partial v}{\partial t} + (\text{grad } v)^T \mathbf{f}(t,\mathbf{x}) ,$$

where

$$\text{grad } v = [\frac{\partial v}{\partial x_1}, \frac{\partial v}{\partial x_2}, \dots, \frac{\partial v}{\partial x_n}]^T .$$

Effective application of $D^+\mathbf{v}$ in the framework of the second Liapunov method is based on the next result by Yoshizawa (1966), which enables calculation of $D^+\mathbf{v}$ without utilizing system motions themselves.

Theorem 4. *Let* $\mathbf{v}$ *be continuous and locally Lipshitzian in* $\mathbf{x}$ *over* $T_\tau \times S$ *and* S *be an open set. Then,*

$$D^+\mathbf{v}(t,\mathbf{x}) = \limsup \{\frac{\mathbf{v}[t+\theta, \mathbf{x}+\theta\mathbf{f}(t,\mathbf{x})] - \mathbf{v}(t,\mathbf{x})}{\theta} : \theta \to 0^+\}$$

holds along solutions χ *of the system (7) at* $(t,\mathbf{x}) \in T_\tau \times S$.

$D^*\mathbf{v}$ will mean that both $D^+\mathbf{v}$ and $D_+\mathbf{v}$ can be used.

I.3.2.4. Stability conditions on time-invariant sets

Following Liapunov (1892), Persidskii (1933), Yoshizawa (1966) and Halanay (1966), the next result is obtained.

Theorem 5. *If* χ *of (7) is continuous on* $T_0 \times R \times N$ *(on* $T_0 \times T_\tau \times N$*) then existence of both an open connected time-invariant neighbourhood* S *of* $\mathbf{x}=\mathbf{0}$ *and a decrescent positive definite function* v *on* S *(on* $T_\tau \times S$*) such that, respectively,* $D^+v(t,\mathbf{x}) \le 0$, $\forall(t,\mathbf{x}) \in R\times S$ $(\forall(t,\mathbf{x}) \in T_\tau \times S)$ *is necessary and sufficient for uniform stability (on* T_τ*) of* $\mathbf{x}=\mathbf{0}$ *of the system (7).*

Proof. Necessity. Let $\mathbf{x}=\mathbf{0}$ of (7) be uniformly stable (on T_τ) and $\epsilon > 0$ be arbitrarily chosen. Let $B_\xi \subseteq N$, $\xi = \min\{\zeta, \delta_M(\epsilon)\}$, $\phi \in K_{[0,\epsilon[}$ and

$$v(t,\mathbf{x}) = \sup\{\phi[\|\chi(t+\sigma;t,\mathbf{x})\|] : \sigma \in [0,+\infty[\} . \qquad (11)$$

Stability of $\mathbf{x}=\mathbf{0}$ and continuity of χ in all arguments together with $\phi \in K_{[0,\epsilon[}$ imply continuity of v in $(t,\mathbf{x}) \in R\times S$ $(T_\tau \times S)$ where $S = B_\xi$.

Stability of $\mathbf{x}=\mathbf{0}$ implies (Proposition 8) :

$$\chi(t;t_o,\mathbf{0}) = \mathbf{0} \ , \ \forall(t,t_o) \in T_o \times R \quad (\forall(t,t_o) \in T_o \times T_\tau) \ .$$

$\phi \in K_{[0,\epsilon[}$ now proves $v(t,\mathbf{0}) = 0$, $\forall t \in R$ $(\forall t \in T_\tau)$.
From (11) it follows :

$$v(t,\mathbf{x}) \geq \phi(\|\mathbf{x}\|) \ , \ \forall(t,\mathbf{x}) \in R \times S \quad (\forall(t,\mathbf{x}) \in T_\tau \times S) \ .$$

Altogether, v is positive definite on S (on $T_\tau \times S$). Uniform stability of $\mathbf{x}=\mathbf{0}$ of (7) and continuity of χ imply [Halanay (1966)] existence of $\pi \in K$ such that $\|\mathbf{x}_o\| < \delta(\epsilon)$ implies $\|\chi(t,t_o,\mathbf{x}_o)\| < \pi(\|\mathbf{x}_o\|)$, $\forall t \in T_o$, $\forall t_o \in R$ $(\forall t_o \in T_\tau)$, $\forall \epsilon > 0$. Hence,

$$v(t,\mathbf{x}) \leq \phi[\pi(\|\mathbf{x}\|)] \ , \ \forall(t,\mathbf{x}) \in R \times S \quad (\forall(t,\mathbf{x}) \in T_\tau \times S) \ ,$$

which proves that v is decrescent on S (on $T_\tau \times S$). Let $\theta > 0$ so that $\mathbf{x}^* = \chi(t+\theta;t,\mathbf{x})$ and $\mathbf{x} = \chi(t;t_o,\mathbf{x}_o)$. Then,

$$\begin{aligned} v(t+\theta,\mathbf{x}^*) &= \sup\{\phi[\|\chi(t+\theta+\sigma;t+\theta,\mathbf{x}^*)\|] : \sigma \in [0,+\infty[\} \\ &= \sup\{\phi[\|\chi(t+\theta+\sigma;t,\mathbf{x})\|] : \sigma \in [0,+\infty[\} \\ &\leq \sup\{\phi[\|\chi(t+\sigma;t,\mathbf{x})\|] : \sigma \in [0,+\infty[\} \\ &= v(t,\mathbf{x}) \ , \end{aligned}$$

$$\forall(t,\mathbf{x}) \in R \times S \quad (\forall(t,\mathbf{x}) \in T_\tau \times S) \ .$$

Hence,

$$D^+v(t,\mathbf{x}) \leq 0 \ , \ \forall(t,\mathbf{x}) \in R \times S \quad (\forall(t,\mathbf{x}) \in T_\tau \times S) \ .$$

Sufficiency. Let the conditions of the theorem be satisfied. Let $\epsilon > 0$ be arbitrarily chosen and ζ be such that $B_\zeta \subseteq N$. Let $\xi = \min\{\epsilon,\zeta\}$ and $\phi_1,\phi_2 \in K_{[0,\zeta[}$ be such (Propositions 11-12) that

$$\phi_1(\|\mathbf{x}\|) \leq v(t,\mathbf{x}) \leq \phi_2(\|\mathbf{x}\|) \ , \ \forall(t,\mathbf{x}) \in R \times B_\zeta \quad (\forall(t,\mathbf{x}) \in T_\tau \times B_\zeta) \ . \quad (12)$$

Now, δ is selected as $\delta(\epsilon) = \phi_2^I[\phi_1(\xi)]$, $\xi = \xi(\epsilon)$. Evidently, $\delta(\epsilon) > 0$. Since D^+v is non-positive and since $V_\eta(t) \subseteq B_\xi$, $\forall t \in R$ $(\forall t \in T_\tau)$, for all $\eta \in]0,\phi_2(\delta)]$, then $\|\mathbf{x}_o\| < \delta(\epsilon)$ implies $v(t_o,\mathbf{x}_o) \leq \phi_2(\|\mathbf{x}_o\|) < \phi_2(\delta(\epsilon)) = \phi_1(\xi)$, and therefore $v[t,\chi(t;t_o,\mathbf{x}_o)] \leq v(t_o,\mathbf{x}_o)$ for all $t \in T_o$ and all $t_o \in R$ $(t_o \in T_\tau)$, which together with (12) yields

$$\phi_1(\|\chi(t;t_o,\mathbf{x}_o)\|) \leq v[t,\chi(t;t_o,\mathbf{x}_o)] \leq v(t_o,\mathbf{x}_o) \leq \phi_2(\|\mathbf{x}_o\|) < \phi_1(\xi) \leq \phi_1(\epsilon) \ ,$$

$$\forall t \in T_o \ , \ \forall t_o \in R \ (\forall t_o \in T_\tau) \ .$$

Altogether, since $\phi_1 \in K_{[0,\zeta[}$, $\|\mathbf{x}_o\| < \delta(\epsilon)$, $\delta(\epsilon) = \phi_2^I[\phi_1(\xi(\epsilon))] > 0$, implies

$$\|\chi(t;t_o,\mathbf{x}_o)\| < \epsilon \ , \ \forall t \in T_o \ , \ \forall t_o \in R \ (\forall t_o \in T_\tau) \ , \ \forall \epsilon > 0 \ . \quad \blacksquare$$

Utilizing $\phi \in KR$ in (11), following the proof of the preceding theorem, Definition 1 and by referring to Barbashin and Krasovskii (1952,1954), it is easy to prove

Theorem 6. *If* χ *(7) is continuous on* $T_0\times R\times R^n$ *(on* $T_0\times T_\tau\times R^n$ *) then existence of a radially unbounded decrescent positive definite in the whole function* v *(on* T_τ *) such that, respectively,* $D^+v(t,x)\le 0$, $\forall(t,x)\in R\times R^n$ $(\forall(t,x)\in T\times R^n)$ *is necessary and sufficient for uniform stability in the whole (on* T_τ *) of* $x=0$ *of the system (7).*

Remark 4. If f is locally Lipschitzian on $R\times N$ (on $T_\tau\times N$) then v in the preceding theorems is also locally Lipschitzian on $R\times N$ (on $T_\tau\times N$) which enables effective calculation of D^+v via Theorem 4.

Remark 5. The preceding theorems hold also when D^+v is replaced by D_+v [McShane (1944) and Lasalle (1976)].

Following Liapunov (1892), Massera (1949,1956), Yoshizawa (1966), Halanay (1966) and Hahn (1967), the next result is obtained.

Theorem 7. *If* χ *is continuous on* $T_0\times R\times N$ *(on* $T_0\times T_\tau\times N$ *) then existence of an open connected time-invariant neighbourhood* S *of* $x=0$ *, a decrescent positive definite function* v *on* S *(on* $T_\tau\times S$ *) and a positive definite function* ψ *on* S *such that, respectively,*

$$D^*v(t,x)\le -\psi(x)\ ,\ \forall(t,x)\in R\times S\quad (\forall(t,x)\in T_\tau\times S)\ , \tag{13}$$

is necessary and sufficient for uniform asymptotic stability of $x=0$ *of the system (7) (on* T_τ *).*

Proof. *Necessity.* Let $x=0$ of (7) be uniformly asymptotically stable (on T_τ). Let $\epsilon>0$ be arbitrarily chosen, ζ be such that $B_\zeta\subseteq N$, $\Delta\in]0,+\infty[$ and $\xi=\min\{\delta_M(\epsilon),\Delta,\zeta\}$. Let $S=B_\xi$, $\phi\in K_{[0,\epsilon[}$, $\alpha\in]1,+\infty[$ and $v(t,x)=\sup\{\phi[\|\chi(t+\sigma;t,x)\|](1+\alpha\sigma)(1+\sigma)^{-1}:\sigma\in[0,+\infty[\}$, $\forall t\in R$. The function v is decrescent and positive definite on S (on $T_\tau\times S$) because χ is continuous in all its arguments, $\phi\in K_{[0,\epsilon[}$, $(1+\alpha\sigma)(1+\sigma)^{-1}$ is also continuous, $\chi(t;t_0,0)\equiv 0$, $\phi(0)=0$, and $\phi(\|x\|)\le v(t,x)\le\phi[\pi(\|x\|)]$, $\forall t\in T_0$, $\forall t_0\in R$ $(\forall t_0\in T_\tau)$, $\forall x\in S$, where $\pi\in K_{[0,\epsilon[}$ is determined in the proof of the necessity part of Theorem 5. Let $x^*=\chi(t+\theta;t,x)$, $x=\chi(t;t_0,x_0)$, $\theta>0$, so that

$$\begin{aligned} v(t+\theta,x^*) &= \sup\{\phi[\|\chi(t+\theta+\sigma;t+\theta,x^*)\|](1+\alpha\sigma)(1+\sigma)^{-1}:\sigma\in[0,+\infty[\} \\ &= \sup\{\phi[\|\chi(t+\theta+\sigma;t,x)\|](1+\alpha\sigma)(1+\sigma)^{-1}:\sigma\in[0,+\infty[\} \\ &= \phi[\|\chi(t+\theta+\sigma^*;t,x)\|](1+\alpha\sigma^*)(1+\sigma^*)^{-1}\ ,\quad \forall t\in R\ . \end{aligned}$$

Let $\nu=\min\{1,\frac{\Delta}{\alpha}\}$. The existence of $\sigma^*\in[0,\tau_u(\Delta,\nu)]$ obeying the last equation is guaranteed by continuity of χ , $\phi\in K_{[0,\epsilon[}$, continuity of $(1+\alpha\sigma)(1+\sigma)^{-1}$ and uniform attraction of $x=0$. Let $\sigma=\theta+\sigma^*$. Then

[Halanay (1966)],

$$\frac{1+\alpha\sigma^*}{1+\sigma^*} = \frac{1+\alpha\sigma}{1+\sigma}\left[1-\frac{(\alpha-1)\,\theta}{(1+\sigma^*)(1+\alpha\sigma)}\right] > 0$$

so that

$$v(t+\theta,\mathbf{x}^*) = \phi\,[\|\chi(t+\sigma;t,\mathbf{x})\|]\,\frac{1+\alpha\sigma}{1+\sigma}\left[1-\frac{(\alpha-1)\,\theta}{(1+\sigma^*)(1+\alpha\sigma)}\right]$$

$$\le v(t,\mathbf{x})\left[1-\frac{(\alpha-1)\,\theta}{(1+\sigma^*)(1+\alpha\sigma)}\right],$$

$$\forall t_o\in R\ \ (\forall t_o\in T_\tau)\ ,\ \forall t\in T_o\,,$$

or

$$\frac{v(t+\theta,\mathbf{x}^*)-v(t,\mathbf{x})}{\theta} \le -\frac{(\alpha-1)\,v(t,\mathbf{x})}{(1+\sigma^*)(1+\alpha\sigma^*+\alpha\theta)}\,,\ \forall t_o\in R\ \ (\forall t_o\in T_\tau)\ ,\ \forall t\in T_o\,.$$

This inequality in limit as $\theta\to 0^+$ takes the form

$$D^*v(t,\mathbf{x}) \le -\psi(\mathbf{x})\ ,\ \forall(t,\mathbf{x})\in R\times S\ \ (\forall(t,\mathbf{x})\in T_\tau\times S)\ ,$$

where

$$\psi(\mathbf{x}) = \frac{(\alpha-1)\,\phi(\|\mathbf{x}\|)}{[1+T(\Delta)][1+\alpha T(\Delta)]}\,,$$

is continuous function, $T(\Delta)\in[\tau_u(\Delta,\nu),+\infty[$, because $\mathbf{x}=0$ is uniformly attractive [Halanay (1966)].
Hence, ψ is positive definite on S $(T_\tau\times S)$.

Sufficiency. Under the conditions of Theorem 7 all conditions of Theorem 5 are fulfilled. Hence, $\mathbf{x}=0$ of (7) is uniformly stable (on T_τ). Its uniform attraction (on T_τ) is proved as follows.
Let ζ be such that $B_\zeta\subseteq S$. Let ϕ_1 and $\phi_2\in K_{[0,\zeta[}$ obey

$$\phi_1(\|\mathbf{x}\|)\le v(t,\mathbf{x})\le\phi_2(\|\mathbf{x}\|)\ ,\ \forall(t,\mathbf{x})\in R\times B_\zeta\ \ (\forall(t,\mathbf{x})\in T_\tau\times B_\zeta)\ . \quad (12)$$

Let

$$\Delta = \phi_2^I\,[\phi_1(\zeta)]\ . \quad (14)$$

As shown in the proof of the sufficiency part of Theorem 5, the conditions (13) and (14) guarantee that $\|\mathbf{x}_o\|<\Delta$ implies

$$v\,[t,\chi(t;t_o,\mathbf{x}_o)]\le v(t_o,\mathbf{x}_o)\ ,\ \forall t\in T_o\ ,\ \forall t_o\in R\ \ (\forall t_o\in T_\tau)\ ,$$

and that v is decreasing in t along motions χ of (7). Let

$$\inf\{v\,[t,\chi(t;t_o,\mathbf{x}_o)]:t\in T_o\}=\nu\ ,\ \forall t_o\in R\ \ (\forall t_o\in T_\tau)\ ,\ \|\mathbf{x}_o\|<\Delta\,. \quad (15)$$

Obviously $\nu\ge 0$. If $\nu>0$ then

$$D^*v\,[t,\mathbf{x}(t;t_o,\mathbf{x}_o)]\le-\gamma$$

where

$$\gamma = \inf\{\psi(\mathbf{x}):\mathbf{x}\in\mathbf{S}\setminus\mathbf{B}_\rho\,,\ \rho=\phi_2^I(\nu)\}\ .$$

Therefore,

$$v\,[t,\chi(t;t_o,\mathbf{x}_o)]\le v(t_o,\mathbf{x}_o)-\gamma(t-t_o)\ ,$$

so that

$$v\,[t,\chi(t;t_o,\mathbf{x}_o)]<\nu \quad \text{for} \quad t\in\left]\frac{v(t_o,\mathbf{x}_o)-\nu}{\gamma}+t_o\,,+\infty\right[\,,$$

which contradicts (15). Hence, $\nu=0$ which together with (13),(15) and positive definiteness of v on S (on $T_\tau \times S$) prove that $\|\mathbf{x}_o\| < \Delta$ implies

$$\lim [\|\chi(t;t_o,\mathbf{x}_o)\| : t \to +\infty] = 0 , \forall t_o \in R \quad (\forall t_o \in T_\tau) ,$$

i.e. that $\mathbf{x}=\mathbf{0}$ is attractive. Let $\rho > 0$ be arbitrary,

$\xi = \min\{\phi_2^I[\phi_1(\zeta)], \phi_2^I[\phi_1(\frac{\Delta}{2})]\}$, $\gamma = \inf\{\psi(\mathbf{x}) : \mathbf{x} \in S \setminus B_\xi\}$, $\gamma = \gamma(\rho)$, $\Delta < +\infty$ and/or $\zeta < +\infty$, and

$$\tau_u(\Delta,\rho) = \frac{\phi_2(\Delta) - \phi_1(\xi)}{\gamma(\rho)} .$$

Then,

$$D^* v(t,\mathbf{x}) \le -\gamma , \forall(t,\mathbf{x}) \in R \times (S \setminus B_\xi) \quad (\forall(t,\mathbf{x}) \in T_\tau \times (S \setminus B_\xi)) ,$$

and for $t = \tau_u(\Delta,\rho) + t_o$, $\tau_u(\Delta,\rho) > 0$, [$\tau_u(\Delta,\rho) = 0$ implies $\|x(t;t_o;\mathbf{x}_o)\| < \rho$, $\forall t \in T_o$, which happens only for $\Delta = \zeta = +\infty$] ,

$$v[t,\chi(t;t_o,\mathbf{x}_o)] \le v(t_o,\mathbf{x}_o) - \gamma(t-t_o) \le \phi_2(\Delta) - \phi_2(\Delta) + \phi_1(\xi) \le \phi_1(\rho) ,$$

so that

$$\phi_1[\|\chi(t;t_o,\mathbf{x}_o)\|] \le \phi_1(\rho) , \forall t_o \in R \quad (\forall t_o \in T_\tau) ,$$

yields

$$\|\chi(t,t_o,\mathbf{x}_o)\| \le \rho \quad \text{at} \quad t = \tau_u(\Delta,\rho) + t_o , \forall \mathbf{x}_o \in B_\Delta .$$

For $t \in]\tau_u(\Delta,\rho),+\infty[$

$$v[t,\chi(t;t_o,\mathbf{x}_o)] < v[t_o + \tau_u(\Delta,\rho) ; \chi(t_o + \tau_u(\Delta,\rho),t_o,\mathbf{x}_o)] \le \phi_1(\rho)$$

so that

$$\|\chi(t;t_o,\mathbf{x}_o)\| < \rho , \quad \forall t \in]t_o + \tau_u(\Delta,\rho),+\infty[, \forall t_o \in R \quad (\forall t_o \in T_\tau) , \forall \mathbf{x}_o \in B_\Delta$$

which proves that attraction of $\mathbf{x}=\mathbf{0}$ is uniform (on T_τ). ∎

Following Barbashin and Krasovskii (1952,1954) and the preceding proof in which we choose $\phi \in KR$ it is easy to prove

Theorem 8. *If* χ *is continuous on* $T_o \times R \times R^n$ *(on* $T_o \times T_\tau \times R^n$ *) then existence of radially unbounded decrescent positive definite in the whole function* v *(on* T_τ *) and positive definite function* ψ *in the whole such that, respectively,*

$$D^* v(t,\mathbf{x}) \le -\psi(\mathbf{x}) , \forall(t,\mathbf{x}) \in R \times R^n \quad (\forall(t,\mathbf{x}) \in T_\tau \times R^n) ,$$

is necessary and sufficient for uniform asymptotic stability in the whole of $\mathbf{x}=\mathbf{0}$ *of the system (7).*

Following Krasovskii (1959) and utilizing $\phi(\zeta) = \zeta^p$ in the proof of Theorem 7 it is easy to prove

Theorem 9. *If* χ *is continuous on* $T_o \times R \times N$ *(on* $T_o \times T_\tau \times N$ *) then existence of a time-invariant neighbourhood* S *of* $\mathbf{x}=\mathbf{0}$ *, a function* v*, positive real numbers* η_1, η_2 *and* η_3 *and a positive integer* p *such that* $v(t,\mathbf{x}) \in C(T_o \times N)$ *and both, respectively,*

(i) $\eta_1 \|\mathbf{x}\|^p \le v(t,\mathbf{x}) \le \eta_2 \|\mathbf{x}\|^p$, $\forall(t,\mathbf{x}) \in R \times S \quad (\forall(t,\mathbf{x}) \in T_\tau \times S)$, *and*

(ii) $D^*v(t,\mathbf{x}) \le -\eta_3\|\mathbf{x}\|^p$, $\forall(t,\mathbf{x}) \in R\times S$ $(\forall(t,\mathbf{x}) \in T_\tau\times S)$

is necessary and sufficient for exponential stability (on T_τ *) of* $\mathbf{x}=0$ *of the system (7).*

Theorem 10. *If* χ *is continuous on* $T_0\times R\times R^n$ *(on* $T_0\times T_\tau\times R^n$ *) then existence of a function* v*, positive real numbers* η_1,η_2 *and* η_3 *and a positive integer* p *such that* $v(t,\mathbf{x}) \in C(T_0\times R^n)$ *and both, respectively,*

(i) $\eta_1\|\mathbf{x}\|^p \le v(t,\mathbf{x}) \le \eta_2\|\mathbf{x}\|^p$, $\forall(t,\mathbf{x}) \in R\times R^n$ $(\forall(t,\mathbf{x}) \in T_\tau\times R^n)$,

and

(ii) $D^*v(t,\mathbf{x}) \le -\eta_3\|\mathbf{x}\|^p$, $\forall(t,\mathbf{x}) \in R\times R^n$ $(\forall(t,\mathbf{x}) \in T_\tau\times R^n)$

is necessary and sufficient for exponential stability in the whole (on T_τ *) of* $\mathbf{x}=0$ *of the system (7).*

Remark 6. If $\mathbf{f}$ is differentiable then v is differentiable on the corresponding sets used in Theorem 5 - Theorem 10.

Remark 7. In the case of stationary system (10) the most powerful extension of the Liapunov method is achieved by the LaSalle invariance principle [see LaSalle (1976)].

I.3.2.5. Stability conditions on time-varying sets

The natural sets on which dynamical properties of time-varying systems should be studied are time-varying sets.

Let $P = 2^{R^n}$ be the power set of R^n . A mapping $S : R \to P$ is a *set-valued function.* Its set value at $t\in R$ is $S(t)$ that will be called *time-varying set.* Iff $S(t) \equiv S$ then it is time-invariant set.
Let $d[\mathbf{x},S(t)]$ - the *distance of* $\mathbf{x}$ *from* $S(t)$ be defined by

$$d[\mathbf{x},S(t)] = \inf[\|\mathbf{x}-\mathbf{y}\| : \mathbf{y} \in S(t)] .$$

Let the *distance between* $S_1(t)$ *and* $S_2(t)$ be defined by [see for example Lee and Markus (1967)] :

$$d[S_1(t),S_2(t)] =$$
$$= \max\{\sup(d[\mathbf{y},S_2(t)] : \mathbf{y} \in S_1(t)) , \sup(d[\mathbf{y},S_1(t)] : \mathbf{y} \in S_2(t))\} .$$

Definition 17.

(i) A set $L \subseteq R^n$ is a *limit set of* S , $S : R\to P$, *at* $t=\tau$, $\lim[S(t) : t\to\tau] = L$, iff $\lim\{d[L,S(t)] : t\to\tau\} = 0$.

(ii) Set valued function S is *continuous at* $t=\tau$ iff $\lim[S(t) : t\to\tau] = S(\tau)$.

(iii) Set valued function S is *continuous* (*on* T_τ) iff it is continuous at every $t \in R$ (at every $t \in T_\tau$), $S(t) \in C(R)$ $(S(t) \in C(T_\tau))$. ■

The set $\{\mathbf{x} : \mathbf{x}=\mathbf{0}\}$ will be denoted by O.

Definition 18. Time-varying set $S(t)$ is *asymptotically contractive* (or *asymptotically contracts to* O) iff both

(i) there is $\tau \in R$ such that $S(t)$ is a neighbourhood of $\mathbf{x}=\mathbf{0}$ for every $t \in T_\tau$

and

(ii) the closure $\bar{S}(t)$ of $S(t)$ obeys $\lim[\bar{S}(t) : t \to +\infty] = O$. ■

This is asymptotic contraction of $S(t)$ as $t \to +\infty$.

Definition 19. A set $S(t)$ is :

(i) *positively invariant with respect to function* v (*on* T_τ) iff for every $t_0 \in R$ $(t_0 \in T_\tau)$ there is $\zeta_0 = \zeta(t_0) \in R$ such that both

a) $V_{\zeta_0}(t_0) \subseteq S(t_0)$

and

b) $V_{\zeta_0}(t) \subseteq S(t)$, $\forall t \in T_0$;

(ii) *invariant with respect to function* v iff there is $\zeta \in R$ such that $V_\zeta(t) \subseteq S(t)$, $\forall t \in R$. ■

The notions of asymptotically contractive sets and (positively) invariant sets with respect to functions were discovered by Grujić (1975). Their importance was shown for stability analysis of non-stationary systems. Let : $\mathbf{N} = \{(t;\mathbf{x}) : t \in R, \mathbf{x} \in N(t)\}$. If R is replaced by T_τ then $\mathbf{N}$ is replaced by $\mathbf{N}_\tau$.

Theorem 11. *If* χ *is continuous in* $(t, t_0, \mathbf{x}) \in T_0 \times \mathbf{N}$ $((t, t_0, \mathbf{x}) \in T_0 \times \mathbf{N}_\tau)$ *and* $N(t)$ *is continuous (on* T_τ*) neighbourhood of* $\mathbf{x}=\mathbf{0}$ *at every* $t \in R$ $(t \in T_\tau)$*, which may be time-invariant,* $N(t) \equiv N$*, or time-varying either asymptotically contractive or not, then existence of functions* v *and* $\phi \in K$*, and continuous neighbourhood* $S(t)$ *of* $\mathbf{x}=\mathbf{0}$ *at every* $t \in R$ $(t \in T_\tau)$ *such that, respectively,*

(i) $v(t,\mathbf{x}) \in C(\mathbf{S})$ $(v(t,\mathbf{x}) \in C(\mathbf{S}_\tau))$,

(ii) $v(t,\mathbf{0}) = 0$, $\forall t \in R$ $(\forall t \in T_\tau)$,

(iii) $v(t,\mathbf{x}) \geq \phi(\|\mathbf{x}\|)$, $\forall (t,\mathbf{x}) \in \mathbf{S}$ $(\forall (t,\mathbf{x}) \in \mathbf{S}_\tau)$,

and

(iv) $D^+v(t,\mathbf{x}) \leq 0$, $\forall (t,\mathbf{x}) \in \mathbf{S}$ $(\forall (t,\mathbf{x}) \in \mathbf{S}_\tau)$,

is necessary for stability (on T_τ*) of* $\mathbf{x}=\mathbf{0}$ *of the system (7).*

Proof. Let $\mathbf{x}=\mathbf{0}$ of (7) be stable (on T_τ), $\epsilon > 0$ and $t_0 \in R$ $(t_0 \in T_\tau)$ be arbitrarily chosen. Let $\nu(t_0)$ be the largest positive number obeying $B_\nu(t_o) \subseteq N(t_o)$ and $\xi(t_o,\epsilon) = \min\{\nu(t_o), \delta_M(t_o,\epsilon)\}$. Continuity of χ in all its arguments implies continuity of $\delta_M(t,\epsilon)$ in $t \in R$ ($t \in T_\tau$) for every fixed $\epsilon > 0$. Continuity of $N(t)$ guarantees continuity of $\nu(t)$ in $t \in R$ ($t \in T_\tau$). Hence, $\xi(t,\epsilon)$ is continuous in $t \in R$ ($t \in T_\tau$) for every fixed $\epsilon > 0$. Let $\phi \in K$ and v be defined for such ϕ by (11) and $S(t) = B_\xi(t,\epsilon)$ that is continuous in $t \in R$ ($t \in T_\tau$) for arbitrary fixed $\epsilon > 0$. Continuity of χ in $(t_o, \mathbf{x}_o) \in \mathbf{S}$ ($(t_o, \mathbf{x}_o) \in \mathbf{S}_\tau$), and $\phi \in K$ imply (i) of the Theorem. Stability of $\mathbf{x} = \mathbf{0}$ (on T_τ) implies $\chi(t; t_o, \mathbf{0}) = \mathbf{0}$, $\forall t \in T_o$, $\forall t_o \in R$ ($\forall t_o \in T_\tau$). Hence, (ii) of the Theorem holdsin view of (11). The condition (iii) of the Theorem is obviously satisfied due to (11). Let $\mathbf{x}^* = \chi(t+\theta; t, \mathbf{x})$, $(t,\mathbf{x}) \in \mathbf{S}$ ($(t,\mathbf{x}) \in \mathbf{S}_\tau$), $\theta \in \,]0,+\infty[$. Then,

$$\begin{aligned} v(t+\theta; \mathbf{x}^*) &= \sup\{\phi[\|\chi(t+\theta+\sigma; t+\theta, \mathbf{x}^*)\|] : \sigma \in [0,+\infty[\} \\ &= \sup\{\phi[\|\chi(t+\theta+\sigma; t, \mathbf{x})\|] : \sigma \in [0,+\infty[\} \\ &\leq \sup\{\phi[\|\chi(t+\sigma_1; t, \mathbf{x})\|] : \sigma_1 \in [0,+\infty[\} \\ &= v(t,\mathbf{x}). \end{aligned}$$

Hence,

$$\frac{v[t+\theta, \chi(t+\theta; t, \mathbf{x})] - v(t,\mathbf{x})}{\theta} \leq 0, \quad \forall (t,\mathbf{x}) \in \mathbf{S} \quad (\forall (t,\mathbf{x}) \in \mathbf{S}_\tau).$$

Taking right upper limit of this inequality we get (iv) of the Theorem. ∎

The conditions of this theorem are only necessary and not sufficient. The form of sufficient conditions essentially depends on that whether $S(t)$ is asymptotically contractive, or not. If it is not asymptotically contractive either as $t \to -\infty$ or $t \to +\infty$ (only as $t \to +\infty$ for stability on T_τ) then there is a time-invariant neighbourhood S of $\mathbf{x} = \mathbf{0}$ such that $S \subseteq S(t)$, $\forall t \in [-\infty, +\infty]$ ($\forall t \in [\tau, +\infty]$). In such cases sufficient conditions that are also necessary were established already by Liapunov (1892). In order to clarify the cause of this phenomenon we state the next result by following Grujić (1975).

Proposition 14. *If S is a time-invariant neighbourhood of $\mathbf{x} = \mathbf{0}$ then it is invariant (positively invariant on T_τ) with respect to every positive definite function (on T_τ), respectively.*
The same is true for every continuous time-varying set $S(t)$ that is a neighbourhood of $\mathbf{x} = \mathbf{0}$ at every $t \in R$ ($t \in T_\tau$) and is not asymptotically contractive as $t \to -\infty$ and $t \to +\infty$ (as $t \to +\infty$), respectively.

However, asymptotically contractive neighbourhood $S(t)$ of $\mathbf{x} = \mathbf{0}$ is neither invariant nor positively invariant with respect to every positive definite function. For example, asymptotically contractive neighbourhoods $S_1(t)$ and $S_2(t)$ of $\mathbf{x} = \mathbf{0}$,

$$S_1(t) = \{\mathbf{x} : \|\mathbf{x}\| < \exp(-t)\}$$

and

$$S_2(t) = \{x : \sum_{i=1}^{n} \alpha_i |x_i| < (1+t^2)^{-1} \ , \ \alpha_i \in]0,+\infty[\ , \ i=1,2,\ldots,n\}$$

are not positively invariant with respect to positive definite functions v_1 and v_2, $v_1(\mathbf{x}) = \mathbf{x}^T H\mathbf{x}$, $H = H^T$ is positive definite, or $v_2(t,\mathbf{x}) = (2+\sin t)\,\mathbf{x}^T H\mathbf{x}$, etc.

Remark 8. If $\mathbf{f}$ is continuous in $(t,\mathbf{x}) \in \mathbf{N}$ $((t,\mathbf{x}) \in \mathbf{N}_\tau)$, and $N(t)$ is continuous (on T_τ) neighbourhood of $\mathbf{x}=\mathbf{0}$ at every $t \in T$ $(t \in T_\tau)$, then the conditions of Theorem 11 are also necessary but not sufficient for asymptotic stability (on T_τ) of $\mathbf{x}=\mathbf{0}$ of the system (7), respectively. If $N(t)$ is asymptotically contractive a number of different sufficient conditions can be obtained for asymptotic stability (on T_τ) of $\mathbf{x}=\mathbf{0}$ of the system (7) (Grujić, 1975).

Example 7 (Grujić, 1975). Let

$$\frac{d\mathbf{x}}{dt} = \frac{1}{2}\,[1+\exp(-t)]^{-1}\,[2\exp(-t) - (1+\exp t)^{-1} + \|\mathbf{x}\|^2]\,\mathbf{x}$$

If $v(t,\mathbf{x}) = [1+\exp(-t)]^2 \|\mathbf{x}\|^2$ is chosen, which is positive definite and differentiable in the whole, then

$$\dot v(t,\mathbf{x}) = -[1+\exp(-t)]\cdot[(1+\exp t)^{-1} - \|\mathbf{x}\|^2]\,\|\mathbf{x}\|^2 \ , \ \forall (t,\mathbf{x}) \in R\times R^n \ .$$

Let

$$S(t) = \{\mathbf{x} : \|\mathbf{x}\|^2 < (1+\exp t)^{-1}\} \ .$$

The set $S(t)$ is continuous neighbourhood of $\mathbf{x}=\mathbf{0}$ at every $t \in R$ and asymptotically contractive. Obviously,

$$\dot v(t,\mathbf{x}) < 0 \ , \ \forall (\mathbf{x} \neq \mathbf{0}) \in S(t) \ , \ \forall t \in R \ .$$

Despite this negativeness of $\dot v$ on $S(t)$ for all $t \in R$ and positive definiteness in the whole of v, the equilibrium state $\mathbf{x}=\mathbf{0}$ of the system is evidently unstable. The cause of this is that the set $S(t)$ is not positively invariant with respect to the function v.

Example 7 can be also helpful for understanding the counterexample by Massera (1949) [see also p.32 in the book by Rouche, Habets and Laloy (1977)] to the following

Hypothesis 1. If there is a time-invariant neighbourhood N of $\mathbf{x}=\mathbf{0}$, $v(t,\mathbf{x}) \in C^{(1,1)}(T_\tau \times N)$ for some $\tau \in R$ and $\phi_i \in K$, $i=1,2$, such that for all $(t,\mathbf{x}) \in T_\tau \times N$ both

(i) $v(t,\mathbf{x}) \geq \phi_1(\|\mathbf{x}\|)$

and

(ii) $\dot v(t,\mathbf{x}) \leq -\phi_2(\|\mathbf{x}\|)$

then $\mathbf{x}=\mathbf{0}$ of the system (7) is asymptotically stable on T_τ . ∎

The hypothesis fails because the function v may be such that for any $\zeta>0$ the set $V_\zeta(t)$ is not asymptotically contractive that should be required because v is not decrescent function.

Definition 20. A function $\omega: R\times R^n \to R^n$ is the *boundary function of a set* $S(t)\subset R^n$ *(on* T_τ *)* iff, respectively, both

(i) $S(t)$ has the boundary $\partial S(t)$ at every $t\in T$ $(t\in T_\tau)$, and there is $\xi\in R$ such that $\omega(t,\mathbf{x})=\xi$, $\forall \mathbf{x}\in\partial S(t)$, $\forall t\in R$ $(\forall t\in T_\tau)$,

and

(ii) for every $t\in R$ $(t\in S_\tau)$, there is $\zeta(t)>0$ such that $\omega(t,\mathbf{x})\neq\xi$, $\forall \mathbf{x}\in N[\zeta(t),\partial S(t)]\setminus\partial S(t)$ where $N[\zeta(t),\partial S(t)]$ is $\zeta(t)$ neighbourhood of $\partial S(t)$. ∎

Definition 21. The boundary function of $S(t)$ is *radially increasing on* $\partial S(t)$ *(on* $\partial S(t)\times T_\tau$ *)* iff for every $t\in R$ $(t\in T_\tau)$, there is $\zeta(t)>0$ such that $\mathbf{x}\in\partial S(t)$ implies

$$\omega[t,(1+\lambda_1)\mathbf{x}] < \xi < \omega[t,(1+\lambda_2)\mathbf{x}] \ , \ \forall|\lambda_1| \ , \ \lambda_2\in\,]0,\zeta(t)[\ , \ \lambda_1<0 \ . \ ∎$$

The next result generalizes Lemma 2.1 by Harrison (1979) to time-varying sets.

Lemma 1. *If* $S(t)$ *is continuous (on* T_τ *) and* χ *is continuous on* $T_o\times R\times R^n$ $(T_o\times T_\tau\times R^n)$ *, then, respectively, both*

(i) the boundary function ω *of* $S(t)$ *is radially increasing on* $\partial S(t)$ *(on* $\partial S(t)\times T_\tau$ *) ,*

and

(ii) $D^+\omega(t,\mathbf{x})<0$, $\forall\mathbf{x}\in\partial S(t)$, $\forall t\in R$ $(\forall t\in T_\tau)$,

are sufficient for $\bar S(t)$ *to be invariant (positively invariant on* T_τ *) with respect to motions of the system (7).*

Proof. Let all conditions of the lemma hold and let there exist $(\sigma,\mathbf{y})\in R\times\partial S(\sigma)$ $((\sigma,\mathbf{y})\in T_\tau\times\partial S(\sigma))$ and $\gamma>0$ such that $\chi(t;\sigma,\mathbf{y})\notin\bar S(t)$, $\forall t\in\,]\sigma,\gamma[$. Hence, $\omega(\sigma,\mathbf{y})=\xi$ and $\omega[t,\chi(t;\sigma,\mathbf{y})]>\xi$, $\forall t\in\,]\sigma,\nu[$, where $\nu\in\,]\sigma,\gamma[$ exists because χ and $S(t)$ are continuous in $t\in R$ $(t\in T_\tau)$. Hence,

$$\frac{\omega[\sigma+\theta;\chi(\sigma+\theta;\sigma,\mathbf{y})]-\omega(\sigma,\mathbf{y})}{\theta} > 0 \ , \ \forall\theta\in\,]0,\nu-\sigma[\ ,$$

which in limit as $\theta\to 0^+$ implies $D^+\omega(\sigma,\mathbf{y})>0$. This result contradicts the condition (ii) of Lemma. The contradiction proves the Lemma. ∎

Notice that $\omega(t,\mathbf{z})>\omega(t,\mathbf{x})$ is admissible even $\mathbf{z}$ is the interior point of $S(t)$ and $\mathbf{x}$ is its boundary point.

Theorem 12. *Let* χ *be continuous on* $T_o \times R \times R^n$ $(T_o \times T_\tau \times R^n)$ *and*

(i) $S(t)$ *be continuous neighbourhood of* $\mathbf{x}=0$ *at every* $t \in R$ $(t \in T_\tau)$*;*

(ii) v *be positive definite on* $\cup[S(t) : t \in R]$ $(\cup[S(t) : t \in T_\tau])$*;*

(iii) *the boundary function* ω *of* $S(t)$ *be radially increasing on* $\partial S(t)$ *(on* $\partial S(t) \times T_\tau$*),* $\omega = v$ *is admissible;*

(iv) $D^+v(t,\mathbf{x}) \leq 0$, $\forall \mathbf{x} \in S(t)$, $\forall t \in R$ $(\forall t \in T_\tau)$ *;*

(v) $D^+\omega(t,\mathbf{x}) < 0$, $\forall \mathbf{x} \in \partial S(t)$, $\forall t \in R$ $(\forall t \in T_\tau)$ *;*

(vi) $S(t)$ *be asymptotically contractive.*

Then, respectively,

(a) *the conditions (i)-(v) are sufficient for stability of* $\mathbf{x}=0$ *of the system (7) (on* T_τ *);*

(b) *the conditions (i), (iii), (v) and (vi) are sufficient for attraction of* $\mathbf{x}=0$ *of the system (7) (on* T_τ *);*

(c) *the conditions (i)-(vi) are sufficient for asymptotic stability of* $\mathbf{x}=0$ *of the system (7) (on* T_τ *).*

Proof.

A. The conditions (i), (iii) and (v) prove that the set $\bar{S}(t)$ is positively invariant (on T_τ) with respect to motions of the system (7), $\mathbf{x}(t_o) \in \bar{S}(t_o) \Rightarrow \chi(t;t_o,\mathbf{x}_o) \in \bar{S}(t)$, $\forall t \in [t_o,+\infty[$, $\forall t_o \in R$ $(\forall t_o \in T_\tau)$, due to Lemma 1.

B. The condition (vi) means that $\lim\,[\bar{S}(t) : t \to +\infty] = 0$.

C. A and B prove (b).

D. Let $t_o \in R$ $(t_o \in T_\tau)$, and $\epsilon > 0$ be arbitrarily chosen. Let $\zeta(t_o,\epsilon) \in \,]0,+\infty[$ be such that $V_{\zeta(t_o,\epsilon)}(t) \subseteq B_\epsilon$, $\forall t \in [t_o;+\infty[$. Existence of the positive number $\zeta(t_o,\epsilon)$ obeying the last condition is assured by both the positive definiteness of v (the condition (ii)) and time-invariance of B_ϵ (Proposition 14).
Let $\delta(t_o,\epsilon) \in \,]0,+\infty[$ be chosen so that $B_{\delta(t_o,\epsilon)} \subseteq V_{\zeta(t_o,\epsilon)}(t_o)$. The condition (iv) and the positive invariance of $\bar{S}(t)$ (A) now prove that $\mathbf{x}_o \in B_{\delta(t_o,\epsilon)}$ implies

$$\chi(t;t_o,\mathbf{x}_o) \in V_{\zeta(t_o,\epsilon)}(t) \subseteq B_\epsilon \ , \ \forall t \in T_o \ , \ \forall t_o \in R \ (\forall t_o \in T_\tau) \ ,$$

which proves stability (on T_τ) of $\mathbf{x}=0$. Hence, (a) holds.

E. The assertion (c) is implied by (a) and (b), which completes the proof. ■

Example 8 (Grujić, 1975). Let n-th order system be defined by

$$\frac{d\mathbf{x}}{dt} = -\mathbf{x} + (4 + 2t + t^2)\|\mathbf{x}\|^2\,\mathbf{x} \ .$$

Let the function v be accepted in the form $v(\mathbf{x}) = \|\mathbf{x}\|^2$ so that it is positive definite in the whole and

$$\dot{v}(\mathbf{x}) = -2\|\mathbf{x}\|^2[1-(4+2t+t^2)\|\mathbf{x}\|^2] .$$

This result suggests that $S(t)$ should be in the form

$$S(t) = \{\mathbf{x} : \|\mathbf{x}\|^2 \le \alpha(4+2t+t^2)^{-1}\} , \quad \alpha \in]0,1] .$$

The boundary $\partial S(t)$ of $S(t)$ is now easily found,

$$\partial S(t) = \{\mathbf{x} : \|\mathbf{x}\|^2 = \alpha(4+2t+t^2)^{-1}\} , \quad \omega(t,\mathbf{x}) = (4+2t+t^2)\|\mathbf{x}\|^2 .$$

ω is the boundary function of the set $S(t)$. It is radially increasing on $\partial S(t)$. Furthermore,

$$\dot{\omega}(t,\mathbf{x}) = -2\|\mathbf{x}\|^2[3+t+t^2-(4+2t+t^2)\|\mathbf{x}\|^2] ,$$

so that on $\partial S(t)$,

$$\dot{\omega}(t,\mathbf{x}) = +2\alpha(4+2t+t^2)^{-1}[(\alpha-1)t^2+(2\alpha-1)t+(4\alpha-3)] .$$

For $\dot{\omega}(t,\mathbf{x})<0$ on $\partial S(t)$ it is sufficient that $\alpha = 0.70$, which implies also

$$\dot{v}(t,\mathbf{x}) < 0 , \quad \forall(\mathbf{x}\neq\mathbf{0}) \in S(t) , \quad \forall t \in R ,$$

and

$$S(t) = \{\mathbf{x} : \|\mathbf{x}\|^2 \le 0.70\,(4+2t+t^2)^{-1}\} .$$

Obviously, the set $S(t)$ is continuous and

$$\lim\,[S(t) : t\to+\infty] = \mathbf{0} .$$

All conditions of Theorem 12 are satisfied. The state $\mathbf{x}=\mathbf{0}$ of the system is asymptotically stable.

Theorem 13. *Let* X *be continuous on* $T_o\times R\times R^n$ $(T_o\times T_\tau\times R^n)$ *and*

(i) $S(t)$ *be continuous neighbourhood of* $\mathbf{x}=\mathbf{0}$ *at every* $t\in R$ $(t\in T_\tau)$;

(ii) v *be positive definite on* $\cup[S(t) : t\in R]$ $(\cup[S(t) : t\in T_\tau])$;

(iii) $S(t)$ *be positively invariant with respect to* v *(on* T_τ*)*;

(iv) $S(t)$ *be asymptotically contractive*

and

(v) $D^+v(t,\mathbf{x}) \le 0$, $\forall \mathbf{x}\in S(t)$, $\forall t\in R$ $(\forall t\in T_\tau)$.

Then, respectively,

(a) *the conditions (i)-(iii) and (v) are sufficient for stability of* $\mathbf{x}=\mathbf{0}$ *(on* T_τ*)*;

(b) *the conditions (i)-(v) are sufficient for asymptotic stability of* $\mathbf{x}=\mathbf{0}$ *(on* T_τ*)*.

Proof.

A. The condition (iii) implies existence of $\zeta(t_o)>0$ such that $V_{\zeta(t_o)}(t_o)\subseteq S(t_o)$ implies

$$V_{\zeta(t_o)}(t) \subseteq S(t) \ , \ \forall t \in T_o \ , \ \forall t_o \in R \ (\forall t_o \in T_\tau) \ .$$

This result together with (ii) and (v) guarantees that $x_o \in V_\zeta(t_o)$ implies

$$\chi(t;t_o,x_o) \in V_\zeta(t) \ , \ \forall t \in T_o \ , \ \forall t_o \in R \ (\forall t_o \in T_\tau) \ , \ \forall \zeta \in]0,\zeta(t_o)[\ .$$

B. Let $t_o \in R$ $(t_o \in T_\tau)$, and $\epsilon > 0$ be arbitrarily chosen. Let $\xi(t_o) \in]0,\zeta(t_o)[$ be such that $V_{\xi(t_o)}(t) \subseteq B_\epsilon$, $\forall t \in T_o$. Such $\xi(t_o)$ exists due to time-invariance of B_ϵ and positive definiteness of v (Proposition 14). Let $\delta(t_o,\epsilon) \in]0,+\infty[$ be such that $B_{\delta(t_o,\epsilon)} \subseteq V_{\xi(t_o)}(t_o)$. Hence, $x_o \in B_{\delta(t_o,\epsilon)}$ implies

$$\chi(t;t_o,x_o) \in B_\epsilon \ , \ \forall t \in T_o \ ,$$

due to the result under A, which proves (a).

C. Let $\Delta(t_o) \in]0,+\infty[$ obey $B_{\Delta(t_o)} \subseteq V_\zeta(t_o)$, $\zeta \in]0,\zeta(t_o)[$. The conditions (i)-(v) guarantee that $x_o \in B_{\Delta(t_o)}$ implies

$$\lim\ [\chi(t;t_o,x_o) : t \to +\infty] = 0 \ , \ \forall t_o \in R \ (\forall t_o \in T_\tau) \ ,$$

due to the result under A. Hence, $x = 0$ is attractive (on T_τ), which together with (a) proves (b). ■

Example 9 (Grujić, 1975). A positive definite function v in the whole, which is not decrescent,

$$v(t,x) = \begin{cases} 2(1+\exp t)\|x\|^2 \ , & \|x\|^2 \le [2(1+\exp t)]^{-1} \\ -2(1+\exp t)\|x\|^4 + (3+2\exp t)\|x\|^2 \ , & [2(1+\exp t)]^{-1} \le \|x\|^2 \le 1 \\ \|x\|^2 \ , & \|x\|^2 \ge 1 \end{cases}$$

is chosen for stability analysis of the n-th order system

$$\frac{dx}{dt} = -\{\tfrac{1}{2}[1+2\exp(-t)] + 2\exp(-2t)\}\, x + [1+2\exp(-t)]^2 \|x\|^2 x \ .$$

Hence,

$$\dot v(t,x) = -2\{1 + 4\exp(-t)[1+\exp(t)][1+2\exp(-t)][1-(2+\exp t)\|x\|^2]\} \ .$$

This result suggests the choice of $S(t)$, $S(t) = \{x : \|x\|^2 < (2+\exp t)^{-1}\}$ so that $S(t)$ is asymptotically contractive, $0 \subset S(t)$, $\forall t \in R$, and $\lim\ [\bar S(t) : t \to +\infty] = 0$.

The set $S(t)$ is positively invariant with respect to the function v because for every $\tau \in R$, the choice of

$$\zeta_M(\tau) = \frac{2+2\exp\tau}{2+\exp\tau} \quad \text{and} \quad \zeta \in]0,\zeta_M(\tau)]$$

imply

$$V_\zeta(t) = \{x : \|x\|^2 < \zeta[2(1+\exp t)]^{-1}\} \subseteq S(t) \ , \ \forall t \in T_\tau \ .$$

All conditions of Theorem 13 are satisfied. Hence, $x=0$ of the system is asymptotically stable.

Example 10 (Grujić, 1975). The n-th order linear system is described by

$$\frac{dx}{dt} = -2^{-1}(\beta + 2\gamma t)(\alpha + \beta t + \gamma t^2)^{-1} x \ , \ 4\alpha\gamma > \beta^2 \ ; \ \alpha, \gamma \in \mathring{R}_+ \ .$$

Let $v(t,x) = (\alpha + \beta t + \gamma t^2)\|x\|^2$. Evidently,

$$v(t,x) \geq (4\alpha\gamma - \beta^2)(4\gamma)^{-1}\|x\|^2 \ .$$

The function v is positive definite in the whole. It is not decrescent. Its total time derivative along system motions obeys $\dot{v}(t,x) = 0$, $\forall x \in R^n$, $\forall t \in R$.

Let $\xi \in]0,+\infty[$ and $S(t) = \{x : \|x\|^2 < \xi(\alpha + \beta t + \gamma t^2)^{-1}\}$. The set $S(t)$ is invariant with respect to v because

$$V_\zeta(t) = \{x : \|x\|^2 < \zeta(\alpha + \beta t + \gamma t^2)^{-1}\} \subseteq S(t) \ , \ \forall t \in]-\infty,+\infty[\ , \ \forall \zeta \in]0,\xi] \ .$$

The set $S(t)$ is obviously continuous and asymptotically contractive for any $\xi \in]0,+\infty[$. Besides, $\dot{v}(t,x) = 0$, $\forall x \in S(t)$, $\forall t \in R$.
All conditions of Theorem 13 are fulfilled. The state $x=0$ of the system is asymptotically stable.

Note that Example 10 illustrates application of Theorem 13 under weaker condition on $\dot{v}$ than that of Hypothesis 1, which is possible because of continuity, asymptotic contraction and (positive) invariance of the set $S(t)$ with respect to the function v . Asymptotic contraction of $S(t)$ and its (positive) invariance with respect to v prove that the set $V_\zeta(t)$, $\zeta \in]0,\xi]$, $\xi < +\infty$, is also asymptotically contractive.

In the case that $\xi = +\infty$ then $S(t) \equiv R^n$, but then the set $V_\zeta(t)$, $\zeta \in]0,+\infty[$, can be accepted as a new set $S(t)$ of Theorem 13. In fact, the next result follows directly from Theorem 13 by relying on the preceding explanation.

Corollary 2 (Barbashin and Krasovskii, 1952; Grujić, 1975). *If there exist a time invariant neighbourhood N of $x=0$, a positive definite function v on N and a real number $\xi \in]0,+\infty[$ such that both*

(i) $\lim [\bar{V}_\zeta(t) : t \to +\infty] = 0$, $\forall \zeta \in]0,\xi]$,

and

(ii) $D^+v(t,x) \leq 0$, $\forall (t,x) \in R \times N$ $(\forall (t,x) \in T_\tau \times N)$,

then the state $x=0$ of the system (7) is asymptotically stable (on T_τ), respectively.

If, in addition, $N = R^n$ and v is radially unbounded (on T_τ) then $x=0$ of the system (7) is asymptotically stable in the whole (on T_τ) .

Notice that in the case when $N=R^n$ this corollary corresponds to Theorem 42.3 by Hahn (1967), which shows that asymptotic stability of **x=0** is then equi-asymptotic stability of it. This assertion can be easily broadened to Corollary 2.

I.4. ON ABSOLUTE STABILITY

I.4.1. Introductory comments

The absolute stability problem has been one of the central problems of the stability theory for its theoretical and engineering importance. Lur'e and Postnikov (1944) showed that mathematical models of hydraulic servo-systems had a particular form referred to as Lur'e-Postnikov system, or in short as Lur'e system. The absolute stability problem has also become the classical problem of the control theory and control engineering.

Since 1944 the problem has been attacked via various approaches. Most of them were aimed to establish sufficient conditions for absolute stability, which was firstly done by Lur'e himself (1951). The Lur'e conditions are purely algebraic and require verification of existence of solutions of non-linear algebraic equations, which has been the key obstacle for their effective application.

Elegant frequency criteria by Popov (1959, 1960, 1961, 1962, 1963, 1964, 1973, 1974) have become most applicable and fruitful for further research in various directions.

The ingenious approach by Popov inspired a great number of scientists for its extensions, whose contributions will be referred to in brief. The crucial extensions of the Popov approach were at first achieved by Yakubovich (1962, 1963, 1964, 1965, 1967, 1968, 1970, 1975), Tsypkin (1962, 1963, 1964), Kalman (1963) and Szöge (1963, 1964). Other important contributions are due to Gelig (1964), Halanay (1964), Ibrahim and Rekasius (1964), Naumov and Tsypkin (1964), Desoer (1965), Jury and Lee (1965), Tokumaru and Saito (1965), Dewey and Jury (1966), Meyer (1965), Anderson (1966), Dumkov (1967), Moore and Anderson (1968), Partovi and Nahi (1969), Bertoni, Bonivento and Sarti (1970), Šiljak and Sun (1971, 1972), Anderson and Moore (1972), Garg and Rabins (1972) and Piatnitskii (1973). Excellent surveys of these and other contributions together with significant original results can be found in the works by Aizerman and Gantmacher (1963), Gantmacher and Yakubovich (1965), Lefschetz (1965), Piatnitskii (1968), Šiljak (1969, 1974) and Narendra and

Taylor (1973).

The problem of the necessary and sufficient conditions for absolute stability in particular cases was solved by Nelepin (1967), Persidskii (1969), Piatnitskii (1970, 1971), Maigarin (1970), Muhametz'anov and Serikbaev (1970).

The necessary and sufficient Liapunov-like conditions for absolute stability were established by Grujić (1978a,b, 1980, 1981b), which show that a functional family should be used rather than a single function whose form is not influenced by the form of system nonlinearities.

I.4.2. Description of Lur'e-Postnikov systems

Non-stationary Lur'e-Postnikov systems are described by

$$\frac{d\mathbf{x}}{dt} = \mathbf{A}(t)\,\mathbf{x} + \mathbf{B}(t)\,\mathbf{f}(t,\Sigma)\ ,\ \Sigma = \mathbf{C}(t)\,\mathbf{x} + \mathbf{D}(t)\,\mathbf{f}(t,\Sigma)\ . \tag{16}$$

It is accepted that the matrix functions $\mathbf{A} \in C(R,R^{n\times n})$, $\mathbf{B} \in C(R,R^{n\times m})$, $\mathbf{C} \in C(R,R^{m\times n})$, $\mathbf{D} \in C(R,R^{m\times m})$ and that they are known. The vector function $\mathbf{f} \in C(R\times R^m,R^m)$ and its real functional form need not be known. However, certain properties of $\mathbf{f}$ are to be known. They will be defined in the sequel. Notice that

$$\mathbf{f} = (\phi_1,\phi_2,\dots,\phi_m)^T \quad \text{and} \quad \Sigma = (\sigma_1,\sigma_2,\dots,\sigma_m)^T \in R^m\ .$$

A matrix function $\mathbf{N} : R\times R^m \to R^{m\times m}$ is associated with $\mathbf{f}$ by

$$\mathbf{N}(t,\mathbf{x};\mathbf{f}) = \operatorname{diag}\left\{\frac{\phi_1(t,\Sigma)}{\sigma_1}, \frac{\phi_2(t,\Sigma)}{\sigma_2}, \dots, \frac{\phi_m(t,\Sigma)}{\sigma_m}\right\}\Bigg|_{\Sigma=\Sigma(t,\mathbf{x},\mathbf{f})}$$

In case $\mathbf{f}(t,\Sigma) \in C^{(1,j)}(R\times R^m)$, $j \in \{0,1\}$, then

$$\mathbf{N}_t(t,\mathbf{x};\mathbf{f}) = \frac{\partial}{\partial t}\mathbf{N}(t,\mathbf{x};\mathbf{f})$$

and in case $\mathbf{f}(t,\Sigma) \in C^{(j,1)}(R\times R^m)$, $j \in \{0,1\}$, then its Jacobian matrix $\mathbf{f}_\Sigma$ is defined by

$$\mathbf{f}_\Sigma(t,\Sigma) = \left[\frac{\partial\phi_i(t,\Sigma)}{\partial\sigma_j}\right] .$$

Let elementwise constant matrices $\mathbf{J},\mathbf{L},\mathbf{M} \in R^{m\times m}$ be of the same type as matrices $\mathbf{N}_t$, $\mathbf{N}$ and $\mathbf{f}_\Sigma$, respectively. For example, the (i,j)-th element of $\mathbf{M}$ is equal to zero iff

$$\frac{\partial\phi_i}{\partial\sigma_j} \equiv 0\ .$$

With the matrices $\mathbf{J}$, $\mathbf{L}$ and $\mathbf{M}$ will be associated sets J , L and M , respectively. It is accepted that $(\mathbf{J}=\mathbf{0}) \in J$ and $(\mathbf{M}=\mathbf{0}) \in M$. These sets may be unbounded. Notice that $M \supseteq L$ holds always.

In case $f(t,\Sigma) \in C^{(1,1)}(R \times R^m)$ then it is accepted that $A(t)$, $B(t)$, $C(t)$ and $D(t)$ are also differentiable in $t \in R$.

The vector nonlinearity $f \in N_o(L)$ iff

(1) $x=0$ is the unique equilibrium state of (16),

and

(2) $N(t,x;f) \in L$, $\forall (t,x) \in R \times R^n$, $x \neq 0$.

Iff in addition to (1) and (2) the function f obeys

(3) $f(t,\Sigma) \in C^{(1,1)}(R \times R^m)$,

(4) $N_t(t,\Sigma) \in J$, $\forall (t,\Sigma) \in R \times R^m$

and

(5) $f_\Sigma(t,\Sigma) \in M$, $\forall (t,\Sigma) \in R \times R^m$

then $f \in N_2(J,L,M)$. Iff J and M are such that some (or all) elements of J, $J \in J$, and M, $M \in M$, can be arbitrarily large (or small when they are negative) then f obeying (1)-(5) belongs to $N_1(L)$.

The sets J, L and M can be prescribed or to be determined. Usually they are fixed and then for the sake of simplicity $N_o(L)$, $N_1(L)$ and $N_2(J,L,M)$ will be denoted by N_o, N_1 and N_2 respectively.

The motion of the system (16) is denoted by

$$\chi(t;t_o,x_o;f), \quad \chi(t_o;t_o,x_o;f) \equiv x_o,$$

where the argument f emphasizes that the form of χ depends on the form of f. Besides, χ is assumed continuous in all its arguments.

I.4.3. Definition of absolute stability

The concept of absolute stability was originally related to time-invariant Lur'e-Postnikov systems and later on it was broadened to time-varying Lur'e-Postnikov systems (16).

Definition 22. The state $x=0$ of the system (16) is

(i) *absolutely stable on* N_j, $j \in \{0,1,2\}$, iff it is asymptotically stable in the whole for every $f \in N_j$;

(ii) *uniformly absolutely stable on* N_j iff it is uniformly asymptotically stable in the whole for every $f \in N_j$;

(iii) *completely uniformly absolutely stable on* N_j iff it is absolutely stable and in addition iff both

(a) for every $\epsilon \in \overset{o}{R}_+$ there is $\delta(\epsilon) > 0$, which is independent of $(t_o,f) \in R \times N_j$, such that $\|x_o\| < \delta(\epsilon)$ implies

$$\|x(t;t_o,x_o;f)\| < \epsilon \ , \ \forall t\in T_o \ , \ \forall t_o\in R \ , \ \forall f\in N_j \ ,$$

and

(b) for every $\Delta \in]0,+\infty[$ and $\rho > 0$ there is $\tau(\Delta,\rho) \in [0,+\infty[$, which is independent of $(t_o,x_o,f) \in R\times R^n\times N_j$, such that $\|x_o\| < \Delta$ implies

$$\|x(t;t_o,x_o;f)\| < \rho \ , \ \forall t \in]t_o + \tau(\Delta,\rho),+\infty[\ , \ \forall t_o\in R \ , \ \forall f\in N_j \ . \blacksquare$$

Notice that complete uniform absolute stability of $x=0$ means its absolute stability uniformly in both initial data $(t_o,x_o) \in R\times R^n$ and nonlinearities $f\in N_j$, $j \in \{0,1,2\}$.

Since Liapunov-like both necessary and sufficient conditions have not been discovered even for asymptotic stability of $x=0$ of time-varying system (7) in general, then such conditions are not known for absolute stability on N_j , $j \in \{0,1,2\}$, of $x=0$ of the system (16).

I.4.4. Liapunov-like conditions for uniform absolute stability

Let $v^f : R\times R^n \to R_+$ be a function whose functional form depends on that of f . Then,

$$F_j = \{v^f : v^f(t,x) \in C^{(i,i)}(R\times R^n)\} \ ,$$

$$i=0 \text{ iff } j=0 \ , \ i=1 \text{ iff } j \in \{1,2\} \ , \ j \in \{0,1,2\} \ ,$$

is a functional family. Besides, the following functional families will be needed,

$$\Phi_j = \{\phi^f : \phi^f \in KR \ , \ f\in N_j\} \ , \quad j \in \{0,1,2\} \ ,$$

$$\Psi_j = \{\psi^f : \psi^f \in K \ , \ f\in N_j\} \ , \quad j \in \{0,1,2\} \ .$$

Theorem 14. *It is necessary and sufficient for uniform absolute stability on* N_j , $j \in \{0,1,2\}$, *of* $x=0$ *of (16) that there exist a function* $\phi \in KR$ *and functional families* F_j , Φ_j *and* Ψ_j *such that both*

(i) *for every* $v^f \in F_j$ *there is* $\phi^f \in \Phi_j$ *obeying*

$$\phi(\|x\|) \le v^f(t,x) \le \phi^f(\|x\|) \ , \ \forall (t,x) \in R\times R^n \ ,$$

and

(ii) *for every* $f \in N_j$ *there are* $v^f \in F_j$ *and* $\psi^f \in \Psi_j$ *satisfying*

$$D^*v^f(t,x) \le -\psi^f(\|x\|) \ , \ \forall (t,x) \in R\times R^n \ .$$

Proof. *Necessity.* Let $x=0$ of (16) be uniformly absolutely stable on N_j , $j \in \{0,1,2\}$. Let $f\in N_j$ be arbitrary but fixed. Let $\phi \in KR$; $\alpha \in]1,+\infty[$, x arbitrary, $x\in R^n$, Δ be arbitrary in $]\|x\|,+\infty[$ and fixed, and

$$v^f(t,x) = \sup\{\phi[\|x(t+\sigma;t,x;f)\|]\ \frac{1+\alpha\sigma}{1+\sigma} : \sigma \in [0,\tau^f(2\Delta,\tfrac{1}{\alpha}\Delta)]\} \ .$$

Hence,

$$v^{f}(t,x) \geq \phi(\|x\|) , \forall (t,x) \in R \times R^{n} , \forall f \in N_{j} .$$

Following now the necessity part of the proof of Theorem 7 it immediately follows that the conditions (i) and (ii) of Theorem 14 are satisfied for $\phi^{f}(\|x\|) = \phi[\pi^{f}(\|x\|)]$ and $\psi^{f} \in K$ such that

$$\psi^{f}(\|x\|) = \frac{(\alpha-1)\,\phi(\|x\|)}{[1+\tau^{f}(2\Delta,\frac{1}{\alpha}\Delta)][1+\alpha\tau^{f}(2\Delta,\frac{1}{\alpha}\Delta)]}$$

where π is defined in the necessity part of the proof of Theorem 7 and superscript f in π^{f} and τ^{f} denotes that π^{f} and τ^{f} are related to $f \in N_{j}$.

Sufficiency. Let $f \in N_{j}$ be arbitrary but fixed. Under the conditions of Theorem 14, all conditions of Theorem 8 are fulfilled. Hence, $x=0$ of (16) is uniformly asymptotically stable in the whole for any $f \in N_{j}$, $j \in \{0,1,2\}$. ∎

The preceding proof and (iii) of Definition 22 show that the next result holds.

Theorem 15. *It is necessary and sufficient for complete uniform absolute stability on* N_{j} , $j \in \{0,1,2\}$, *of* $x=0$ *of the system (16) that there exist a functional family* F_{j} *and functions* ϕ_{i} , $\phi_{i} \in KR$, $i=1,2$, *and* $\psi \in K$ *such that both*

(i) $\phi_{1}(\|x\|) \leq v^{f}(t,x) \leq \phi_{2}(\|x\|)$, $\forall (t,x) \in R \times R^{n}$, $\forall v^{f} \in F_{j}$,

and

(ii) $D^{*}v^{f}(t,x) \leq -\psi(\|x\|)$, $\forall (t,x) \in R \times R^{n}$, $\forall f \in N_{j}$.

Theorems 14, 15 show that a functional family F_{j} is to be used for studying absolute stability rather than a single function v whose form is not influenced by the form of the nonlinearity $f \in N_{j}$.

I.4.5. Criteria for absolute stability of time-varying systems

The problem of algorithmic construction of a function v^{f} obeying the corresponding requirements for a stability property of a given system (16) even when f is fixed and known has not been solved in general. Therefore, the form of v^{f} can be accepted and then conditions are to be found under which it will satisfy the corresponding conditions of Theorem 14 or Theorem 15.

Definition 23. A functional family F_{j} is a *Liapunov functional family on* N_{j} , $j \in \{0,1,2\}$, of the system (16) iff it obeys Theorem 14. ∎

Let

$$v^{f}(t,x) = (x^T, f^T)\, H(t)(x^T, f^T)^T + 2 \int_0^{\Sigma(t,x)} f^T(t,\Sigma)\, \Theta d\Sigma \qquad (17)$$

where

$$H(t) \equiv H(t) \in C^{(1)}(R, R^{(n+m)\times(n+m)})$$

and

$$\Theta \in R^{m\times m} \ , \ \Theta = \mathrm{diag}\,\{\theta_1, \theta_2, \dots, \theta_m\} \ .$$

In order to establish conditions under which $F_2 = \{v^f : v^f(17)\ , f \in N_2\}$ is a Liapunov functional family on N_2 of the system (16) the following notation will be used :

$$E(t,L) = [I_m - D(t)\, L]^{-1}\, C(t) \quad \text{iff} \quad \det [I_m - D(t)\, L] \neq 0 \ , \ \forall (t,L) \in R\times L \ ,$$

$$F(t,L) = A(t) + B(t)\, LE(t,L) \ ,$$

$$R(t,L) = \begin{bmatrix} I_n \ , & 0 \\ 0 \ , & E^T(T,L)\, L \end{bmatrix} H(t) \begin{bmatrix} I_n \ , & 0 \\ 0 \ , & LE(t,L) \end{bmatrix} + E^T(t,L)\, L\Theta E(t,L)$$

$$Q(t,J,L,M) = [I_m - D(t)\, M]^{-1} \{\dot{C}(t) + C(t)\, A(t) + $$
$$+ [\dot{D}(t) + C(t)\, B(t)\, L + D(t)\, J]\, E(t,L)\}$$
$$\text{iff} \quad \det [I_m - D(t)\, M] \neq 0 \ , \ \forall (t,M) \in R\times M \ ,$$

$$T(t,J,L,M) = \dot{H}(t) + F^T(t,L)\, H(t) + H(t)\, F(t,L) + E^T(t,L)\, L\Theta Q(t,J,L,M)$$
$$+ Q^T(t,J,L,M)\, \Theta LE(t,L) + E^T(t,L)\, J\Theta E(t,L) \ .$$

Here, I_m is m×m identity matrix.

Theorem 16. *If* L *is open set then for functional family*

$$F_2 = \{v^f : v^f(17)\ , f \in N_2(J,L,M)\}$$

to be Liapunov functional family on $N_2(J,L,M)$ *it is necessary and sufficient that*

(i) *the matrix function* $R(t,L)$ *be positive definite on* $R\times L$,

(ii) $\sup\{\|R(t,L)\| : (t,L) \in R\times L\} < +\infty$,

and

(iii) *the matrix function* $T(t,J,L,M)$ *be negative definite on* $R\times J\times L\times M$.

Proof. *Necessity.* Let $F_2 = \{v^f : v^f(17)\ , f \in N_2(J,L,M)\}$ be a Liapunov functional family on $N_2(J,L,M)$ of the system (16). Let $f(t,\Sigma) = L\Sigma$, $L \in L$. Hence, $f \in N_2(J,L,M)$. Now, $f(t,\Sigma) = L\Sigma$ implies (i) and (ii) of Theorem 16 due to (ii) of Definition 22 and (i) of Theorem 14. Along motions of (16) it follows for any $f \in N_2(J,L,M)$,

$$\frac{d}{dt} v^{\mathbf{f}}(t,\mathbf{x}) = \dot{v}^{\mathbf{f}}(t,\mathbf{x},\mathbf{N},\mathbf{f}_\Sigma, \int_0^\Sigma \Sigma^T \mathbf{N}_t \Theta d\Sigma) .$$

Since L is open and $\dot{v}^{\mathbf{f}}$ is continuous in all arguments it follows that for every $(t,\mathbf{J},\mathbf{L},\mathbf{M}) \in R\times J\times L\times M$ there is $\mathbf{f} \in N_2(J,L,M)$ such that

$$\frac{d}{dt} v^{\mathbf{f}}(t,\mathbf{x}) = \mathbf{x}^T \mathbf{T}(t,\mathbf{J},\mathbf{L},\mathbf{M})\, \mathbf{x} . \tag{18}$$

This result, (ii) of Theorem 14 and (ii) of Definition 22 imply necessity of (iii) of Theorem 16.

Sufficiency. The following are obviously true :

(a) for every $(t,\mathbf{x},\mathbf{f}) \in R\times R^n\times N_2(J,L,M)$ there are $\mathbf{L}_i = \mathbf{L}_i(t,\mathbf{x},\mathbf{f}) \in L$, $i = 1,2$, such that

$$\mathbf{x}^T\mathbf{R}(t,\mathbf{L}_1)\, \mathbf{x} \le v^{\mathbf{f}}(t,\mathbf{x}) \le \mathbf{x}^T\mathbf{R}(t,\mathbf{L}_2)\, \mathbf{x} ,$$

and

(b) for every $(t,\mathbf{x},\mathbf{f}) \in R\times R^n\times N_2(J,L,M)$ there are $(\mathbf{J},\mathbf{L},\mathbf{M}) \in J\times L\times M$ such that (18) holds.

Now, (i) and (ii) of Theorem 16 and (a) imply (i) of Theorem 14, and (iii) of Theorem 16 together with (b) implies (ii) of Theorem 14, which completes the proof. ∎

Remark 9. Notice that the matrix function $\mathbf{R}$ is positive definite on $R\times L$ iff there is $\epsilon > 0$ such that the matrix $[\mathbf{R}^T(t,\mathbf{L}) + \mathbf{R}(t,\mathbf{L}) - \epsilon\, \mathbf{I}_n]$ is positive definite for every $(t,\mathbf{L}) \in R\times L$.

Remark 10. Conditions of Theorem 16 are purely linear system stability conditions. In case that L is compact then these conditions are sufficient for uniform absolute stability on N_2 of $\mathbf{x} = \mathbf{0}$.

Example 11 (Grujić, 1981b). Let the second order Lur'e-Postnikov system possess different time-varying terms described by

$$\frac{d\mathbf{x}}{dt} = \begin{bmatrix} -1-t^2 , & 0.5\sin t \\ 0.5\cos t , & -1-t^2 \end{bmatrix} \mathbf{x} + \begin{bmatrix} t , & 0 \\ 0 , & t \end{bmatrix} \begin{bmatrix} \phi_1(t,\sigma_1) \\ \phi_2(t,\sigma_2) \end{bmatrix} ,$$

$$\Sigma = \begin{bmatrix} \sigma_1 \\ \sigma_2 \end{bmatrix} = \begin{bmatrix} 0.1\, t , & 0.01 \sin t \\ 0.01\cos t , & 0.1\, t \end{bmatrix} \mathbf{x} .$$

The vector nonlinearity $\mathbf{f} = (\phi_1,\phi_2)^T$ is time-dependent and $\mathbf{f} \in N_1(L)$, where now

$$L = \{L : 0 \le L \le 9I_2\} \ , \ L = \mathrm{diag}\,\{\ell_1,\ell_2\} \ ,$$
$$J = \{J : J = \mathrm{diag}\,\{j_1,j_2\} \ , \ j_i \in \bar{R} \ , \ i = 1,2\}$$
$$M = \{M : M = \mathrm{diag}\,\{m_1,m_2\} \ , \ m_i \in \bar{R} \ , \ i = 1,2\} \ .$$

In this case $\mathbf{D}(t) \equiv \mathbf{0}$. Hence, $\mathbf{E}(t,\mathbf{L}) = \mathbf{C}(t)$ and

$$\mathbf{F}(t,\mathbf{L}) = \begin{bmatrix} -1-(1-0.1\,\ell_1)\,t^2 & , \ (0.5+0.01\,\ell_1 t)\sin t \\ (0.5+0.01\,\ell_2 t)\cos t & , \ -1-(1-0.1\,\ell_2)\,t^2 \end{bmatrix}$$

It can be easily verified that $[\mathbf{F}^T(t,\mathbf{L}) + \mathbf{F}(t,\mathbf{L}) + \epsilon\,\mathbf{I}_2]$, $\epsilon \in\,]0,1.8[$ is negative definite for every $(t,\mathbf{L}) \in R \times L$, which suggests the choice of $\mathbf{H}(t) \equiv \mathbf{I}_2$ and $\Theta = \mathbf{0}$. Hence, $\mathbf{R}(t,\mathbf{L}) \equiv \mathbf{I}_2$, $\|\mathbf{R}(t,\mathbf{L})\| \equiv 1$, and $\mathbf{T}(t,\mathbf{J},\mathbf{L},\mathbf{M}) \equiv \mathbf{F}^T(t,\mathbf{L}) + \mathbf{F}(t,\mathbf{L})$ is negative definite on $R \times J \times L \times M$. All conditions of Theorem 16 are fulfilled. The state $\mathbf{x} = \mathbf{0}$ of the system is uniformly absolutely stable on $N_2(J,L,M) = N_1(L)$. In this example,

$$F_1 = \{v^{\mathbf{f}} : v^{\mathbf{f}}(t,\mathbf{x}) = \mathbf{x}^T\mathbf{x} \ , \ \forall \mathbf{f} \in N_1(L)\}$$

is one member family. Derivative of v obeys, along system motions,

$$\frac{dv}{dt} \le -\epsilon\,\mathbf{x}^T\mathbf{x} \ , \ \epsilon \in\,]0,1.8[\ , \ \forall(t,\mathbf{x}) \in R \times R^n \ , \ \forall \mathbf{f} \in N_1(L) \ .$$

Hence $\mathbf{x} = \mathbf{0}$ is completely uniformly absolutely stable on $N_1(L)$ due to Theorem 15.

Example 12 (Grujić, 1978a). The second order non-stationary Lur'e-Postnikov system is described by

$$\frac{d\mathbf{x}}{dt} = \begin{bmatrix} -\dfrac{20+30t^2}{1+t^2} & , \ 2t \\ -2t & , \ -\dfrac{30+20t^2}{1+t^2} \end{bmatrix} \mathbf{x} + \begin{bmatrix} 0 \\ 1 \end{bmatrix} \mathbf{f}(t,\Sigma) \ ,$$

$$\Sigma = [1,t^2]\,\mathbf{x} + (-t^2)\,\mathbf{f}(t,\Sigma) \ .$$

The nonlinearity f is scalar and $\mathbf{f} \in N_0(L)$, where $L = [0,1]$. Since $\mathbf{D}(t) = [-t^2]$, then $\mathbf{I}_1 - \mathbf{D}(t)\,L = [1 + Lt^2]$, $L \in L$, obeys the condition $\det\,[\mathbf{I}_1 - \mathbf{D}(t)\,\mathbf{L}] \ne 0$ on $R \times L$. Hence,

$$\mathbf{E}(t,L) = [\mathbf{I} - \mathbf{D}(t)\,L]^{-1}\,\mathbf{C}(t) = \frac{1}{1+Lt^2}\,[1,t^2] \ ,$$

and finally,

$$\mathbf{F}(t,L) = \mathbf{A}(t) + \mathbf{B}(t)\, L\mathbf{E}(t,L)$$

$$= \begin{bmatrix} -\dfrac{20+30t^2}{1+t^2} \,, & 2t \\ -2t + \dfrac{L}{1+Lt^2} \,, & -\dfrac{30+(29L+20)\,t^2+19Lt^4}{(1+t^2)(1+Lt^2)} \end{bmatrix} .$$

The matrix $[\mathbf{F}^T(t,L) + \mathbf{F}(t,L) + \mathbf{I}_2]$ is negative definite for every $(t,L) \in \mathbf{R}\times L$, which suggests $\mathbf{H}(t) \equiv \mathbf{I}_2$ and $\Theta = \mathbf{0}$ so that $\mathbf{R}(t,L) \equiv \mathbf{I}_2$. Hence, $\|\mathbf{R}(t,L)\| \equiv 1$ and $\mathbf{T}(t,L) = \mathbf{F}^T(t,L) + \mathbf{F}(t,L)$.
The matrix function $\mathbf{T}$ is negative definite on $\mathbf{R}\times L$. Therefore, $v(\mathbf{x}) = \mathbf{x}^T\mathbf{x}$ and

$$\frac{dv}{dt} \le -v(\mathbf{x}) \ , \ \forall (t,\mathbf{x}) \in \mathbf{R}\times\mathbf{R}^n \ , \ \forall \mathbf{f} \in N_0(L) \ ,$$

which show that all conditions of Theorem 15 are satisfied. It follows that $\mathbf{x}=\mathbf{0}$ of the system is completely uniformly absolutely stable on $N_0([0,1])$.

In special cases $\mathbf{D}(t) \equiv \mathbf{0}$ and $\mathbf{f}$ is independent of $t \in \mathbf{R}$. Then the system (16) takes the following form

$$\frac{d\mathbf{x}}{dt} = \mathbf{A}(t)\,\mathbf{x} + \mathbf{B}(t)\,\mathbf{f}(\Sigma) \ , \ \Sigma = \mathbf{C}(t)\,\mathbf{x} \ . \tag{19}$$

Following Grujić (1977a), we state

Theorem 17. *For complete uniform absolute stability on* $N_0([0,\mathbf{K}])$ *of* $\mathbf{x}=\mathbf{0}$ *of the system (19) it is sufficient that :*

(i) *there exists a real number* $\mu > 0$ *for which real parts of all eigenvalues of* $[\mathbf{A}(t) + \mu\mathbf{I}_n]$ *are negative for every* $t \in \mathbf{R}$ *,*

(ii) *there exist a positive definite matrix* $\mathbf{H} = \mathbf{H}^T \in \mathbf{R}^{n\times n}$ *and non-negative diagonal matrix* $\Theta \in \mathbf{R}^{\ell\times\ell}$ *obeying both*

(a) the matrix function $[2\mathbf{K}^{-1} - \Theta\mathbf{C}\mathbf{B} - \mathbf{B}^T\mathbf{C}^T\Theta]$ *is positive definite on* $\mathbf{R}$*, and*

(b) the matrix function

$$[\mathbf{A}^T\mathbf{H} + \mathbf{H}\mathbf{A} + (\mathbf{A}^T\mathbf{C}^T\Theta + \mathbf{C}^T + \mathbf{H}\mathbf{B})(2\mathbf{K}^{-1} - \Theta\mathbf{C}\mathbf{B} - \mathbf{B}^T\mathbf{C}^T\Theta)^{-1}(\mathbf{A}^T\mathbf{C}^T\Theta + \mathbf{C}^T + \mathbf{H}\mathbf{B})^T]$$

is negative definite on $\mathbf{R}$.

Application of this theorem will be shown in the framework of singularly perturbed large-scale systems stability analysis.

I.4.6. Criteria for absolute stability of time-invariant systems

In the case A, B, C and D are matrices with real valued elements rather than matrix functions and the vector nonlinearity f is time-independent then the Lur'e-Postnikov system is time-invariant,

$$\frac{dx}{dt} = Ax + Bf(\Sigma) \ , \ \Sigma = Cx + Df(\Sigma) \ . \tag{20}$$

In some cases the linear part of the system is not stable.
Let such Lur'e-Postnikov systems be described by

$$\frac{d\hat{x}}{dt} = \hat{A}\hat{x} + \hat{B}\hat{f}(\hat{\Sigma}) \ , \ \hat{\Sigma} = \hat{C}\hat{x} \ , \tag{21}$$

where $\hat{f} \in N_0(\hat{L})$ and $\hat{L} = (\hat{\Gamma},\hat{K})$ is such a diagonal matrix sector that the system (21) is stable for $\hat{f} = L\hat{\Sigma}$ and every $L \in \hat{L}$. Then, by following Tsypkin (1962b), we define

$$A = \hat{A} + \hat{B}\hat{\Gamma}\hat{C} \ , \ B = \hat{B} \ , \ C = \hat{C} \ ,$$

provided that $\hat{\Gamma}$ is such that A is stable, $x = \hat{x}$, and $\Sigma = \hat{\Sigma}$, $f(\Sigma) = \hat{f}(\Sigma) - \hat{\Gamma}\Sigma$. Hence $K = \hat{K} - \hat{\Gamma}$, $L = (0,K)$ and $D = 0$.

The preceding notation transforms the system (21) into the form of the system (20). It is clear that without losing in generality it may be accepted that A is stable matrix.

Notice that $(\hat{\Gamma},\hat{K})$ means that this sector can be closed, $(\hat{\Gamma},\hat{K}) = [\hat{\Gamma},\hat{K}]$, or semi-open, $(\hat{\Gamma},\hat{K}) = [\hat{\Gamma},\hat{K}[$. Iff there is an infinite diagonal entry of K then the set $L = [\Gamma,K[$ or $L =]\Gamma,K[$, where Γ is a diagonal matrix, $\Gamma \le K$ elementwise.

Let $W(s) = C(A - sI_n)^{-1}B - D$ be the transfer matrix of the linear part of the system (20) with $Y_i = f$ as input and with $Y_o = -\Sigma$ as its output. Then

$$\mathrm{He}\, W(j\omega) = \frac{1}{2}\,[W(j\omega) + W^T(-j\omega)]$$

is the hermitian part of $W(j\omega)$.

Theorem 18. *It is sufficient for absolute stability on* $N_0[(0,K)]$ *of* $x=0$ *of the system (20) that*

(i) *the pair* (A,B) *is completely controllable :*

$$\mathrm{rank}\,(B, AB, \dots, A^{n-1}B) = n \ ,$$

(ii) *the pair* (A,C) *is observable :*

$$\mathrm{rank}\,(C^T, A^TC^T, \dots, (A^T)^{n-1}C^T) = n \ ,$$

(iii) *the matrix* A *is stable :*

$$\mathrm{Re}\,\lambda_i(A) < 0 \ , \ \forall i = 1,2,\dots,n \ ,$$

and that

(iv) *there exists positive diagonal matrix* $T \in R^{m \times m}$ *and diagonal matrix* $\Theta \in R^{m \times m}$ *such that*

$$\mathrm{He}\,\{T\,[C(A - j\omega I_n)^{-1} B - D + K^{-1}] + j\omega\Theta C(A - j\omega I_n)^{-1} B\}$$

is positive semi-definite for all $\omega \in [0,+\infty[$ *and for* $\omega \to +\infty$.

Proof of this theorem can be found in the book by Narendra and Taylor (1973). The requirements for controllability and observability of the pair (A,B) and (A,C), respectively, can be removed by following Popov (1960) [see also Tokumaru and Saito (1965)]. ∎

Theorem 19. *In the case* $D = 0$ *it is sufficient for absolute stability on* $N_0((0,K))$ *of* $x = 0$ *of the system (20) that both*

(i) A *is stable matrix,* $\mathrm{Re}\,\lambda_i(A) < 0$, $\forall i = 1,2,\dots,n$,

and

(ii) *there exists diagonal matrix* $\Theta \in R^{m \times m}$ *such that the matrix* $K^{-1} + \mathrm{He}\,[(I_m + j\omega\Theta)\,W(j\omega)]$ *is positive definite for all* $\omega \in [0,+\infty[$ *and for* $\omega \to +\infty$.

Remark 11. The Popov-like criteria (Theorems 18,19) reduce absolute stability analysis to positive (semi)-definiteness test of $m \times m$ matrix dependent on ω for all $\omega \geq 0$. This is difficult computational problem. Greater number m of nonlinearities, more difficult computational problem. Greater order n of A, again more difficult computational problem even for digital computer calculation. It is evident that for higher-order systems new approaches to their stability analysis are needed [see Šiljak (1971)].

I.5. ON STABILITY PROPERTIES OF SINGULARLY PERTURBED SYSTEMS

I.5.1. Introductory comments

A physical system can be composed of subsystems with different speeds of their transient responses. Each of them has its own natural time scale. If all of them were disconnected from others then a dynamical property of each would be studied in the corresponding time scale. It appears meaningful to use such information and to explore whether under simple additional conditions on interactions and subsystems it is possible to infer the over-all system property from that of all disconnected subsystems. This phenomenon of different time scales related to disconnected subsystems is mathematically expressed by existence of arbitrarily small positive parameters μ_i at the highest derivatives in differential equations. When all μ_i are formally set equal to zero, then the number of differential equations is decreased. This is singular case.

The singular perturbation approach to systems analysis and design has been successfully used for model-order reduction and separation of time scales. A number of results obtained on various singular perturbation problems were excellently reviewed by Kokotović, O'Malley Jr. and Sannuti (1975,1976).

Liapunov stability properties of singularly perturbed systems were studied by Gradshtein (1951), Tichonov (1952), Klimushev and Krasovskii (1962), Hoppensteadt (1966,1967,1968,1974), Desoer and Shensa (1970), Shensa (1971), Wilde and Kokotović (1972), Šiljak (1972), Zien (1973), Porter (1974,1976,1977a,b), Habets (1974a,b) Geraschenko and Geraschenko (1975), Grujić (1976a,b , 1977c,d , 1978c , 1979a,b,c , 1981a), Suzuki and Miura (1976), Young, Kokotović and Utkin (1977), Javid (1978), Chow (1978), Chow and Kokotović (1978), Khalil and Kokotović (1979), and Martynyuk and Gutowski (1979).

I.5.2. System description

A singularly perturbed system S to be considered is governed by two non-linear vector differential equations

$$\frac{d\mathbf{x}}{dt} = \mathbf{f}(t,\mathbf{x},\mathbf{y},\mu) \ , \tag{22a}$$

$$\mu\frac{d\mathbf{y}}{dt} = \mathbf{g}(t,\mathbf{x},\mathbf{y},\mu) \ . \tag{22b}$$

The state vector of the whole system is $(\mathbf{x}^T,\mathbf{y}^T)^T$, where $\mathbf{x}\in R^n$ and $\mathbf{y}\in R^m$, $\mathbf{f}(t,\mathbf{x},\mathbf{y},\mu)\in C(R\times R^n\times R^m\times R_+,R^n)$ and $\mathbf{g}(t,\mathbf{x},\mathbf{y},\mu)\in C(R\times R^n\times R^m\times R_+,R^m)$. The parameter μ is positive and can be arbitrarily small. It is accepted that $\mu\in]0,1]$.

Open connected neighbourhoods $N_x\subseteq R^n$ and $N_y\subseteq R^m$ of $\mathbf{x}=\mathbf{0}$ and $\mathbf{y}=\mathbf{0}$, respectively, are related to $\mathbf{f}$ and $\mathbf{g}$ so that $(\mathbf{x}^T,\mathbf{y}^T)^T=\mathbf{0}$ is the unique equilibrium state of the system (22) in $N_x\times N_y$ - the cartesian product set of N_x and N_y , for every $\mu\in]0,1]$.

If μ is formally set equal to zero, then the system (22) degenerates into the reduced-order (or, the degenerate) system S_0 (subscript zero due to $\mu=0$)

$$\frac{d\mathbf{x}}{dt} = \mathbf{f}(t,\mathbf{x},\mathbf{y},0) \ , \tag{23a}$$

$$\mathbf{0} = \mathbf{g}(t,\mathbf{x},\mathbf{y},0) \ . \tag{23b}$$

It is accepted that $\mathbf{g}(t,\mathbf{x},\mathbf{y},0)$ becomes zero for every $t \in R$ and $\mathbf{x} \in N_x$ iff $\mathbf{y}=\mathbf{0}$. The justification for this demand is based on Hoppensteadt's effective usage (Hoppensteadt, 1966) of Liapunov's transformation of coordinates (Liapunov, 1892, and Comment 2 herein) in the framework of singularly perturbed systems. Hence, the reduced-order system can be put in the form

$$\frac{d\mathbf{x}}{dt} = \mathbf{f}(t,\mathbf{x},\mathbf{0},0) , \tag{24}$$

which is suitable for the stability analysis of the equilibrium state.

If $\mu > 0$ is small enough then the system (22) is composed of a slow part and a fast part. The fast (or, the boundary layer) system S_τ is obtained from (22) after changing the time scale by introducing τ , $\tau = \frac{t-t_0}{\mu}$. Then the boundary-layer system is obtained in the form

$$\frac{d\mathbf{y}}{d\tau} = \mathbf{g}(\alpha,\mathbf{b},\mathbf{y},0) . \tag{25}$$

Here, α and $\mathbf{b}$, $\mathbf{b} = (\beta_1,\beta_2,\dots,\beta_n)^T$, are real valued scalar and vector parameters, respectively. They are introduced instead of t and $\mathbf{x} \in N_x$, respectively.

It is assumed that $\mathbf{g}$ becomes zero for every $t \in R$, $\mathbf{x} \in N_x$ and $\mu \in [0,1]$ iff $\mathbf{y}=\mathbf{0}$.

Separation of time scales is advantageous from the stability viewpoint owing to the possibility of a separate analysis of the degenerate system $S_0(24)$ and the fast system $S_\tau(25)$, and for the order reduction of the stability problem. This means that the following problem can be solved.

Under what qualitative properties of $\mathbf{f}$ and $\mathbf{g}$, which hold on $N_x \times N_y$, can uniform asymptotic stability of the equilibrium state of (22) be inferred from such a property of (24) and (25) ?

A solution for this problem will be presented by following Grujić (1981a).

I.5.3. Liapunov-like conditions for asymptotic stability

Let

$$N_{xo} = \{\mathbf{x} : \mathbf{x} \in N_x , \mathbf{x} \neq \mathbf{0}\} , \quad N_{yo} = \{\mathbf{y} : \mathbf{y} \in N_y , \mathbf{y} \neq \mathbf{0}\} ,$$

and for $v(\alpha,\mathbf{b},\mathbf{y}) \in C^{(1,1,1)}(R \times R^n \times R^m, R)$,

$$v_\alpha = \frac{\partial v}{\partial \alpha} , \quad v_b = \left(\frac{\partial v}{\partial \beta_1}, \frac{\partial v}{\partial \beta_2}, \dots, \frac{\partial v}{\partial \beta_n}\right)^T .$$

The problem solution is based on the following two assumptions that relate some positive definite functions θ and v to (24) and (25).

Assumption 1. There exists $\theta \in C^{(1,1)}(R\times N_{xo}, R_+)$ that is decrescent and positive definite on N_x and, in addition, radially unbounded as soon as $N_x = R^n$, and there exist non-negative numbers ζ_1 and ζ_2 and functions $\phi \in C(R^n, R_+)$ and $\Phi \in C(R^m, R_+)$ that are positive definite on N_x and N_y, respectively, so that $\zeta_1 < 1$ and

(a) $\theta_t(t,\mathbf{x}) + \theta_{\mathbf{x}}^T(t,\mathbf{x})\, \mathbf{f}(t,\mathbf{x},\mathbf{0},0) \le -\phi(\mathbf{x})$, $\forall (t,\mathbf{x}) \in R\times N_{xo}$,

(b) $\theta_{\mathbf{x}}^T(t,\mathbf{x})[\mathbf{f}(t,\mathbf{x},\mathbf{y},\mu) - \mathbf{f}(t,\mathbf{x},\mathbf{0},0)] \le \zeta_1\phi(\mathbf{x}) + \zeta_2\Phi(\mathbf{y})$,

$\forall (t,\mathbf{x},\mathbf{y},\mu) \in R\times N_{xo}\times N_y \times]0,1]$. ∎

The hypotheses of Assumption 1 and the condition (a) guarantee uniform asymptotic stability of $\mathbf{x}=\mathbf{0}$ of (24), which is global as soon as $N_x = R^n$. The second condition (b) is the required qualitative property of $\mathbf{f}$ on $N_x\times N_y$.

Assumption 2. There exists either $v(t,\mathbf{x},\mathbf{y}) \in C^{(1,1,1)}(R\times N_x\times N_y, R_+)$ that is decrescent and positive definite on $N_x\times N_y$ and radially unbounded in $\mathbf{y}$ uniformly in $\mathbf{x}\in N_x$ as soon as $N_y = R^m$, or $\{v(t,\mathbf{y}) \in C^{(1,1)}(R\times N_{yo}, R_+)$ that is decrescent and positive definite on N_y and radially unbounded as soon as $N_y = R^m\}$, and there exist non-negative numbers ξ_1, ξ_2, ξ_3 and ξ_4 and an integer $\pi \ge 1$ such that $\xi_1 < 1$, $\xi_2 < 1$ and

(a) $v_{\mathbf{y}}^T\, \mathbf{g}(\alpha,\mathbf{b},\mathbf{y},0) \le -\Phi(\mathbf{y})$,

$\forall(\alpha,\mathbf{b},\mathbf{y}) \in R\times N_x\times N_y$, or $\{\forall(\alpha,\mathbf{b},\mathbf{y}) \in R\times N_x\times N_{yo}\}$, respectively;

(b) $v_{\mathbf{y}}^T[\mathbf{g}(\alpha,\mathbf{b},\mathbf{y},\mu) - \mathbf{g}(\alpha,\mathbf{b},\mathbf{y},0)] \le \xi_1\mu^\pi\phi(\mathbf{b}) + \xi_2\Phi(\mathbf{y})$,

$\forall(\alpha,\mathbf{b},\mathbf{y},\mu) \in R\times N_x\times N_y\times]0,1]$, or $\{\forall(\alpha,\mathbf{b},\mathbf{y},\mu) \in R\times N_x\times N_{yo}\times]0,1]\}$, respectively;

(c) $v_\alpha + v_{\mathbf{b}}^T\, \mathbf{f}(\alpha,\mathbf{b},\mathbf{y},\mu) \le \xi_3\phi(\mathbf{b}) + \xi_4\Phi(\mathbf{y})$,

$\forall(\alpha,\mathbf{b},\mathbf{y},\mu) \in R\times N_x\times N_y\times]0,1]$, or $\{\forall(\alpha,\mathbf{b},\mathbf{y},\mu) \in R\times N_x\times N_{yo}\times]0,1]\}$, respectively. ∎

The numbers ζ_1, ζ_2, ξ_1, ξ_2, ξ_3 and ξ_4 of Assumptions 1 and 2 should be chosen as small as possible.

If v is independent of $\mathbf{x}$, which is admissible, then it should be positive definite on N_y only. If it is also time-invariant, then condition (c) vanishes in Assumption 2.

Let

$$\tilde{\mu} = \frac{1-\xi_2}{\zeta_2 + \xi_4} .$$

It is a lower estimate of the upper bound of allowable μ as expressed by

Theorem 20. *If Assumptions 1 and 2 hold, then the equilibrium state* $(x^T,y^T)^T = 0$ *of (22) is uniformly asymptotically stable for every* $\mu \in]0,\tilde{\mu}[$ *and for* $\mu \to 0$ *as soon as* $1 > \zeta_1 + \xi_1 \tilde{\mu}^{\pi-1} + \xi_3$.
If, in addition, $N_x \times N_y = R^{m+n}$ *then the equilibrium state is uniformly asymptotically stable in the whole for every* $\mu \in]0,\tilde{\mu}[$ *and for* $\mu \to 0$.

Proof. Let ν be defined by $\nu = \theta + v$. Then,

$$\nu(t,x,y) \in C^{(1,1,1)}(R \times N_{xo} \times N_{yo})$$

and it is decrescent and positive definite on $N_x \times N_y$ due to Assumptions 1 and 2. Its eulerian derivative $\dot{\nu}[t,x(t),y(t),\mu]$ along a motion of (22) is found in the following form for $z(t) = (x^T(t),y^T(t))^T \neq 0$ $\{z(t) = 0 , t \in [t_o,+\infty[$, means the equilibrium state is reached and, therefore, it will not be considered henceforth$\}$:

$$\dot{\nu} = \theta_t + \theta_x^T f + v_t + v_x^T f + \frac{1}{\mu} v_y^T g$$

due to (22). The right-hand side of this equation is transformed as follows :

$$\dot{\nu} = \theta_t + \theta_x^T f(t,x,0,0) + \theta_x^T [f(t,x,y,\mu) - f(t,x,0,0)] + v_t + v_x^T f(t,x,y,\mu)$$
$$+ \frac{1}{\mu} v_y^T g(t,x,y,0) + \frac{1}{\mu} v_y^T [g(t,x,y,\mu) - g(t,x,y,0)] .$$

The conditions (a), (b) of Assumption 1 and (a)-(c) of Assumption 2 yield

$$\dot{\nu}(t,x,y,\mu) \le -(1 - \zeta_1 - \xi_1 \mu^{\pi-1} - \xi_3)\,\phi(x) - \frac{1}{\mu}[1 - \xi_2 - \mu(\zeta_2 + \xi_4)]\,\Phi(y) ,$$
$$\forall \mu \in]0,\tilde{\mu}[\text{ and } \mu \to 0 , \ \forall (t,x,y) \in R \times N_{xo} \times N_{yo} . \qquad (26)$$

Let

$$N_{ox} = \{z : x=0 , y \in N_{yo}\} , \ N_{oy} = \{z : x \in N_{xo} , y=0\} , \ N_o = N_{ox} \cup N_{oy} .$$

It is evident that

$$N_x \times N_y = N_{xo} \times N_{yo} \cup N_o \cup \{z : z=0\} .$$

Let ν_M be the maximal positive number for which the largest connected neighbourhood $V_{\nu_M}(t)$ of $z=0$ obeying

$$\nu(t,x,y) \in [0,\nu_M[, \ \forall (x,y) \in V_{\nu_M}(t) , \ \forall t \in R ,$$

is a subset of $N = N_x \times N_y$ for every $t \in R$. Such a $\nu_M > 0$ exists because ν is positive definite on N and N is time-invariant neighbourhood of $z=0$ (Proposition 14).

Let τ_i and τ_i^* , $t_o \le \tau_i < \tau_i^* \le +\infty$, denote instants such that

$$z(t) \in V_{\nu_M}(t) \setminus N_o , \ \forall t \in]\tau_i, \tau_i^*[, \ \tau_i > t_o ,$$

and $\quad z(t) \in N_o \ , \ \forall t \in [\tau^*_{i-1}, \tau_i]$.

If $z(t_o) \in V_{\nu_M}(t_o) \setminus N_o$ then $i=0$, $\tau_o = t_o$, $[\tau_o, \tau^*_o[\; = [t_o, \tau^*_o[$ is the first interval to be considered and the next one is $[\tau^*_o, \tau_1]$.

If $z(t_o) \in N_o$, then $i=1$, $\tau^*_o = t_o$ and $[\tau^*_o, \tau_1]$ is the first interval to be considered, and the next one is $]\tau_1, \tau^*_1[$.

In what follows, $i \geq 0$ is an integer.

Let

$$\zeta(t;t_o,z_o;\mu) = (\chi^T(t;t_o,z_o;\mu), \eta^T(t;t_o,z_o;\mu))^T \ , \quad \zeta(t;t_o,z_o;\mu) \equiv z_o \ ,$$

be the motion of the system (22) through z_o at $t = t_o$ for $\mu > 0$.

Proposition 15. *The function* ν *is strictly decreasing in* $t \in [\tau^*_{i-1}, \tau_i]$ *along motions* $\zeta(t;t_o,z_o;\mu)$ *of (22) for every* $\mu \in]0,\tilde{\mu}[$ *and for* $\mu \to 0$.

Proof. *Part 1.* Let $\exists \hat{t} \in [\tau^*_{i-1}, \tau_i[$ such that

$$\nu[t, x(t), y(t)] \leq \nu[\hat{t}, x(\hat{t}), y(\hat{t})]$$

for some $t \in]\tau_{i-1}, \tau^*_{i-1}[$. If $\hat{t} = \tau^*_{i-1}$ then $\exists \bar{\tau}_1, \bar{\tau}_2 \in]\tau_{i-1}, \tau^*_{i-1}[$, $\bar{\tau}_1 < \bar{\tau}_2$, such that

$$\nu[\bar{\tau}_1, x(\bar{\tau}_1), y(\bar{\tau}_1)] \leq \nu[\bar{\tau}_2, x(\bar{\tau}_2), y(\bar{\tau}_2)]$$

due to the continuity of ν and ζ in $t \in T_o$, $\forall t_o \in R$, which is implied by continuity of **f** and **g**, and due to continuity of ν . Hence,

$$\exists \tau_3 \in [\bar{\tau}_1, \bar{\tau}_2] \quad \text{for which} \quad \left. \frac{d\nu(t)}{dt} \right|_{t=\tau_3} \geq 0 \ .$$

However, this contradicts (26) due to positive definiteness of ϕ and Φ and

$$(1 - \zeta_1 - \xi_1 \mu^{\pi-1} - \xi_3) > 0 \quad , \quad \frac{1}{\mu}[1 - \xi_2 - \mu(\zeta_2 + \xi_4)] > 0 \ , \ \forall \mu \in]0,\tilde{\mu}[\ .$$

Hence, $\hat{t} = \tau^*_{i-1}$ is impossible, and $\hat{t} \in]\tau^*_{i-1}, \tau_i]$ is to be considered.

Let $T_1 \subseteq [\tau^*_{i-1}, \tau_i[$ be the set of all instants t such that $x(t) = 0$, and let $T_2 \subseteq [\tau^*_{i-1}, \tau_i[$ be the set of all instants obeying $y(t) = 0$. Since $z(t) = 0$ is excluded, $\forall t \in [t_o, +\infty[$, and since the system motions are continuous, then either $T_1 = [\tau^*_{i-1}, \tau_i[$ or $T_2 = [\tau^*_{i-1}, \tau_i[$.

Let $T_1 = [\tau^*_{i-1}, \tau_i[$ for the sake of preciseness. Then,

$$\theta[t, x(t)] = \theta(t, 0) \ , \ \forall t \in T_1 \ , \ \text{and} \quad \nu[t, x(t), y(t)] = \nu[t, 0, y(t)] \ .$$

Further,

$$\frac{d}{dt}\nu[t, 0, y(t)] = \frac{d}{dt} v[t, 0, y(t)] \leq -\frac{1}{\mu}(1 - \xi_2 - \mu\xi_4)\,\Phi[y(t)] \ ,$$
$$t \in T_1 \ , \ \forall \mu \in]0,\tilde{\mu}[\ \text{and} \ \mu \to 0 \ . \tag{27}$$

This disproves $\hat{t} \in T_1$. Let now $T_2 = [\tau^*_{i-1}, \tau_i[$. Then $y(t) = 0$, $\forall t \in T_2$.

Hence, $\nu[t,x(t),y(t)] = \nu[t,x(t),0]$, $\forall t \in T_2$,

and $\frac{d}{dt}\nu[t,x(t),0] \le -(1-\zeta_1-\xi_3)\,\phi[x(t)]$, $\forall t \in T_2$,

which disproves $\hat{t} \in T_2$. Altogether, there is not $\hat{t} \in [\tau^*_{i-1},\tau_i[$ as defined above.

Part 2. Results (26), (27), Part 1, the definition of $\tilde{\mu}$, the conditions $1 > \zeta_1 + \xi_1\tilde{\mu}^{\pi-1} + \xi_3$, $\zeta_2 > 0$, $\xi_3 > 0$ and the positive definiteness of Φ and ϕ , prove that ν is strictly decreasing over every $[\tau^*_{i-1},\tau_i[$, $\tau^*_{i-1} \ge t_o$, $\forall i \ge 1$.

Part 3. Let $\exists\hat{t} \in [\tau^*_{i-1},\tau_i]$ such that

$$\nu[t,x(t),y(t)] \ge \nu[\hat{t},x(\hat{t}),y(\hat{t})]$$

for some $t \in]\tau_i,\tau^*_i[$. Hence $\exists\bar{\tau}_1,\bar{\tau}_2 \in]\tau_i,\tau^*_i[$, $\bar{\tau}_1 < \bar{\tau}_2$, such that

$$\nu[\bar{\tau}_1,x(\bar{\tau}_1),y(\bar{\tau}_1)] \le \nu[\bar{\tau}_2,x(\bar{\tau}_2),y(\bar{\tau}_2)]$$

due to continuity of both $\nu[t,x(t),y(t)]$ and ζ in t , and Part 2. Hence, $\exists\bar{\tau}_3 \in [\bar{\tau}_1,\bar{\tau}_2[$ such that

$$\frac{d}{dt}\nu[t,x(t),y(t)]\bigg|_{t=\tau_3} \ge 0$$

which contradicts (26) as explained in Part 1.

Now, Parts 1-3 prove Proposition 15, which together with positive definiteness and decrescency of ν prove (Hahn, 1967) uniform stability of $z=0$ of (22), $\forall\mu \in]0,\mu[$ and for $\mu \to 0$. Further, the positive definiteness of Φ and ϕ , $(1-\zeta_1-\xi_1\mu^{\pi-1}-\xi_3) > 0$ and $(1-\zeta_1-\xi_1\mu^{\pi-1}) > 0$, $\forall\mu \in]0,\tilde{\mu}[$, and for $\mu \to 0$ due to the definition of $\tilde{\mu}$, prove that $\dot{\nu}$ is less than a negative definite function on $N_{xo}\times N_{yo}$, on N_{ox} and on N_{oy} . This result, positive definiteness and decrescency of ν prove that $z=0$ is uniformly attractive which completes the proof of the first statement of the theorem. In the case $N_x\times N_y = R^{m+n}$ then ν is also radially unbounded, which together with the preceding result proves the second statement of the theorem. ■

Application of this theorem will be shown in the framework of absolute stability analysis of singularly perturbed Lur'e-Postnikov systems.

I.5.4. Singularly perturbed Lur'e-Postnikov systems

Let system (22) be of the Lur'e form (Grujić, 1981a) :

$$\dot{x} = A_{11}x + q_1\Phi_1(\sigma_1) + A_{12}y \quad , \quad \sigma_1 = c^T_{11}x + c^T_{12}y \tag{28a}$$

$$\mu\dot{y} = \mu A_{21}x + A_{22}y + q_2\Phi_2(\sigma_2) \quad , \quad \sigma_2 = \mu c^T_{21}x + c^T_{22}y \ . \tag{28b}$$

The matrices $A_{(\cdot)}$ and vectors $c_{(\cdot)}$ and $q_{(\cdot)}$ are of the appropriate order. The non-linearities Φ_i , i=1,2 , are one-one, continuous, $\Phi_i(0) = 0$, and in Lur'e sectors, Φ_i in $[0,k_i]$, $k_i \in]0,+\infty[$, $\frac{\Phi_i(\sigma_i)}{\sigma_i} \in [0,k_i]$, i=1,2 , $\forall \sigma_i \in]-\infty,+\infty[$.

In addition, only those non-linearities Φ_i are considered for which $x=0$ and $y=0$ are the unique equilibrium states of the degenerate system

$$\frac{dx}{dt} = A_{11}x + q_1\Phi_1(\sigma_1^o) \ , \ \sigma_1^o = c_{11}^T x \ , \tag{29}$$

and of the boundary-layer system, respectively,

$$\frac{dy}{d\tau} = A_{22}y + q_2\Phi_2(\sigma_2^o) \ , \ \sigma_2^o = c_{22}^T y \ . \tag{30}$$

These assumptions are guaranteed by $c_{ii}^T A_{ii}^{-1} q_i > 0$, i=1,2 .

We accept that A_{11} is a stable matrix (Barnett and Storey, 1970), pair (A_{11},q_1) - controllable, and that there exist numbers $\psi_1 \in [0,+\infty[$ and $\epsilon_1 \in]0,+\infty[$ such that

$$k_1^{-1} + \mathrm{Re}\,(1 + j\psi_1\omega)\, c_{11}^T(A_{11} - j\omega I_n)^{-1} q_1 - \epsilon_1 q_1^T(A_{11}^T + j\omega I_n)^{-1}(A_{11} - j\omega I_n)^{-1} q_1 \geq 0 \ , \ \forall\omega \in [0,+\infty] \ .$$

Then,

$$\theta(x) = (x^T H x + \psi_1 \int_0^{\sigma_1^o} \Phi_1(\sigma_1^o)\, d\sigma_1)^{1/2}$$

is a Liapunov function of the degenerate subsystem (29) for every Φ_1 in $[0,k_1]$, where H_1 is the solution (Yakubovich, 1962; Kalman, 1963; Popov, 1973; Narendra and Taylor, 1973) of

$$A_{11}^T H_1 + H_1 A_{11} + g_1 g_1^T = -\epsilon_1 I_1 \ , \ h_1 + H_1 q_1 = -\sqrt{\gamma_1}\, g_1 \tag{31}$$

for

$$\gamma_1 = k_1^{-1} - \psi_1\, c_{11}^T q_1 \ , \ h_1 = \tfrac{1}{2}(\psi_1\, A_{11}^T c_{11} + c_{11}) \ . \tag{32}$$

The assumptions 1,2 can be now verified.

Test of Assumption 1. Let $\theta(x)$ and H_1 be determined as above. Hence, $\theta(x)$ is decrescent, positive-definite on R^n and radially unbounded.

Let condition (a) be tested first.

(a) In this case $\theta_t = 0$ and

$$\theta_x^T(x)\, f(x,0,0) \leq -\tfrac{1}{2}\, \epsilon_1 \eta_2^{-1} \|x\| \ , \ \forall(x \neq 0) \in R^n$$

where

$$\eta_2 = \Lambda^{1/2}(H_1 + \tfrac{1}{2}\psi_1 k_1 c_{11} c_{11}^T)$$

and $\Lambda(\cdot)$ is the maximal eigenvalue of the matrix $(\cdot)$. Hence,

$$\phi(x) = \eta_3 \|x\| \ , \ \eta_3 = \frac{1}{2}\epsilon_1 \eta_2^{-1} \ ,$$

and

$$\theta_t + \theta_x^T f(x,0,0) \le -\phi(x) \ , \ \forall (x \ne 0) \in R^n \ ,$$

which implies $N_x = R^n$, $N_{xo} = \{x : x \ne 0 \ , x \in R^n\}$.

(b) The accepted choice of $\theta(x)$ yields

$$\theta_x^T [f(x,y,\mu) - f(x,0,0)] =$$

$$= \frac{1}{2\theta(x)} x^T \left(2H_1 + \psi_1 \frac{\Phi_1(\sigma_1^o)}{\sigma_1^o} c_{11} c_{11}^T\right) \{A_{12}y + q_1[\Phi_1(\sigma_1) - \Phi_1(\sigma_1^o)]\}$$

$$\le \zeta_1 \phi(x) + \zeta_2 \Phi(y) \ , \ \forall x \in N_{xo} \ , \ \forall y \in R^m \ , \ \forall \mu \in]0,1] \ ,$$

for

$$\Phi(y) = \rho_3 \|y\| \ ,$$

$$\zeta_1 = k_1 (\eta_1 \eta_3)^{-1} \cdot \eta_2 \|q_1\| \cdot \|c_{11}\| \ ,$$

$$\zeta_2 = (\eta_1 \rho_3)^{-1} \eta_2 (k_1 \|c_{12}\| \cdot \|q_1\| + \|A_{12}\|) \ ,$$

$$\eta_1 = \lambda^{1/2}(H_1) \ ,$$

where $\lambda(\cdot)$ denotes the minimal eigenvalue of the matrix $(\cdot)$. The value of $\rho_3 > 0$ will be determined in the sequel. The numbers ζ_1 and ζ_2 and functions θ , ϕ and Φ satisfy Assumption 1.

Test of Assumption 2. Let $v(y) = \|y\|$ be a tentative v function. This choice is accepted in order to show another alternative for construction of a Liapunov function. Hence, v is decrescent, positive-definite on R^n and radially unbounded.

(a) In order to test this condition of Assumption 2, we present the boundary-layer system in the Rosenbrock form :

$$\frac{dy}{dt} = D_{22}(\alpha_2)\, y$$

where

$$D_{22}(\alpha_2) = A_{22} + \alpha_2(\sigma_2^o)\, q_2 c_{22}^T \quad , \quad \alpha_2(\sigma_2^o) = \frac{\Phi_2(\sigma_2^o)}{\sigma_2^o} \ .$$

Referring to Grujić (1977e), the matrix $\hat{D}_{22}(\alpha_2) = D_{22}^T(\alpha_2) + D_{22}(\alpha_2)$ is negative definite for every $(\sigma, \phi_2) \in R \times N_o([0,k_2])$ iff both $\hat{D}_{22}(0)$ and $\hat{D}_{22}(k_2)$ are negative definite, which is accepted in the analysis. Finally, $\Phi(y) = \rho_3 \|y\|$ and $v_y^T g(\alpha,b,y,0) \le -\Phi(y)$, $\forall (y \ne 0) \in R^m$ verify condition (a).

(b) $$v_y^T [g(\alpha,b,y,\mu) - g(\alpha,b,y,0)] = \frac{1}{v} y^T \{\mu A_{21} b + q_2[\Phi_2(\sigma_2) - \Phi_2(\sigma_2^o)]\} \ , \quad \forall (y \ne 0) \in R^m \ .$$

Let

$$\zeta_1 = 2\epsilon^{-1} \eta_2 \sup_{\alpha \in [0,k_2]} \|A_{21} + \alpha q_2 c_{21}^T\| \quad , \quad \xi_2 = k_2 \|q_2 c_{21}^T\| \rho_3^{-1} \ .$$

It is assumed that $\xi_2 < 1$. Then

$$v_{\mathbf{y}}^T[\mathbf{g}(\alpha,\mathbf{b},\mathbf{y},\mu) - \mathbf{g}(\alpha,\mathbf{b},\mathbf{y},0)] \le \xi_1\mu\phi(\mathbf{b}) + \xi_2\Phi(\mathbf{y}) ,$$
$$\forall(\alpha,\mathbf{b},\mathbf{y},\mu) \in R\times R^n\times R^m\times]0,+\infty[,$$

which implies $\pi = 1$ in (b) of Assumption 2.

(c) In this case $v_\alpha \equiv 0$ and $v_{\mathbf{b}} \equiv \mathbf{0}$, which yield $\xi_3 = 0$, $\xi_4 = 0$. The lower estimate of the upper bound of allowable μ is given by

$$\tilde{\mu} = \frac{1-\xi_2}{\zeta_2} .$$

Now, $1 > \zeta_1 + \xi_1$ implies absolute stability of $\mathbf{z} = (\mathbf{x}^T,\mathbf{y}^T)^T = \mathbf{0}$ of the system (28) for every $\mu \in]0,\tilde{\mu}[$ on $N_0(L)$, $L = [\mathbf{0},\mathbf{K}]$, $\mathbf{K} = \mathrm{diag}\{k_1,k_2\}$.

Example 13 (Grujić, 1979c). Let

$$\mathbf{A}_{11} = \begin{bmatrix} 0 , & 1 \\ -1 , & -2 \end{bmatrix} , \quad \mathbf{q}_1 = \begin{bmatrix} 0 \\ 1/10 \end{bmatrix} , \quad \mathbf{c}_{11} = \begin{bmatrix} -1/100 \\ 0 \end{bmatrix} ,$$

$$\mathbf{A}_{12} = \mathbf{I} , \quad \mathbf{c}_{12} = \begin{bmatrix} 1 \\ 1 \end{bmatrix} , \quad k_1 = 2 ,$$

and

$$\mathbf{A}_{21} = 10^{-3}\mathbf{I}_2 , \quad \mathbf{c}_{21} = \begin{bmatrix} 10^{-3} \\ 0 \end{bmatrix} , \quad k_2 = 1 ,$$

$$\mathbf{A}_{22} = \begin{bmatrix} -4 , & 1 \\ 1 , & -4 \end{bmatrix} , \quad \mathbf{q}_2 = \begin{bmatrix} 1 \\ 1 \end{bmatrix} , \quad \mathbf{c}_{22} = \begin{bmatrix} 1 \\ 0 \end{bmatrix} .$$

We select $\psi_1 = 1$ and $\epsilon_1 = 1/10$ so that

$$\frac{1}{k_1} + \mathrm{Re}\,(1 + j\psi_1\omega)\,\mathbf{c}_{11}^T(\mathbf{A}_{11} - j\omega\mathbf{I}_2)^{-1}\mathbf{q}_1 - \epsilon_1\mathbf{q}_1^T(\mathbf{A}_{11}^T + j\omega\mathbf{I}_2)^{-1}(\mathbf{A}_{11} - j\omega\mathbf{I}_2)^{-1}\mathbf{q}_1$$
$$\equiv \frac{1}{k_1} = \frac{1}{2} > 0 .$$

Hence,

$$\mathbf{g}_1 = \begin{bmatrix} 0 \\ 0 \end{bmatrix} \quad \text{and} \quad \mathbf{H}_1 = \begin{bmatrix} h_{11} , & h_{12} \\ h_{12} , & h_{22} \end{bmatrix}$$

is determined from

$$\begin{bmatrix} 0 , & -1 \\ 1 , & -2 \end{bmatrix} \begin{bmatrix} h_{11} , & h_{12} \\ h_{12} , & h_{22} \end{bmatrix} + \begin{bmatrix} h_{11} , & h_{12} \\ h_{12} , & h_{22} \end{bmatrix} \begin{bmatrix} 0 , & 1 \\ -1 , & -2 \end{bmatrix} = -\frac{1}{10} \begin{bmatrix} 1 , & 0 \\ 0 , & 1 \end{bmatrix}$$

in the form

$$\mathbf{H}_1 = \frac{1}{20} \begin{bmatrix} 3 , & 1 \\ 1 , & 1 \end{bmatrix} .$$

Therefore, $\eta_1 = 0.16$ and $\eta_2 = 0.45$.

The matrix $\hat{\mathbf{D}}_{22}(\alpha_2)$ is obtained as

$$\hat{\mathbf{D}}_{22}(\alpha_2) = \begin{bmatrix} -8 + 2\alpha_{22} , & 2 + \alpha_{22} \\ 2 + \alpha_{22} , & -8 \end{bmatrix} .$$

Hence, $\hat{\mathbf{D}}_{22}(0)$ and $\hat{\mathbf{D}}_{22}(1)$ are both negative-definite.

Finally, $\zeta_1 = 0.05$, $\zeta_2 = 1.88$, $\xi_1 = 0.02$ and $\xi_2 = 0.002$. Hence, $\tilde{\mu} = 0.52$.

Since $\zeta_1 + \xi_1 = 0.53$ is less than 1 , it follows that $\mathbf{z} = (\mathbf{x}^T, \mathbf{y}^T)^T = \mathbf{0}$ of the system defined in this example is absolutely stable for every $\mu \in]0,\tilde{\mu}[$, i.e. $\mu \in]0,0.52[$, on $N_0(L)$, $L = [\mathbf{0},\mathbf{K}]$, $\mathbf{K} = \text{diag}\{2,1\}$.

The advantage of the separation of time scales is the achieved order problem reduction. Instead of the straightforward analysis of the 4th-order system, two 2nd-order systems were considered and the condition $1 > \zeta_1 + \xi_1$ was tested. Another advantage of this approach to the stability analysis is the more effective way for a system Liapunov function construction.

However, if the dimensions m and n of the reduced-order system (24), (29) and of the boundary-layer system (25), (30) are high, then a further reduction of the order problem is needed.

COMMENTS ON REFERENCES

1. In 1881-1886, Poincaré had published his four memoirs under a general title : "On the Curves Defined by Differential Equations", in which an idea on necessity to study functions, defined by differential equations directly and not by means of their reduction to the more simple functions, was realized successively.
In Liapunov's work : "General Problem of Motion Stability" (Liapunov, 1892), his doctoral dissertation, an evolution of "the fundamental

variant of the stability theory" (Moiseyev, 1949) ascending to Aristotel and having arisen twenty-two centuries ago, is completed. This is the main merit of Liapunov in front of the science.

2. The precise definition of uniform stability was given by Persidskii (1933), and uniform asymptotic stability by Malkin (1954). Statements of the corresponding theorems on stability also belong to them.

3. Definition 1.a-e) is based on those of Liapunov (1892), Persidskii (1933), Chetayev (1946), Massera (1949,1956), Barbashin and Krasovskii (1952), Malkin (1954), Antosiewicz (1958) [see also Krasovskii (1959); Kalman and Bertram (1960); Nemytskii and Stepanov (1960)], Zubov (1964) [see also Coppel (1965), Yoshizawa (1966), Bhatia and Szegö (1967)], Demidovich (1967), Hahn (1967) [see also Šiljak (1969,1974a), Barbashin (1970)], Narendra and Taylor (1973), Grujić (1975,1977) [see also Rouch, Habets and Laloy (1980)], Martynyuk and Gutowski (1979), Matrosov, Anapolsky and Vasilyev (1980). A historical survey of development of stability definiitons introduced till 1975 can be found in works by Yoshizawa (1966), and Rouch, Habets and Laloy (1980).

4. Conditions for solutions existence and uniqueness can be found in the following books : Nemytskii and Stepanov (1960), Codington and Levinson (1955), Hartman (1970), Gutowski (1971), Pontryagin (1961) [see also the paper by Kalman and Bertram (1960)].

5. Duboshin (1950) illustrates that Poincaré's known theorem, which states possibility of expansion of a system of differential equations solutions in series in powers of a small parameter, is a special case of Liapunov's (1892) general theorem, which was the basis of his first method of motion stability analysis. As it is known, Poincaré proved his theorem for a case of a system of the order not greater than the third order, having mentioned in Notes a possibility of its generalization to a system of an arbitrary order. In 1894, Picard proved Poincaré's theorem by another method and also for a special case. In 1911, i.e. after 20 years of Liapunov's book (1892) publication and after some years of its publication in French, Moulton proved Poincaré's theorem for a system of an arbitrary order involving one parameter only (in Liapunov's theorem there can be several (p) parameters).

6. Survey of works on solution of the Liapunov matrix equation is given in the book by Barnett and Storey (1970). An effective method for its solution is also given in the work by Karpinskii and Larin (1976).

7. Liapunov (1892) proposed to denote auxiliary functions of his second method by V; in the present book we denote scalar functions by V and v.

8. The concept of asymptotically contractive sets and positively invariant sets with respect to a function is introduced in the paper by Grujić (1975). The meaning of these concepts is illustrated in the nonstationary systems analysis.

9. Popov's ingenious approach has inspired a great number of scientists, who contributed much to the development of the theory of absolute stability. Yakubovich (1962-1965,1967,1968,1970,1975) and Tsypkin (1962-1964), Kalman (1963) and Szegö (1963,1964) were the first to develop Popov's method.
Considerable results were also obtained by Gelig (1964), Halanay (1964), Ibrahim and Rekasius (1964), Naumov and Tsypkin (1964), Desoer (1965), Jury and Lee (1965), Tokumaru and Saito (1965), Meyer (1965), Dewey and and Jury (1966), Anderson (1966), Dymkov (1967), Moore and Anderson (1968), Partovi and Nahi (1969), Bertoni, Bonivento and Sarti (1970), Šiljak and Sun (1971,1972), Šiljak (1972), Anderson and Moore (1972), Garg and Robbins (1972), Piatnitskii(1970). The sufficiently complete outlines together with considerable original results can be found in the works by Aizerman and Gantmakher (1963), Gantmakher and Yakubovich (1968), Lefschetz (1965), Piatnitskii (1968), Šiljak (1969,1974), Narendra and Taylor (1973).
For some particular cases the necessary and sufficient conditions of absolute stability are stated by Nelepin (1967), Persidskii (1969), Piatnitskii (1970,1971), Mukhametzyanov and Serikbayev (1970). Necessary and sufficient Liapunov like conditions for absolute stability of and Aizerman conjecture for any Lur'e-Postnikov system are due to Grujić (1978-1981).

10. Numerous results, obtained for the singularly perturbed systems, are well described in the following works : Kokotović, O'Malley and Sannuti (1975,1976). Stability of the singularly perturbed systems according to Liapunov was studied by Gradshtein (1951), Tikhonov (1952). Flatto and Levinson (1955), Pontryagin (1957), Mischenko (1959), Klimushev and Krasovskii (1962), Razumikhin (1963), Hoppensteadt (1966, 1967,1968,1974), Desoer and Shensa (1970), Shensa (1971), Wilde and Kokotović (1972), Šiljak (1972), Vasiljeva and Butuzov (1973), Zien (1973), Porter (1974,1976,1977a-b), Habets (1974a-b), Geraschenko and Geraschenko (1975), Mischenko and Rozov (1975), Grujić (1976a-b , 1977c, 1978c , 1979a-b-c , 1981a), Suzuki and Miura (1976), Kuzmina (1977), Young, Kokotović and Utkin (1977), Javid (1978), Chow (1978), Chow and Kokotović (1978), Khalil and Kokotović (1979), Martynyuk and Gutowski (1979).

Aizerman, M.A., and F.R. Gantmacher (1963), *Absolute Stability of Control Systems*. A.N. SSSR, Moscow (in Russian).

Anderson, B.D.O. (1966), Stability of control systems with multiple nonlinearities. *J. Franklin Institute*, *282*, No.3, 155-160.

Anderson, B.D.O. and J.B. Moore (1972), Construction of Liapunov functions for non-stationary systems containing non-inertial nonlinearities. *Avtom. i Telem*, No.5, 14-21 (in Russian).

Antosiewicz, H.A. (1958), A survey of Liapunov's second method. In *Contributions to the Theory of Nonlinear Oscillations*, edited by S. Lefschetz, Vol. IV, Princeton University Press, Princeton, 141-166.

Barbashin, Ye.A. (1967), *Introduction to the Theory of Stability*. M. Nauka, 223 p. (in Russian).

Barbashin, Ye.A. (1970), *The Liapunov functions*. Nauka, Moscow (in Russian).

Barbashin, Ye.A.,and N.N. Krasovskii (1952), On the stability of motion in the large. *Dokl. Akad. Nauk SSSR*, *86*, No.3, 453-456 (in Russian).

Barbashin, Ye.A.,and N.N. Krasovskii (1954), On the existence of Liapunov functions in the case of asymptotic stability in the whole. *Prikl. Mat. Meh.*, *XVIII*, 345-350.

Barnett, S., and C. Storey (1970), *Matrix Methods in Stability Theory*. Nelson, London.

Bertoni, G., C. Bonivento, and E. Sarti (1970), A graphical method for investigating the absolute stability of time-varying systems. *Ricerche di Automatica*, *1*, No.1, 102-111.

Bhatia, N.P., and G.P. Szegö (1967), *Dynamical Systems : Stability Theory and Applications*. Springer-Verlag, Berlin.

Chetaev, N.G. (1946), *Stability of Motion*. OGIZ, Moscow (in Russian).

Chow, J.H. (1978), Asymptotic stability of a class of non-linear singularly perturbed systems. *J. Franklin Inst.*, *305*, 275-281.

Chow, J.H., and P.V. Kokotović (1978), Near-optimal feedback stabilization of a class of nonlinear singularly perturbed systems. *SIAM J. Control and Optimization*, *16*, No.5, 756-770.

Coddington, E.A., and N. Levinson (1955), *Theory of Ordinary Differential Equations*. McGraw Hill, New York. (Russian translation published in 1958 by Inostranoi Literaturi, Moscow.)

Coppel, W.A. (1965), *Stability and Asymptotic Behaviour of Differential Equations*. D.C. Heath and Co., Boston.

Demidovich, B.P. (1967), *Lectures on the Mathematical Theory of Stability*. Nauka, Moscov (in Russian).

Desoer, C.A. (1965), A generalization of the Popov criterion. *IEEE Trans. A.C.*, *AC-10*, No.2, 182-185.

Desoer, C.A., and M.J. Shensa (1970), Network with very small and very large parasitics : natural frequencies and stability. *Proc. IEEE*, *58*, 1933-1938.

Dewey, A.G., and E.I. Jury (1966), A stability inequality for a class of non-linear feedback systems. *IEEE Trans. A.C.*, *AC-11*, No.1, 54-62.

Duboshin, G.N. (1950), On one Poincaré's theorem. *Vestn. Mosk. un-ta, Ser. Yestestv. nauk* (Natural Sciences), No.12, 35-38 (in Russian).

Dymkov, V.I. (1967), On absolute stability of frequency-modulated systems. *Avtom. i Telem*, 109-114 (in Russian).

Flatto, L., and N. Levinson (1955), Periodic solutions of singularly perturbed systems. *J. Rational Mech. Anal., 4*, No.6, 943-950.

Gaiduk, A.R. (1976), Absolute stability of control systems with multiple nonlinearities. *Avtom. i Telemeh*, No.6, 5-11 (in Russian).

Gantmacher, F.R. (1974a), *The Theory of Matrices. Vol. 1*. Chelsea Publ. Co., New York.

Gantmacher, F.R. (1974b), *The Theory of Matrices. Vol. 2*. Chelsea Publ. Co., New York.

Gantmacher, F.R., and V.A. Yakubovich (1965), Absolute stability of nonlinear control systems. In *Analytical Mechanics, Stability of Motions and Space Balistics*. Nauka, Moscow, 30-63 (in Russian).

Garg, D.P., and M.J. Rabins (1972), Stability bounds for nonlinear systems designed via frequency domain stability criteria. *Trans. of the ASME J. of Dynamic Systems, Meas. and Control*, 262-265.

Gelig, A.H. (1964), Investigation of stability of non-linear discontinuous automatic regulating systems with non-unique equilibrium states. *Avtom. i Telemeh, XXV*, No.2, 153-160 (in Russian).

Gelig, A.H., and Komarnitskaya (1966), Absolute stability of non-linear systems with non-unique equilibrium state in critical cases. *Avtom. i Telem.*, No.8, 5-14 (in Russian).

Geraschenko, E.I. and S.M. Geraschenko (1975), *Method of Motion Decomposition and Optimization of Non-Linear Systems*, Nauka, Moscow (in Russian).

Gradshtein, U.S. (1951), Application of Liapunov's stability theory to the theory of differential equations with small multipliers in the derivative terms. *Dokl. AN SSSR, 81*, No.6, 985-986 (in Russian).

Grujić, Lj.T. (1975), Novel development of Lyapunov stability of motion. *Int. J. Control, 22*, No.4, 525-549.

Grujić, Lj.T. (1976a), General stability analysis of large-scale systems. *Proc. IFAC Symp. on Large-Scale Systems Theory and Applications*, 203-213.

Grujić, Lj.T. (1976b), Vector Liapunov functions and singularly perturbed large-scale systems. *Proc. 1976 JACC*, 408-416.

Grujić, Lj.T. (1977a), Un lemme matriciel réciproque; application à la stabilité absolue. *C.R. Acad. Sci.*, Paris, Ser.A, *284*, 1409-1412.

Grujić, Lj.T. (1977b), Is the stability theory consistent and complete ? *First World Conf. on Mathematics at the Service of Man*, Barcelona, July 11-16, 20.

Grujić, Lj.T. (1977c), Stability theory of sets and singularly perturbed large-scale systems. *Ibidem*, 22.

Grujić, Lj.T. (1977d), Converse Lemma and singularly perturbed large-scale systems. *Proc. 1977 JACC*, 1107-1112.

Grujić, Lj.T. (1977e), Stability and instability of product sets. *Systems Science, 3*, No.1, 14-31.

Grujić, Lj.T. (1978a), Absolute stability of non-stationary systems : resolutions and applications. *Proc. 1978 JACC*, Philadelphia, 327-337.

Grujić, Lj.T. (1978b), Solutions for the Lur'e-Postnikov and Aizerman problems. *Int. J. Systems Sci., 9*, No.12, 1359-1372.

Grujić, Lj.T. (1978c), Singular perturbations, uniform asymptotic stability and large-scale systems. *Proc. 1978 JACC*, 339-347.

Grujić, Lj.T. (1979a), Singular perturbations, large-scale systems and asymptotic stability of invariant sets. *Int. J. Systems Sci.*, *10*, No.12, 1323-1341.

Grujić, Lj.T. (1979b), Sets and singularly perturbed systems. *Systems Sci.*, *5*, No.4, 327-338.

Grujić, Lj.T. (1979c), Singular perturbations and large-scale systems. *Int. J. Control*, *29*, No.1, 159-169.

Grujić, Lj.T. (1980), Necessary and sufficient Liapunov-like conditions for absolute stability and Aizerman conjecture. *Math. Physics*, *28*, 7-20 (in Russian).

Grujić, Lj.T. (1981a), Uniform asymptotic stability of non-linear singularly perturbed general and large-scale systems. *Int. J. Control*, *33*, No.3, 481-504.

Grujić, Lj.T. (1981b), On absolute stability and the Aizerman conjecture. *Automatica*, *17*, No.2, 335-349.

Grujić, Lj.T., P. Borne, and J.C. Gentina (1979), Matrix approaches to the absolute stability of time-varying Lur'e-Postnikov systems. *Int. J. Control*, *30*, 967-980.

Habets, P. (1974a), *Stabilité asymptotique pour des problèmes de perturbations singulières*. Centro Internazionalle Mathematico Estivo, Ed. Cremonese, Roma.

Habets, P. (1974b), A consistency theory of singular perturbations of differential equations. *SIAM J. Appl. Math.*, *26*, 136-152.

Hahn, W. (1967), *Stability of Motion*, Springer-Verlag, Berlin.

Halanay, A. (1964), Absolute stability of certain non-linear regulating systems with time-lags. *Avtom. i Telemeh*, *XXV*, No.3, 290-301.

Halanay, A. (1966), *Differential Equations*. Academic Press, New York.

Harrison, G.W. (1979), Persistent sets via Lyapunov functions. *Nonlinear Analysis*, *3*, No.1, 73-80.

Hartmann, P. (1964), *Ordinary Differential Equations*. John Wiley and Sons, New York. (Russian translation published in 1970 by MIR, Moscow.)

Hoppensteadt, F. (1966), Singular perturbations on the infinite interval. *Trans. Amer. Math. Scr.*, *123*, 521-535.

Hoppensteadt, F. (1967), Stability in systems with parameter. *J. Math. Anal. Appl.*, *18*, 129-134.

Hoppensteadt, F. (1968), Asymptotic stability in singular perturbation problems. *J. Diff. Eq.*, *4*, 350-358.

Hoppensteadt, F. (1974), Asymptotic stability in singular perturbation problems. II : Problems having matched asymptotic expansion solutions. *J. Diff. Eq.*, *15*, 510-521.

Ibrahim, E.S., and Z.V. Rekasius (1964), A stability criterion for nonlinear feedback systems. *IEEE Trans. on Aut. Cont.*, *AC-9*, No.2, 151-159.

Javid, S.H. (1978), Uniform asymptotic stability of linear time-varying singularly perturbed systems. *J. Franklin Inst.*, *305*, 27-37.

Jury, E.I., and V.V. Lee (1965), Absolute stability of systems with many nonlinearities. *Avtom. i Telemeh.*, *26*, No.6, 945-965 (in Russian).

Kalman, R.E. (1963), Liapunov functions for the problem of Lur'e in automatic control. *Proc. Nat. Acad. Sci. U.S.A.*, *49*, No.2, 201-205.

Kalman, R.E., and J.E. Bertram (1960), Control system analysis and design via the "second method" of Lyapunov, I. *Trans. of ASME : J. Basic Eng.*, *82*, 371-393.

Karpinsky, F.G., and V.B. Larin (1967), On algebraic and differential Riccati's equations solutions. *Mat. Fizika*, *19*, 36-41 (in Russian).

Khalil, H.K., and P.V. Kokotovic (1979), D-stability and multi-parameter singular perturbation. *SIAM J. Control Optim.*, *17*, 56-65.

Klimushev, A.I., and N.N. Krasovskii (1962), Uniform asymptotic stability of systems of differential equations with a small parameter in the derivative terms. *J. Appl. Math. Mech.*, *25*, 1011-1025.

Kokotović, P.V., R.E. O'Malley Jr., and P. Sannuti (1975), Singular perturbations and order reduction in control theory - An overview. *Prepr. 6th IFAC World Congress*, IC, 51.3.

Kokotović, P.V., R.E. O'Malley Jr., and P. Sannuti (1976), Singular perturbations and order reduction in control theory - An overview. *Automatica*, *12*, 123-132.

Krasovskii, N.N. (1959), *Certain Problems of the Theory of Stability of Motion*. FIZMATGIZ, Moscow (in Russian).

Kuzmina, L.K. (1977), On solutions stability of some systems of differential equations with a small parameter for derivatives. *Prikl. Matem. i Mekhanika*, No.3, 567-573 (in Russian).

LaSalle, J.P. (1976), *The Stability of Dynamical Systems*. SIAM, Philadelphia.

LaSalle, J., and S. Lefschetz (1961), *Stability by Liapunov's Direct Method*. Academic Press, New York.

Lee, E.B., and L. Markus (1967), *Foundations of Optimal Control Theory*. John Wiley and Sons Inc., New York.

Lefschetz, S. (1965), *Stability of Nonlinear Control Systems*. Academic Press, New York.

Liapunov, A.M. (1892), *General Problem of Stability of Motion*. Harkov Math. Soc. (Published in *Collected Papers*, *2*, Ac. Sci. USSR, Moscow-Leningrad, 1956, 5-263.) (in Russian).

Lur'e, A.I. (1951), *Certain Non-linear Problems of the Theory of Automatic Control*. Gostehizdat, Moscow (in Russian).

Lur'e, A.I., and V.N. Postnikov (1944), On the stability theory of control systems. *Prikl. Mat. Meh.*, *VIII*, No.3, 246-248 (in Russian).

Maigarin, B.Z. (1970), Certain criteria for absolute stability of control systems. *Avtom. i Telemh.*, No.1, 188-191 (in Russian).

Malkin, I.G. (1954), On the question of the reciprocal Liapunov's theorem on asymptotic stability of control systems. *Avtom. i Telemeh.*, No.1, 188-191 (in Russian).

Malkin, I.G. (1968), *Motion Stability Theory*. Nauka, Moscow (in Russian).

Martynyuk, A.A., and R. Gutowski (1979), *Integral Inequalities and Stability of Motion*. Naukova dumka, Kiev (in Russian).

Massera, J.L. (1949), On Liapunov's conditions of stability. *Ann. of Math.*, *50*, 705-721.

Massera, J.L. (1956), Contributions to stability theory. *Ann. of Math.*, *64*, 182-206.

Matrosov, V.M. (1962), To the theory of stability of Motion. *Prikl. Matem. i Mekhanika, 25*, No.5, 885-895 (in Russian).

Matrosov, V.M., Yu.L. Anapolsky, and S.N. Vasilyev (1980), *Comparison in Mathematical Theory of Systems*. Nauka, Novosibirsk, 479 p. (in Russian).

McShane, E.J. (1944), *Integration*. Princeton University Press, Princeton.

Meyer, K.R. (1965), Liapunov functions for the problem of Lur'e. *Proc. Nat. Acad. Sci. U.S.A., 53*, 501-503.

Mishchenko, Ye.F. (1959), Asymptotic methods in the theory of relaxational oscillations. *Uspekhi Matem. Nauk, 14*, No.6, 229-236 (in Russian).

Mishchenko, Ye.F., and N.Kh. Rozov (1975), *Differential Equations with a small parameter and relaxation oscillations*. Nauka, Moscow (in Russian).

Moiseyev, N.D. (1949), *Essays on Development of the Theory of Stability*. GITTL, Moscow-Leningrad (in Russian).

Moore, J.B., and B.D.O. Anderson (1968), A generalization of the Popov criterion. *J. Franklin Institute, 285*, No.6, 488-492.

Moulton, F. (1902), *An introduction to Celestial Mechanics*. MacMillan, New York.

Mukhametzianov, I.A., and S.S. Serikbayev (1970), On necessary and sufficient conditions for absolute stability of certain nonlinear systems. *Avtom. i Telemeh., 11*, 11-18 (in Russian).

Narendra, K.S., and J.H. Taylor (1973), *Frequency Domain Criteria for Absolute Stability*. Academic Press, New York.

Naumov, V.N., and Y.Z. Tsypkin (1964), Frequency criteria for absolute stability of processes in non-linear automatic control systems. *Avtom. i Telem., XXV*, No. 6, 852-867 (in Russian).

Nelepin, R.A. (1967), On the problem of exact boundary of the region of absolute stability of control systems. *Avtom. i Telemeh.*, No.4, 30-37 (in Russian).

Nemytskii, V.V., and V.V. Stepanov (1960), *Qualitative Theory of Differential Equations*. Princeton University Press, Princeton.

Partovi, S., and N.E. Nahi (1969), Absolute stability of dynamic systems containing non-linear functions of several state variables. *Automatica, 5*, No.4, 465-473.

Persidskii, K.P. (1933), On stability of motion at first approximation. *Mat. Sb., 40*, 284-293 (in Russian).

Persidskii, S.K. (1969), On the problem of absolute stability. *Avtom. i Telemeh.*, No.12, 5-11 (in Russian).

Piatnitskii, E.S. (1968), New investigations on absolute stability of automatic control systems (Survey). *Avtom. i Telemeh.*, No.6, 5-36 (in Russian).

Piatnitskii, E.S. (1970), Absolute stability of nonstationary nonlinear systems. *Avtom. i Telemeh.*, No.1, 5-15 (in Russian).

Piatnitskii, E.S. (1973), On existence of absolutely stable systems which do not obey the criterion by Popov. *Avtom. i Telemeh.*, No.1, 30-37 (in Russian).

Poincaré, H. (1881-1882), Sur les courbes définies par une équation différentielle. *Journal de Mathématiques*, série 3, *7*, 375-422; *8*, 251-296.

Pontryagin, L.S. (1957), Asymptotic behaviour of systems of differential equations solutions with a small parameter under the highest derivatives. *Izv. AN SSSR*, Ser. Matem., *21*, No.5, 605-626 (in Russian).

Pontryagin, L.S. (1958), Systems of ordinary differential equations with small parameters for high derivatives. In *Proc. of the 3rd All-Union Math. Congress, Moscow*. Izd-vo AN SSSR, *3*, 570-577 (in Russian).

Pontryagin, L.S. (1970), *Ordinary Differential Equations*. Nauka, Moscow (in Russian).

Popov, V.M. (1959), Criterii de stabilitate pentru sistemele neliniare de reglare automata, bazate pe utilizarea transformatei Laplace. *Studii si cerretari de energetica*, Acad. R.P.R. anul, *IX*, No.4, 647-680.

Popov, V.M. (1960), Noi criterii de stabilitate pentru sistemele automate neliniare. *Studii si cercetari de energetica*, Acad. R.P.R., anul., *X*, No.3, 73-88.

Popov, V.M. (1961), On absolute stability of nonlinear automatic control systems. *Avtom. i Telem.*, *XXII*, No.8, 961-979 (in Russian).

Popov, V.M. (1962), On a critical case of the absolute stability. *Avtom. i Telemeh.*, *XXIII*, No.1, 3-24 (in Russian).

Popov, V.M. (1963), Solution of a new stability problem of control systems. *Avtom. i Telem.*, *XXIV*, No.1, 7-28 (in Russian).

Popov, V.M. (1964), On an absolute stability theory problem of control systems. *Avtom. i Telem.*, *XXV*, No.9, 1257-1262 (in Russian).

Popov, V.M. (1973), *Hyperstability of Control Systems*. Springer Verlag, New York.

Popov, V.M. (1974), Dichotomy and stability by frequency-domain method. *Proc. of IEEE*, *62*, No.5, 548-562.

Porter, B. (1974), Singular perturbation methods in the design of stabilizing feedback controllers for multivariable linear systems. *Int. J. Control*, *20*, No.4, 689-692.

Porter, B. (1976), Design of stabilizing feedback controllers for a class of multivariable linear systems with slow and fast modes. *Int. J. Control*, *23*, No.1, 49-54.

Porter, B. (1977a), Singular perturbation methods in the design of stabilizing state-feedback controllers for multivariable linear systems. *Int. J. Control*, *26*, 583-587.

Porter, B. (1977b), Singular perturbation methods in the design of full-order observers for multivariable linear systems. *Int. J. Control*, *26*, 589-594.

Razumikhin, B.S. (1963), On stability of the systems of differential equations solutions with a small parameter by multipliers for the derivatives. *Sib. Matem. Zhurnal*, *4*, No.1, 225-230 (in Russian).

Rouche, N., P. Habets, and M. Laloy (1977), *Stability Theory by Liapunov's Direct Method*. Springer-Verlag, New York. (Russian translation published by Miz.)

Shensa, M.J. (1971), Parasitics and the stability of equilibrium points of non-linear networks. *IEEE Trans. Circ. Theory*, CT-18, 181-484.

Suzuki, M., and M. Miura (1976), Stabilizing feedback controllers for singularly perturbed linear constant systems. *IEEE Trans. A.C.*, AC-21, 123-124.

Szegö, G.P. (1963), On the absolute stability of sampled-data control systems. *Proc. Nat. Acad. Sci. U.S.A.*, *50*, 558-560.

Szegö, G.P. (1964), On the absolute stability of sampled-data systems : the indirect control case. *IEEE Trans. A.C.*, AC-9, 160-163.

Szegö, G.P., and R. Kalman (1963), Sur la stabilité absolue d'un système d'équations aux différences finies. *C.R. Acad. Sci.* (Paris), *257*, 388-390.

Šiljak, D. (1969), *Nonlinear Systems*. John Wiley and Sons, New York.

Šiljak, D.D. (1971), New algebraic criteria for positive realness. *J. Franklin Institute*, *291*, No.2, 109-120.

Šiljak, D.D. (1972), Singular perturbation of absolute stability. *IEEE Trans. AC*, 720.

Šiljak, D. (1974), *Stability of Control Systems*. Faculty of Electrical Engineering, Belgrade (in Serbo-Croatian).

Šiljak, D.D., and K.C. Sun (1971), Exponential absolute stability of discrete systems. *Z. Angew. Math. Mech.*, 271-275.

Šiljak, D.D., and K.C. Sun (1972), On exponential absolute stability. *Int. J. Control*, *16*, 1003-1018.

Tikhonov, A.N. (1952), Systems of differential equations containing small parameters in the derivatives. *Math. Sbor.*, *31*, *73*, No.3, 576-586 (in Russian).

Tokumaru, H., and N. Saito (1965), On the absolute stability of automatic control systems with many nonlinear characteristics. *Memoirs of the Faculty of Engineering*, Kyoto University, *27*, No.3, 347-379.

Tsypkin, Y.Z. (1962a), On stability in the whole of non-linear and sampled-data automatic control systems. *Dokl. AN SSSR*, *145*, No.1, 52-55 (in Russian).

Tsypkin, Y.Z. (1962b), On certain properties of absolutely stable nonlinear sampled-data automatic systems. *Avtom. i Telem.*, *XXIII*, No.12, 1565-1570 (in Russian).

Tsypkin, Y.Z. (1963), Absolute stability of equilibrium states and processes in non-linear sampled-data automatic systems. *Avtom. i Telem.*, *XXIV*, No.12, 1601-1615 (in Russian).

Tsypkin, Y.Z. (1964), Frequency criteria for absolute stability of nonlinear sampled-data systems. *Avtom. i Telem.*, *XXV*, No.3, 281-289 (in Russian).

Vasilyeva, A.B., and V.F. Butuzov (1973), *Asymptotic expansion of the singularly perturbed equations solutions*. Nauka, Moscow (in Russian).

Wilde, R.R., and P. Kokotović (1972), Stability of singularly perturbed systems and networks with parasitics. *IEEE Trans. A.C.*, AC-17, 245-246.

Yakubovich, V.A. (1962), Resolution of certain matrix inequalities appearing in the theory of automatic control. *Dokl. AN SSSR*, *143*, No.6, 1304-1307 (in Russian).

Yakubovich, V.A. (1963a), Absolute stability of non-linear regulating systems in critical cases. I. *Avtom. i Telem.*, *XXIV*, No.3, 293-303 (in Russian).

Yakubovich, V.A. (1963b), *Ibidem*. II. *Avtom. i Telem.*, *XXIV*, No.6, 717-731 (in Russian).

Yakubovich, V.A. (1963c), Frequency conditions for absolute stability of regulating systems with hysteresis nonlinearities. *Dokl. AN SSSR*, *149*, No.2, 288-291 (in Russian).

Yakubovich, V.A. (1964a), Absolute stability of non-linear regulating systems in critical cases. III. *Avtom. i Telem.*, *XXV*, No.5, 601-612 (in Russian).

Yakubovich, V.A. (1964b), Method of matrix inequalities in the stability theory of nonlinear regulating systems. I. *Avtom. i Telem.*, *XXV*, No.7, 1017-1029 (in Russian).

Yakubovich, V.A. (1965), *Ibidem.* III. *Avtom. i Telem.*, *XXVI*, No.4, 577-590 (in Russian).

Yakubovich, V.A. (1967a), Frequency conditions for absolute stability of control systems with several non-linear or linear non-stationary blocks. *Avtom. i Telem.*, *XXVIII*, No.6, 5-30 (in Russian).

Yakubovich, V.A. (1967b), Absolute stability of sampled-data systems with several non-linear or linear non-stationary blocks. I. *Avtom. i Telem.*, *XXVIII*, No.9, 59-72 (in Russian).

Yakubovich, V.A. (1968), *Ibidem.* II. *Avtom. i Telem.*, *XXVIX*, 81-101 (in Russian).

Yakubovich, V.A. (1970), Absolute instability of non-linear control systems. I : General frequency criteria. *Avtom. i Telem.*, *XXXI*, No.12, 5-14 (in Russian).

Yakubovich, V.A. (1975), Frequency conditions for oscillations in non-linear regulating systems with single one-one or hysteresis non-linearity. *Avtom. i Telem.*, *XXXVI*, 51-64 (in Russian).

Yoshizawa, T. (1966), *Stability Theory by Liapunov's Second Method.* The Mathematical Society of Japan, Tokyo.

Young, K.K.D., P.V. Kokotović, and V.I. Utkin (1977), A singular perturbation analysis of high-gain feedback systems. *IEEE Trans. A.C.*, AC-22, 931-938.

Zien, L. (1973), An upper bound for the singular parameter in a stable, singularly perturbed system. *J. Franklin Inst.*, *295*, 373-381.

Zubov, V.I. (1964), *Methods of Liapunov and their Applications.* P. Noordhoff Ltd., Groningen.

Zypkin, J.S. (1963), Die absolute Stabilität nichtlinearer Impulsregelsysteme. *Regelungstechnik*, No.4, 145-148.

CHAPTER II

THE STABILITY THEORY OF COMPARISON SYSTEMS

II.1. INTRODUCTORY NOTES

II.1.1. Original concepts of the comparison method

An idea of a comparison has led to a number of fundamental results. We shall deal with numerous problems concerning the topic of this chapter.

In Liapunov's memoir (1892) (Liapunov , 1935) the stability problem in the first approximation was solved by comparing solutions of the comparison system

$$\frac{dx}{dt} = Ax + X(x) \quad , \quad \forall x \in N$$

where A is an $n \times n$ constant matrix, X is a holomorphic vector-function in x , components of which contain in their decompositions the terms not smaller than that of the second order, with solutions of a system of equations of the first approximation

$$\frac{dx}{dt} = Ax$$

for the defined functions X . Namely, Liapunov had stated the following (Liapunov, 1935, pp. 95-96).

"**Theorem 1.** When the defining equation corresponding to a system of differential equations of the perturbed motion has only roots with the negative real parts, the unperturbed motion is stable and in such a way that any perturbed motion, for which perturbations are small enough, will asymptotically approach the unperturbed one."

"**Theorem 2.** When the defining equation has roots with the negative real parts, then, nevertheless which would be its remaining roots, a known conditional stability will exist for an unperturbed motion. Namely, in case of existence of k such roots, the motion will be stable for perturbations subordinate to some equations of the form

$$F_j(a_1,a_2,\dots,a_n) = 0 \ , \ (j = 1,2,\dots,n-k)$$

in which F_j are the holomorphic functions of the initial values a_s of functions x_s, turning into zero, when all a_s vanish, and which allow us to represent all these values as holomorphic functions of independent values."

"**Theorem III.** When among roots of the defining equations we find such ones that their real parts are positive, then the unperturbed motion is unstable."

V.V. Nemytskii and V.V. Stepanov (Nemytskii, Stepanov, 1949, p.168) wrote :

"Stability analysis and also defining of the families of 0-curves can be mainly done via the comparison method. Essentially, this method is the following.
Let there be given

$$\frac{dx_i}{dt} = f_i(t,x_1,\dots,x_n) \ , \ i = 1,2,\dots,n \ , \qquad \text{(A)}$$

which we call the comparison system, and let the analyzed system have the form :

$$\frac{dx_i}{dt} = f_i(t,x_1,\dots,x_n) + X_i(t,x_1,\dots,x_n) \ , \qquad \text{(B)}$$

where X_i are "small" in different senses; then we try to make conclusions on behaviour of the integral curves of the system (B) by means of that of integral curves of the system (A). Particularly, much have been done in supposition that the comparison system is a linear system with the constant or variable coefficients."

Concretizing systems (A) and (B), A.A. Martynyuk and R. Gutowski (1979, p. 79) considered the Cauchy problem for equations

$$\frac{d\mathbf{y}}{dt} = \mathbf{P}(t,\mathbf{y}) + \mathbf{F}(t,\mathbf{y}) \overset{\Delta}{=} \mathbf{Y}(t,\mathbf{y}) \ , \ \mathbf{y}(t_o) = \mathbf{y}_o \ , \qquad \text{(B1)}$$

$$\frac{d\mathbf{x}}{dt} = \mathbf{P}(t,\mathbf{x}) \ , \ \mathbf{x}(t_o) = \mathbf{y}_o = \mathbf{x}_o \ . \qquad \text{(A1)}$$

The system (A1) is referred to herein as a comparison system.

Under the *Problem I*, we shall mean the following : the algorithm's construction and the proof of statements, allowing us to determine the properties of solutions of the system (B1) on the basis of the system (A1).

Let $y_r = \chi(t;t_o,y_o)$ be a solution of the system (A1). We denote by $y = r+\chi$ and transform the system (B1) to the form :

$$\frac{dr}{dt} = Y(t,r+\chi) - Y(t,\chi) = \phi(t,r) .$$

Let us denote

$$\left.\frac{\partial\phi}{\partial r}\right|_{r=0} = A(t) \ , \ F(t,r) = \phi(t,r) - A(t)r .$$

Further on we shall consider system of equations

$$\frac{dr}{dt} = A(t)r + F(t,r) \ , \ r(t_o) = r_o \ , \tag{B2}$$

$$\frac{ds}{dt} = A(t)s \ , \ s(t_o) = s_o = r_o \ . \tag{A2}$$

Similarly, under the *Problem II* we shall mean the following : the algorithms' construction and the proof of statements which allow us to determine the properties of the (B2) system solutions on the basis of the (A2) system analysis.

As far as the comparison systems (A)-(A2) are concerned, we suppose that they are either completely integrable or more accessible for analysis.

Grebenikov and Ryabov (1979, p.38) considered a system

$$\frac{dz}{dt} = Z(t,z,\mu) \ , \ z(0) = z_o \ , \tag{B3}$$

where $z \in G$, $G \subset R^n$, $\mu \in M$ is a small parameter. Alongside with the system (B3) they used a system of differential equations :

$$\frac{dw}{dt} = W(t,w,\mu) \ , \ w(0) = z_o \ , \tag{A3}$$

which is called the comparison system for the system (B3).

In concrete cases the right part of the system (A3) can be chosen so that it should not depend on t or so that the system (A3) has a number of first integrals.

In this interpretation of the comparison method a great number of results on the qualitative behaviour of solutions, particularly, of systems (A2), (B2), was obtained (Bellman, 1954; Demidovich, 1967, Česari, 1964, et al.).

Another approach in which the idea of a comparison is utilized, is based on the Liapunov method of functions. Depending on the nature of Liapunov's functions - scalar, vector or matrix ones - we come to a scalar equations usually set in a vector form or matrix comparison medium, respectively. It is important to note the following : an application of the scalar or vector Liapunov function leads to a comparison equation (system) of the smaller order than that of the initial system, but the dimension of the Liapunov comparison matrix-functions (Martynyuk, 1984b) is admitted to exceed that of the state space of the initial system. At the same time the comparison matrix mediums can be more appropriate for the analysis since solutions of the corresponding systems of equations are partially ordered by a cone of non-negative definite matrices.

II.1.2. The Liapunov functions and comparison equations generated by them

The Liapunov functions defined in I.3.2.2 are applied in the trends of the classical development of the second method, basing, in fact, on the simpliest differential inequalities

$$D^{*}u(t,\mathbf{x}) \le 0 \ , \ D^{*}u(t,\mathbf{x}) \le -a(u(t,\mathbf{x})) \ .$$

These inequalities were applied by Liapunov himself and they were initial ones until a consideration of the more complicated inequalities (Melnikov, 1956; Antosiewicz, 1958; Opial, 1960).

Corduneanu (1960) , Antosiewicz (1958) , Lakshmikantham (1963, 1964) have found quite a general principle being called the comparison principle with a functional of a Liapunov-like function.

The stability theory development by means of the comparison equations application is shown in a number of papers [14].

Assumption 1. For an autonomous system (10), Ch.I, a comparison equation

$$\frac{du}{dt} = \omega(u) \ , \ u \in R_+ \ ,$$

exists then and only then, when there exist :

1) a positive definite on N (in the whole) function

$$u : N \to R \quad (u : R^n \to R) \ ;$$

2) a continuous, non-negative function ω , $\omega(0) = 0$,

$$\omega(u) \in \mathrm{Lip}_u(N) \quad (\omega(u) \in \mathrm{Lip}_u(R^n))$$

such that

$$D^{+}u(\mathbf{x}) \le \omega(u(\mathbf{x})) \ , \ \forall(\mathbf{x} \ne \mathbf{0}) \in N \quad (\forall(\mathbf{x} \ne \mathbf{0}) \in R^n) \ .$$

Assumption 2. For a non-stationary system (7) of Ch.I, a comparison equation

$$\frac{du}{dt} = \omega(t,u) \ , \ t \in T_\tau \ , \ u \in R_+ \ , \ \tau \in R \tag{2}$$

[14] See 1) of Comments on References to Ch. II.

exists then and only then when there exist :

1) a positive definite on T_τ , $\tau \in R$, with respect to N (in the whole) function $v : R \times N \to R$ ($v : R \times R^n \to R$) ;
2) a continuous, non-negative in the domain $T_\tau \times R_+$ function ω : $T_\tau \times R_+ \to R_+$, $\omega(t,0) = 0$, $\omega(t,u) \in \mathrm{Lip}_u(N)$, ($\omega(t,u) \in \mathrm{Lip}_u(R^n)$) such that

$$D^+ v(t,\mathbf{x}) \le \omega(t, v(t,\mathbf{x})) , \quad \forall t \in T_\tau , \quad \forall(\mathbf{x} \neq 0) \in N \quad (\forall(\mathbf{x} \neq 0) \in R^n) . \blacksquare$$

Item 2 of the present chapter is devoted to the analysis of stability conditions of the solution $u=0$ of equations (1),(2) and to the study of the link with stability conditions for the state $\mathbf{x}=0$ of the systems (7),(10) of Chapter I.

II.1.3. Vector-functions and comparison systems

Let us consider the totality of real functions $v_i : R^n \to R$, $i=1,2,\dots,m$, being the components of a vector $\mathbf{v} = [v_1, v_2, \cdots, v_m]^T$, for which we define the function $\bar{V}(\mathbf{x}) = \max\{v_i(\mathbf{x}) , i=1,2,\dots,m\}$.

Definition 1. Vector-function $\mathbf{v} : R^n \to R^m$ is to be called :

1) *positive semi-definite*, iff a time-invariant neighbourhood N of the point $\mathbf{x}=0$, $N \subseteq R^n$, exists such that
   a) v_i are continuous on N , $v_i(\mathbf{x}) \in C(N)$, $\forall i \in [1,m]$;
   b) $\bar{V}$ is non-negative on N , $\bar{V}(\mathbf{x}) \ge 0$, $\forall \mathbf{x} \in N$;
   c) v_i vanishes at the point $\mathbf{x}=\mathbf{0}$, $v_i(\mathbf{0})=0$, $\forall i \in [1,m]$;
2) *positive semi-definite on the neighbourhood* S of the point $\mathbf{x}=\mathbf{0}$, iff condition 1) of the Definition 1 is fulfilled for $N=S$;
3) *positive semi-definite in the whole*, iff conditions 1) of the Definition 1 are fulfilled for $N=R^n$;
4) *negative semi-definite (on the neighbourhood* S *of the point* $\mathbf{x}=\mathbf{0}$ *or in the whole)*, iff $(-\mathbf{v})$ is positive semi-definite (on the neighbourhood S or in the whole) respectively. ∎

Definition 2. Vector-function $\mathbf{v} : R^n \to R^m$ is to be called :

1) *positive definite* iff there exists a time-invariant neighbourhood N , $N \subseteq R^n$, of the point $\mathbf{x}=\mathbf{0}$ on which it is positive semi-definite and $\bar{V}(\mathbf{x}) > 0$, $\forall(\mathbf{x} \neq \mathbf{0}) \in N$;
2) *positive definite on the neighbourhood* S of the point $\mathbf{x}=\mathbf{0}$, iff conditions 1) of the Definition 2 are fulfilled for $N=S$;
3) *positive definite in the whole*, iff conditions 1) of the Definition 2 are fulfilled for $N=R^n$;

4) *negative definite* (*on the neighbourhood* S of the point $\mathbf{x}=\mathbf{0}$ or *in the whole*) iff $(-\mathbf{v})$ is positive definite (on the neighbourhood S or in the whole). ■

Proposition 1. In order that the vector-function $\mathbf{v}$ be positive definite on the neighbourhood N of the point $\mathbf{x}=\mathbf{0}$, it is necessary and sufficient that there exist functions $\varphi_i \in K_{[0,\alpha[}$, $i=1,2$, where $\alpha = \sup\{\|\mathbf{x}\| : \mathbf{x}\in N\}$, such for which conditions $v_i(\mathbf{x}) \in C(N)$, $\forall i\in[1,m]$, and $\varphi_1(\|\mathbf{x}\|) \le \bar{V}(\mathbf{x}) \le \varphi_2(\|\mathbf{x}\|)$, $\forall \mathbf{x}\in N$, are fulfilled.

Assumption 3. For an autonomous system (10) of Chapter I, a comparison system (an aggregate system)

$$\frac{d\mathbf{u}}{dt} = \Omega(\mathbf{u}), \quad \mathbf{u}\in R^m, \tag{3}$$

exists iff

1) there exists a positive definite on N (in the whole) vector-function $\mathbf{v} = [v_1, v_2, \dots, v_m]^T$, $v_i : R^n \to R$, $v_i \in C(N)$ ($v_i \in C(R^n)$),
2) a continuous with respect to $\mathbf{u}$ in $M \subseteq R^m$ vector-function $\Omega(\mathbf{u}) = (\omega_1(\mathbf{u}), \omega_2(\mathbf{u}), \dots, \omega_m(\mathbf{u}))^T$, $\omega_i \in \mathrm{Lip}_u(M)$, $i\in[1,m]$, exists and $\omega_k(\mathbf{u}) \le \omega_k(\mathbf{u}')$ for $u_k = u_k'$, $u_\nu \le u_\nu'$, $\nu \ne k$, $(\mathbf{u},\mathbf{u}')\in M$ (or $(\mathbf{u},\mathbf{u}')\in R^m$) $\forall(\nu,k)\in[1,m]$;
3) an inequality

$$D^+ \mathbf{v}(\mathbf{x}) \le \Omega(\mathbf{v}(\mathbf{x})), \quad \forall(\mathbf{x}\ne\mathbf{0})\in N \quad (\forall(\mathbf{x}\ne\mathbf{0})\in R^n),$$

is fulfilled. ■

Together with a system of equations (7) Ch. I, we shall consider a vector-function $\mathbf{v}$, $\mathbf{v}(t,\mathbf{x}) = [v_1(t,\mathbf{x}), \dots, v_m(t,\mathbf{x})]^T$ the components of which are $v_i : T_\tau \times R^n \to R$, $\forall i\in[1,m]$. As it was shown above we determine $\bar{V}(t,\mathbf{x}) = \max\{v_i(t,\mathbf{x}),\ \forall i\in[1,m]\}$ (Matrosov, 1968) and formulate a number of definitions.

Definition 3. A vector-function $\mathbf{V} : R\times R^n \to R^m$ is to be called :

1) *positive semi-definite on* T_τ, $\tau\in R$, iff there exists a connected time-invariant neighbourhood N of the point $\mathbf{x}=\mathbf{0}$, $N \subseteq R$, such that
   a) v_i are continuous with respect to $(t,\mathbf{x})\in T_\tau\times N$, $v_i \in C(T_\tau\times N)$, $\forall i\in[1,m]$;
   b) $\bar{V}(t,\mathbf{x})$ is non-negative on N : $\bar{V}(t,\mathbf{x}) \ge 0$, $\forall(t,\mathbf{x})\in T_\tau\times N$;
   c) v_i vanishes for $\mathbf{x}=\mathbf{0}$, $v_i(t,\mathbf{x}) = 0$, $\forall t\in T_\tau$, $\forall i\in[1,m]$;
2) *positive semi-definite on* $T_\tau\times S$ iff conditions 1) of the Definition 3 are fulfilled for $N=S$;
3) *positive semi-definite* in the whole on T_τ iff conditions 1) of the Definition 3 are fulfilled for $N=R^n$;

4) *negative semi-definite (in the whole) on* T_τ (*on* $T_\tau \times N$) iff $(-\mathbf{v})$ is positive semi-definite (in the whole) on T_τ (on $T_\tau \times N$).

In the definition given above the expression "*on* T_τ" can be omitted iff its conditions are fulfilled for all $\tau \in R$. ■

Definition 4. A vector-function $\mathbf{v}: R \times R^n \to R^m$ is to be called :

1) *positive definite on* T_τ, $\tau \in R$, iff there exists a connected time-invariant neighbourhood N of the point $\mathbf{x}=\mathbf{0}$, $N \subseteq R$, such that $\mathbf{v}$ is positive semi-definite on $T_\tau \times N$ and there exists a positive definite function w on N, $w: R^n \to R$, satisfying the condition

$$w(\mathbf{x}) \le \bar{V}(t,\mathbf{x}) \ , \ \forall (t,\mathbf{x}) \in T_\tau \times N \ ;$$

2) *positive definite on* $T_\tau \times S$, iff conditions 1) of the Definition 4 are fulfilled for $N=S$;
3) *positive definite in the whole on* T_τ iff conditions 1) of the Definition 4 are fulfilled for $N=R^n$;
4) *negative definite (in the whole) on* T_τ (*on* $T_\tau \times N$) iff $(-\mathbf{v})$ is positive definite (in the whole) on T_τ (on $T_\tau \times N$).

In the present definition the expression "*on* T_τ" can be omitted iff its conditions are fulfilled for all $\tau \in R$. ■

Proposition 2. In order that the vector-function $\mathbf{v}: R \times R^n \to R^m$ be positive definite on $T_\tau \times N$, where N is a time-invariant neighbourhood of the point $\mathbf{x}=\mathbf{0}$, it is necessary and sufficient that the following conditions be fulfilled :

a) $v_i(t,\mathbf{x}) \in C(T_\tau \times N)$, $\forall i \in [1,m]$;

b) $v_i(t,\mathbf{0}) = 0$, $\forall t \in T_\tau$, $\forall i \in [1,m]$;

c) there exists a function $\varphi \in K_{[0,\alpha[}$, where $\alpha = \sup\{\|\mathbf{x}\| : \mathbf{x} \in N\}$ and $\varphi(\|\mathbf{x}\|) \le \bar{V}(t,\mathbf{x})$, $\forall (t,\mathbf{x}) \in T_\tau \times N$.

Assumption 4. For a non-stationary system (7) Ch.I, a comparison system (an aggregate system)

$$\frac{d\mathbf{u}}{dt} = \Omega(t,\mathbf{u}) \ , \ (t,\mathbf{u}) \in T_\tau \times M \ (T_\tau \times R^m) \tag{4}$$

exists then and only then when

1) there exists a positive definite on $T_\tau \times N$ (in the whole) vector-function $\mathbf{v}$, $\mathbf{v}(t,\mathbf{x}) = (v_1(t,\mathbf{x}), v_2(t,\mathbf{x}), \ldots, v_m(t,\mathbf{x}))^T$, $v_i: R \times R^n \to R$;
2) there exists a continuous with respect to $(t,\mathbf{u}) \in T_\tau \times M$ function $\Omega(t,\mathbf{u}) = (\omega_1(t,\mathbf{u}), \omega_2(t,\mathbf{u}), \ldots, \omega_m(t,\mathbf{u}))^T$, $\omega_k(t,\mathbf{0}) = 0$, $\omega_k \in \mathrm{Lip}_u(M)$, $\omega_k(t,\mathbf{u}) \le \omega_k(t,\mathbf{u}')$ for $u_k = u_k'$, $u_\nu \le u_\nu'$, $k \ne \nu$, $\forall (\mathbf{u},\mathbf{u}') \in M$ (or $\forall (\mathbf{u},\mathbf{u}') \in R^m$), $k,\nu \in [1,m]$;

3) an inequality

$$D^+\mathbf{v}(t,\mathbf{x}) \le \Omega(t,\mathbf{v}(t,\mathbf{x})) \ , \ \forall(t,\mathbf{x}) \in T_T \times N \ \ (T_T \times R^m)$$

is fulfilled. ■

These systems are analyzed somewhat later in Section II.3. [15]

II.1.4. Matrix-functions

The concept of the Liapunov matrix-function generalizes the concept of a vector-function. Its content is the following (Martynyuk, 1984a).

Let us consider a totality of functions $u_{ij} : R^n \to R$, $i = 1,2,\dots,m$. $j = 1,2,\dots,m$:

$$\begin{matrix} u_{11}(\mathbf{x}) & u_{12}(\mathbf{x}) & \dots & u_{1m}(\mathbf{x}) \\ u_{21}(\mathbf{x}) & u_{22}(\mathbf{x}) & \dots & u_{2m}(\mathbf{x}) \\ \dots\dots & & \dots & \dots\dots \\ u_{m1}(\mathbf{x}) & u_{m2}(\mathbf{x}) & \dots & u_{mm}(\mathbf{x}) \end{matrix} \ .$$

We shall denote this totality by $\mathbf{B}(\mathbf{x})$ and call it *the Liapunov matrix-function* [16] under the conditions given in what follows.

Let

$$\bar{B}(\mathbf{x}) = \max\{u_{ij}(\mathbf{x}) \ , \ i\in[1,m] \ ; \ j\in[1,m]\} \ .$$

We shall determine a derivative of the matrix-function $\mathbf{B}$ along solutions of system (10) Ch.I by formula

$$\overset{\circ}{\mathbf{B}}(\mathbf{x}) = \operatorname*{Lim\,inf}_{\theta\to 0^+} \frac{1}{\theta} \left(\mathbf{B}(\mathbf{x}(\theta;\mathbf{x})) - \bar{B}(\mathbf{x})\mathbf{U}\right) \ ,$$

where $\mathbf{U} = (U_{ij})$ is an $m\times m$ - matrix, $U_{ij} = 1$, $\forall(i,j) \in [1,m]$.

If elements $u_{ij}(\mathbf{x})$ of the matrix-function $\mathbf{B}$ are continuous, then

$$\overset{\circ}{\mathbf{B}}(\mathbf{x}) = \begin{cases} -\infty & \text{iff } (i,j) \notin T(\mathbf{x}) \ ; \\ D_+\mathbf{B}(\mathbf{x}) & \text{iff } (i,j) \in T(\mathbf{x}) \ , \end{cases}$$

where

$$T(\mathbf{x}) = \{(i,j)\in[1,m] : u_{ij}(\mathbf{x}) = \bar{B}(\mathbf{x})\} \ .$$

Definition 5. The matrix-function $\mathbf{B} : R^n \to R^{m\times m}$ is to be called :

1) *positive semi-definite* iff a time-invariant neighbourhood N of the point $\mathbf{x}=0$, $N \subseteq R^n$, exists such that :
   a) u_{ij} are continuous on N ; $u_{ij}(\mathbf{x}) \in C(N)$, $\forall(i,j)\in[1,m]$;
   b) $\bar{B}(\mathbf{x})$ is non-negative on N ; $\bar{B}(\mathbf{x})\ge 0$, $\forall \mathbf{x}\in N$;
   c) u_{ij} vanishes for $\mathbf{x}=0$, $\forall(i,j)\in[1,m]$;

[15] See 2) of Comments on References to Ch. II.
[16] See 3) of Comments on References to Ch. II.

2) *positive semi-definite on the neighbourhood* S of the point $x=0$ iff all conditions 1) of the Definition 5 are fulfilled for $N=S$;
3) *positive semi-definite in the whole*, iff all conditions 1) of the Definition 5 are fulfilled for $N=R^n$;
4) *negative semi-definite* (*on the neighbourhood* S of the point $x=0$ or *in the whole*) iff $-B$ is positive semi-definite (on the neighbourhood S or in the whole).

Definition 6. The matrix-function $B : R^n \to R^{m\times m}$ is to be called :

1) *positive definite* iff there exists a time-invariant neighbourhood N, $N\subseteq R^n$, of the point $x=0$, on which B is positive semi-definite and and $\bar{B}(x)>0$, $\forall(x\neq 0)\in N$;
2) *positive definite on the neighbourhood* S of the point $x=0$ iff conditions 1) of the Definition 6 are fulfilled for $N=S$;
3) *positive definite in the whole* iff conditions 1) of the Definition 6 are fulfilled for $N=R^n$;
4) *negative definite (on the neighbourhood* S of the point $x=0$ or in the whole) iff $-B$ is positive definite (on the neighbourhood S of the point $x=0$ or in the whole). ∎

Proposition 3. In order that the matrix-function B be positive definite it is necessary and sufficient that in a neighbourhood N of the point $x=0$ there exist functions $\varphi_i\in K_{[0,\alpha[}$, $i=1,2$, where $\alpha = \sup\{\|x\| : x\in N\}$, such that $u_{ij}(x)\in C(N)$, $\forall(i,j)\in[1,m]$, and $\varphi_1(\|x\|) \le \bar{B}(x) \le \varphi_2(\|x\|)$, $\forall x\in N$. ∎

Let us consider a totality of functions $u_{ij}(t,x) : T_\tau\times R^n \to R$, $\forall(i,j) \in[1,m]$ and determine

$$\bar{B}(t,x) = \max\{u_{ij}(t,x),\ i\in[1,m],\ j\in[1,m]\} .$$

The derivative of the matrix-function $B : T\times R^n\to R^{m\times m}$ along solutions of the system (7) Ch.I we define by

$$\overset{\circ}{B}(t,x) = \lim_{\theta\to 0^+} \inf \frac{1}{\theta}\{B[x(t+\theta, x+\theta f(t,x))] - \bar{B}(t,x)U\} ,$$

where $U=(U_{ij})$ is an $m\times m$-matrix, $U_{ij}=1$, $\forall(i,j)\in[1,m]$.

If elements $u_{ij}(t,x)$ are continuous for all $(t,x)\in T_\tau\times N$ $(\forall(t,x) \in T_\tau\times R^n)$ then

$$\overset{\circ}{B}(t,x) = \begin{cases} -\infty & \text{iff } (i,j)\in T(t,x) ; \\ D_+B(t,x) & \text{iff } (i,j)\in T(t,x) , \end{cases}$$

where

$$T(t,x) = \max\{(i,j)\in[1,m] : u_{ij}(t,x) = \bar{B}(t,x)\} .$$

Definition 7. The matrix-function $B : T_\tau \times R^n \to R^{m\times m}$ is to be called :

1) *positive semi-definite on* T_τ, $\tau \in R$, iff there exists a connected time-invariant neighbourhood N of the point $x=0$, $N \subseteq R^n$, such that :
   a) u_{ij} are continuous with respect to $(t,x) \in T_\tau \times N$, $u_{ij}(t,x) \in C(T_\tau \times N)$, $\forall (i,j) \in [1,m]$;
   b) $\bar{B}(t,x)$ is non-negative on N, $\bar{B}(t,x) \geq 0$, $\forall (t,x) \in T_\tau \times N$;
   c) u_{ij} vanishes for $x=0$, $u_{ij}(t,0) = 0$, $\forall t \in T_\tau$, $\forall (i,j) \in [1,m]$;
2) *positive semi-definite on* $T_\tau \times S$ iff conditions 1) of the Definition 7 are fulfilled for $N=S$;
3) *positive semi-definite in the whole on* T_τ iff conditions 1) of the Definition 7 are fulfilled for $N=R^n$;
4) *negative semi-definite (in the whole) on* T_τ *(on* $T_\tau \times N$*)* iff $(-B)$ is positive semi-definite (in the whole) on T_τ (on $T_\tau \times N$).

In the Definition 7 the expression "*on* T_τ" can be omitted iff its conditions are fulfilled for all $\tau \in R$. ∎

Definition 8. The matrix-function $B : T_\tau \times R^n \to R^{m\times m}$ is to be called :

1) *positive definite on* T_τ, $\tau \in R$, iff there is a connected time-invariant neighbourhood N of the point $x=0$, $N \subseteq R^n$, such that B is positive semi-definite on $T_\tau \times N$ and there exists a positive definite function w on N; $w : R^n \to R$, satisfying the condition
$$w(x) \leq \bar{B}(t,x), \quad \forall (t,x) \in T_\tau \times N ;$$
2) *positive definite on* $T_\tau \times S$ iff conditions 1) of the Definition 8 are fulfilled for $N=S$;
3) *positive definite in the whole on* T_τ iff conditions 1) of the Definition 8 are fulfilled for $N=R^n$;
4) *negative definite (in the whole) on* T_τ *(on* $T_\tau \times N$*)* iff $(-B)$ is positive definite (in the whole) on T_τ (on $T_\tau \times N$), respectively.

In the Definition 8 the expression "*on* T_τ" can be omitted iff its conditions are fulfilled for all $\tau \in R$. ∎

Proposition 4. In order that the matrix-function $B : R \times R^n \to R^{m\times m}$ be positive definite on $T_\tau \times N$, where N is a time-invariant neighbourhood of the point $x=0$, it is necessary and sufficient that the following conditions be fulfilled :

a) $u_{ij}(t,x) \in C(T_\tau \times N)$;
b) $u_{ij}(t,0) = 0$, $\forall t \in T_\tau$;

c) there exists a function $\varphi \in K_{[0,\alpha[}$, where $\alpha = \sup\{\|\mathbf{x}\| : \mathbf{x} \in N\}$, such that

$$\varphi(\|\mathbf{x}\|) \le \bar{B}(t,\mathbf{x}) \quad , \quad \forall (t,\mathbf{x}) \in T_\tau \times N \ . \ ■$$

Remark 1. (Martynyuk, Gutowski, 1979). The matrix-functions naturally appear in case there is need for a calculation of derivatives of higher order than the first one of the vector-Liapunov functions. Really, we consider a vector-function $\mathbf{v}$,

$$\mathbf{v}(t,\mathbf{x}) = (v_1(t,\mathbf{x}), v_2(t,\mathbf{x}), v_m(t,\mathbf{x}))^T \ ,$$

where $v_j(t,\mathbf{x}) \in C^{\ell,\ell}_{t,x}(T_\tau \times N)$, and successively calculate the full derivatives $v_j^{(k)}(t,\mathbf{x})$ and its components up to ℓ-th -order in virtue of system (7) Ch.I by formulas

$$v_j^{(1)}(t,\mathbf{x}) = \partial v_j/\partial t + (\nabla v_j)^T \mathbf{f}(t,\mathbf{x}) \quad , \ j \in [1,m]$$

$$\cdots\cdots\cdots\cdots\cdots\cdots\cdots\cdots$$

$$v_j^{(\ell)}(t,\mathbf{x}) = \partial v_j^{(\ell-1)}/\partial t + (\nabla v_j^{(\ell-1)})^T \mathbf{f}(t,\mathbf{x}) \quad , \ j \in [1,m] \ .$$

Naturally, we assume here that system (7) Ch.I has a sufficiently smooth right part together with functions v_j, $j \in [1,m]$.

Further on we can make a table

$$\begin{array}{cccc} v_1(t,\mathbf{x}) & v_2(t,\mathbf{x}) & \cdots & v_m(t,\mathbf{x}) \ ; \\ v_1^{(1)}(t,\mathbf{x}) & v_2^{(1)}(t,\mathbf{x}) & \cdots & v_m^{(1)}(t,\mathbf{x}) \ ; \\ \cdots & \cdots & \cdots & \cdots \\ v_1^{(\ell)}(t,\mathbf{x}) & v_2^{(\ell)}(t,\mathbf{x}) & \cdots & v_m^{(\ell)}(t,\mathbf{x}) \ , \end{array}$$

from which the Liapunov matrix-function can be constructed if the corresponding conditions are fulfilled (Definition 21, Section 4.1).

II.2. THE LIAPUNOV FUNCTIONS AND COMPARISON EQUATIONS

We shall study stability of the state $\mathbf{x}=\mathbf{0}$ of the system described by equation (7) Ch.I via the analysis of the comparison equation (2) for which the point $u=0$ is the equilibrium state (the zero solution) (see I.2.3).

Further on we shall have need for some results of the theory of differential equations.

II.2.1. On monotonicity and solutions estimations

According to the general concept of the comparison method, the relationship between stability of the state $\mathbf{x}=\mathbf{0}$ of the system (7) Ch.I and the solution $u=0$ of the solution system (2) lead to the necessity of obtaining stability conditions for the solution $u=0$ of the system (2). The simpliest way of their determination is to integrate equation (2) in quadratures. However, such cases are, in reality, exceptions. A necessity to study properties of the equation (2) solutions arises in this connection.

Proposition 5. Let functions $\omega(t,u)$ and $\theta(t,u)$ be defined in a rectangle $\mathbf{D}: t_o \le t \le t_o+a$, $|u-u_o| \le b$; $a,b \in \check{R}_+$

$$\omega(t,u) \in C^{(1,1)}(\mathbf{D}) \quad , \theta(t,u) \in C^{(1,1)}(\mathbf{D}) \quad ,$$

and satisfy an inequality

$$\omega(t,u) < \theta(t,u) \quad , \forall (t,u) \in \mathbf{D} \ .$$

If functions $u(t)$, $\tilde{u}(t)$ are continuous and differentiable on the interval $[t_o, t_o+\delta]$ $(0<\delta\le a)$ and satisfy equations

$$\frac{du}{dt} = \omega(t,u) \quad , \frac{d\tilde{u}}{dt} = \theta(t,\tilde{u}) \quad ,$$

with the initial conditions $u(t_o) = u_o$, $\tilde{u}(t_o) = u_o$, then

$$u(t;t_o,u_o) < \tilde{u}(t;t_o,u_o) \quad \text{for} \quad t_o < t < t_o+\delta \ .$$

Proof. We shall consider a function $z(t) = \tilde{u}(t) - u(t)$ and its Euler derivative. It is obvious that

$$z(t_o) = 0 \quad \text{and} \quad \left.\frac{dz}{dt}\right|_{t=t_o} = \left.\frac{d\tilde{u}}{dt}\right|_{t=t_o} - \left.\frac{du}{dt}\right|_{t=t_o} = \theta(t_o,u_o) - \omega(t_o,u_o) > 0 \ .$$

On account of continuity of $z(t)$ there exists a neighbourhood of the point t_o where the function z is positive. Let at an interval $[t_o, t_o+\delta]$ of continuity and differentiability of functions u and $\tilde{u}$ there exist a value t_1 $(t_o < t_1 < t_o+\delta)$, where $z(t_1) = 0$. In this connection $z(t) > 0$ for $t_o < t < t_1$. Besides, at the point $t = t_1$ there is

$$\left.\frac{dz}{dt}\right|_{t=t_1} = \theta(t_1, \tilde{u}(t_1;t_o,u_o)) - \omega(t_1, u(t_1;t_o,u_o)) \le 0$$

$$\tilde{u}(t_1;t_o,u_o) = u(t_1;t_o,u_o) \quad ,$$

which is impossible due to the condition

$$\theta(t_1,u) - \omega(t_1,u) > 0$$

of the Proposition 5. It follows from this contrast that $z(t_1) > 0$ and, hence, the Proposition 5 is proved. ∎

Definition 9. The solution $r(t)$ defined on an interval $[t_o, t_o+\delta]$ is called an *upper (lower) solution of the equation (2)* iff

$$u(t;t_o,u_o) < r^+(t) \quad (u(t;t_o,u_o) > r^-(t)) \ , \ \forall t \in \,]t_o, t_o+\delta[$$

for any solution u of the equation (2) with the initial value u_o, $(t_o,u_o) \in \mathbf{D}$.

Theorem 1. (The comparison principle). Let a function $\omega : R \times R_+ \to R_+$ be continuous and let there exist a unique solution $u(t;t_o,u_o)$ passing through $(t_o,u_o) \in R \times R_+$ of the Cauchy problem

$$\frac{du}{dt} = \omega(t,u) \ , \ u(t_o) = u_o \ .$$

Let us suppose that an upper solution $u^+(t;t_o,u_o)$ is defined over $[t_o, t_o+\delta_1[$ and there is a differentiable function $v : [t_o, t_o+\delta_1[\to R$ such that $v(t_o) \le u_o$ for $t = t_o$ and

$$\frac{dv}{dt} \le \omega(t,v) \ , \ \forall t \in [t_o, t_o+\delta_1[\ , \tag{5}$$

then

$$v(t) \le u^+(t;t_o,u_o) \quad \text{for} \quad t \in [t_o, t_o+\delta_1[\ . \ \blacksquare \tag{6}$$

The theorem remains correct if we set the sign "$\ge$" instead of "$\le$" in inequalities (5) and (6) and consider a lower $u^-(t;t_o,u_o)$ solution in estimation (6) instead of an upper $u^+(t;t_o,u_o)$ one.

The proof of the theorem is illustrated in the book by Rouche, Habets and Laloy (1980) [17] .

Proposition 6. Let functions $\theta(t,u)$ and $\omega(t,u)$ be continuous and defined on an open domain $\mathbf{D}' \subset \mathbf{D}$ and satisfy an inequality

$$\theta(t,u) < \omega(t,u) \ , \ \forall (t,u) \in \mathbf{D}' \ ,$$

and also

$$|\theta(t,u)| \le m \ , \ |\omega(t,u)| \le m \ , \ m \in \breve{R}_+ \ .$$

Suppose that $\delta = \min(a, b/m)$ and let us denote by $u^+(t;t_o,u_o)$ and $u^-(t;t_o,u_o)$ upper and lower solution of an equation

$$\frac{du}{dt} = \omega(t,u) \ , \ u(t_o) = u_o \ ,$$

and by $\tilde{u}^+(t;t_o,u_o)$, $\tilde{u}^-(t;t_o,u_o)$ upper and lower solution of an equation

$$\frac{d\tilde{u}}{dt} = \theta(t,\tilde{u}) \ ,$$

passing through the point (t_o,u_o) .

Then on a joint interval of existence of the considered solutions we obtain

$$\tilde{u}^-(t;t_o,u_o) \le \tilde{u}^+(t;t_o,u_o) < u^-(t;t_o,u_o) \le u^+(t;t_o,u_o) \ . \ \blacksquare$$

---

[17] See 4) of Comments on References to Ch. II.

The proof of the Proposition 6 follows from the Proposition 1 and Theorem 1.

Let us consider a question of Chaplygin's theorem application in connection with Propositions 5,6 and Luzin's results (Luzin, 1951).

Suppose that we wouldn't be able to obtain a solution of the comparison equation (2).

We shall try to give such two functions $g(t,u)$ and $\tilde{g}(t,u)$ which satisfy the relationship

$$g(t,u) < \omega(t,u) < \tilde{g}(t,u) \ , \ \forall (t,u) \in \mathbf{D} \ ,$$

in the bounds of the domain $\mathbf{D}$, moreover, the differential equations connected with them,

$$\frac{d\psi}{dt} = g(t,\psi) \ , \ \frac{d\eta}{dt} = \tilde{g}(t,\eta) \ , \ \psi(t_o) = \eta(t_o) = u_o \ ,$$

would be integrated exactly. Then an, unknown to us, solution of the Cauchy problem connected with the equation (2) can be found between solutions $\psi(t;t_o,u_o)$ and $\eta(t;t_o,u_o)$ passing through the point u_o for $t=t_o$, i.e.

$$\psi(t;t_o,u_o) < u(t;t_o,u_o) < \eta(t;t_o,u_o)$$

on a general interval of existence of the considered solutions.

Thus, Propositions 5,6 justify the comparison equation (2) analysis by means of its substitution by a simpler one, admitting the exact integration.

The proof of the Proposition 6 follows from the Proposition 5 and Theorem 1.

Let us continue analysis of inequality (5) together with an integral inequality connected with it. Namely, let the following conditions be fulfilled :

(H1). There exist functions $w(t)$, $k(t)$ continuous and non-negative on T_o ;

(H2). There exists a function $c(t,u)$, continuous with respect to $(t,u) \in \mathbf{D}$ and non-negative for $u \geq 0$;

(H3). There exists a continuous, non-negative and non-decreasing function $g(z)$ for $z \geq 0$, moreover, $g(z_o) > 0$ for $z_o \in]0,a]$, $a \in \check{R}_+$;

(H4). A representation

$$\omega(t,u) = w(t)u + c(t,u) \ , \ \forall (t,u) \in \mathbf{D} \ .$$

is admissible.

Theorem 2. Let

1) assumptions (H1)-(H3) be fulfilled;
2) inequalities

a) $$u(t) \le a + \int_{t_o}^{t} [w(s)u(s) + c(s,u(s))]\, ds ,$$

b) $$c\,[t, z\exp(\int_{t_o}^{t} w(s)ds)]\,\exp(-\int_{t_o}^{t} w(s)ds) \le k(t)\, g(z) , \quad t\in T_o ,$$

hold.

Then for all $t \in [t_o, \tau[$ the following estimation is valid

$$u(t) \le G^{-1}\,[G(a) + \int_{t_o}^{t} k(s)\,ds]\,\exp(\int_{t_o}^{t} w(s)\,ds) , \tag{7}$$

where G^{-1} is a function, inverse to that of

$$G(w) = \int_{u_o}^{w} \frac{ds}{g(s)} , \quad 0 < u_o \le a \le u ,$$

and the value τ can be determined from a relationship

$$\int_{u_o}^{\infty} \frac{ds}{g(s)} = G(a) + \int_{t_o}^{\tau} k(s)\,ds .$$

If to the enumerated conditions we add one more :

3) there exists a constant $\tilde{a}\in \check{R}_+$ such that

$$\int_{t_o}^{\infty} k(s)\,ds \le \int_{\tilde{a}}^{\infty} \frac{ds}{g(s)} ,$$

then the inequality (5) is fulfilled for all $t\in T_o$ and $\forall a \in]0,\tilde{a}[$.

Proof. We assume that

$$z = u\exp(-\int_{t_o}^{t} w(s)\,ds)$$

and note that

$$c\,[t, z\exp(\int_{t_o}^{t} w(s)\,ds)] \le k(t)\exp(\int_{t_o}^{t} w(s)\,ds)\, g(z) , \quad \forall t\in T_o .$$

Hence

$$c(t,u) \le u(t)\exp(\int_{t_o}^{t} w(s)\,ds)\;\; g(u\exp[-\int_{t_o}^{t} w(s)\,ds]) .$$

We transform the inequality 2.a) of the Theorem 2 to the form

$$u(t) \le a + \int_{t_o}^{t} \Big[w(s)\,u(s) + k(s)\exp(\int_{t_o}^{t} w(s)\,ds)\, g(u(s)\exp[-\int_{t_o}^{t} w(s)\,ds])\Big]\, ds$$

$$= R(t)\exp\left(\int_{t_o}^{t} w(s)\,ds\right) .$$

It is not difficult to find out that

$$\frac{dR}{dt}\exp\left(\int_{t_o}^{t} w(s)\,ds\right) + R(t)\,w(t)\exp\left(\int_{t_o}^{t} w(s)\,ds\right) \le$$

$$\le w(t)\,R(t)\exp\left(\int_{t_o}^{t} w(s)\,ds\right) + k(t)\exp\left(\int_{t_o}^{t} w(s)\,ds\right) g(R(t)) .$$

Hence

$$\frac{dR}{dt} \le k(t)\,g(R(t)) .$$

As far as $a>0$ in virtue of the Condition (H3) of the Theorem 2, we obtain

$$\int_{a}^{R(t)} \frac{dr}{g(r)} \le \int_{t_o}^{t} k(s)\,ds .$$

Further on we shall obtain

$$G(R(t)) \le G(a) + \int_{t_o}^{t} k(s)\,ds$$

taking into account the notation $G(w)$.

Let us pay our attention to the fact that in a general case function $R(t)$ can make inadmissible an unbounded continuation to the right side. Let us define β , for which $R(t)=\infty$ from an equation

$$G(\infty) = \int_{t_o}^{\infty} \frac{dr}{g(r)} = G(a) + \int_{t_o}^{\beta} k(s)\,ds .$$

The condition of unbounded continuation of the function $R(t)$ is given by an inequality

$$G(a) + \int_{t_o}^{\infty} k(s)\,ds \le \int_{u_o}^{\infty} \frac{dr}{g(r)} ,$$

or

$$\int_{t_o}^{\infty} k(t)\,dt \le -\int_{u_o}^{a} \frac{dr}{g(r)} + \int_{u_o}^{\infty} \frac{dr}{g(r)} = \int_{a}^{\infty} \frac{dr}{g(r)} .$$

This inequality will be correct for those $a \in]0,\tilde{a}]$, for which

$$\int_{t_o}^{\infty} k(t)\,dt \le \int_{\tilde{a}}^{\infty} \frac{dr}{g(r)} < \int_{a}^{\infty} \frac{dr}{g(r)}$$

for $a<\tilde{a}$ i.e. we have $\beta=\infty$ for $a \in]0,\tilde{a}[$.

For values $t \in [t_o,\beta[$ an inequality

$$R(t) \le G^{-1}[G(a) + \int_{t_o}^{t} k(s)\,ds]$$

and estimation

$$u(t) \le R(t)\exp\left(\int_{t_o}^{t} w(s)\,ds\right) \le G^{-1}[G(a) + \int_{t_o}^{t} k(s)\,ds]\exp\left(\int_{t_o}^{t} w(s)\,ds\right)$$

will be correct. ■

Example 1. Let $w(t) = 1$ and $c(t,u) = u^2$ for all $t \ge t_o$. Having fulfilled conditions of the Theorem 2 we obtain from the inequality

$$u(t) \le a + \int_{t_o}^{t} [u(s) + u^2(s)]\,ds \quad , \ t \ge t_o \ ,$$

an estimation

$$u(t) \le \frac{a\exp(t-t_o)}{1 + a[1 - \exp(t-t_o)]} \quad \text{for} \quad t \in [t_o, \tau[\ ,$$

where

$$\tau = t_o + \ln\frac{1+a}{a} \ .$$

Really, functions k and g mentioned in estimation 2 b) of the Theorem 2 are the following

$$k(t) = \exp(t-t_o) \ , \ g(z) = z^2 \ .$$

As far as

$$G(u) = \int_{u_o}^{u} \frac{dr}{r^2} = -\frac{1}{u} + \frac{1}{u_o} = \lambda \ , \ u = G^{-1}(\lambda) = \frac{1}{-\lambda + \frac{1}{u_o}} \ ,$$

then

$$\rho \triangleq G(a) + \int_{t_o}^{t} k(s)\,ds = \int_{u_o}^{u}\frac{dr}{r^2} + \int_{t_o}^{t}\exp(s-t_o)\,ds = -\frac{1}{a} + \frac{1}{u_o}\exp(t-t_o) - 1 \ .$$

Further on

$$G^{-1}(\rho) = \frac{1}{-[G(a) + \int_{t_o}^{t} u(s)\,ds] + \frac{1}{u_o}} = \frac{a}{1 + a[1 - \exp(t-t_o)]} \ .$$

Now, basing on inequality (7), we obtain the desired estimation. Value τ can be defined from an equation

$$\int_{u_o}^{\infty}\frac{dr}{r^2} = \int_{u_o}^{a}\frac{dr}{r^2} + \int_{t_o}^{\tau}\exp(s-t_o)\,ds \ .$$

Theorem 3. Let :

1) conditions (H1)-(H3) be fulfilled, moreover, in condition (H1) the function $w(t) = -\alpha(t)$ holds for all $t \ge t_o$;
2) inequalities

a) $u(t) \le a + \int_{t_o}^{t} [-\alpha(s)\, u(s) + c(s,u(s))]\, ds$,

b) $c\,[t\,,\, z \exp(-\int_{t_o}^{t} \alpha(s)\, ds)]\, \exp(\int_{t_o}^{t} \alpha(s)\, ds) \le k(t)\, g(z)$, $t \ge t_o$, $z \ge 0$

hold.

Then for all $t \in [t_o, \tau[$ an estimation

$$u(t) \le G^{-1}[G(a) + \int_{t_o}^{t} k(s)\, ds]\, \exp(-\int_{t_o}^{t} \alpha(s)\, ds) \tag{8}$$

is valid and the value τ can be defined from

$$\int_{u_o}^{\infty} \frac{dr}{g(r)} = G(a) + \int_{t_o}^{\tau} k(s)\, ds \, .$$

If to the enumerated conditions of Theorem 3 we add one more :

3) there exists a constant $\tilde{a} \in \check{R}_+$ such that

$$\int_{t_o}^{\infty} k(s)\, ds \le \int_{\tilde{a}}^{\infty} \frac{dr}{g(r)} \, ,$$

then the estimation (8) holds for all $t \ge t_o$ and all $a \in]0, \tilde{a}[$. ∎

The proof of the Theorem 3 is similar to that of the Theorem 2.

Let us note that the inequality 2 a) must be fulfilled together with the condition :

$$\frac{du}{dt} \le -\alpha(t)\, u + c(t,u) \, , \quad \forall t \in T_o \, , \quad u(t_o) = a \, .$$

II.2.2. Special cases of the general comparison equations

Theorems 2,3 can be utilized for obtaining various estimations of comparison equations solutions. Let us consider several cases.

Case 1. Let all conditions of the Theorem 2 be fulfilled. Suppose that

$$\int_{t_o}^{\infty} k(t)\, dt < \infty \quad \text{and} \quad \int_{u_o}^{\infty} \frac{dr}{g(r)} < \infty \, , \quad t_o \ne 1 \, ,$$

and there exists a constant $\tilde{a} \in]0, +\infty[$ such that

$$\int_{t_o}^{\infty} k(t)\, dt = \int_{\tilde{a}}^{\infty} \frac{dr}{g(r)} \, .$$

Then there exists a constant $M(a,t_o)$ such that

$$u(t) \le M(a,t_o) < +\infty$$

for all $t \in [t_o,\infty[$ and $a \in]0,\tilde{a}[$.

Example 2. Let

$$w(t) = \frac{1}{t^2} \quad \text{and} \quad c(t,u) = \frac{1}{t^2}\exp[-(\frac{1}{t}-\frac{1}{t_o}) + u\exp(\frac{1}{t}-\frac{1}{t_o})] .$$

In case 1 from an inequality

$$u(t) \le a + \int_{t_o}^{t} (\frac{1}{s^2}u(s) + \frac{1}{s^2}\exp[-(\frac{1}{s}-\frac{1}{t_o}) + u(s)\exp(\frac{1}{s}-\frac{1}{t_o})])\,ds \ , \quad t \ge t_o ,$$

an estimation

$$u(t) \le \exp[-(\frac{1}{t}-\frac{1}{t_o})]\ln\frac{1}{\frac{1}{t} + [\exp(-a) - \frac{1}{t_o}]} \tag{10}$$

is fulfilled for all $t \in [t_o,+\infty[$ and $a \in]0,\tilde{a}[$, where $\tilde{a} = \ln t_o$.

Functions k and g of the Theorem 2 have the form

$$k(t) = \frac{1}{t^2} \ , \quad g(z) = \exp(z)$$

as far as

$$c(t\ , z\exp[-(\frac{1}{t}-\frac{1}{t_o})])\exp(\frac{1}{t}-\frac{1}{t_o}) = \frac{1}{t^2}\exp[-(\frac{1}{t}-\frac{1}{t_o}) + z]\exp(\frac{1}{t}-\frac{1}{t_o})$$

$$= \frac{1}{t^2}\exp(z) .$$

Functions G and G^{-1}

$$G(u) = \int_{u_o}^{u}\exp(-r)\,dr = -\exp(-u) + \exp(u_o) ;$$

$$-u = \ln[-\lambda + \exp(-u_o)] ;$$

$$u = G^{-1}(\lambda) = -\ln[-\lambda + \exp(-u_o)] .$$

Further on,

$$\rho = \int_{u_o}^{a}\exp(-r)\,dr + \int_{t_o}^{t}\frac{dr}{r^2} = -\exp(-a) + \exp(u_o) + (-\frac{1}{t} + \frac{1}{t_o}) ,$$

$$G^{-1}(\rho) = -\ln[-(G(a) + \int_{t_o}^{t}k(s)\,ds) + \exp(-u_o)] = -\ln[\frac{1}{t} - \frac{1}{t_o} + \exp(-a)] .$$

Taking into account all this, we obtain the inequality (10) from the estimation (7).

Value $\tilde{a}$ can be defined from condition (9)

$$\int_{t_o}^{\infty}k(t)\,dt = \frac{1}{t_o} < \infty \ , \quad \int_{\tilde{a}}^{\infty}\frac{dr}{g(r)} = \exp(-\tilde{a}) < \infty ,$$

$$\frac{1}{t_o} = \exp(-\tilde{a}) \ , \ \tilde{a} = \ln t_o \ .$$

Case 2. Let all conditions of the Theorem 2 be fulfilled. Suppose that

$$\int_{t_o}^{\infty} [k(t) + w(t)]\, dt < \infty \ , \ \int_0^{(\cdot)} \frac{dr}{g(r)} < \int_{(\cdot)}^{\infty} \frac{dr}{g(r)} = \infty \ . \qquad (11)$$

Then there exists a constant $M(a,t_o)$ such that

$$u(t) \le M(a,t_o) < \infty$$

for all $t \in [t_o,\infty[$ and $a \in]0,\tilde{a}[$, where $\tilde{a}$ can be arbitrarily large.

Example 3. Let

$$w(t) = \frac{1}{t^2} \quad \text{and} \quad c(t,u) = \frac{1}{t^2} u^{1/3} \exp(-\frac{2}{3t}) \quad \text{for all} \quad t \ge t_o \ .$$

In case 2 from an inequality

$$u(t) \le a + \int_{t_o}^{t} [\frac{1}{s^2} u(s) + \frac{1}{s^2} u^{1/3}(s) \exp(-\frac{2}{3s})]\, ds \ , \ t \ge t_o \ ,$$

an estimation

$$u(t) \le [(a \exp(\frac{1}{t_o}))^{2/3} - \frac{2}{3}(\frac{1}{t} - \frac{1}{t_o})]^{3/2} \exp(-\frac{1}{t}) \qquad (12)$$

and value

$$M(a,t_o) = [(a \exp(\frac{1}{t_o}))^{2/3} + \frac{2}{3t_o}]^{3/2}$$

are fulfilled for all $t \in [t_o,\infty[$ and $a \in]0,\tilde{a}[$, where $\tilde{a}$ can be arbitrarily large. Let us prove it. It is the fact that

$$c[t, 2\exp(\int_{t_o}^{t} w(s)\, ds)] \exp(-\int_{t_o}^{t} w(s)\, ds) =$$

$$= c[t, z\exp(-(\frac{1}{t} - \frac{1}{t_o}))] \exp(\frac{1}{t} - \frac{1}{t_o}) = \frac{1}{t^2} z^{1/3} \exp(-\frac{2}{3t_o}) \ .$$

For functions k and g we determine expressions in the form

$$k(t) = \frac{1}{t^2} \exp(-\frac{2}{3t_o}) \ , \ g(z) = z^{1/3} \ .$$

Conditions (11) are fulfilled as far as

$$\int_{t_o}^{\infty} [w(s) + k(s)]\, ds = \int_{t_o}^{\infty} \frac{1}{t^2} \exp(-\frac{2}{3t_o})\, dt + \int_{t_o}^{\infty} \frac{dt}{t^2} = \frac{1}{t_o} \exp(-\frac{2}{3t_o}) + \frac{1}{t_o} < \infty \ ;$$

$$\int_0^A \frac{dr}{g(r)} = \int_0^A r^{-1/3}\, dr = \frac{3}{2} A^{2/3} < \infty \ ; \ \int_A^{\infty} \frac{dr}{g(r)} = \int_A^{\infty} r^{-1/3}\, dr = \frac{3}{2} r^{2/3} = \infty \ .$$

Further on

$$G(u) = \int_{u_o}^{u} r^{-1/3}\,dr = \frac{3}{2}(u^{2/3} - u_o^{2/3}) = \lambda \quad ;$$

$$u = G^{-1}(\lambda) = (\frac{2}{3}\lambda + u_o^{2/3})^{3/2} \quad ;$$

$$\rho = \int_{u_o}^{a} r^{-1/3}\,dr + \int_{t_o}^{t} \frac{1}{r^2}\exp(-\frac{2}{3t_o})\,dr = \frac{3}{2}(a^{2/3} - u^{2/3}) + (\frac{1}{t} - \frac{1}{t_o})\exp(-\frac{2}{3t_o}) \quad ;$$

$$G^{-1}(\rho) = [\frac{2}{3}(G(a) + \int_{t_o}^{t} k(s)\,ds) + u^{2/3}]^{3/2} = [a^{2/3} - \frac{2}{3}(\frac{1}{t} - \frac{1}{t_o})\exp(-\frac{2}{3t_o})]^{3/2} \quad .$$

On the basis of estimation (7) we obtain the inequality (12).

Case 3. Let all conditions of the Theorem 2 be fulfilled. Suppose that

$$\int_{t_o}^{\infty} [w(t) + k(t)]\,dt < \infty \quad , \quad \int_{(\cdot)}^{\infty} \frac{dr}{g(r)} < \int_{0}^{(\cdot)} \frac{dr}{g(r)} = \infty$$

and there exists a constant $\tilde{a} \in]0,\infty[$ such that

$$\int_{t_o}^{\infty} k(t)\,dt = \int_{\tilde{a}}^{\infty} \frac{dr}{g(r)} \quad .$$

Then there exists a constant $M(a,t_o)$ such that

$$u(t) \le M(a,t_o) < \infty$$

for all $t \in [t_o,\infty[$ and $a \in]0,\tilde{a}[$, where $M(a,t_o) \to 0$ for $a \to 0$, $t_o > 0$.

Example 4. Let

$$w(t) = \frac{1}{t^2} \quad , \quad c(t,u) = u^2 \frac{1}{t^2}\exp(\frac{1}{t})$$

for all $t \ge t_o > 0$.

In Case 3 from the inequality

$$u(t) \le a + \int_{t_o}^{t} [\frac{1}{s^2}u(s) + \frac{1}{s^2}u^2(s)\exp(\frac{1}{s})]\,ds \quad , \quad t \ge t_o$$

an estimation

$$u(t) \le \frac{a\exp[-(\frac{1}{t} - \frac{1}{t_o})]}{1 + a(\frac{1}{t} - \frac{1}{t_o})\exp(\frac{1}{t_o})}$$

is fulfilled for all $t \in [t_o,\infty)$ and $a \in]0,\tilde{a}[$, where $\tilde{a} = t_o\exp(-\frac{1}{t_o})$ and the constant $M(a,t_o)$ can be defined by an expression

$$M(a,t_o) = \frac{a\,t_o\exp(\frac{1}{t_o})}{t_o - a\exp(\frac{1}{t_o})} \quad .$$

To convince ourselves in that, we must determine functions, which are contained in estimation (7). From the fact that

$$c\,[t\,,\,z\exp(-(\tfrac{1}{t}-\tfrac{1}{t_o}))]\exp(\tfrac{1}{t}-\tfrac{1}{t_o}) = \frac{1}{t^2}\exp(\tfrac{1}{t})\,z^2\exp(-\tfrac{2}{t}+\tfrac{2}{t_o})\exp(\tfrac{1}{t}-\tfrac{1}{t_o})$$
$$= z^2\,\frac{1}{t^2}\exp(\tfrac{1}{t_o})$$

we determine

$$k(t) = \frac{1}{t^2}\exp(\tfrac{1}{t_o})\ ,\quad g(z) = z^2\ .$$

Functions ρ and G^{-1} have the form

$$\rho = \int_{u_o}^{a}\frac{dr}{r^2} + \int_{t_o}^{t}\frac{1}{r^2}\exp(\tfrac{1}{t_o})\,dr = -\frac{1}{a}+\frac{1}{u_o}+(-\frac{1}{t}+\frac{1}{t_o})\exp(\tfrac{1}{t_o})\ ;$$

$$G^{-1}(\rho) = \frac{a}{1+a(\frac{1}{t}-\frac{1}{t_o})\exp(\frac{1}{t_o})}\ .$$

By combining these functions into inequality (7), we obtain estimation (13). An expression $\tilde{a}$ can be defined from an expression

$$\int_{t_o}^{\infty}\frac{1}{t^2}\exp(\tfrac{1}{t_o})\,dt = \int_{\tilde{a}}^{\infty}\frac{dr}{r^2}\ .$$

Case 4. Let all conditions of the Theorem 2 be fulfilled. Suppose that

$$\int_{t_o}^{\infty}[w(t)+k(t)]\,dt < \infty\ ,\quad \int_{0}^{(\cdot)}\frac{dr}{g(r)} = \int_{(\cdot)}^{\infty}\frac{dr}{g(r)} = \infty\ .$$

Then there exists a constant $M(a,t_o)$ such that

$$u(t) \le M(a,t_o) < \infty$$

for all $t\in[t_o,\infty[$ and $a\in\,]0,\tilde{a}[$, where $\tilde{a}$ can be arbitrarily large. Besides, $M(a,t_o)\to 0$ for $a\to 0\ ,\ t_o>0$.

Example 5. Let $w(t)=\frac{1}{t^2}$, $c(t,u)=\frac{1}{t^2}u$ for all $t\ge t_o>0$. In Case 4 from the inequality

$$u(t) \le a + 2\int_{t_o}^{t}\frac{1}{s^2}u(s)\,ds\ ,\quad t\ge t_o\ ,$$

an estimation

$$u(t) \le a\exp[\,2(\tfrac{1}{t_o}-\tfrac{1}{t})] \tag{14}$$

is fulfilled for all $t\in[t_o,\infty[$ and $a\in\,]0,\tilde{a}[$, where $\tilde{a}$ can be arbitrarily large; the constant $M(a,t_o)$ can be defined by an expression

$$M(a,t_o) = a\exp(\tfrac{2}{t_o})\ ,\quad t_o>0\ .$$

To obtain estimation (14) we must act similarly as with the previous one. We define

$$w(t) = \frac{1}{t^2} \; , \; g(z) = z$$

and further on

$$\int_{t_o}^{\infty} [w(t) + k(t)]\, dt = \int_{t_o}^{\infty} \frac{2}{t^2}\, dt = \frac{2}{t_o} < \infty \; ,$$

$$\int_{o}^{A} \frac{ds}{s} = \ln s \Big|_{o}^{A} = \infty \; , \; \int_{A}^{\infty} \frac{ds}{s} = \ln s \Big|_{A}^{\infty} = \infty \; .$$

Functions ρ and G^{-1} have the form

$$\rho = \ln \frac{a}{u_o} + \left(-\frac{1}{t} + \frac{1}{t_o}\right) \; , \; G^{-1}(\rho) = a \exp\left(\frac{1}{t_o} - \frac{1}{t}\right) \; .$$

On the basis of inequality (7) and the calculated functions we obtain estimation (14). It is obvious that $\lim M(a,t_o) = 0$ for $a \to 0$ and $t_o > 0$.

Case 5. Let all conditions of the Theorem 3 $(w(t) = -\alpha(t) \; , \forall t \geq t_o > 0)$ be fulfilled. Suppose that

$$\int_{u_o}^{\infty} \frac{dr}{g(r)} < \infty \; .$$

Besides, there exists a constant $\tilde{a} \in]0,\infty[$, such that

$$\int_{t_o}^{\infty} k(t)\, dt = \int_{\tilde{a}}^{\infty} \frac{dr}{g(r)}$$

Then there exists a constant $M(a,t_o) < \infty$ such that

$$u(t) \leq M(a,t_o) \exp\left(-\int_{t_o}^{t} \alpha(s)\, ds\right)$$

for all $t \in [t_o,\infty[$ and $a \in]0,\tilde{a}[$.

Case 6. Let all conditions of the Theorem 3 $(w(t) = -\alpha(t) \; , \forall t \geq t_o > 0)$ be fulfilled. Suppose that

$$\int_{t_o}^{\infty} k(t)\, dt < \infty \; , \; \int_{o}^{(\cdot)} \frac{dr}{g(r)} < \int_{(\cdot)}^{\infty} \frac{dr}{g(r)} = \infty \; .$$

Then there exists a constant $M(a,t_o) < \infty$ such that

$$u(t) \leq M(a,t_o) \exp\left(-\int_{t_o}^{t} \alpha(s)\, ds\right)$$

for all $t \in [t_o,\infty[$ and $a \in]0,\tilde{a}[$, where $\tilde{a}$ can be arbitrarily large.

Case 7. Let all conditions of the Theorem 3 $(w(t) = -\alpha(t), \forall t \geq t_o > 0)$ be fulfilled. Suppose that

$$\int_{t_o}^{\infty} k(t)\,dt < \infty \qquad \int_{(\cdot)}^{\infty} \frac{dr}{g(r)} < \int_{0}^{(\cdot)} \frac{dr}{g(r)} = \infty .$$

Besides, there exists a constant $\tilde{a} \in]0,\infty[$ such that

$$\int_{t_o}^{\infty} k(t)\,dt = \int_{\tilde{a}}^{\infty} \frac{dr}{g(r)} .$$

Then there exists a constant $M(a,t_o) < \infty$ such that

$$u(t) \leq M(a,t_o) \exp\left(-\int_{t_o}^{t} \alpha(s)\,ds\right)$$

for all $t \in [t_o,\infty[$ and $a \in]0,\tilde{a}[$, where $M(a,t_o) \to 0$ for $a \to 0$ and $t_o > 0$.

Case 8. Suppose that

$$\int_{t_o}^{\infty} k(t)\,dt < \infty , \quad \int_{0}^{(\cdot)} \frac{dr}{g(r)} = \int_{(\cdot)}^{\infty} \frac{dr}{g(r)} = \infty .$$

Then there exists a constant $M(a,t_o) < \infty$ such that

$$u(t) \leq M(a,t_o) \exp\left(-\int_{t_o}^{t} \alpha(s)\,ds\right)$$

for all $t \in [t_o,\infty[$ and $a \in]0,\tilde{a}[$, where $\tilde{a}$ can be arbitrarily large; and $M(a,t_o) \to 0$ for $a \to 0$ and $t_o > 0$.

Case 9. Let in Condition (H4) there be the function $w \equiv p$ and $c(t,u) = 0$. According to (Rouche, Habets and Laloy, 1980) a solution of a comparison equation

$$\frac{du}{dt} = p(t)\,u \tag{15}$$

a) is stable if

$$(\forall t_o \in T_o)(\exists a > 0)(\forall t \geq t_o) \int_{t_o}^{t} p(s)\,ds \leq a ;$$

b) is uniformly stable if

$$(\exists a > 0)(\forall t_o \in T_o)(\forall t \geq t_o) \int_{t_o}^{t} p(s)\,ds \leq a ;$$

c) is equi-asymptotically stable if

$$(\forall t_o \in T_o) \int_{t_o}^{t} p(s)\,ds \to -\infty \quad \text{for} \quad t \to +\infty$$

Case 10. Let in Condition (H4) there be the function $w(t)=0$ and $c(t,u)=a\exp[-ku]+\varphi(t)-a$, where $\varphi: R\to R$ is a continuous function, $a,k\in\check{R}_+$.

A solution of the comparison equation

$$\frac{dv}{dt}=a\exp[-kv]+\varphi(t)-a\ ,\quad v(t_o)=c\ ,\quad t_o=0\ , \tag{16}$$

has the form

$$v(t)=\psi(t)+k^{-1}\ln\{\exp(kc)+ak\int_0^t\exp[-k\,\psi(s)]\,ds\}$$

and is defined for all $t\in T_o$, $c\in\,]0,\tilde{c}[$, where

$$\psi(t)=\int_0^t\varphi(s)\,ds-at\ ,\quad \tilde{c}<h\ ,\quad h\in\check{R}_+\ .$$

Let us pay our attention to the equation (16) and consider it under some hypothesis on the function $\varphi(t)$:

a) if $\varphi(t)=0$, then

$$v(t)=-at+k^{-1}\ln\{\exp(kc)+\exp(kat)-1\}$$

is defined for all $t\in T_o$, $c\in\,]0,\tilde{c}[$, $\tilde{c}<h$, $h\in\check{R}_+$;

b) if $\varphi(t)=b$, $b\in\check{R}_+$ is a small value, then

$$v(t)=(b-a)\,t+k^{-1}\ln\ \exp(kc)-\frac{a}{b-a}\,[\exp(-k(b-a)t)-1]$$

is defined for all $t\in T_o$, $c\in\,]0,\tilde{c}[$, $\tilde{c}<h$, $h\in\check{R}_+$, $b\neq a$;

c) if $v(t_o)=c=0$, then

$$v(t)=\psi(t)+k^{-1}\ln\{1+ak\int_0^t\exp(-k\psi(s))\,ds\}$$

is defined for all $t\in T_o$ [18].

II.2.3. General stability theorems on the basis of scalar comparison equations

In the comparison method, a connection between stability of the state and stability of the solution of equation (2) is based on estimations which follow from the comparison principle (Liapunov, 1935; Chaplygin, 1919).

G.I. Melnikov's (1956) result is the basis of the statement given in the sequel.

---

[18] See 5) of Comments on References to Ch. II.

Proposition 7. If the vector-function $\mathbf{f}$ in system (7), Ch. I, is continuous on $T \times N$ (on $T_\tau \times N$) and, besides :

a) there exists a decrescent positive definite function v on S (on $T_\tau \times S$), $v(t,\mathbf{x}) \in C^{(1,1)}(S)$, $[v(t,\mathbf{x}) \in C^{(1,1)}(T_\tau \times S)]$;

b) the right-hand Dini derivative of the function v in virtue of system (7), Ch. I, is negative definite on $\mathbf{N}$ (on $T_\tau \times N$), then

we can illustrate the method for the comparison equation construction, which is included into the Assumption 1.

This statement can be proved by the direct construction of comparison equations (see II.5).

Proposition 8. In order that the state $u = 0$ of the comparison equation (1) be asymptotically stable (in the whole) it is necessary and sufficient :

a) that the function $\omega(u)$ satisfies the condition 2) of the Assumption 1;

b) that $\omega(u) < 0$ is fulfilled for $u > 0$ $(u \in \mathbf{R}_+)$. ■

(For the Proof see the Section 3.5.)

Theorem 4. If the vector-function $\mathbf{f}$ in system (7), Ch. I, is continuous on $T \times \mathbf{N}$ (on $T_\tau \times \mathbf{N}$), then iff 1)-4) hold,

1) there exists an open connected time-invariant neighbourhood S of the state $\mathbf{x} = \mathbf{0}$;

2) there exists a decrescent positive definite function v on $\mathbf{S}$ (on $T_\tau \times S$) and a function a, which belongs to the class K, such that

$$a(\|\mathbf{x}\|) \le v(t,\mathbf{x}) \ , \ \forall(t,\mathbf{x}) \in T\times S \quad (\forall(t,\mathbf{x}) \in T_\tau \times S)$$

3) there exists a function ω which satisfies conditions a)-b) of the Proposition 7 and the estimation

$$D^+v(t,\mathbf{x}) \le \omega(v(t,\mathbf{x})) \ , \ \forall(t,\mathbf{x}) \in T\times S \quad (\forall(t,\mathbf{x}) \in T_\tau \times S)$$

is fulfilled;

4) conditions a)-b) of the Proposition 8 are fulfilled,

the state $\mathbf{x}=\mathbf{0}$ of system (7), Ch.I, is asymptotically stable.

Theorem 5. If the vector-function $\mathbf{f}$ in system (7), Ch. I, is continuous on $T\times S$ (on $T_\tau \times S$) and there exist :

1) an open connected time-invariant neighbourhood S of the state $\mathbf{x} = 0$;

2) a decrescent positive definite function v on S (on $T_\tau \times S$) and a function a belonging to the class K such that

$$a(\|x\|) \le v(t,x) \ , \ \forall (t,x) \in T\times S \quad (\forall (t,x) \in T_\tau \times S)$$

3) a function $\omega : R\times R_+ \to R_+$, $(t,u) \to \omega(t,u)$, is continuous and it ensures existence of the unique solution $u(t;t_o,u_o)$ of the equation (2), passing through the point $(t_o,u_o) \in \operatorname{int}(T\times R_+)$; $\omega(t,0) = 0$, $\forall t \in T$, and

$$D^+v(t,x) \le \omega(t,v(t,x)) \ , \ \forall (t,x) \in T\times S \quad (\forall (t,x) \in T_\tau \times S)$$

then

a) stability of $u = 0$ of the equation (2) implies stability of $x = 0$ of system (7), Ch. I ;
b) asymptotic stability of $u=0$ of the equation (2) implies equi-asymptotic stability of $x=0$ of the system (7), Ch.I.

If in addition to 1)-2) of Theorem 5, one more condition is fulfilled :

4) a function b, belonging to the class K obeys

$$v(t,x) \le b(\|x\|) \ , \ \forall (t,x) \in T\times S \quad (\forall (t,x) \in T_\tau \times S)$$

then

c) uniform stability of $u=0$ of the equation (2) implies uniform stability of $x=0$ of the system (7), Ch.I;
d) uniform asymptotic stability of $u=0$ of the equation (2) implies uniform asymptotic stability of $x=0$ of the system (7), Ch.I.

For a proof see Theorems 5, 7, Ch.I, and the book by Rouche, Habets and Laloy (1980, p.66).

Example 6. In the problem of stability of a body of the variable mass in the sense of Liapunov (Aminov, 1959; Matrosov, 1962; Martynyuk, Gutowski, 1979) we consider equations of the perturbed motion in a pseudo-linear form (also called : the Rosenbrock form)

$$\frac{dx}{dt} = A(t,x)\,x \ , \ x \in R^n \ .$$

For this system we shall consider a positive definite function v,

$$v(t,x) = x^T B(t)\,x$$

with the derivative in virtue of equations of the perturbed motion

$$D^*v(t,x) = x^T \frac{dB}{dt} x \ ,$$

where

$$A^T(t,x)\,B(t) + B(t)\,A(t,x) = 0 \quad \text{for all} \quad (t,x) \in T_\tau \times S \ ,$$

and 0 is a zero-matrix.

According to the Theorem 1, p. 85, the positive definiteness of the function $v(t,\mathbf{x})$ and a constant negativeness of the derivative $D^*v(t,\mathbf{x})$ are sufficient conditions for stability. These conditions do not coincide with the necessary ones (Aminov, 1959) for bodies of a variable mass. The situation can be somewhat changed by the scalar comparison equations application. For the expression $D^*v(t,\mathbf{x})$ we have an estimation

$$D^*v(t,\mathbf{x}) \le \left\|\frac{d\mathbf{B}}{dt}\right\| \rho(v(t,\mathbf{x})) ,$$

where $\rho(v) = (a^{-1}(v))^2$. Here the function $a(\cdot)$, which belongs to the class K , is such that

$$a(\|\mathbf{x}\|) \le v(t,\mathbf{x}) , \quad \forall (t,\mathbf{x}) \in T_\tau \times S .$$

We consider a comparison equation

$$\frac{du}{dt} = \left\|\frac{d\mathbf{B}}{dt}\right\| \rho(u) .$$

As far as $v(t,\mathbf{x})$ is a positive definite form, then

$$\int_0^\infty \frac{dr}{\rho(r)} = +\infty$$

and the condition for stability of the solution $u=0$ is

$$\text{A)} \qquad \int_0^\infty \left\|\frac{d\mathbf{B}}{dt}\right\| dt < \infty .$$

According to Theorem 5 : 1) we conclude stability of the solution $\mathbf{x}=\mathbf{0}$ of the initial system. As far as (Matrosov, 1962) for bodies of a variable mass

$$\left|\int_0^\infty \frac{db_{ij}}{dt} dt\right| = |b_{ij}(\infty) - b_{ij}(0)| < \infty$$

holds, then the condition A) can be fulfilled, if derivatives db_{ij}/dt $\forall (i,j) \in [1,m]$ on a semi-axis $[0,\infty[$ change the sign the finite number times. This situation is quite probable in numerous practical problems for bodies of a variable mass.

Thus, on the basis of a scalar comparison equation we define the unique sufficient condition of stability : the positive-definiteness of the quadratic form $v(t,\mathbf{x})$. As it was stated by Aminov (1959) this condition was often also necessary.

II.2.4. The generalized comparison equation

Alongside with the system of equations (7), Ch.I, we shall consider a scalar equation

$$\frac{du}{dt} = \omega(t,u,\mathbf{x}) \ , \ u(t_o) = u_o \ , \tag{17}$$

determined on $T_o \times R_+ \times N$ $(N \subseteq R^n)$. Let the function $\omega(t,u,\mathbf{x})$ be defined, continuous and non-decreasing with respect to $(u,\mathbf{x})$ for fixed t from the domain T_o.

Let us note that for a solution of the problem (17) the function $\mathbf{x}(\cdot)$ can be considered as given for values $t \in T_o$.

Proposition 9. Suppose that the function $\omega : T_o \times R_+ \times N \to R_+$, $(t,u,\mathbf{x}) \to \omega(t,u,\mathbf{x})$ is continuous and non-decreasing with respect to $(u,\mathbf{x})$ for fixed t, which ensures the unique solution $u(t;t_o,u_o)$ of the problem (17), passing through some points $(t_o,u_o) \in \text{int}(T_o \times R_+)$ for $\mathbf{x}(t) \in N$. By $[t_o,\beta[$ $(\beta > t_o)$ we denote the maximal interval of existence of the solution $u(t;t_o,u_o)$. Let the function $r(t)$ be continuous and differentiable, $r : [t_o,\beta[\to R_+$ such that

1) $r(t_o) \le u_o$;
2) $\frac{dr}{dt} \le \omega(t,r,\mathbf{x})$ on $[t_o,\beta[$.

Then

$$r(t) \le u(t;t_o,u_o) \ , \ \forall t \in [t_o,\beta[\text{ and } x(t) \in N \ . \ \blacksquare$$

(For a proof see paragraph 3 of the present Chapter).

Theorem 5 is generalized in case of the comparison equation (17) as follows.

Theorem 6. If the vector-function $\mathbf{f}$ of the system (7), Ch.I, is continuous on $T \times N$ (on $T_\tau \times N$) and there exist :

1) an open connected time-invariant neighbourhood S of the state $\mathbf{x} = \mathbf{0}$;
2) a decrescent positive definite function v on S (on $T_\tau \times S$) and a function a, which belongs to the class K, such that

$$a(\|\mathbf{x}\|) \le v(t,\mathbf{x}) \ , \ \forall(t,\mathbf{x}) \in T \times S \quad (\forall(t,\mathbf{x}) \in T_\tau \times S) \ ;$$

3) a function

$$\omega : T \times R_+ \times S \to R_+ \ , \ (t,u,\mathbf{x}) \to \omega(t,u,\mathbf{x}) \ , \ \omega(t,0,\mathbf{0}) = 0 \ , \ \forall t \in T \ ,$$

continuous, non-decreasing with respect to $(u,\mathbf{x})$ and ensuring existence of the unique solution $u(t;t_o,u_o)$ of the Cauchy problem for the system (17) provided $(t_o,u_o) \in \text{int}(T \times R_+)$ and

$$D^+ v(t,\mathbf{x}) \le \omega(t,\|\mathbf{x}\|,\mathbf{x}) \ , \ \forall(t,\mathbf{x}) \in T \times S \quad (\forall(t,\mathbf{x}) \in T_\tau \times S) \ ;$$

4) a solution of the system

$$\frac{dx}{dt} = f(t,x) \quad , \quad \frac{du}{dt} = \omega(t,a^{-1}(u),x) \tag{18}$$

for all $t \in T$ (for all $t \in T_\tau$), then

a) u-stability of the zero-solution of the system (18) implies stability of the state $x=0$ of the system (7), Ch. I ;

b) asymptotic u-stability of the zero-solution of the system (18) implies equi-asymptotic stability of the state $x=0$ of the system (7), Ch. I .

If in addition to 1)-4) one more condition is fulfilled :

5) for a function b , which belongs to the class K ,

$$v(t,x) \le b(\|x\|) \quad , \quad \forall (t,x) \in T \times N \quad (\forall (t,x) \in T_\tau \times S) \ ,$$

then

c) uniform u-stability of the zero-solution of the system (18) implies uniform stability of the state $x=0$ of the system (7), Ch. I ;

d) uniform asymptotic u-stability of the zero solution of the system (18) implies uniform asymptotic stability of the state $x=0$ of the system (7), Ch. I .

Proof. From the fact that the function $\omega(t,u,x)$ is non-decrescent, we define

$$\omega(t,\|x\|,x) \le \omega(t,a^{-1}(v(t,x)),x)$$

for $\|x\| \le a^{-1}(v(t,x))$. Hence,

$$D^+v(t,x) \le \omega(t,a^{-1}(v(t,x)),x) \ .$$

A comparison equation has the form :

$$\frac{du}{dt} = \omega(t,a^{-1}(u),x) \quad , \quad u(t_o) = u_o \ .$$

In view of Theorem 1 for any $(t_o,x_o) \in T_i \times S$ and $(t_o,u_o) \in T_i \times R_+$ and any $t \in T_\tau$, for which $x(t;t_o,x_o)$ and $u(t;t_o,u_o)$ are defined, we determine an estimation

$$a(\|x\|) \le v(t,x(t;t_o,x_o)) \le u(t;t_o,v(t_o,x_o)) \ . \tag{19}$$

On the basis of estimation (19) and conditions of the system (18) u-stability, stability of the state $x=0$ of the system (7), Ch. I , can be determined similarly as in the Theorem 6 [19]. ■

Example 7 (Hatvani, 1975). We consider a mechanical system, which is described by a differential equation

[19] See 6) of Comments on References to Ch. II.

$$\frac{d^2y}{dt^2} + a(t)\frac{dy}{dt} + b(t)\,g(y) = 0 \quad , \quad t \geq 0$$

where the function a (damping coefficient) is continuous and b (a generalized rigidity) is continuously differentiable on $]0,+\infty[$; function g is continuous on $]-\infty,+\infty[$ and

$$\int_0^{x_1} g(r)\,dr > 0 \quad \text{for} \quad x_1 \neq 0 \ .$$

The perturbed motions are considered over the interval $[0,+\infty[$ for sufficiently small initial perturbations $y(0)$, $\dot{y}(0)$. By using the substitution $y = x_1$, $\dot{y} = x_2$ it is obtained that

$$\frac{dx_1}{dt} = x_2 \ , \ \frac{dx_2}{dt} = -b(t)\,g(x_1) - a(t)\,x_2 \ .$$

We introduce the notations

$$[a]_+ = \begin{cases} a & \text{if } a \geq 0 \ ; \\ 0 & \text{if } a < 0 \ , \end{cases} \qquad [a]_- = \begin{cases} 0 & \text{if } a \geq 0 \ ; \\ -a & \text{if } a < 0 \ . \end{cases}$$

Let us consider a positive definite function v ,

$$v(t,x_1,x_2) = \frac{x_2^2}{2} + q(t)\int_0^{x_1} g(s)\,ds \ ,$$

where q is a continuously differentiable function on T_o such that $q(t) \geq q_o > 0$ for $t \geq 0$, $q_o = \text{const}$ and

$$\int_0^{\infty} \left[\frac{\dot{q}(t)}{q(t)}\right]_+ dt < \infty \quad , \quad \int_0^{\infty} |q(t) - b(t)| \exp[2t]\,dt < \infty \ .$$

System (18) has the form

$$\frac{dx_1}{dt} = x_2 \ , \ \frac{dx_2}{dt} = -b(t)\,g(x_1) - a(t)\,x_2 \ ,$$

$$\frac{du}{dt} = p(t)\,u + (q(t) - b(t))\,g(x_1)\,x_2$$

where

$$p(t) = \left[\frac{\dot{q}(t)}{q(t)}\right]_+ + 2\,[a(t)]_- \ .$$

Having fulfilled the requirement

$$2\,|q(x_1)\,x_2| \leq (1 + u^2)\|\mathbf{x}\|^2 \ , \ u = \text{const} \ ,$$

and taking into account that $\|\mathbf{x}(t)\| \leq K\|\mathbf{x}(0)\| \exp(t)$, we define u-stability of the zero-solution and, hence, stability of the motion $y = \dot{y} = 0$, if

$$\left|\frac{dg(y)}{dy}\right| \leq u \quad ; \quad -\infty < y < \infty \quad ; \quad \int_0^{\infty} (|a(s)| + |b(s)|)\,ds < \infty \ .$$

II.2.5. The scalar comparison equation construction

We consider a system of equations of the perturbed motion

$$\frac{dx}{dt} = Px + q(x) + h(t,x) \tag{20}$$

in which $x \in R^n$; P is an $n \times n$ - constant matrix; $q(x)$ contains polynomials of smallness not more than of the second order, or series absolutely converging in the neighbourhood N of the state $x = 0$; $h(t,x)$ contains bounded functions on $T_o \times N$, which satisfy the estimation

$$|h_s(t,x)| \le f_s(t) \ , \ \forall t \in T_o \ , \forall x \in N \ , \ \forall s \in [1,n] \ , \tag{21}$$

where f_s are some non-negative, continuous and bounded functions of time.

By applying a linear non-singular transformation $y = Ax$ $(\det A \neq 0)$ to the system of equations (20) we reduce the matrix P to the diagonal form :

$$\frac{dy}{dt} = \Lambda y + w(y) + p(t) \ , \tag{22}$$

where $y \in R^n$; Λ is a diagonal matrix; w is a vector function with the components which take into account the vector q ; p is a vector function, connected with estimations (21).

Following the results of Tikhonov's work (1969) we shall consider expressions for variables

$$y_s = r_s \exp(i\theta_s) \ , \ \bar{y}_s = r_s \exp(-i\theta_s) \ .$$

Let us note that for real values λ_s variables y_s are real, and for complex values λ_s variables y_s are complex. As far as moduli r_s of the complex conjugate variables coincide, then r_s of different moduli will be only n_1 $(n_1 < n)$. For a function

$$v = \sum_{s=1}^{n_1} r_s$$

we define

$$\frac{dv}{dt} = \sum_{s=1}^{n_1} \alpha_s r_s + \frac{1}{2} \sum_{s=1}^{n_1} (w_s e^{-i\theta_s} + \bar{w}_s e^{i\theta_s}) + \frac{1}{2} \sum_{s=1}^{n_1} (p_s e^{-i\theta_s} + \bar{p}_s e^{i\theta_s}) \ . \tag{23}$$

It is obvious that v is a positive definite function.

An estimation of terms of the right part of expression (23) will be realized in the domain $N = \{y : |y_s| < \hbar \ , \hbar \in \check{R}_+\}$.

The first term can be estimated in the following way

$$\sum_{s=1}^{n_1} \alpha_s r_s \le \alpha v \ , \ \alpha = \max\{\alpha_s \ , s \in [1,n_1]\} \ . \tag{24}$$

For the second term it will be

$$\left|\frac{1}{2}\sum_{s=1}^{n_1}(w_s e^{-i\theta_s}+\bar{w}_s e^{i\theta_s})\right| \le v^2\sum_{s=1}^{n_1}\beta_s = \beta v^2 \tag{25}$$

where

$$\beta = \sum_{s=1}^{n_1}\beta_s \quad \text{and} \quad \beta_s = \sum_{\substack{m_1+\dots+m_n=k\\ k=2,3,\dots}} |a_s^{(m_1\dots m_n)}|\, \hbar^{k-2}$$

if

$$w_s = \sum_{\substack{m_1+m_2+\dots+m_n=k\\ k=2,3,\dots}} a^{(m_1\dots m_n)}\, y_1^{m_1} y_2^{m_2}\dots y_n^{m_n}$$

are the convergent series in the domain N.

Taking into account estimation (21) the third term can be estimated as follows :

$$\left|\frac{1}{2}\sum_{s=1}^{n_1}(p_s e^{-i\theta_s}+\bar{p}_s e^{i\theta_s})\right| \le \sum_{s=1}^{n_1}|p_s(t)| \le \varphi(t) \tag{26}$$

where

$$\varphi(t) = \sum_{s=1}^{n_1}\sum_{j=1}^{n}|a_{sj}|\, p_s(t)$$

is a bounded non-negative function.

From estimations (24)-(26) we obtain a differential inequality

$$\frac{dv}{dt} \le \alpha v + \beta v^2 + \varphi(t) \ , \tag{27}$$

determined for all $t\in T_o$ and $v<\hbar$, $\hbar\in \check{R}_+$.

A comparison equation, corresponding to the inequality (27)

$$\frac{du}{dt} = \alpha u + \beta u^2 + \varphi(t) \ , \ u(t_o) = v_o \ , \tag{28}$$

is an equation of the Rikkati type and, in general, it is not integrated in quadratures.

Remark 1. If

$$(\forall t_o \in T_i)\ (\forall t\in T_o) \qquad v_o + \int_{t_o}^{t}\varphi(s)\,ds \le a \ , \quad a\in\check{R}_+$$

then an inequality

$$v(t) \le a + \int_{t_o}^{t}(\alpha v(s) + \beta v^2(s))\,ds$$

can be analyzed on the basis of Theorems 2,3.

Remark 2. Possibility of the approximate analysis of equation (28) can be connected with the exponential majorization of the right part of correlation

$$\alpha v + \beta v^2 \le \ell(e^{-kv} - 1) ,$$

where values ℓ and k are chosen to obey the condition

$$\max_{v \le h^*} [(\alpha v + \beta v^2) - (\ell e^{-kv} - 1)] \to \min \quad \text{for} \quad h^* < \hbar .$$

The obtained comparison equation

$$\frac{du}{dt} = \ell(e^{-ku} - 1) + \varphi(t)$$

can be completely analyzed (see 2.2). As it is shown in Tikhonov's paper (1969), values ℓ and k can be defined by formulas

$$\ell = -\frac{\alpha}{k} , \quad k = \frac{3}{2h^*} \left(1 - \sqrt{1 + \frac{8}{3} \frac{\beta h^*}{\alpha}}\right) .$$

The constant $h^* > 0$ is determined from the condition that $k \in R$,

$$h^* \le -\frac{3}{8} \frac{\alpha}{\beta} .$$

Finally,

$$h^* = \min\left\{\hbar , -\frac{3}{8} \frac{\alpha}{\beta}\right\} , \quad \alpha < 0 .$$

Example 8 (Tikhonov, 1974). We consider a motion of the rigid body with the lengthwise symmetry plane in the vertical plane; this motion is defined by the initial data of motion. The condition for realization of such a lengthwise motion is the absence (at the initial moment and in the process of motion) of the perturbations which violate the symmetry of motion. With the aim of simplification of the problem we consider a non-controlled motion of the body and the perturbed motion in such bounds of the kinematic parameters of the motion variation, that we can neglect the density of the medium variation. Equations of the perturbed motion have the form

$$\frac{dx_s}{dt} = \sum_{k=1}^{4} p_{sk} x_k + \sum_{\substack{m_1+m_2+m_3+m_4=m \\ m=2,3}} P_s^{(m_1 \dots m_4)} x_1^{m_1} \dots x_4^{m_4} + X_s , \quad s \in [1,4] ,$$

where p_{sk} , $P_s^{(m_1 \dots m_4)}$ are constants, X_s are holomorphic functions initiating in their decomposition with the terms not lower than of the fourth order, and converging in the neighbourhood $N \subset \{|x_s| \le \hbar ,$ $s \in [1,4]\}$. We assume that the roots of a characteristic equation

$$\det(P - \lambda I) = 0 , \quad P = [p_{sk}]_{s,k=1}^{4} ,$$

are simple; moreover, λ_1 and λ_2 have "small" and λ_3 , λ_4 "comparatively large" (over their absolute value) negative real parts (see Vedrov, 1938).

By means of the non-singular transformation we reduce the initial system to a diagonal form

$$\frac{dy_s}{dt} = \lambda_s y_s + \sum_{\substack{m_1+\ldots+m_4=m \\ m=3,4}} P_s^{(m_1\ldots m_4)} y^{m_1}\ldots y^{m_4} + Y_s \ , \quad s \in [1,4] \ .$$

Further on we introduce variables

$$z_s = y_s - \sum_{\substack{m_1+m_2=m \\ m=2,3}} B_s^{(m_1 m_2)} y_1^{m_1} y_2^{m_2} \ , \quad s = 3,4 \ ,$$

so that they could satisfy inequalities

$$\frac{dz_s}{dt} = \lambda_s z_s + \sum_{k=3}^{4} \Big[\sum_{i=1}^{4} b_{sk}^{(i)} y_i + \sum_{m_1+\ldots+m_4=2} b_{sk}^{(m_1\ldots m_4)} y_1^{m_1}\ldots y_4^{m_4} \Big] z_k + Z_s \ ,$$

$s=3,4$, where $b_{sk}^{(i)}$, $b_{sk}^{(m_1\ldots m_4)}$ are constants, Z_s are nonlinear functions not lower than of the fourth order. It is not difficult to verify that solutions z_3 , z_4 (if we neglect the terms higher than those of the third order) quickly enough tend to zero. Really, let us consider the function

$$v = \sum_{s=3}^{4} |z_s|^2 \ .$$

In the neighbourhood $N_y \subset R^2$ this function satisfies a differential inequality

$$\frac{dv}{dt} \le 2(-a + h^* M)\, v \ ,$$

where $a = \max(\lambda_3, \lambda_4)$;

$$M = \sum_{s=3}^{4} \sum_{k=3}^{4} \Big(\sum_{i=1}^{4} |b_{sk}^{(i)}| + h^* \sum_{m_1+\ldots+m_4=2} |b_{sk}^{(m_1\ldots m_4)}| \Big) \ .$$

Hence, for $v_o = v(t_o)$ we have

$$|z_s(t)| < \sqrt{v_o} \exp[-(a - h^* M)\, t] \ , \quad s = 3,4 \ ,$$

and as far as the value a is large enough according to the condition, then $|z_s|$ quickly decreases and, beginning with $t_o > 0$ become negligibly small . It means that it is possible to use approximate correlations

$$y_s = \sum_{\substack{m_1+m_2=m \\ m=2,3}} B_s^{(m_1 m_2)} y_1^{m_1} y_2^{m_2} \ , \quad s = 3,4$$

for a transformation of the two remained equations to the form

$$\frac{dy_s}{dt} = \lambda_s y_s + \sum_{\substack{m_1+m_2=m \\ m=2,3}} Q_s^{(m_1 m_2)} y_1^{m_1} y_2^{m_2} + \widetilde{Y}_s \ , \quad s = 3,4 \ .$$

Here $\tilde{Y}_s$ are nonlinear functions with power higher than those of the third order, $|y_s| \le h$, $Q_s^{(m_1 m_2)}$ are constants which can be calculated. The obtained equations by means of a substitution

$$z_s = y_s + \sum_{\substack{m_1+m_2=m\\ m=2,3}} C_s^{(m_1 m_2)} y_1^{m_1} y_2^{m_2} , \quad s = 1,2$$

under a suitable choice of $C_s^{(m_1 m_2)}$ are reduced to the form

$$\frac{dz_1}{dt} = \lambda_1 z_1 + C_1^{(21)} z_1^2 z_2 ;$$

$$\frac{dz_2}{dt} = \lambda_2 z_2 + C_2^{(12)} z_1 z_2^2 ,$$

where $C_1^{(21)}$, $C_2^{(12)}$ are calculated.
Further on we introduce the notations

$$\alpha_1 = \frac{1}{2} (C_1^{(21)} + C_2^{(12)}) , \quad \beta = \frac{1}{2i} (C_1^{(21)} - C_2^{(12)})$$

and take into account that $\lambda_{1,2} = -\alpha \pm i\beta$.
We represent variables z_1 and $z_2 = \bar{z}_1$ in the form

$$z_1 = \rho \exp(i\theta) , \quad z_2 = \rho \exp(-i\theta) .$$

It is not difficult to determine that

$$\rho = \left(C_1 \exp(2\alpha t) + \frac{\alpha_1}{2}\right)^{-1/2} ,$$

$$\theta = \beta t - \frac{\beta_1}{2\alpha_1} \ln \left|\exp(-2\alpha t) + \frac{C_1 \alpha}{\alpha_1}\right| + C_2 ,$$

where

$$C_1 = \left(\frac{1}{\rho_o^2} - \frac{\alpha_1}{\alpha}\right) \exp(-2\alpha t_o) , \quad C_2 = \theta_o - \beta t_o + \frac{\beta}{2\alpha_1} \ln \left|\exp(-2\alpha t_o) + \frac{C_1 \alpha}{\alpha_1}\right| .$$

Thus, the solutions z_1 and z_2 are obtained in the form

$$z_{1,2} = \rho(t;t_o,\rho_o) \exp[\pm i\theta(t;t_o,\theta_o)] .$$

Analysis of the solution $z_{1,2}$ for :
a) $\alpha_1 < 0$ $(\alpha > 0)$
b) $\alpha_1 > 0$ $(\alpha > 0)$
illustrates that ρ tends to zero either over the whole domain of the initial perturbations or under too small initial perturbations, $\rho_o < \sqrt{\alpha/\alpha_1}$.

II.2.6. A refined method of comparison equations construction

In the course of construction of comparison equations we often use estimations of modulus of components of the right part in expression of the total derivative of the Liapunov function. All this, as a rule, simplifies construction of these equations, but at the same time it leads to more rough their forms. The illustrated in II.2.5 method of the comparison equation construction has a more precise definition (Tikhonov, 1965, 1969; Melnikov, 1975).

We consider a special case of a zero root. Let there be given equations of the perturbed motion

$$\frac{dy_s}{dt} = \lambda_s y_s + \mu_{s-1} y_{s-1} + p_\nu^s y^\nu + Y_s(t,y) \;, \quad s \in [1,k] \quad (\nu \in [2,m]) \qquad (29)$$

where p_ν^s are constant nonlinear terms (y^ν) having the form of polynomials; functions $Y_s(t,y)$ have the sense of constantly effecting perturbations. We assume that system (29) satisfies the condition of existence of the continuous solution of the corresponding Cauchy problem, connected with system (29).

By means of a nonlinear transformation

$$y_s = z_s + \sum_{k=2}^{m} \sum_{m_1+\dots+m_2=k} b_s^{(m_1\dots m_n)} z_1^{m_1} z_2^{m_2} \dots z_n^{m_n}$$

with the constant coefficients $b_s^{(m_1\dots m_n)}$, which are chosen in conformity with the condition of invariability of a linear part of the system (29), the system of equations (29) is reduced to a nonlinear canonical form. In this connection constants p_ν^s can be different from zero only for the special values of the indices ν, i.e. when divisors $\nu\lambda - \lambda_s \approx 0$. In the considered case of a special root, for $\lambda_1 = \alpha_1 \approx 0$, $\alpha_{n_1} \le \alpha_{n_1-1} \le \dots \le \alpha_2 < 0$ the first equation of the system (29) takes the form

$$\frac{dy_1}{dt} = \lambda_1 y_1 + \sum_{k=2}^{m} g_k y_1^k + Y_1(t,y) \;. \qquad (30)$$

In the remained equations of the system (29) each nonlinear term with the coefficient p_ν^s different from zero obligatory contains as a multiplier one of the "non-special" variables $y_2, y_3, \dots, y_n$ in the first or higher power. Assume also that $p_{\nu,0\dots0}^s = 0$ for $s \ge 2$.

Let us determine the function

$$v = \sum_{s=1}^{n_1} r_s$$

in the form $v = v_1 + v_o$, where

$$v_o = \sum_{s=2}^{n_1} r_s$$

corresponds to non-special variables. In expression of the total derivative of the function v in virtue of the system (29) we isolate some terms, corresponding to $s=1$, namely

$$\frac{dv}{dt} = \sum_{s=1}^{n_1} \alpha_s r_s + \frac{1}{2} \sum_{s=2}^{n_1} \mu_{s-1}(y_{s-1} e^{-i\theta_s} + \bar{y}_{s-1} e^{i\theta_s}) + g_k y_1^{k-1} r_1 + $$
$$+ \frac{1}{2} \sum_{s=2}^{n_1} (p_\nu^s y^\nu e^{-i\theta_s} + \bar{p}^s \bar{y}^\nu e^{i\theta_s}) + \frac{1}{2} \sum_{s=1}^{n_1} (Y_s e^{-i\theta_s} + \bar{Y}_s e^{i\theta_s}) \tag{31}$$

We estimate the terms of the right part of expression (31) as follows :

* the first term :

$$\sum_{s=1}^{n_1} \alpha_s r_s = \alpha_1 \sum_{s=1}^{n_1} r_s + \sum_{s=2}^{n_1} (\alpha_s - \alpha_1) r_s \le \alpha_1 v + (\alpha_2 - \alpha_1) v_o \ ;$$

* the second term :

$$\operatorname{Re} \sum_{s=2}^{n_1} \mu_{s-1} y_{s-1} e^{-i\theta_s} \le \sum_{s=1}^{n_1-1} \mu r_s \le \mu v \ , \ \mu_{s-1} \le \mu \ , \ s \in [2,n] \ ,$$

where μ is a sufficiently small constant;

* the third term :

$$\sum_{k=2}^{m} g_k y_1^{k-1} v_1 \le \sum_{k=2}^{m} g_k^* v^k + \sum_{k=2}^{m} k|g_k| v_o v^{k-1} ,$$

where

$$g_k^* = \begin{cases} g_k & \text{for } k \text{ odd} , \\ |g_k| & \text{for } k \text{ even} ; \end{cases}$$

* the fourth term :

$$\frac{1}{2} \sum_{s=2}^{n_1} (p_\nu^s y^\nu e^{-i\theta_s} + \bar{p}_\nu^s \bar{y}^\nu e^{i\theta_s}) \le \sum_{s=2}^{n_1} |p_\nu^s| r^{|\nu|-1} v_o = \sum_{k=2}^{m} c_k v^{k-1} v_o ,$$

$$c_k = \sum_{|\nu|=k} \sum_{s=2}^{n_1} |p_\nu^s| \ ;$$

* the fifth term :

$$\operatorname{Re} \sum_{s=1}^{n_1} (Y_s e^{-i\theta_s}) \le \sum_{s-1}^{n_1} |Y_s| \le \varphi(t) \ .$$

Having taken into account the illustrated estimations, the relationship (31) takes the form

$$\frac{dv}{dt} \le (\alpha_1 + \mu) v + \sum_{k=2}^{m} g_k^* v^k + v^o [\alpha_2 - \alpha_1 + \sum_{k=2}^{m} (c_k + k|g_k|) v^{k-1}] + \varphi(t) \ .$$

Taking into account the fact that the root α_2 is negative and its absolute value is not small, we can show that the domain $v \le h^*$, $h^* < \hbar$ is such that

$$\alpha_2 = \alpha_1 + \sum_{k=2}^{m} (c_k + k|g_k|) v^{k-1} .$$

In this connection

$$\frac{dv}{dt} \le (\alpha_1 + \mu)\, v + \sum_{k=2}^{m} g_k^* v^k + \varphi(t) \ , \qquad (32)$$

in the domain $v \le h^*$ and for all $t \in T_o$.

A comparison equation, corresponding to the inequality (32)

$$\frac{du}{dt} = (\alpha_1 + \mu)\, u + \sum_{k=2}^{m} g_k^* u^k + \varphi(t) \qquad (33)$$

has the peculiarity that coefficients g_k^* for odd values of the index k , can be negative.

Remark 3. In case of a pair of special roots $\alpha_1 \approx 0$, $\alpha_{n_1} \le \alpha_{n_1 - 1} \le \ldots \le \alpha_2 < 0$, a differential inequality

$$\frac{dv}{dt} \le (\alpha_1 + \mu)\, v + \sum_{k=3}^{m} g_k^* v^k + \varphi(t) \quad \text{for} \quad v \le h^* \ , \ k\text{-odd} \ , \qquad (34)$$

holds, which is constructed similarly as the inequality (32). A comparison equation corresponding to it is

$$\frac{du}{dt} = (\alpha_1 + \mu)\, u + \sum_{k=3}^{m} g_k^* u^k + \varphi(t) \quad , \ k\text{-odd} \ . \qquad (35)$$

Inequalities (32), (34) can be analyzed on the basis of Theorems 2, 3.

II.2.7. Several applications of scalar comparison equations

A. Suppose that for a time-invariant system of equations of the perturbed motion

$$\frac{dx}{dt} = Px + q(x) \qquad (36)$$

the following conditions are fulfilled :

1) matrix P is stable, $\mathrm{Re}\, \lambda_i(P) < -b^2 < 0$, $\forall i \in [1,n]$;
2) in $\bar{N} = \{x : \|x\| \le r\}$ an estimation $\|q(x)\| \le C_r \|x\|$ is fulfilled, where C_r is constant;
3) through each point N there passes at least one solution $x(t;t_o,x_o)$ bounded for all $t \in T_o$, and belonging to N .

Let us assume the following notations (Nemytskii, 1955) :

$$\|P\| = \Big(\sum_{i,k=1}^{n} p_{ik} \Big)^{1/2} \ ,$$

m is a number of components q_i of a vector $(q_1, q_2, \ , q_n)^T$ which are different from zero;

$$\epsilon_M = \frac{n-1}{n} \, \frac{1}{n\sqrt{n-1}} \, \frac{b^2}{\|P\|} \ .$$

By extending the domain of determination of the system (36) right part on the whole complex plane and fulfilling a unitary transformation of the system (36) linear part, after a substitution of variables $y_k = \epsilon^{k-1} z_k$, $\epsilon \in]0,1]$, we obtain a system

$$\frac{dz_1}{dt} = \lambda_1 z_1 + \epsilon b_{12} z_2 + \dots + \epsilon^{n-1} b_{1n} z_n + \psi_1 ,$$

$$\dots\dots\dots\dots\dots\dots\dots\dots\dots\dots$$

$$\frac{dz_n}{dt} = \lambda_n z_n + \frac{1}{\epsilon^{n-1}} \psi_n .$$

Here the functions $\psi_1,\dots,\psi_n$ take into account the transformation of components of nonlinearities $q_1(x),\dots,q_n(x)$.

If motion $x(t;t_o,x_o)$ remaining in $N \subseteq R^n$, $\forall t \in T_o$, for $t \to +\infty$ approaches zero, we shall call it the 0^+-motion.

Theorem 7 (Nemytskii, 1955). In order that the state $x=0$ of the system (36) be uniformly attractive from $\bar{N}_M = \{x : \|x\| \le \epsilon_M^{n-1} r\}$ it is sufficient to fulfill conditions 1)-3) with the constant in condition 2) which satisfies an inequality

$$c_r < b^2 n^{-2} m^{-1/2} \epsilon_M^{n-1} .$$

Proof. By means of the function $v = |z|^2$ for an estimation

$$\frac{dv}{dt} \le 2 \left[-b^2 + n\sqrt{n-1}\, \|A\|\, \epsilon + \frac{C_r}{\epsilon^{n-1}} n\sqrt{m} \right] v$$

we construct a comparison equation

$$\frac{du}{dt} = 2 \frac{\varphi(\epsilon)}{n\sqrt{m}} u , \quad u(0) = u_o , \tag{37}$$

where

$$\frac{\varphi(\epsilon)}{n\sqrt{m}} = \frac{b^2 \epsilon^{n-1} - n\sqrt{n-1}\, \|P\|\, \epsilon^n}{n\sqrt{m}} .$$

As far as

$$u(t;t_o,u_o) = u_o \exp\left[2 \frac{\varphi(\epsilon)}{n\sqrt{m}} (t-t_o)\right] , \quad \forall t_o \in T_i , \quad \forall t \in T_o ,$$

and having fulfilled conditions of the theorem, $\varphi(\epsilon)/n\sqrt{m} < 0$, we can see that the statement of the theorem is evident. ∎

B. We consider system (36) for a special type of nonlinearities, namely, we assume that the following conditions are fulfilled :

1') matrix P is stable, $\operatorname{Re} \lambda_i(P) < \frac{1}{2}\kappa < 0$, $\forall i \in [1,n]$;

2') $q_i(x) = x_i\, \Pi(x)$, where $\Pi(x)$ is a quadratic form of variables;

3') a solution $x(t;t_o,x_o)$ of the system (36) is defined for $(t_o,x_o) \in T_i \times R^n$.

The next theorem is due to Martynyuk and Verbitskii (1982).

Theorem 8. In order that the state $\mathbf{x}=0$ of the system (36) be uniformly attractive from

$$N = \left\{ \mathbf{x} : v(\mathbf{x}) < - \frac{\kappa\lambda}{\Sigma \mu_i |\nu_i|} \right\} ,$$

it is sufficient to fulfill conditions 1')-3'). ■

Here

$$\mu_i = p_{ii} + \sqrt{\sum_{k=1}^{n} p_{ik}^2} \quad \text{and} \quad \nu_i$$

are maximal in modulus eigenvalues of the matrices of both

$$\Phi_i(\mathbf{x}) = \frac{\partial v}{\partial x_i} x_i \quad \text{and} \quad \Pi_i(\mathbf{x}) ,$$

respectively; λ is a minimal eigenvalue of the matrix $\mathbf{B}$ of the function $v(\mathbf{x}) = \mathbf{x}^T \mathbf{B} \mathbf{x}$.

Proof. For an estimation of the total derivative of the quadratic scalar Liapunov function $v(\cdot)$,

$$D^* v(\mathbf{x}) \le \kappa v(\mathbf{x}) + \sum_i \Phi_i(\mathbf{x}) \Pi_i(\mathbf{x}) ,$$

the following estimation is valid :

$$D^* v(\mathbf{x}) \le \kappa v(\mathbf{x}) + \sum_i \frac{\mu_i |\nu_i|}{\lambda^2} v^2(\mathbf{x}) \tag{38}$$

as far as

$$\lambda r^2 \le v(\mathbf{x}) \quad \text{and} \quad \frac{|\nu_i|}{\lambda} v(\mathbf{x}) > \Pi_i(\mathbf{x}) , \quad |\nu_i| R^2 \ge \Pi_i(\mathbf{x}) ,$$

$$\frac{\mu_i}{\lambda} v(\mathbf{x}) > \Phi_i(\mathbf{x}) , \quad \mu_i R^2 \ge \Phi_i(\mathbf{x}) , \quad R = \dim N = |N| .$$

A comparison equation

$$\frac{du}{dt} = \kappa u + \beta u^2 , \quad \kappa < 0 , \quad \beta > 0 , \quad u(0) = u_o ,$$

which corresponds to the differential inequality (38), has the solution

$$u(t;u_o) = \left(-\frac{\beta}{\kappa} + C_1 \exp[-\kappa t] \right)^{-1} .$$

From the condition $u(t;u_o) \ge 0$, $\forall t \in T_o$ we have $C_1 \ge \delta > 0$. The initial value u_o is subordinate to the condition of $0 < u_o < \kappa/\beta$. Hence, the Theorem 8 is proved. ■

C. Let us estimate the influence of nonlinearity of the pseudo-sliding law on the domain of attraction of an unperturbed motion of the model of biaxial rail vehicle on the basis of the Theorem 8.

According to the work by Lazaryan, Korotenko, Demin, Radchenko (1969) the investigated model of the rail vehicle is a mechanical system which consists of three rigid bodies, connected by visco-elastic elements (Figure 5). For a determination of interaction forces between wheels and rails we apply a nonlinear theory of pseudo-sliding. Variation of radii of the surface of wheels rolling depending on their transverse movements with respect to the rail threads linearly (a conic wheels profile). The model is described by a system of nonlinear differential equations of the 14th order. The zero solution of the system corresponds to an unperturbed motion. For some values of parameters an unperturbed motion of the vehicle is asymptotically stable, though the motion of a separately taken wheeled couple with conic wheels is always unstable.

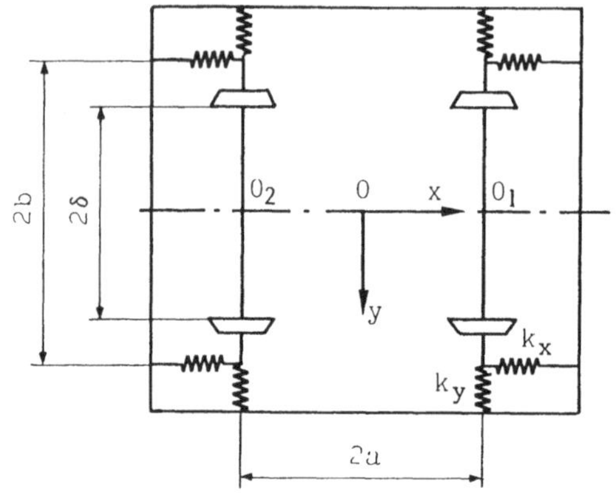

Figure 5

Equations of a rectilinear motion of the model have the form :

$$\begin{aligned} &m\ddot{y} + 2k_y(2y - 2H\theta - y_1 - y_2) = 0 , \\ &J_x\ddot{\theta} + 4k_z b^2\theta + 4\beta_z b^2\dot{\theta} - 2k_y H(2y - 2H\theta - y_1 - y_2) = 0 , \\ &J_z\ddot{\psi} + 2k_y a(2a\psi - y_1 + y_2) + 2k_x b^2(2\psi - \psi_1 - \psi_2) = 0 , \\ &m\ddot{y}_j + 2k_y(y_j - y + (-1)^j a\psi + H\theta) = Q_{y_j} , \\ &J_{z_1}\ddot{\psi}_j + 2k_x b^2(\psi_j - \psi) = Q_{\psi_j} , \quad j = 1,2 . \end{aligned} \tag{39}$$

Here

$$Q_{yj} = \sum_{i=1}^{2} Y_{ji},$$

i is a number of the wheels of the j-th wheeled couple;
$Q_{\psi_j} = (X_{j1} - X_{j2})d$, X_{ji} is a lengthwise component of pseudo-sliding force at the point of contact of the i-th wheel; Y_{ij} is a transverse component of the pseudo-sliding force at the point of contact of the i-th wheel. The pseudo-sliding force as a function of relative sliding

can be determined by formula

$$F = -\frac{k_1 \epsilon}{\sqrt{1 + (\frac{\epsilon k_1}{k_f p})^2}} ,$$

where k_1 is a coefficient of pseudo-sliding, k_f is a coefficient of Coulomb's friction; p is a loading on the wheel; ϵ is a relative sliding, $\epsilon = v_*/v$; v is the velocity of an unperturbed motion; v_* is the velocity of sliding at the point of the wheel contact. We denote

$$k_3 = \frac{1}{2} \frac{k_1^3}{(k_f p)^2}$$

and define the first two terms of an expansion in powers of ϵ of the function F :

$$F = -k_1 \epsilon + k_3 \epsilon^3 = -k_1 \frac{v_*}{v} + k_3 (\frac{v_*}{v})^3 .$$

The lengthwise and transverse components of pseudo-sliding force are determined by expressions :

$$X = F \frac{v_{*x}}{v_*} = -k_1 \frac{v_{*x}}{v} + k_3 (\frac{v_{*x}}{v})^3 + k_3 \frac{v_{*x} v_{*y}^2}{v^3} ,$$

$$Y = F \frac{v_{*y}}{v_*} = -k_1 \frac{v_{*y}}{v} + k_3 (\frac{v_{*y}}{v})^3 + k_3 \frac{v_{*y} v_{*x}^2}{v^3} ,$$

in which the lengthwise and transverse components of the velocity of sliding are determined respectively as follows :

$$(i=1) \quad (v_{*x})_{j_1} = d\dot{\psi}_j + \frac{v}{r_o} \gamma y_j \; ; \; (v_{*y})_{j_1} = \dot{y}_j - v\psi_j ,$$

$$(i=2) \quad (v_{*x})_{j_2} = -d\dot{\psi}_j - \frac{v}{r_o} \gamma y_j \; ; \; (v_{*y})_{j_2} = \dot{y}_j - v\psi_j .$$

Here r_o is a radius of the middle circle of rolling; γ is a conicity of the wheel.

With this in view

$$X_{j1} = -X_{j2} = X_j \; ; \; Y_{j1} = Y_{j2} = Y_j \; ;$$
$$Q_{yj} = 2Y_j \; ; \; Q_{\psi j} = 2X_j d . \qquad (40)$$

Taking into account expressions of forces (40) and applying a substitution

$$p_j = u_j - v\psi_j \; ; \; q_j = \delta w_j - \frac{v\gamma}{r_o} y_j , \; j = 1,2 ,$$

we can write down the system (39) in new variables in the form of

$$\frac{dy}{dt} = u \; ; \; m \frac{du}{dt} = -4k_y y + 4k_y H\theta + 2k_y (y_1 + y_2) ,$$

$$\frac{d\theta}{dt} = \chi \; ; \; J_x \frac{d\chi}{dt} = 4k_y H_y - 4(k_z b^2 + k_y H^2)\theta + 4\beta_z b^2 \chi - 2k_y H(y_1 + y_2) \; ;$$

$$\frac{d\psi}{dt} = \Omega \; ; \quad J_z \frac{d\Omega}{dt} = -4(k_y a^2 + k_x b^2)\psi + 2k_y a(y_1 - y_2) + 2k_x b^2(\psi_1 + \psi_2) \; ;$$

$$\frac{dy_j}{dt} = p_j + v\psi_j \; ; \quad j = 1,2 \; ;$$

$$m_1 \frac{dp_j}{dt} = 2k_y(y - H\theta - (-1)^j a\psi) + (m_1 \frac{v^2\gamma}{\delta r_o} + 2k_y) y_j - \\ - \frac{2k_1}{v} p_j - \frac{\gamma m_1}{\delta} q_j + \frac{2k_3 p_j}{v^3}(p_j^2 + q_j^2) \; ;$$

$$\frac{d\psi_j}{dt} = -\frac{v\gamma}{\delta r_o} y_j + \frac{1}{\delta} q_j \; ;$$

$$\frac{1}{\delta} J_{z1} \frac{dq_j}{dt} = 2k_x b^2 \psi + \frac{J_{z1} v\gamma}{\delta r_o} p_j + (\frac{J_{z1} v^2 \gamma}{dr_o} - 2k_x b^2)\psi_j - \\ - 2\frac{k_1\delta}{v} q_j + 2\frac{k_3\delta q_j}{v^3}(p_j^2 + q_j^2) \quad , \quad j = 1,2 \, . \tag{41}$$

For a linear approximation of system (41) we construct numerically a quadratic Liapunov function $v(\mathbf{x}) = \mathbf{x}^T \mathbf{A} \mathbf{x}$, $\mathbf{x} \in R^{14}$. Assuming that $J_{z1} = m_1 d^2$ for a full derivative of the function (in view of the system (41)) we have an estimation

$$D^* v(\mathbf{x}) \le \kappa v + \frac{2k_3}{m_1 v^3} \left[(p_1^2 + q_1^2)(\frac{\partial v}{\partial p_1} p_1 + \frac{\partial v}{\partial q_1} q_1) + (p_2^2 + q_2^2)(\frac{\partial v}{\partial p_2} p_2 + \frac{\partial v}{\partial q_2} q_2)\right] . \tag{42}$$

Let us denote

$$\Pi_j = p_j^2 + q_j^2 \; ; \quad \phi_j = \frac{\partial v}{\partial p_j} p_j + \frac{\partial v}{\partial q_j} q_j \; ;$$

$\Lambda(\Pi_j)$, $\Lambda(\phi_j)$ are maximal eigenvalues of Π_j, ϕ_j , $j = 1,2$; $\Lambda(\mathbf{A})$ is the maximal eigenvalue of the matrix $\mathbf{A}$. In this connection $\nu_i = 1$. The estimation (42) is reduced to the form :

$$D^* v(\mathbf{x}) \le \kappa v + \alpha(\Pi_1 \phi_1 + \Pi_2 \phi_2) \le \kappa v + \frac{\alpha(\mu_1 + \mu_2)}{\Lambda(\mathbf{A})} v^2 \, , \tag{43}$$

where

$$\alpha = \frac{2k_3}{m_1 v^3} \, .$$

The comparison equation given in **B**, which is easily integrated, corresponds to the inequality (43). The domain of attraction of the system (41) state $\mathbf{x} = \mathbf{0}$ can be given by an expression

$$N = \left\{ \mathbf{x} : v(\mathbf{x}) < -\frac{\Lambda^2(\mathbf{A})}{\alpha(\mu_1 + \mu_2)} \right\} \subset R^{14}$$

and any motion initiated inside this domain, will not leave the sphere of radius

$$R = \sqrt{-\frac{\kappa \Lambda^2(\mathbf{A})}{\alpha(\mu_1 + \mu_2)}} \, .$$

II.3. STABILITY OF THE COMPARISON SYSTEMS SOLUTIONS

II.3.1. The non-degeneracy of monotonicity. Definition

We consider a system of differential equations

$$\frac{du}{dt} = \Omega(t,u) \ , \tag{44}$$

where $u \in S \subset R^m$, $\Omega : T_o \times S \to R^m$, $(t,u) \to \Omega(t,u)$. Let R^m be a vector space, an order in which is given by a cone of non-negative elements $K = \{u \in R^m , u_i \geq 0 , i = 1,2,\ldots,m\}$ with interior $\overset{o}{K}$. By α we denote a set $\{i_1,\ldots,i_p\}$ ordered by means of correlations $1 \leq i_1 < i_2 < \ldots < i_p \leq m$, $|\alpha| = p < m$; $\hat{\alpha}$ is another set which is obtained from the set $\alpha = \{i_1,\ldots,i_p\}$ by crossing out the value $i_j \in [i_1,i_p]$; y_α is a vector with coordinates $y_{i_1},\ldots,y_{i_p}$; K_α is a cone in R^m which is defined by the formula $K_\alpha = \{y \in R^m , y_\alpha \geq 0 , y_{\hat{\alpha}} = 0 , |\alpha| = p\}$; its interior is $\overset{o}{K}_\alpha$, $y_\alpha > 0$; a cone segment is $<a,b> - \{y : y \in R^m , a_i \leq y_i \leq b_i , i \in [1,n]\}$. We denote by $e \in R^m$ a vector $e = 1,1,\ldots,1^T$ and suppose that $S \supset$ $\supset \{u : 0 \leq u \leq e\}$.

Definition 10. (Martynyuk, Obolenskii, 1978). If for any set β , for which $|\beta| \leq |\alpha|$, it follows from $u-v \in \overset{o}{K}_\beta$ that

$$\Omega_\beta(t,u) - \Omega_\beta(t,v) \in K_\beta \setminus 0 \ , \ \forall (u,v) \in S \ , \ \forall t \in T_o \ ,$$

where $\Omega_\beta(\cdot)$ is a vector, constructed by means of the set β from functions $\Omega_1(t,u),\ldots,\Omega_m(t,u)$, then we say that the vector field, which is induced by system (44), satisfies the $W^{|\alpha|}$-condition.

Corrolary 1. If in the domain $S \times T_o$ for components of a vector function $\Omega(t,u)$ the following inequalities are fulfilled

$$\begin{gathered} \Omega_s(t,u') \leq \Omega_s(t,u'') \quad \text{for} \quad u_s' = u_s'' \ , \ u_\nu' \neq u_\nu'' \quad (\nu \neq s) \\ u = (u_1,\ldots,u_m)^T \in S \subset R^m \ , \quad t \in T_o \ , \quad s \in [1,m] \end{gathered} \tag{45}$$

then the system (44) satisfies the W^o-condition. ∎

Inequalities (45) are known as Ważewski's condition (1949).

Definition 11. A system of ordinary differential equations (44), the right parts of which satisfy the condition of uniform non-singular monotonicity (Ważewski's condition), is to be called *Chaplygin-Ważewski's system*. ∎

II.3.2. The basic statements of the comparison principle

The following important property of the system (44) can be derived from the comparison principle (see II.2). Let for the right part of the system of differential equations (44) the two vector-functions $\mathbf{g}(t,\mathbf{u})$ and $\tilde{\mathbf{g}}(t,\mathbf{u})$; $\mathbf{g} : T_o \times S \to R^m$ be given, which satisfy the Ważewski condition, and such that

a) $\mathbf{g}(t,\mathbf{u}) < \Omega(t,\mathbf{u}) < \tilde{\mathbf{g}}(t,\mathbf{u})$, $\forall (t,\mathbf{u}) \in T_o \times S$;

b) systems of equations

$$\frac{d\mathbf{w}}{dt} = \mathbf{g}(t,\mathbf{w}) , \tag{46}$$

$$\frac{d\mathbf{v}}{dt} = \tilde{\mathbf{g}}(t,\mathbf{v}) \tag{47}$$

are integrable and their solutions $\mathbf{w}(t)$, $\mathbf{v}(t)$ with the initial conditions $\mathbf{w}(t_o) = \mathbf{v}(t_o) = \mathbf{u}_o$ for $t = t_o$ are defined on $T_\beta \subset T_o$. In addition, all solutions of the system (44), starting from the initial point $(t_o, \mathbf{u}_o) \in T_o \times S$, lie between the two special solutions of the systems (46),(47) (the minimal one of the system (46) and the maximal-of the system (47)), i.e.

$$\mathbf{w}^-(t;t_o,\mathbf{u}_o) < \mathbf{u}(t;t_o,\mathbf{u}_o) < \mathbf{v}^+(t;t_o,\mathbf{u}_o) \tag{48}$$

on a general interval of existence of the considered solutions.

This holds also in case functions $\mathbf{g}$ and $\tilde{\mathbf{g}}$ do not depend on $t \in T_o$, i.e. the following is valid :

a) $\mathbf{g}(\mathbf{u}) < \Omega(t,\mathbf{u}) < \tilde{\mathbf{g}}(\mathbf{u})$, $\forall (t,\mathbf{u}) \in T_o \times S$;

b) systems of equations

$$\frac{d\mathbf{w}}{dt} = \mathbf{g}(\mathbf{w}) , \tag{49}$$

$$\frac{d\mathbf{v}}{dt} = \tilde{\mathbf{g}}(\mathbf{v}) \tag{50}$$

are integrable and their solutions are determined as it was given above; hence the estimation (48) is valid for the minimal and maximal solutions of autonomous systems (49),(50).

Estimation (48) allows us to analyze stability (instability) of the state $\mathbf{u} = \mathbf{0}$ of the non-integrable (at all) system (44) on the basis of the two systems (46),(47) (or (49),(50), which are either integrable or autonomous).

Let us recollect the known concepts.

The solution $\mathbf{u}^+ : [t_o, \beta[\to R^m$ of the system (44) with the initial conditions $(t_o, \mathbf{u}_o)$ is called the (right) maximal solution if any other one $\mathbf{u} : [t_o, \tilde{\beta}[\to R^m$ with the same initial conditions satisfies an inequality

$$\mathbf{u}^+(t;t_o,\mathbf{u}_o) \geq \mathbf{u}(t;t_o,\mathbf{u}_o) , \quad \forall t \in [t_o,\beta[\cap [t_o,\tilde{\beta}[.$$

Similarly we define the (right) minimal solution $u^- : [t_o,\beta[\to R^m$ as a solution which satisfies the inequality $u^-(t;t_o,u_o) \le u(t;t_o,u_o)$, $\forall t \in [t_o,\beta[\cap [t_o,\tilde{\beta}[$. In this connection solutions u^- and u are considered for one and the same initial condition (t_o,u_o) .

The proof of estimation (48) is based on two statements, which we illustrate by utilizing lemmas of the items 2,3, Ch. IX of the book by Rouch, Habets and Laloy (1980) and the statement 5.29 of Gutowski's monograph (Gutowski, 1976).

Proposition 10. Let functions $\Omega,\tilde{g} : T_o\times S \to R^m$ be continuous, satisfying an inequality $\Omega < \tilde{g}$ for all $(t,u) \in T_o\times S$ and let $\tilde{g}$ satisfy the W^o - condition; the function $v^+ : [t_o,\beta[\to R^m$ be a maximal solution of the system (47) passing through the point $(t_o,u_o) \in T_i\times S$, $u : [t_o,\tilde{\beta}[\to R^m$ be any solution of the system (44) passing through the point $(t_o,u_o) \in T_i\times S$. Then $u(t;t_o,u_o) < v^+(t,t_o,u_o)$, $\forall t \in [t_o,\beta[\cap [t_o,\tilde{\beta}[$.

Proposition 11. Let functions $\Omega,g : T_o\times S \to R^m$ be continuous, satisfying an inequality $\Omega > g$ for all $(t,u) \in T_o\times S$ and let g satisfy the W^o - condition, the function $w^- : [t_o,\beta[\to R^m$ be a minimal solution of the system (46) passing through the point $(t_o,u_o) \in T_i\times S$, $u : [t_o,\tilde{\beta}[\to R^m$ be any solution of the system (44) passing through the point $(t_o,u_o) \in T_i\times S$. Then $u(t;t_o,u_o) > w^-(t;t_o,u_o)$, $\forall t \in [t_o,\beta[\cap [t_o,\tilde{\beta}[$.

II.3.3. Definitions of the comparison system stability

For the system (44) we can consider non-negative solutions $u(t;t_o,u_o)$, $\forall t \in T_o$. It allows us to formulate the corresponding concepts in the following form.

Definition 12. The state $u=0$ of the system (44) is :

a) *stable with respect to* T_i iff for any $t_o\in T_i$ and any $\epsilon\in]0,\infty[$ there exists $\delta(t_o,\epsilon)$ such that for any u_o , $0 \le u_o \le \delta e$ an estimation $u^+(t;t_o,u_o) < \epsilon e$ is fulfilled for all $t\in T_o$;

b) *uniformly stable with respect to* T_i iff conditions of the Definition 12.a) are fulfilled and for every $\epsilon \in]0,\infty[$ the corresponding maximal δ denoted by $\delta_M(t,\epsilon)$ obeys :

$$\inf [\delta_M(t,\epsilon) : t\in T_i] > 0 ;$$

c) *stable in the whole with respect to* T_i iff conditions of the Definition 12.a) are fulfilled and $\delta_M(t,\epsilon) \to +\infty$, $\forall \epsilon \to +\infty$, $\forall t\in T_i$;

d) *uniformly stable in the whole with respect to* T_i iff conditions of the Definitions 12.b),c) are fulfilled. ∎

Let us note that the explanation to the Definition 1 (I.2.4) is also valid for the Definition 12.

Definition 13. The state $\mathbf{u}=\mathbf{0}$ of the system (44) is :

a) *attractive with respect to* T_i iff for any $t_o \in T_i$ there exists $\Delta(t_o) > 0$ and for any $\zeta > 0$ and for $\mathbf{u}_o : 0 \le \mathbf{u}_o \le \Delta(t_o)\,\mathbf{e}$ there is $\tau(t_o, \mathbf{x}_o, \zeta) \in [0, +\infty[$ such that an estimation $\mathbf{u}^+(t; t_o, \mathbf{u}_o) < \zeta\mathbf{e}$ is fulfilled for all $t \in]t_o + \tau(t_o, \mathbf{x}_o, \zeta), +\infty[$;

b) $\mathbf{u}_o$ *- uniformly attractive with respect to* T_i iff conditions under a) are fulfilled and for any $t_o \in T_i$ and any $\eta \in]0, +\infty[$ there exist $\Delta(t_o) > 0$ and $\tau_u[t_o, \Delta(t_o), \eta] \in [0, +\infty[$ such that

$$\sup\,[\tau_m(t, \mathbf{u}_o, \eta) : 0 \le \mathbf{u}_o \le \Delta(t_o)\,\mathbf{e}\,] = \tau_u[t_o, \Delta(t_o), \eta] \; ;$$

c) t_o *- uniformly attractive with respect to* T_i iff conditions of a) are fulfilled and for any $\eta \in]0, +\infty[$ there exist $\Delta > 0$ and $\tau_u(\mathbf{u}_o, \eta) \in [0, +\infty[$ such that

$$\sup\{\tau_m(t_o, \mathbf{u}_o, \eta) : (t_o, \mathbf{u}_o) \in T_i\} = \tau_u(\mathbf{u}_o, \eta) \; ;$$

d) *uniformly attractive with respect to* T_i iff conditions of the Definition 13.a)-c) are fulfilled and for any $\eta \in]0, +\infty[$ there exist $\Delta > 0$ and $\tau_u(\Delta, \eta) \in [0, +\infty[$ such that

$$\sup\{\tau_m(t_o, \mathbf{u}_o, \eta) : (t_o, \mathbf{u}_o) \in T_i \times [0 \le \mathbf{u}_o \le \Delta\mathbf{e}]\} = \tau_u(\Delta, \eta) \; . \; \blacksquare$$

Definitions 13.a)-d) are modified into those of *"attractive in the whole"* if conditions of the Definition 13.a) are fulfilled for any $\Delta(t_o) \in]0, +\infty[$ and $t_o \in T_i$.

Definition 14. The state $\mathbf{u}=\mathbf{0}$ of the system (44) is :

a) *asymptotically stable with respect to* T_i iff it is stable with respect to T_i and attractive with respect to T_i ;

b) *equi-asymptotically stable with respect to* T_i iff it is stable with respect to T_i and $\mathbf{u}_o$ - uniformly attractive with respect to T_i ;

c) *quasi-uniformly asymptotically stable with respect to* T_i iff it is uniformly stable with respect to T_i and t_o - uniformly attractive with respect to T_i ;

d) *uniformly asymptotically stable with respect to* T_i iff it is uniformly stable with respect to T_i and uniformly attractive with respect to T_i ;

e) *exponentially stable with respect to* T_i iff there exist $\Delta > 0$ and real values $\alpha \ge 1$ and $\beta > 0$ such that for $0 \le \mathbf{u}_o \le \Delta\mathbf{e}$ the inequality

$$\mathbf{u}(t; t_o, \mathbf{u}_o) \le \alpha\mathbf{u}_o \exp[-\beta(t - t_o)] \; , \quad \forall t \in T_o \; , \quad \forall t_o \in T_i \; ,$$

is valid. $\blacksquare$

Definitions 14.a)-d) become the corresponding definitions of *asymptotic stability in the whole* provided both the corresponding type of stability in the whole and attraction in the whole hold. In Definitions 12-14, the expression "*with respect to* T_i" can be omitted iff $T_i = R$.

II.3.4. Linear comparison systems

We consider an autonomous linear system of differential equations

$$\frac{d\mathbf{u}}{dt} = \mathbf{P}\mathbf{u} , \tag{51}$$

where $\mathbf{u} \in R^m$; $\mathbf{P}$ is an $m \times m$ - matrix, $\mathbf{P} = (p_{ij})$

Definition 15 (Newman, 1959). A real $m \times m$ - matrix W is to be called Metzler's matrix (an M-matrix) if (i,j)-th element w_{ij} of $\mathbf{W}$ satisfies

$$w_{ij} \begin{cases} < 0 , & i = j , \\ \geq 0 , & i \neq j , \ \forall (i,j) \in [1,m] . \end{cases}$$

Definition 16. An autonomous linear system (51) is to be called a comparison system iff $p_{ij} \geq 0$, $i \neq j$, $\forall (i,j) \in [1,m]$. ■

Proposition 12. Let $\mathbf{P}$ be an $(m \times m)$-matrix such that $p_{ij} \geq 0$, $\forall (i,j) \in [1,m]$, $i \neq j$. If $\mathbf{w} : [t_o, \beta[\to R^m$ satisfies an inequality

$$\frac{d\mathbf{w}}{dt} \leq \mathbf{P}\mathbf{w} ,$$

and $\mathbf{v} : [t_o, \tilde{\beta}[\to R^m$ is a solution of the Cauchy problem

$$\frac{d\mathbf{v}}{dt} = \mathbf{P}\mathbf{v} , \quad \mathbf{v}(t_o) = \mathbf{v}_o ,$$

then an estimation

$$\mathbf{w}(t; t_o, \mathbf{w}_o) \leq \mathbf{v}(t; t_o, \mathbf{v}_o) , \quad \forall t \in [t_o, \beta[\cap [t_o, \tilde{\beta}[,$$

is valid if only $\mathbf{w}_o \leq \mathbf{v}_o$. ■

The proof of this proposition can be found in Barbashin's work (1970, p.87).

Proposition 13. Let $\mathbf{P}$ in system (51) be an M-matrix. Then the following statements are equivalent :

a) matrix $\mathbf{P}$ is stable (all $\operatorname{Re} \lambda_j(\mathbf{P}) < -\delta$, where λ_j are roots of an equation $\det(\mathbf{P} - \lambda \mathbf{I}) = 0$) ;

b) the following inequalities hold :

$$(-1)^k \begin{vmatrix} p_{11} & p_{12} & \cdots & p_{1k} \\ p_{21} & p_{22} & \cdots & p_{2k} \\ \cdots & \cdots & \cdots & \cdots \\ p_{k1} & p_{k2} & \cdots & p_{kk} \end{vmatrix} > 0 , \quad k = 1,2,\dots,m ;$$

(this condition is called the Sevyastyanov-Kotelyanskii condition or Hick's condition);

c) for any $c > 0$, $c = (c_1, c_2, \dots, c_m)^T$, there is always $b > 0$, $b = (b_1, b_2, \dots, b_m)^T$, obeying

$$P^T b = -c \ ;$$

d) a diagonal matrix B with the positive diagonal exists such that the matrix $P^T B + BP$ is negative definite. ■

In special case, when $B = I$ (I is the unit $m \times m$-matrix), the matrix $P^T + P$ is negative definite [20].

Theorem 9. The state $u = 0$ of the comparison system (51) is asymptotically stable then and only then when there exists a positive vector $u_o = (u_{10}, u_{20}, \dots, u_{mo})^T$ such that

$$P^T u_o < 0 \ , \ u_o > 0 \ .$$

Proof. Necessity. Let the state $u = 0$ of system (51) be asymptotically stable. Then the matrix P is stable (i.e. roots λ_j of the characteristic equation $\det(P - \lambda I) = 0$ satisfy the condition $\operatorname{Re} \lambda_j(P) < 0$, $\forall j \in [1, m]$). The consequence of the equivalence between the statements a) and b) of the Proposition 13 is that in this case there exists a vector $u_o > 0$ such that $P^T u_o < 0$ $(u_o \in \check{R}^m_+)$.

Sufficiency. Together with the system (51) we shall consider the function $v(u) = u^T B u$ $(u \geq 0)$. After some transformations it is not difficult to determine that

$$D^+ v = u^T (P^T B + BP) u \quad , \quad \forall u \in R^m_+ \ .$$

As far as P is an M-matrix obeying c) of Proposition 13 due to the condition of Theorem 9, then the matrix $P^T B + BP$ is negative definite and hence $D^+ v < 0$ holds, $\forall (u \neq 0) \in R^m_+$

This proves asymptotic stability of the state $u = 0$ of the system (51) [21]. ■

Proposition 13 together with the Theorem 9 can be defined by the effectively verified conditions of asymptotic stability of the state $x = 0$ of nonlinear autonomous equations of the perturbed motion (10), Ch. I.

Example 9 (Martynyuk, Obolenskii, 1980). We consider a system of proportional-integral control, reduced to the form

---

[20] See 7) of Comments on References to Ch. II.
[21] See 8) of Comments on References to Ch. II.

$$\frac{dx_i}{dt} = -\rho_i x_i + \sigma \ , \quad i = 1,2,\dots,n \ ,$$

$$\frac{d\sigma}{dt} = \sum_{i=1}^{n} a_i x_i - p\sigma - f(\sigma) \ , \tag{52}$$

where $\rho_i > 0$, $p > 0$, $\sigma f(\sigma) > 0$ for $\sigma \neq 0$, $f(0) = 0$.

Let us illustrate estimations of the domain of the parameters values for which the state $(\mathbf{x} = \mathbf{0} \ , \ \sigma = 0)$ of the system (52) is asymptotically stable. By means of a substitution

$$y_i = \frac{1}{2} x_i^2 \ , \quad i \in [1,n] \ , \quad z = \frac{1}{2} \sigma^2 \ ,$$

we shall reduce the system (52) to the form

$$\frac{dy_i}{dt} = -\rho_i x_i + x_i \sigma \le \frac{1}{2} \rho_i x_i^2 + \frac{1}{2} \frac{\sigma^2}{p_i} \ , \quad i \in [1,n] \ ,$$

$$\frac{dz}{dt} = \sum_{i=1}^{n} a_i x_i \sigma - p\sigma^2 - \sigma f(\sigma) \le \sum_{i=1}^{n} |a_i| \rho_i \frac{x_i^2}{2} - (2p - \sum_{i=1}^{n} \frac{|a_i|}{\rho_i}) \frac{\sigma^2}{2} - \sigma f(\sigma) \ .$$

Further on asymptotic stability of the state $(\mathbf{x} = \mathbf{0} \ , \sigma = 0)$ according to the Proposition 13 and the Theorem 9 will follow from conditions for asymptotic stability of the state $\mathbf{y} = \mathbf{0}$, $z = 0$ of a system

$$\frac{dy_i}{dt} = -\rho_i y_i + \frac{1}{\rho_i} z \ ,$$

$$\frac{dz}{dt} = \sum_{i=1}^{n} |a_i| \rho_i y_i - (2p - \sum_{i=1}^{n} \frac{|a_i|}{\rho_i}) z - g(z) \ .$$

The state $(\mathbf{y} = \mathbf{0} \ , z = 0)$ is asymptotically stable then and only then when the system of inequalities

$$y_i^o > \frac{z}{\rho_i^2} \ , \quad y_i^o > 0 \ , \quad i = 1,2,\dots,n \ ,$$

$$(2p - \sum_{i=1}^{n} \frac{|a_i|}{\rho_i}) z^o + g(z^o) > \sum_{i=1}^{n} |a_i| \rho_i y_i^o \ , \quad z^o > 0 \ .$$

is joint.

After some transformations of this system we define an estimation

$$\sum_{i=1}^{n} \frac{|a_i|}{\rho_i} \le p \tag{53}$$

which determines restrictions on parameters for which the state $(\mathbf{y} = \mathbf{0}$, $z = 0)$ is asymptotically stable.

Remark 4. For the system of equations (52) for $n = 4$ Piontkovskii and Rutkovskaya (1967) determined the following estimation of the domain of parameters

$$\frac{1}{(\min_i (\rho_i))^2} \sum_{i=1}^{4} |a_i|^2 < (\frac{p}{4})^2 \ . \tag{54}$$

It is obvious that both for $n=4$ and $n>4$ the estimation (53) defines a larger domain compared with that ensured by the estimation (54).

Let us give a criterion for exponential stability of the state $u=0$ of the system (51). We shall present the matrix P of the system (51) in the form $Q-R=P$, where $Q=\frac{1}{2}(P-P^T)$, $R=-\frac{1}{2}(P^T+P)$. According to the matrices Q and R we shall construct an $m\times m^2$-matrix defined by $H=(R,QR,Q^2R,\dots,Q^{m-1}R)$ the columns of which are matrices $R,QR,Q^2R,\dots,Q^{m-1}R$. Let us agree to write $R>0$ iff all main diagonal minors of the matrix R are positive :

$$\Delta_k = \begin{vmatrix} r_{11} & \dots & r_{1k} \\ \dots & \dots & \dots \\ r_{k1} & \dots & r_{kk} \end{vmatrix} > 0 \ , \ k=1,2,\dots,m \ ,$$

and $R\geq 0$ iff $\Delta_k \geq 0$, $k=1,2,\dots,m$.

Theorem 10. An isolated state $u=0$ of the comparison system (51) is exponentially stable then and only then when $R>0$ or $R\geq 0$ and the rank $H=m$.

Proof. As a consequence of the Proposition 13.d) the matrix P^TB+BP is negative definite, as far as $p_{ij}\geq 0$, $i\neq j$, $\forall (i,j)\in[1,m]$. In this connection $R>0$, or $R\geq 0$ and the rank $H=m$, are necessary and sufficient. Besides, we can give the function $v(u)=u^TBu$ $(u\geq 0)$ for which $D^+v<-\delta v$ guarantees exponential stability of the state $u=0$, $\delta = \lambda_m(-P^TB-BP)\,\Lambda_M^{-1}(B)$. ∎

II.3.5. Nonlinear systems with an isolated equilibrium state

We consider an autonomous nonlinear system

$$\frac{du}{dt} = \Omega(u) \ , \quad u\in R^m \ . \tag{55}$$

Assumption 5. Let :

1) the right part of system (55) satisfy the W^o-condition; it is continuous and a solution of the Cauchy problem for any $u_o\in R^m$ is locally unique;
2) there exist a neighbourhood $\mathcal{U}$ of the state $u=0$ such that for all $u\in\mathcal{U}$, $u\neq 0$, $\Omega(u)\neq 0$ and $\Omega(u)=0$ for $u=0$.

Theorem 11. An isolated state $u=0$ of the comparison system (55) is asymptotically stable then and only then when conditions of the Assumption 5 are fulfilled and there exists a positive vector $u_o\in \mathring{K}\cap\mathcal{U}$ such that the system of inequalities

$$\Omega_i(u_{10},\dots,u_{mo}) < 0 \ , \ \forall i\in[1,m]$$

is joint. ∎

The proof illustrated in the paper by Martynyuk and Obolenskii (1980) is based on a number of additional statements.

Proposition 14. If conditions of the Assumption 5 are fulfilled, then the set $A_r = \{u : u \in B_r \;\&\; \gamma \in R \to \Omega(u) = \gamma u\}$ is compact and not empty, and the real function $\gamma(u) = (\Omega(u), u)/|u|^2$ is continuous and not equal to zero on $A_r \cap U$, where $|u| = |u_1| + |u_2| + \dots + |u_m|$ and

$$B_r = \{u : u \in K \;;\; \sum_{i=1}^{m} u_i^2 = r^2 \;,\; r \in \check{R}_+\} \;.$$

Proposition 15. If conditions of the Assumption 5 are fulfilled and

$$\hat{\gamma}(\bar{u}) = \max_{u \in A_r} \frac{(\Omega(u), u)}{|u|^2} < 0 \quad , \quad \bar{u} \in \partial B_r \;,$$

then in $\overset{o}{K}$ there exists a vector u_o such that the system of inequalities

$$\Omega_i(u_{1o}, \dots, u_{mo}) < 0 \;,\; \forall i \in [1,m] \;,$$

is joint.

Proposition 16. Let $H^T(u)$ be a local semi-group of transformations of the conic segment. If conditions of the Assumption 5 are fulfilled and if there exists a vector $\tilde{u} \in B_r \cap U$ such that the system of inequalities

$$\Omega\; (\tilde{u}_1, \dots, \tilde{u}_m) \geq 0 \;,\; \forall i \in [1,m] \;,$$

is joint, then we can find such $\delta > 0$ that for all $0 < t < \delta$ an inequality

$$H^T(\tilde{u}) \geq \tilde{u} \;,$$

is fulfilled, if only for one component $\tilde{u}_i$ a strict inequality holds.

Proposition 17. If the conditions of the Assumption 5 are fulfilled and if there exists a sequence u_s such that $u_s \in K \cap U$, $\lim_{s \to +\infty} u_s = 0$ and for any vector u_s an inequality

$$\Omega_i(\tilde{u}_s) \geq 0 \;,\; i = 1,2,\dots,m \;,$$

is fulfilled for $i = 1,2,\dots,m$, then for a given s we can find the value τ_s such that $\{u : u = u(t;0,u_s) \;,\; \forall t > \tau_s\} \cap \bar{U} = \phi$.

Proposition 18. If conditions of the Proposition 5 are fulfilled and there exists a positive vector $u_o \in \overset{o}{K} \cap U$ for which a system of inequalities

$$\Omega_i(u_{1o}, \dots, u_{mo}) < 0 \;,\; \forall i \in [1,m] \;,$$

is joint, then the state $u = 0$ of the system (55) is asymptotically stable. ∎

Let us note that Propositions 14-17 prove the necessity of conditions of the Theorem 11, while the Proposition 18 proves sufficiency of its conditions.

Theorem 12. The state $\mathbf{u}=\mathbf{0}$ of the comparison system (55) is unstable if :

a) conditions of the Assumption 5 are fulfilled;

b) there exists a sequence $\mathbf{u}_s \in K$, $\lim_{s\to+\infty} \mathbf{u}_s = \mathbf{0}$, such that for each vector $\tilde{\mathbf{u}}_s$ the system of inequalities

$$\Omega_i(\tilde{u}_1,\tilde{u}_2,\dots,\tilde{u}_m) \geq 0 \ , \ i \in [1,m] \ ,$$

is joint, if only for one $j \in [1,m]$ the inequality is strictly fulfilled;

c) there exists a neighbourhood U of the state $\mathbf{u}=\mathbf{0}$ such that on a set $K_{\mathbf{u}_s} \cap U$ a vector field $\Omega(\mathbf{u})$ does not vanish. ■

The proof of the Theorem 12 is settled by Propositions 16, 17.

II.3.6. The theorem of Zaidenberg-Tarsky and algebraic solvability of the stability problem

In contrast to the general stability problem there is algebraically solvable one for a comparison system on the basis of the theorem by Zaidenberg-Tarskii (Lopatinskii, 1980).

Assumption 6. A comparison system

$$\frac{du_i}{dt} = \Omega_i(u_1,u_2,\dots,u_m,\beta) \ , \ i = 1,2,\dots,m \ , \tag{56}$$

satisfies the following conditions :

1) functions $\Omega_i(u_1,u_2,\dots,u_m,\beta)$ are polynomials, $\Omega_i(0,0,\dots,0,\beta)=0$, $(u_1,u_2,\dots,u_m)$ are variables, $m \geq 2$, $\beta = (\beta_1,\dots,\beta_p)$ are variable parameters from the domain E, $p \geq 0$;

2) polynomials

$$\frac{\partial\Omega_i}{\partial u_j}(u_1,u_2,\dots,u_m,\beta) \ , \ i \neq j \ , \ \forall(i,j) \in [1,m] \ ,$$

are non-negative for all $\beta \in E$;

3) a general criterion for asymptotic stability (instability) of the comparison system (56) is contained in the finite system of equations of the following types

$$\begin{aligned} &f(u_1,\dots,u_m,\beta) = 0 \ , \ g(u_1,\dots,u_m,\beta) > 0 \ , \ h(u_1,\dots,u_m,\beta) < 0 \ , \\ &k(u_1,\dots,u_m,\beta) \neq 0 \ , \ \ell(u_1,\dots,u_m,\beta) \geq 0 \ , \ m(u_1,\dots,u_m,\beta) \leq 0 \ . \end{aligned} \tag{57}$$

Functions f , g , k , h , ℓ , m which are constructed according to the right part of system (56), ensure fulfillment of a system of inequalities

$$\Omega_i(u_1,\dots,u_m,\beta) < 0 \ , \ u_i > 0 \ ;$$

$$\Omega_i(u_1,\dots,u_m,\beta) > 0 \ , \ u_i < 0 \ ,$$

and also an isolation of the equilibrium state $\mathbf{u}=\mathbf{0}$. The determination of the values of the parameters $\bar{\beta} \in E$ for which these inequalities would ensure asymptotic stability (instability) of the state $\mathbf{u}=\mathbf{0}$ of the system (56), is the problem of analysis of (57).

Theorem 13. By means of the finite number of algebraic operations on system (57) we can define the finite number of systems $\Sigma_1,\Sigma_2,\dots,\Sigma_r$ of equalities and inequalities of the same form, which contain β , but do not contain $u_1,\dots,u_m$,and which possess the following property : in order that for the real parameter values $\beta_1=\bar{\beta}_1 , \dots , \beta_p=\bar{\beta}_p$ the system (57) have only one real solution $u_1=\xi_1 , u_2=\xi_2 , \dots , u_m=\xi_m$ it is necessary and sufficient that values $\bar{\beta}_1,\dots,\bar{\beta}_p$ (substituted instead of $\beta_1,\dots,\beta_p$) satisfy only one of the systems of equations and inequalities $\Sigma_1,\Sigma_2,\dots,\Sigma_r$. ■

Let us illustrate Theorem 13 via a number of examples (Martynyuk, Obolenskii, 1978, 1980).

Example 10. Let a linear comparison system

$$\frac{d\mathbf{u}}{dt} = \mathbf{P}\mathbf{u} \ , \ \mathbf{P} = [p_{ij}]_1^m \ , \ p_{ij} \geq 0 \ , \ i \neq j \ , \ \forall (i,j) \in [1,m] \ , \qquad (58)$$

be given. The collection illustrated in Theorem 13 is widely known and it is represented by Sevastyanov-Kotelyanskii's inequalities

$$(-1)^k \begin{vmatrix} p_{11} & p_{12} & \cdots & p_{1k} \\ \cdots & \cdots & \cdots & \cdots \\ p_{k1} & p_{k2} & \cdots & p_{kk} \end{vmatrix} > 0 \ , \ k = 1,2,\dots,m \ . \qquad (59)$$

As it is known these inequalities (Proposition 13.b)) are the necessary and sufficient conditions for asymptotic stability of the state $\mathbf{u}=\mathbf{0}$ of system (58).

Example 11. Inequalities (59) are necessary and sufficient for asymptotic stability of the state $\mathbf{u}=\mathbf{0}$ of the comparison system

$$\frac{du_i}{dt} = \Big(\sum_{j=1}^{m} a_{ij}\, u_j^{2\ell_j+1}\Big)^{2m_i+1} \ , \ i,j = 1,2,\dots,m \ ,$$

where $a_{ij} \geq 0$ for $i \neq j$; ℓ_i , m_i are natural numbers.

Example 12. Let a system

$$\frac{du_i}{dt} = -\rho_i^{2\ell_i+1} u_i^{2\ell_i+1} + u_{i+1}^{2\ell_i+1} \quad , \quad i = 1,2,\dots,m-1 \; ,$$
$$\frac{du_m}{dt} = \sum_{i=1}^{m-1} |a_i| \, u_i^{2\ell_i+1} - \rho_m u_m^{2\ell_m+1} \; , \qquad (60)$$

be given, where $\rho_i > 0$; $\ell_i, \forall i \in [1,m]$, are the arbitrary natural numbers; $|a_1| > 0$. If

$$\rho_m = \sum_{i=1}^{m-1} |a_i| \left(\prod_{s=1}^{m-1} \frac{1}{\rho_s} \right)^{2\ell_m+1} ,$$

then the "straight line" given by an equation

$$u_i = \left(\prod_{s=1}^{m-1} \frac{1}{\rho_s} \right) u_m \quad , \quad i = 1,2,\dots,m-1 \; ,$$

is asymptotically stable.

The state $\mathbf{u} = 0$ of the system (60) is asymptotically stable if the perturbations of parameters ρ_k and $|a_i|$, $\forall i \in [1,m-1]$ are subordinate to the restriction

$$\sum_{i=1}^{m-1} |a_i| \left(\prod_{s=1}^{m-1} \frac{1}{\rho_s} \right) < \rho_m$$

Let in the system (60) there be $m = 2$. Then,

$$\frac{du_1}{dt} = a_{11} u_1^{2\ell_1+1} + a_{12} u_2^{2\ell_1+1} \; ,$$
$$\frac{du_2}{dt} = a_{21} u_1^{2\ell_2+1} + a_{22} u_2^{2\ell_2+1} \; , \qquad (61)$$

where $a_{12}, a_{21} \geq 0$; ℓ_1, ℓ_2 are the natural numbers.

The necessary and sufficient conditions for asymptotic stability of the state $u_1 = 0$, $u_2 = 0$ of the system (61) have the form :

$$a_{11} < 0 \quad , \quad \begin{vmatrix} a_{11}^{2\ell_1+1} & a_{12}^{2\ell_2+1} \\ a_{21}^{2\ell_1+1} & a_{22}^{2\ell_2+1} \end{vmatrix} > 0 \; .$$

II.3.7. Nonlinear autonomous comparison systems with a non-isolated singular point

We consider a comparison system

$$\frac{du_i}{dt} = \Omega_i(u_1, u_2, \dots, u_m) \quad , \quad i = 1,2,\dots,m \; . \qquad (62)$$

Assumption 7. The system (62) satisfies the conditions :

1) the right parts of the system (62) are continuous and the solution of the corresponding Cauchy problem for $(u_{10},\dots,u_{m_o}) \in S \subset R^m$ is locally unique;
2) a vector field, generated by the system (62), satisfies the $W^{|\alpha|}$ - condition, where $2|\alpha| + 4 \geq m$;
3) for every point $u \in M$, where

$$M = \{u \in R^m \,, \sum_{i=1}^{m} \Omega_i^2(u) = 0\} \,,$$

there exists a strictly monotonous continuous curve $c(s) \in M$, determined for $s \in]-\delta(u), \delta(u)[$, where $\delta(u) > 0$, $c(0) = 0$.

Theorem 14. Many fixed points of the system (62) (a set M) are asymptotically stable if all conditions of the Assumption 7 are fulfilled.

The proof of the theorem is illustrated in the paper (Martynyuk and Obolenskii, 1980) and is based on two auxiliary statements. ■

Proposition 19. If conditions of the Assumption 7 are fulfilled, then the intersection of the set M with an arbitrary cube

$$Q_h = \{u \in R^m \,, |u_i| < h \,, h \in \check{R}_+ \,, i \in [1,m]\}$$

is compact and there exists a neighbourhood N of the set M, invariant relative to the semi-group of transformations $H^t(u_o)$ $(u_o \in Q_h$, $t > 0)$, which is generated by a vector field, corresponding to system (62).

Proposition 20. If conditions of the Assumption 7 are fulfilled, then for an arbitrary point $u_o \in Q_h$ an ω - limit point of the system (62) solution belongs to M .

II.3.8. Several applications of nonlinear comparison systems

We consider a large-scale system of equations of the perturbed motion

$$\frac{dx_i}{dt} = A_i x_i + h_i(x_1, x_2, \dots, x_m) \,, \quad i \in [1,m] \,, \tag{63}$$

where $x_i \in R^{n_i}$, A_i are $n_i \times n_i$ - matrices; $h_i : R^n \to R^{n_i}$ are functions of interactions among the independent (free) subsystems

$$\frac{dx_i}{dt} = A_i x_i \,, \quad i \in [1,m] \,. \tag{64}$$

Let us suppose that the roots of characteristic equations

$$\det(A_i - \lambda_i I_i) = 0 \,, \quad i \in [1,m] \,, \tag{65}$$

where I_i is an $n_i \times n_i$ -unit matrix, are simple ones with the negative real parts. We apply the method of the comparison equation construction, illustrated in II.2.5 and by means of the non-singular linear transformation $x_i = B_i z_i$ we reduce the system (63) to the form

$$\frac{d\rho_i}{dt} = \alpha_i \rho_i + \frac{1}{2}(F_i e^{-i\theta_i} + \bar{F}_i e^{i\theta_i}) , \quad i \in [1,m] \tag{66}$$

where

$$\alpha_i = \mathrm{Re}\,\{\mathrm{diag}(\lambda_i^1,\dots,\lambda_i^{n_i})\} , \quad F_i = B_i^{-1} H_i(B_1 z_1,\dots,B_m z_m) .$$

With the system (63) we associate a tentative vector Liapunov function

$$\mathbf{V}(\rho_1,\dots,\rho_m) = (\sum_{j=1}^{n_1^*} \rho_{1j} , \dots , \sum_{j=1}^{n_m^*} \rho_{mj})^T$$

where $n_i^* < n_i$, $i \in [1,m]$ is a number of various roots of the characteristic equations (65).

Assumption 8. The large-scale system (63) is such that ;

1) the roots of the characteristic equations (65) are simple, with the negative real parts;
2) there exists a vector-function $\mathbf{\Omega(u)} = (\Omega_1(u_1,\dots,u_m) ,\dots, \Omega_m(u_1,\dots,u_m))^T$ which satisfies the W^o-condition and inequalities

$$\left|\frac{1}{2}\sum_{j=1}^{n_i^*} (F_i^j e^{-i\theta_s^j} + \bar{F}_i^j e^{i\theta_s^j})\right| \le \Omega_i(u_1,\dots,u_m) , \quad \forall i \in [1,m] .$$

Theorem 15. If conditions of the Assumption 8 are fulfilled and there exists a vector $\mathbf{u^*} \in K$ such that the system of inequalities

$$\alpha_1^o u_1^* + \Omega_1(u_1^*,\dots,u_m^*) < 0 , \quad (u_1^* > 0,\dots,u_m^* > 0) ,$$
$$\dots\dots\dots\dots\dots\dots\dots\dots\dots\dots\dots\dots$$
$$\alpha_m^o u_m^* + \Omega_m(u_1^*,\dots,u_m^*) < 0 , \quad (u_1^* > 0,\dots,u_m^* > 0) ,$$

is solvable, then the state $\mathbf{x} = \mathbf{0}$ of system (63) is asymptotically stable.

Proof. Under the conditions of the Theorem 15 we obtain a system of equations

$$\frac{d\mathbf{u}}{dt} = \alpha^* \mathbf{u} + \Omega(\mathbf{u}) , \quad \mathbf{u} \in R^m ,$$

where

$$\alpha^* = \mathrm{diag}\{\alpha_1^o,\alpha_2^o,\dots,\alpha_m^o\} , \quad \alpha_s^o = \max\{\alpha_j , j \in [1,n_s^*]\} ,$$

which is a comparison system for the equations (63). Together with the Theorem 11 this completes the proof of the Theorem 15. ■

Example 13 (Martynyuk, Nikitina, 1982). We consider equations of a motion of the self-oriented wheel of chassis with a weightless stand.

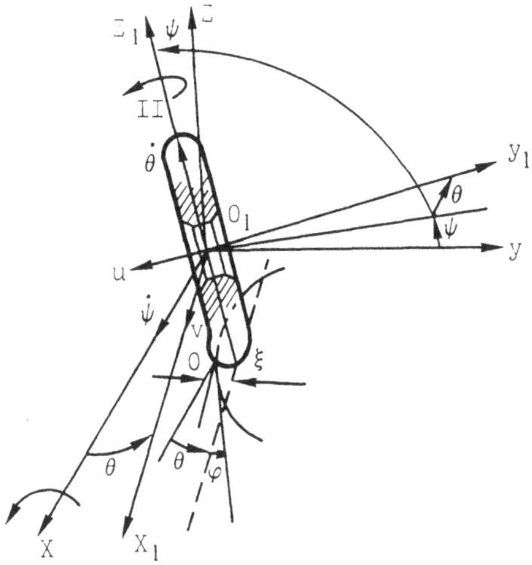

Figure 6

We assume that the conditions are fulfilled, for which the theory of rolling of an elastic pneumatic, being developed by N.V. Keldysh (1945), is valid. Let x, y, z be the Cartesian coordinates of the crossover point of the straight line of the largest slope, which is drawn in the middle plane of the wheel, through its centre, with the plane of the road. We locate the beginning of the non-stationary system of coordinates $Cx_1y_1z_1$ connected with the moving mechanical system into the centre of masses of the wheel with a weightless stand (Figure 6).

The position of an elastic wheel in the inertial system of coordinates $Oxyz$ will be determined by an angle of rotation θ, coordinates x, y and also by deformations ξ, φ. Let us introduce the following simplifying proposition : we shall consider a wheel rigid with respect to the rotation about the axis Ox.

Equations of the perturbed motion of the elastic wheel at a constant velocity of an unperturbed motion $v = \text{const}$ have the form :

$$m\dot{u} + h_u u + h_\xi(u + v\theta + v\varphi) - b_y y - a\xi = 0 \,, \tag{67}$$

$$J\ddot{\theta} + h_\theta\dot{\theta} + H_\theta\dot{\theta}|\dot{\theta}| + b_\theta\theta - b\varphi = 0 \,, \tag{68}$$

where u is a velocity of the transverse removal of the wheel :

$$u = -\dot{y} \,, \tag{69}$$

a, b, h_ξ are parameters, which are contained in the expressions of forces, stipulated by a deformation of pneumatic; b_y, b_θ are coefficients of the stand rigidity relative to the flexion and the torsion; H_θ is a characteristic of a nonlinear damp; h_θ, h_u are characteristics of linear friction; M, J are a mass and a moment of inertia relative to the axis Cx of the wheel, respectively.

Equations (67),(68), of non-holonomic connections

$$u + \dot{\xi} + v(\theta + \varphi) = 0 \quad , \quad \dot{\varphi} + \dot{\theta} - (\alpha\xi - \beta\varphi)v = 0$$

and an expression (69) constitute a closed system for the determination of $y(t), \theta(t), \xi(t), \varphi(t)$, $\forall t \in T_o$.

In variables $y_1 = \theta$, $y_2 = \dot{\theta}$, $y_3 = \varphi$, $y_4 = \xi$, $y_5 = y$, $y_6 = u$ the system of differential equations of the perturbed motion takes the form

$$\dot{y}_1 = y_2 \ , \quad \dot{y}_2 = -\frac{b_\theta}{J} y_1 - \frac{h_\theta}{J} y_2 + \frac{b}{J} y_3 - \frac{H_\theta}{J} y_2 |y_2| \ , \tag{70}$$

$$\dot{y}_3 = -\beta v y_3 + \alpha v y_4 - y_2 \ , \quad \dot{y}_4 = -v y_3 - v y_1 - y_6 \ , \tag{71}$$

$$\dot{y}_5 = -y_6 \ , \quad \dot{y}_6 = \frac{b_y}{m} y_5 - \frac{(h_u + h_\xi)}{m} y_6 - \frac{h_\xi}{m} v(y_1 + y_3) + \frac{a}{m} y_4 \ . \tag{72}$$

Further on in subsystems (70)-(72) we shall pass on to the canonical variables z_k $(k = 1,2,\dots,6)$.

Let us choose the vector Liapunov function with the following components, corresponding to subsystems (70)-(72) :

$$v_k = \rho_k \ , \ \rho_k = z_{2k-1} e^{-i\theta_k} \ , \ \rho_k = z_{2k} e^{i\theta_k} \ , \ k = 1,2,3 \ .$$

After several transformations we obtain

$$\begin{gathered} \dot{v}_1 = -a_{11} v_1 + a_{12} v_2 - \gamma v_1^2 \ , \quad \dot{v}_2 = a_{21} v_1 - a_{22} v_2 + a_{23} v_3 \ , \\ \dot{v}_3 = a_{31} v_1 + a_{32} v_2 - a_{33} v_3 \ , \end{gathered} \tag{73}$$

where

$$a_{11} = h_\theta / 2J \ ; \ a_{12} = -bS_2 \sin\theta_1 / (vJk_1) \ ; \ \gamma = 2H_\theta S_1 |S_1| \sin\theta_1 / (Jk_1) \ ;$$

$$a_{21} = v(Q\cos\theta_1 - S_1 \sin\theta_2) / k_2 \ ; \ a_{23} = QS_3/k_3 \ ; \ a_{33} = h/(2m) \ ;$$

$$a_{31} = -vh_\xi \sin\theta_3 \cos\theta_1 / (mk_3) \ ; \ a_{32} = \sin\theta_3 (a\cos\theta_2 - h_\xi S_2) / (mk_3) \ ;$$

$$k_1 = \sqrt{4b_\theta J - h_\theta^2} / (2J) \ ; \ k_2 = v\sqrt{4\alpha - \beta^2} / 2 \ ; \ k_3 = \sqrt{4b_y m - h^2} / (2m) \ ;$$

$$Q = \alpha_2 \sin\theta_2 - k_2 \cos\theta_2 \ ; \ S_1 = \alpha_1 \cos\theta_1 + k_1 \sin\theta_1 \ ;$$

$$S_2 = \alpha_2 \cos\theta_2 + k_2 \sin\theta_2 \ ; \ S_3 = \alpha_3 \cos\theta_3 + k_3 \sin\theta_3 \ ;$$

$$\alpha_1 = h_\theta / (2\theta) \ ; \ \alpha_2 = \beta v / 2 \ ; \ \alpha_3 = h/(2m) \ ; \ h = h_\xi + h_u \ .$$

For fixed values θ_k $(k = 1,2,3)$ estimations of the derivatives in expressions (73) are made so that the Ważewski conditions are fulfilled.

The domain of attraction of the singular point can be determined by inequalities

$$y_1^o > \frac{D_2 a_{12} - a_{11} D_1}{\gamma D_1} \ , \tag{74}$$

$$y_2^o > y_1^o D_2 / D_1 \ , \quad y_3^o > y_1^o D_3 / D_1 \ , \tag{75}$$

where

$$D_1 = a_{22} a_{33} - \bar{a}_{23} \bar{a}_{32} \ ; \ D_2 = \bar{a}_{21} a_{33} + \bar{a}_{23} \bar{a}_{31} \ ; \ D_3 = \bar{a}_{21} a_{32} + \bar{a}_{31} a_{22} \ . \tag{76}$$

Here the bar over letters in the formulas (76) means the corresponding coefficient after the estimation. Existence of the domain of attraction of the non-zero singular point of the comparison system in case of

$$D_2\bar{a}_{12} > a_{11}D_1$$

indicates boundedness of the system solution. The stability domain in the parameters plane $h_u v$ can be defined from the inequality

$$D_1 > 0 .$$

We constructed in Figure 7 the stability domains in the parameter plane $h_u v$ for the following values of a: 10^4, 2.10^4, 5.10^4 (the curves 1-3), and we constructed in Figure 8 the attraction domains which were obtained according to (74),(75), depending on the parameter v for the following values of the remaining parameters :

$m = 105$ kg.sec^2/m	$h_u = 200$ kg.sec/m
$T = 90$ kg.sec^2	$h_\xi = 100$ kg.sec/m
$a = 10^4$ kg/m	$H_\theta = 1000$ kgm.sec^2
$b = 0,5.10^4$ kg/m	$b_\theta = 2.10^6$ kgm
$\beta = 5,3$ m^{-2}	$b_u = 2.10^6$ kg/m
$h_u = 4,5$ m^{-1}	$h_\theta = 200$ kg.sec/m

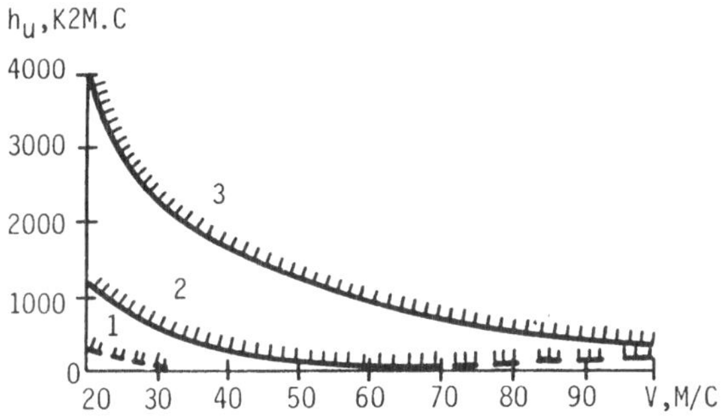

Figure 7

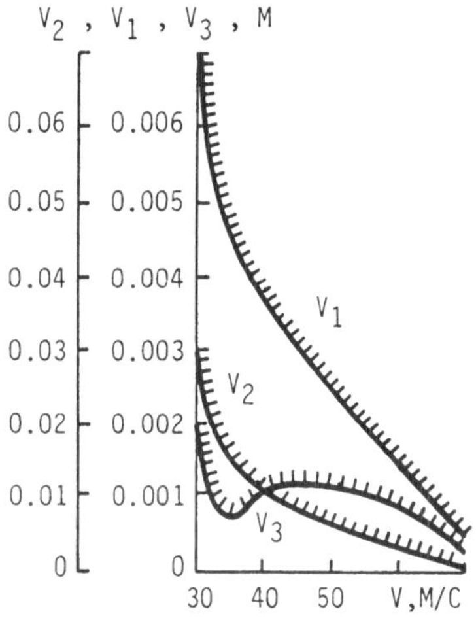

Figure 8

Theorem 11 is utilized in the problem of estimation of the asymptotic stability domain. Let $u = u^*$ be an isolated equilibrium state of the system (55).

For asymptotic stability in the whole of the state $u = u^*$ it is necessary and sufficient that the same state is both stable in the whole and attractive in the whole (see the Definition 3, Ch.I).

Example 14. Let us construct an estimation of the attraction domain of the equilibrium state of an inertial part of the double wheeled chassis. According to the theory of Keldysh (1945) the equations of motion of such a mechanical system, together with the description of the damper and connections, have the form (Keldysh, 1945; Goncharenko, Lobas, Nikitina, 1981)

$$J\ddot{\psi} + h_\psi\dot{\psi} + [b_{11} + 2(\rho N + \sigma N_r + C_b d^2)]\,\psi + 2\,Iv\dot{\theta}/r + (b_{12} + 2IV/r)\,\theta + b_{13}z - $$
$$- d_{11}\chi - 2h_\xi r\dot{\xi} - 2(ar + \sigma N)\,\xi - h_\psi\dot{\chi}\sin\lambda = 0 \;,$$
$$J^*\ddot{\theta} + h_\theta\dot{\theta} + b_{22}\theta - 2IV/r\,(1 + d^2/r^2)\,\dot{\psi} + (b_{21} - 2fC_b d^2)\,\psi + b_{22}z -$$
$$- d_{21}\chi - 2b\varphi - h_\theta\dot{\chi}\cos\chi = 0 \;,$$
$$m\ddot{z} + b_{33}z + (b_{31} + 2\sigma N)\,\psi + b_{32}\theta - d_{31}\chi - 2h_\zeta\dot{\zeta} - 2a\zeta = 0 \;,$$
$$H\dot{\chi}_1^2\,\mathrm{sign}\,\dot{\chi}_1 = k(\chi - \chi_1) \;,$$
$$\dot{\zeta}_1 + \dot{z} + r\dot{\psi} + V\theta + V\varphi = 0 \;,$$
$$\dot{\varphi} + \dot{\theta} - \alpha V\zeta + \beta V\varphi - \gamma V\psi = 0 \;,$$
$$k(\chi - \chi_1) = e_{11}\psi + e_{12}\theta + e_{13}z - g_{11}\chi \;. \tag{77}$$

Here θ, ψ, z are inertial, and ζ, φ, χ_1, χ are inertia-free coordinates. It is not difficult to find out that in the system (77) there is an excessive coordinate. Let us introduce a notation $\sigma = \chi - \chi_1$, and reduce a system of equations and correlations (77) to the form

$$\frac{dx_i}{dt} = \sum_{k=1}^{8} \alpha_{ik}x_k + b_i\sqrt{|\sigma|}\,\mathrm{sign}\,\sigma + e_i\sigma \;, \quad i = 1,2,\dots,8 \;,$$
$$\frac{d\sigma}{dt} = \sum_{k=1}^{8} a_k x_k - b_\sigma\sqrt{|\sigma|}\,\mathrm{sign}\,\sigma \;, \quad b_\sigma > 0 \;. \tag{78}$$

We introduce functions

$$v_1 = z_a e^{-i\theta_a} \;, \quad v_2 = \bar{z}_a e^{i\theta_a} \;, \quad v_3 = z_e\,\mathrm{sign}\,z_e \;, \quad \bar{W} = \sigma\,\mathrm{sign}\,\sigma \tag{79}$$

where z_a are canonical variables after reduction of the system (78) to the canonical form with respect to a part of the variables.

On the basis of (79) we compose a comparison system

$$\frac{du_i}{dt} = -\alpha_i u_i + B_i\sqrt{u} + s_i u \;,$$
$$\frac{du}{dt} = \sum_{j=1}^{n-N} k_j u_j - b_\sigma\sqrt{u} \;, \quad i = 1,2,\dots,n-N \;. \tag{80}$$

Here $-\alpha_i = \mathrm{Re}(\lambda_i)$; λ_i - are roots of the characteristic polynomial of the first n equations of the system (78) among which there are N complex conjugate ones; all other coefficients are obtained by an ordinary method in the course of construction of the comparison system (80).

Let us stipulate the following : for the zero value of their arguments derivatives of the functions (79) are not considered; the local uniqueness of the Cauchy problem (80) solution corresponds to a choice of only one branch of the solution, which will be an estimation for other solutions from above.

Values of the parameters must satisfy the inequality

$$b_\sigma \alpha_1\alpha_2 \dots \alpha_{n-N} >$$
$$B_1 k_1 \alpha_2\alpha_3 \dots \alpha_{n-N} + B_2 k_2 \alpha_1\alpha_3 \dots \alpha_{n-N} + \dots + B_{n-N} k_{n-N} \alpha_1\alpha_2 \dots \alpha_{n-N-1} .$$

The vector of an invariant cone $\mathbf{u}$ has a restriction from above

$$\mathbf{e}^T\mathbf{u} < A^2/B^2 , \quad \mathbf{e} = 1,1,\dots,1^{\;T} .$$

The attraction domain of the state $\mathbf{u}=0$ can be estimated by inequalities

$$u_{io} > (B_i \sqrt{u_o} + s_i u_o)/\alpha_i , \quad i = 1,2,\dots,n-N .$$

II.3.9. Reducible comparison systems

The necessary and sufficient conditions for asymptotic stability of the state $\mathbf{u}=0$ of the systems (51) or (55) do not admit the direct generalization to non-stationary comparison systems [22]. At the same time some non-stationary comparison systems can be analyzed, in particular, reducible one in Liapunov's sense.

Definition 17 (Šiljak, 1975). Matrix $\mathbf{P}(t)$ is called a *non-autonomous M-matrix*, iff the following conditions are fulfilled

$$p_{ij}(t) \begin{cases} < 0 , & \forall t \in T_o , \; i = j ; \\ \geq 0 , & \forall t \in T_o , \; i \neq j , \; \forall (i,j) \in [1,m] . \end{cases} \quad \blacksquare$$

Definition 18. Non-autonomous linear system

$$\frac{d\mathbf{u}}{dt} = \mathbf{P}(t)\,\mathbf{u} \tag{81}$$

is called *reducible comparison system* if $\mathbf{P}(t)$ is the non-autonomous M-matrix and if by means of Liapunov's transformation it can be reduced to the system

$$\frac{d\mathbf{u}}{dt} = \mathbf{B}\mathbf{u} \tag{82}$$

[22] See 9) of Comments on References to Ch. II.

with the constant M-matrix $\mathbf{B}$. ■

Let us recollect that we call Liapunov's transformation a linear one defined by $\mathbf{u} = \mathbf{L}(t)\,\mathbf{v}$, where $\mathbf{L}(t)$ is an $m\times m$ - Liapunov matrix.

Definition 19. Matrix $\mathbf{L}(t) \in C^1(T_0)$ with complex, in general, elements is called *the Liapunov matrix* iff the following conditions are fulfilled :

a) $\mathbf{L}(t)$ and $\dot{\mathbf{L}}(t)$ are bounded for all $t\in T_0$, i.e.

$$\sup\{\|\mathbf{L}(t)\| : \forall t\in T_0\} < +\infty \quad , \quad \sup\{\|\dot{\mathbf{L}}(t)\| : \forall t\in T_0\} < +\infty \; ,$$

b) $|\det \mathbf{L}(t)| \geq a > 0$, $a\in \check{R}_+$, $\forall t\in T_0$. ■

The matrix $\mathbf{L}(t)$ properties can be found in the book by Demidovich (1967).

Theorem 16 (Erugin, 1946). The linear differential system (81) is reducible then and only then when any of its fundamental matrices $\mathbf{X}(t)$ can be represented in the form

$$\mathbf{X}(t) = \mathbf{L}(t)\exp[\mathbf{B}t] \tag{83}$$

Proof. Let us note that for a matrix $\mathbf{B}$, satisfying the condition $b_{ij} \geq 0$, $i \neq j$, $(i,j) \in [1,m]$, elements of the matrix $\exp[\mathbf{B}t]$ are non-negative. It follows from the fact that

$$\exp[\mathbf{B}t] = \left(\exp[\frac{\mathbf{B}t}{N}]\right)^N \quad , \quad N \text{ is a positive value}$$

and

$$\exp[\frac{\mathbf{B}t}{N}] = \mathbf{I} + \mathbf{B}\frac{t}{N} + \mathbf{B}^2\frac{t^2}{2N^2} + \ldots \quad ■$$

Theorem 17. An isolated state $\mathbf{u}=\mathbf{0}$ of the non-autonomous system (81) is asymptotically stable if

a) system (81) is reducible;
b) a positive vector $\mathbf{u}^\circ = u_1^\circ,\ldots,u_m^\circ{}^T$ exists such that $\mathbf{B}^T\mathbf{u}^\circ < 0$, $\mathbf{u}^\circ > 0$. ■

Theorem 18. The isolated state $\mathbf{u}=\mathbf{0}$ of the non-autonomous linear system (81) is stable, if the matrix $\mathbf{B}$ of the reduced system is such that $b_{ij} \geq 0$, $i \neq j$, $(i,j) \in [1,m]$, and the Liapunov matrix $\mathbf{L}(t)$ satisfies an estimation

$$\|\mathbf{L}(t)\| \leq a\,\|\exp[\mathbf{B}t]\|^{-1} \quad , \quad \forall t\in T_0 \quad , \quad a\in\check{R}_+ \; .$$

Proof. Statement of the Theorem 18 follows from the presentation of the system (81) solution in the form $\mathbf{u}(t) = \mathbf{X}(t)\,\mathbf{u}_0$ and from (83) for the fundamental matrix $\mathbf{X}(t)$. ■

Definition 20. Non-autonomous M-matrix $\mathbf{P}$ with the continuous bounded elements $p_{ij}(t)$ *has a quasi-dominant diagonal* iff constant d_i exist such that

$$d_j|p_{jj}(t)| > \sum_{\substack{i=1\\ i\neq j}}^{m} d_i|p_{ij}(t)| \ , \quad \forall t\in T_o \ , \ \forall j\in[1,m] \ . \ \blacksquare$$

Theorem 19. The isolated state $\mathbf{u}=\mathbf{0}$ of the non-autonomous system (81) is exponentially stable if the M-matrix $\mathbf{P}$ has continuous, bounded elements and a quasi-dominant diagonal.

Proof. We utilize Šiljak's method (Šiljak, 1978). For a function

$$\nu(\mathbf{u}) = \sum_{i=1}^{m} d_i u_i$$

it is not difficult to find an estimation $\dot{\nu}(\mathbf{u}) \le -\pi\nu(\mathbf{u})$, which holds for every $t\in T_o$. Here $\pi\in\check{R}_+$ and it is determined from the condition

$$|p_{jj}(t)| - d_j^{-1}\sum_{\substack{i=1\\ i\neq j}}^{m} d_i|p_{ij}(t)| \ge \pi \ , \quad \forall t\in T_o \ ,$$

which is valid because of the fact that the M-matrix $\mathbf{P}$ has a quasi-dominant diagonal. Taking into account the fact that $\|\mathbf{u}\| \le |\mathbf{u}|$, where

$$\|\mathbf{u}\| = \Big(\sum_{i=1}^{m} u_i^2\Big)^{1/2} \quad \text{and} \quad |\mathbf{u}| = \sum_{i=1}^{m}|u_i|$$

we define

$$|\mathbf{u}(t)| \le \alpha|\mathbf{u}_o| \exp[-\pi(t-t_o)] \ .$$

Here

$$\alpha = d_M d_m^{-1} \ ; \ d_M = \max\{d_i : i\in[1,m]\} \ ; \ d_m = \min\{d_i : i\in[1,m]\} \ .$$

II.4. MATRIX-FUNCTIONS APPLICATION TO THE STABILITY ANALYSIS

II.4.1. Main properties of matrix-functions

In II.1.4 we introduced the matrix-function $\mathbf{B}$ in $\mathbf{x}$ for an autonomous system (10), Ch.I, and the function $\mathbf{B}$ in $(t,\mathbf{x})$ for a non-autonomous system (7), Ch. I. In this section we consider only the autonomous system

$$\frac{d\mathbf{x}}{dt} = \mathbf{g}(\mathbf{x}) \ , \tag{84}$$

where the state vector $\mathbf{x}\in R^n$, and $\mathbf{g}(\mathbf{x})$ is continuous and satisfies the Lipschitz condition with respect to $\mathbf{x}$ in N; the solution $\mathbf{X}(t;\mathbf{x}_o)$ is unique and it continuously depends on t_o and $\mathbf{x}_o$, $t_o\ge 0$, $\mathbf{x}_o\in \overset{o}{N}$.

Let $\mathcal{U} \subseteq \mathcal{N}$ and $\mathbf{B} : \mathbf{R}^n \to \mathbf{R}^{m\times m}$ be the matrix Liapunov function defined on $\mathcal{U}$ as follows.

Definition 21. A matrix function $\mathbf{B}$ in $\mathbf{x}$ is to be called the *matrix Liapunov function for the system* (84) *on* $\mathcal{U}$ iff the conditions

1) $\mathbf{B}(\mathbf{x}) \in C^1(\mathcal{U})$;
2) $\overset{\circ}{\mathbf{B}}(\mathbf{x}) \leq 0$, $\forall(\mathbf{x} \neq \mathbf{0}) \in \mathcal{U}$, along motions of (84),

are fulfilled. ■

Let us state the properties of the matrix-function $\mathbf{B}$ similar to those which are characteristic to the vector-functions and let us formulate LaSalle's Invariance principle for the given case.

Proposition 21. Let $\mathbf{x},\mathbf{y} \in \mathcal{N}$ and $\mathbf{B}$ be a matrix-function. Then :

a) $\|\mathbf{B}(\mathbf{y}) - \mathbf{B}(\mathbf{x})\| \leq \max\{|u_{ij}(\mathbf{y}) - u_{ij}(\mathbf{x})| : i,j \in [1,m]\}$;

b) $\bar{B}$ is continuous at the point $\mathbf{x}^* \in \mathcal{N}$ if $\mathbf{B}(\mathbf{x})$ is continuous at this point;

c) existence of a neighbourhood $\mathcal{L}(\mathbf{x}) = \mathcal{L}$ of the point $\mathbf{x} \in \mathcal{N}$ such that for $\mathbf{y} \in \mathcal{L}$ there holds $\mathcal{T}(\mathbf{y}) \subset \mathcal{T}(\mathbf{x})$ follows from the continuity of $\mathbf{B}(\mathbf{x})$ on $\mathcal{N}$;

d) $\bar{B}(\mathbf{y}) - \bar{B}(\mathbf{x}) = \max\limits_{(i,j) \in \mathcal{T}(\mathbf{x})} (u_{ij}(\mathbf{y}) - u_{ij}(\mathbf{x}))$ holds from the continuity of $\mathbf{B}$ on $\mathcal{N}$ for $\mathbf{y}$ sufficiently close to $\mathbf{x}$;

e) from the fact that $\bar{B}(\mathbf{x}) \in C^1(\mathcal{U})$ it follows that

$$\dot{\bar{B}}(\mathbf{x}) = \max_{(i,j) \in \mathcal{T}(\mathbf{x})} u_{ij}(\mathbf{x}) = \max_{(i,j) \in \mathcal{T}(\mathbf{x})} \overset{\circ}{u}_{ij}(\mathbf{x}) \ . \ ■$$

Let us denote $\mathcal{E} = \{\mathbf{x} : \dot{B}(\mathbf{x}) \nless 0 , \mathbf{x} \in \bar{\mathcal{U}} \cap \mathcal{N}\}$ and let $\mathcal{M}$ be the largest invariant set in $\mathcal{E}$ and $\mathbf{B}_{\bar{c}} = \{\mathbf{x} : \bar{B}(\mathbf{x}) < c\}$.

Theorem 21 (The Invariance Principle). Let for the system (84) there exist the Liapunov matrix-function $\mathbf{B}$ defined on $\mathcal{U}$ and the solution $\mathbf{x}(t) = \chi(t;\mathbf{x})$ of the system (84) be precompact on $\mathcal{U}$ for all $t \geq 0$. Then $\chi(t;\mathbf{x}) \to \mathcal{M} \cap \mathbf{B}_{\bar{c}}$ for any $0 < c < +\infty$ and $t \to +\infty$. ■

The proof is similar to that given by LaSalle for the vector Liapunov functions (LaSalle, 1976).

II.4.2. Theorems of direct method based on matrix-functions

In order to formulate and prove theorems on stability and instability we state the following

Proposition 22. Let $\Phi(t)$ be a continuous matrix-function on an interval $]a,b[$. Then $\Phi(t)$ is a decreasing on $]a,b[$ matrix-function if $\dot{\Phi}(t) \le 0$ for all $t \in]a,b[$.

Proof. It is sufficient to prove that all elements of the matrix-function $\Phi(t)$ are decreasing functions on $]a,b[$. Really, if $(i,j) \notin T(x)$ then the derivative $D_+u_{ij}(t) < 0$. In case when $(i,j) \in T(x)$, then $D_+u_{ij}(t) \le 0$. In both cases functions $u_{ij}(t)$ will be decreasing on $]a,b[$. ■

Let us state and prove the theorems on stability and instability for the system (84) utilizing the Liapunov matrix-functions.

Theorem 22. If there exists a positive definite Liapunov matrix-function $B : N \to R^{m\times m}$ such that for any $x \in N$ its derivative in virtue of system (84) is $\overset{o}{B}(x) \le 0$, then $x = 0$ of system (84) is stable.

Proof. As far as B is continuous and positive definite on N , the function $\bar{B}$ is also continuous and positive definite on N ; moreover $\bar{B}(0) = 0$. Besides, there exists a function $\varphi_1 \in K_{[o,\alpha[}$ such that $\bar{B}(x) \ge \varphi_1(\|x\|)$. Let there be given $\epsilon > 0$. As far as $\bar{B}(x)$ is continuous and $\bar{B}(0) = 0$, we can find $\delta = \delta(\epsilon) > 0$ such that $\bar{B}(x_o) < \varphi_1(\epsilon)$ for all $x_o \in S_\delta = \{x \in N : \|x\| < \delta\}$. As far as $\overset{o}{B}(\chi(t;x_o)) \le 0$, then, in view of the Proposition 22, $B(\chi(t;x_o))$ is a decreasing matrix-function. Hence, on the basis of the theorem on the Dini derivative, $\dot{\bar{B}}(\chi(t;x_o)) \le 0$ (Rouche, Habets, Laloy, 1977). Then for any $t \ge 0$ we shall obtain

$$\varphi_1(\|\chi(t;x_o)\|) \le \bar{B}(\chi(t;x_o)) \le \bar{B}(x_o) < \varphi_1(\epsilon) ,$$

where $\chi(t;x_o)$ is a solution of the system (1) with the initial condition x_o . As far as $\varphi_1 \in K_{[o,\alpha[}$, then $\|\chi(t;x_o)\| < \epsilon$ for all $t \ge 0$. ■

Theorem 23. If there exists a positive definite Liapunov matrix-function $B : N \to R^{m\times m}$ such that for any $(x \ne 0) \in N$ its derivative in view of the system (84) is $\overset{o}{B}(x) < 0$, then the solution of the system (84) is asymptotically stable.

Proof. Since conditions of the theorem are the intensification of those of the Theorem 1, the solution of system (84) is stable. Let us illustrate that for $x_o \in \overset{o}{N}$ an equality

$$\lim_{t \to +\infty} \|\chi(t;x_o)\| = 0 ,$$

is valid, where $\chi(t;x_o)$ is a non-trivial solution of the system (84).

According to the condition of the theorem there exist functions $\varphi_k \in K_{[o,\alpha[}$, $k = 1,2$ such that

$$\varphi_1(\|x\|) \le \bar{B}(x) \le \varphi_2(\|x\|) \ , \ \forall x \in N \ .$$

Since $\overset{\circ}{B}(x(t,x_o)) < 0$, then $\dot{\bar{B}}(x(t;x_o)) < 0$ and hence, the function $\bar{\Phi}(t) = \bar{B}(x(t;x_o))$ is monotonously decreasing, and since it is bounded from below, it has a finite limit

$$\lim_{t\to+\infty} \bar{\Phi}(t) = \inf_t \bar{\Phi}(t) = \alpha \ge 0 \ . \tag{85}$$

Let us illustrate that the number $\alpha = 0$. In order to show it, we suppose that $\alpha > 0$. Then, in view of continuity of the function $\bar{\Phi}$ the solution $x(t;x_o)$ satisfies an inequality

$$\|x(t;x_o)\| \ge \beta > 0 \quad \text{for} \quad 0 \le t < +\infty \ , \tag{86}$$

where $\beta = \text{const}$. But as far as $\dot{\bar{\Phi}}(t) < 0$, it follows from (86) that $\dot{\bar{\Phi}}(t) \le -b$, $b = \text{const}$. Hence, for all $t > t_o \ge 0$ an inequality

$$\bar{\Phi}(t) = \bar{\Phi}(t_o) + \int_{t_o}^{t} \dot{\bar{\Phi}}(s)\, ds \le \bar{\Phi}(t_o) - b(t-t_o)$$

will be fulfilled. That is impossible, because the right part of the inequality is negative for $t \to +\infty$, which contradicts the positiveness of the matrix $B(x)$. Thus,

$$\lim_{t\to+\infty} \bar{\Phi}(t) = \lim_{t\to+\infty} \bar{B}(x(t;x_o)) = 0 \ .$$

Hence, in consequence of the property of having fixed sign of $\bar{B}(x)$ we obtain the inequality (85). ■

Let us state the theorem on instability similar to that of Chetayev.

Theorem 24. If for the system (84) there exists such a matrix Liapunov function B that :

1) the domain of its positiveness $P = \{x : \bar{B}(x) > 0 \ , x \in N\}$ adjoins the origin of coordinates;
2) an inequality $\bar{B}(x) = 0$ is fulfilled on the set $\partial P \cap N$;
3) $\bar{B}(x)$ is bounded on the domain P ;
4) on the domain P there holds $\overset{\circ}{\bar{B}}(x) \ge \varphi(\bar{B}(x))$, where $\varphi \in K_{[o,\alpha[}$ is some function of Hahn's class K, where $0 < \alpha < +\infty$,

then the unperturbed motion is unstable.

Proof. Let us prove that during the evolution of t one solution of the system (84) will leave the domain N.

In view of condition 1) of the theorem, we can find $x_o \in P \cap S_\delta$ such that $B(x_o) > 0$, where $S_\delta = \{x : x \in N \ , \|x\| < \delta\}$ is an open sphere with the centre in the origin. We consider the solution $x(t;x_o)$. As far as $\bar{B}(\mathbf{x})$ is bounded on the domain P, the function $\bar{\Phi}(t) = \bar{B}(\chi(t;\mathbf{x}_o))$ is also bounded on this domain, i.e. $\bar{\Phi}(t) \le \alpha < +\infty$. Hence, on the basis of the condition 4), the inequality

$$\alpha \ge \bar{\Phi}(t) = \bar{\Phi}(t_o) + \int_{t_o}^{t} \bar{\Phi}(s)\, ds \ge \bar{\Phi}(t_o) + \varphi(\bar{\Phi}(t_o))(t-t_o)$$

is fulfilled as long as $\chi(t;\mathbf{x}_o) \in P$. That is why $\chi(t;\mathbf{x}_o)$ must leave the domain P with the increasing of t. But $\chi(t;\mathbf{x}_o)$ cannot leave P through $\partial P \cap N$ due to the condition 2), hence it leaves N. ∎

Example 15. We consider a system of equations of the perturbed motion

$$\begin{aligned}
\frac{dz_1}{dt} &= -\rho_1^3(z_1^2 + \sin^2(z_2-1) - a)^3 + (z_2 + z_3 + \cos^2(z_2-1) - b)^3 - \\
&\quad - \sin 2(z_2-1)\, \varphi(z_1,z_2,z_3) \\
\frac{dz_2}{dt} &= \varphi(z_1,z_2,z_3) \\
\frac{dz_3}{dt} &= \rho_2^5(z_2 + z_3 + \cos^2(z_2-1) - b)^5 + (z_1 + z_3)^5 + \\
&\quad + (\sin 2(z_2-1) - 1)\, \varphi(z_1,z_2,z_3) - \rho_3(z_1 + z_3)^7
\end{aligned} \tag{87}$$

where

$$\begin{aligned}
\varphi(z_1,z_2,z_3) &= -\rho_2^5(z_2 + z_3 + \cos^2(z_2-1) - b)^5 + (z_1 + z_3)^5 - \\
&\quad - \rho_4(z_1 + \sin^2(z_2-1) - a)^7 - \rho_5(z_2 + z_3 + \cos^2(z_2-1) - b)^7 - \\
&\quad - \rho_1^3(z_1 + \sin^2(z_2-1) - a)^3 + (z_2 + z_3 + \cos^2(z_2-1) - b)^3 ;
\end{aligned}$$

$$a = \sin^2(-1), \quad b = \cos^2(-1), \quad \rho_i > 0, \quad i = 1,2,\dots,5 .$$

We determine restrictions on parameters ρ_i for which the state $z_1 = z_2 = z_3 = 0$ of the system (87) will be stable or asymptotically stable.

We introduce a matrix-function in the form

$$\mathbf{B}(\mathbf{z}) = \{u_{ij}(\mathbf{z})\}_{i,j=1,2,3} , \tag{88}$$

where

$$\begin{aligned}
u_{ij}(\mathbf{z}) &= 0 \qquad \text{for } i \ne j , \; i,j = 1,2,3 ; \\
u_{11}(\mathbf{z}) &= \frac{1}{2}(z_1 + \sin^2(z_2-1) - a)^2 ; \\
u_{22}(\mathbf{z}) &= \frac{(1+\epsilon)^2}{2\rho_1^2}(z_2 + z_3 + \cos^2(z_2-1) - b)^2 ; \\
u_{33}(\mathbf{z}) &= \frac{(1+\epsilon)^4}{2\rho_1^2\rho_2^2}(z_1 + z_3)^2 .
\end{aligned}$$

We consider sets

$$\Omega_1 = \{\mathbf{z} ; \mathbf{z} \in R^3 : (u_{11}^{1/2} \ge u_{22}^{1/2} \wedge u_{33}) \wedge (u_{11} \ge u_{22} \wedge u_{11} \ge u_{33})\} ,$$

$$\Omega_2 = \{\mathbf{z} ; \mathbf{z} \in \mathbf{R}^3 : (u_{11}^{1/2} \le u_{22}^{1/2} \wedge u_{11}^{1/2} \le u_{33}^{1/2}) \wedge (u_{11} \ge u_{22} \wedge u_{11} \ge u_{33})\} ,$$

$$\Omega_3 = \{\mathbf{z} ; \mathbf{z} \in \mathbf{R}^3 : (u_{22}^{1/2} \ge u_{11}^{1/2} \wedge u_{22}^{1/2} \ge u_{33}^{1/2}) \wedge (u_{22} \ge u_{11} \wedge u_{22} \ge u_{33})\} ,$$

$$\Omega_4 = \{\mathbf{z} ; \mathbf{z}\ \mathbf{R}^3 : (u_{22}^{1/2} \le u_{11}^{1/2} \wedge u_{22}^{1/2} \le u_{33}^{1/2}) \wedge (u_{22} \ge u_{11} \wedge u_{22} \ge u_{33})\} ,$$

$$\Omega_5 = \{\mathbf{z} ; \mathbf{z}\ \mathbf{R}^3 : (u_{33}^{1/2} \ge u_{11}^{1/2} \wedge u_{33}^{1/2} \ge u_{22}^{1/2}) \wedge (u_{33} \ge u_{11} \wedge u_{33} \ge u_{22})\} ,$$

$$\Omega_6 = \{\mathbf{z} ; \mathbf{z}\ \mathbf{R}^3 : (u_{33}^{1/2} \le u_{11}^{1/2} \wedge u_{33}^{1/2} \le u_{22}^{1/2}) \wedge (u_{33} \ge u_{11} \wedge u_{33} \ge u_{22})\} ,$$

$$\Omega_7 = \{\mathbf{z} ; \mathbf{z}\ \mathbf{R}^3 : (u_{11}^{1/2} \ge u_{22}^{1/2})\} ,$$

$$\Omega_8 = \{\mathbf{z} ; \mathbf{z}\ \mathbf{R}^3 : (u_{11}^{1/2} \le u_{22}^{1/2})\} .$$

On the set $\Omega_1 \cup \Omega_2$, $\bar{B}(\mathbf{z}) = u_{11} > 0$, and the total derivative of $\mathbf{B}(\mathbf{z})$ is

$$\overset{\circ}{\mathbf{B}}(\mathbf{z}) = \begin{bmatrix} \dot{u}_{11} & -\infty & -\infty \\ -\infty & -\infty & -\infty \\ -\infty & -\infty & -\infty \end{bmatrix} .$$

On the set $\Omega_3 \cup \Omega_4$, $\bar{B}(\mathbf{z}) = u_{22} > 0$, and the total derivative of $\mathbf{B}(\mathbf{z})$ is

$$\overset{\circ}{\mathbf{B}}(\mathbf{z}) = \begin{bmatrix} -\infty & -\infty & -\infty \\ -\infty & \dot{u}_{22} & -\infty \\ -\infty & -\infty & -\infty \end{bmatrix} .$$

On the set $\Omega_5 \cup \Omega_6$, $\bar{B}(\mathbf{z}) = u_{33} > 0$, and the total derivative of $\mathbf{B}(\mathbf{z})$ is

$$\overset{\circ}{\mathbf{B}}(\mathbf{z}) = \begin{bmatrix} -\infty & -\infty & -\infty \\ -\infty & -\infty & -\infty \\ -\infty & -\infty & \dot{u} \end{bmatrix} .$$

It is not difficult to notice that for derivatives $\dot{u}_{ii}(\mathbf{z})$, $i = 1,2,3$, the following estimations hold

$$\dot{u}_{11}(\mathbf{z}) \le (z_1 + \sin^2(z_2-1) - a)^4 \frac{(1 - (1+\epsilon)^3)\, \rho_1^3}{(1+\epsilon)^3} , \quad \forall z \in \Omega_1 ;$$

$$\dot{u}_{22}(\mathbf{z}) \le (z_2 + z_3 + \cos^2(z_2-1) - b)^6 \frac{(1 - (1+\epsilon)^5)\, \rho_2^5}{\rho_1^2 (1+\epsilon)^3} , \quad \forall z \in \Omega_3 ;$$

$$\dot{u}_{33}(z) \le (z_1+z_3)^8 \frac{(1+\epsilon)^4}{\rho_1^2\rho_2^2}\left(\rho_4\frac{(1+\epsilon)^{14}}{\rho_1^7\rho_2^7}+\rho_5\frac{(1+\epsilon)^7}{\rho_2^7}-\rho_3\right), \quad \forall z\in\Omega_5$$

$$\dot{u}_{11}(z) \le \frac{1+\epsilon}{\rho_1}(z_2+z_3+\cos^2(z_2-1)-b)^4(1-(1+\epsilon)^3), \quad \forall z\in\Omega_2$$

$$\dot{u}_{22}(z) \le \frac{(1+\epsilon)^3}{\rho_1^2\rho_2}(z_1+z_3)^6(1-(1+\epsilon)^5), \quad \forall z\in\Omega_4$$

$$\dot{u}_{33}(z) \le \frac{(1+\epsilon)^4}{\rho_1\rho_2}\left(\rho_4+\rho_5\frac{\rho_1^7}{(1+\epsilon)^7}-\rho_3\frac{\rho_1^7\rho_2^7}{(1+\epsilon)^{14}}\right)(z_1+\sin^2(z_2-1)-a)^8, \quad \forall z\in\Omega_7\cap\Omega_6;$$

$$\dot{u}_{33}(z) \le \frac{(1+\epsilon)^4}{\rho_1^2\rho_2}\left(\frac{\rho_4}{\rho_1^7}(1+\epsilon)^6+\frac{\rho_5}{1+\epsilon}-\rho_3\frac{\rho_2^7}{(1+\epsilon)^8}\right)(z_2+z_3+\cos^2(z_2-1)-b)^8, \quad \forall z\in\Omega_8\cap\Omega_6. \tag{89}$$

Let us denote

$$\psi = \frac{\rho_4}{\rho_1^7\rho_2^7}+\frac{\rho_5}{\rho_2^7}. \tag{90}$$

Taking into account that

$$\bigcup_{i=1}^{6}\Omega_i = R^3,$$

it follows from estimations (89) that cases $\rho_3>\psi$ and $\rho_3=\psi$ correspond to the asymptotic stability and stability of the equilibrium state $z=0$. Namely, if $\rho_3>\psi$ we can determine $\epsilon>0$ such that $B(z)>0$ and $\mathring{B}(z)<0$, $\forall z\in R^3$. If $\rho_3=\psi$, then, assuming that $\epsilon=0$, we obtain $B(z)>0$ and $\mathring{B}(z)=0$, $\forall z\in R^3$.

II.4.3. The scalar Liapunov function construction on the basis of matrix-functions

Let for the system (84) there be constructed a matrix Liapunov function

$$B(x) = \{u_{ij}(x)\}_{i,j=1}^{m}, \quad u_{ij}(\mathbf{0})=0, \quad \forall(i,j)\in[1,m], \tag{91}$$

and

a) $$\bar{B}(x) = \max_{i,j} u_{ij}(x), \quad \forall(i,j)\in[1,m]; \tag{92}$$

b) the derivative of B along solutions of the system (84) obeys

$$\mathring{B}(x) = \liminf_{t\to0^+}\frac{1}{t}\,(B(\chi(t;x))-\bar{B}(x)\,U), \tag{93}$$

where U is an $m\times m$-matrix with the elements $U_{ij}=1$, $\forall(i,j)\in[1,m]$;

c) the function

$$v(x) = c^T B(x)\,c, \tag{94}$$

where $c = c_1,\dots,c_m{}^T\in R^m$, $c_i\neq0$ are constant scalars, $i=1,2,\dots,m$;

d) the derivative of the function (84) has the form

$$D_+v(x) = c^T\mathring{B}(x)\,c, \tag{95}$$

where $\overset{\circ}{B}(x)$ is determined by the formula (93).

For a statement of the basic theorems on stability of the state $x=0$ of the system (84) on the basis of the matrix-function (91) we have the following possibility : at first, the immediate application of the function (91) and its derivative (93). Secondly, application of the function (94) and its derivative (95) permits us, trying to preserve the general form of the well-known criteria of stability, to consider a matrix-function to be an initial one, i.e. to take the statement of criteria of stability in terms of existence of the matrix-function B .

II.4.3.1. Theorems on stability (stability in the whole)

The criterion of stability, the basis of which constitutes the first theorem on stability of Liapunov's direct method, is widely-known.

Basing on the concept of the matrix-function we state for the given case the following

Theorem 25. Let the system (84) be such that :

1) the matrix-function $B : N \to R^{m\times m}$ exists on the domain N;
2) there exists an m-dimensional vector $c \neq 0$ (componentwise);
3) the function v (94) is positive definite on N;
4) the total derivative D_+v (95) is :
   a) negative semi-definite on N; or
   b) identically equal to zero on N,

   along the system (84).

Then the equilibrium state $x=0$ of the system (84) is stable.

If all conditions of the Theorem 25 are fulfilled for $N=R^n$ and the function (94) is radially unbounded, then the equilibrium state $x=0$ of system (84) is stable in the whole. ■

The proof of the theorem is similar to the known ones (Krasovskii, 1959; Hahn, 1967).

Let there exist a domain $\Omega = \{x : x \in N , 0 \leq v(x) < a , a \in \check{R}_+\} \subseteq N$ over which $D_+v(x) \leq 0$. We denote by M the largest invariant set in Ω on which $D_+v(x) = 0$.

Theorem 26. Let the system (84) be such that :

1) conditions 1)-3) of Theorem 25 are fulfilled;
2) on the set Ω the total derivative (95) satisfies the condition $D_+v(x) \leq 0$.

Then the set M is attractive with respect to the domain Ω , i.e. all motions originating from the domain Ω tend to the set M for $t \to +\infty$. ∎

The proof of the theorem is similar to that of Theorem 26.1 (Hahn, 1967) taking into account the fact that the function (94) is scalar.

II.4.3.2. Theorems on asymptotic stability

Theorem 27. Let the system (84) be such that along its solutions the following holds :

1) conditions 1)-3) of the Theorem 25 are fulfilled;
2) the total derivative (95) is negative definite.

Then the equilibrium state $x = 0$ of the system (84) is asymptotically stable.

If all conditions of the Theorem 27 are fulfilled for $N = R^n$ and the function (94) is radially unbounded, then the equilibrium state $x = 0$ of the system (84) is asymptotically stable in the whole.

Proof. In view of the condition 1) of Theorem 27 for the system (84) we construct the function v in $\mathbf{x}$, which is positive definite on N and its total derivative (95) is negative definite. Hence, all conditions of Theorem 25.2 (Hahn, 1967) are fulfilled, which proves the statement of Theorem 27. ∎

Let G be the largest *invariant set* in R^n , which does not contain the entire trajectories of the system (84), except the point $x = 0$.

Theorem 28. Let the system (84) be such that :

1) conditions 1)-3) of Theorem 25 are fulfilled for $N = R^n$;
2) function (94) is radially unbounded;
3) along the solutions of the system (84) the total derivative (95) is
   a) negative definite outside G ;
   b) negative semi-definite on G ;

Then the equilibrium state $x = 0$ of the system (84) is asymptotically stable in the whole. ∎

II.4.3.3. Theorems on instability

At first we state two theorems on instability, generalizing Liapunov's theorems on instability (Liapunov, 1935).

Suppose that there exist positive constants g' , g and the function w such that

$$D_+ v(x) = g' \, c^T B(x) \, c \tag{96}$$

or
$$D_+v(x) = g\, c^T B(x)\, c + w(x) \; . \tag{97}$$

Functions v, $v(x) = c^T B(x) c$, and w are not semi-definite and of the opposite sign iff $\forall (x \neq 0) \in N$ the following two relationships

$$(v(x) \geq 0 \;\&\; w(x) \leq 0) \quad ; \quad (v(x) \leq 0 \;\&\; w(x) \geq 0)$$

do not hold.

Theorem 29. Let the system (84) be such that :

1) the conditions 1)-2) of Theorem 25 are fulfilled;
2) the derivative (95) along the solutions of the system (84) has a fixed sign;
3) the function v (94) is not semi-definite on N of the sign opposite to D_+v (95).

Then the equilibrium state $x = 0$ of the system (84) is unstable. ∎

Theorem 30. Let the system (84) be such that :

1) conditions 1)-2) of Theorem 25 are fulfilled;
2) there exists a positive constant g and a function w such that the derivative D_+v (95) of the function v (94) has the form (97); moreover the function w either identically vanishes or it is semi-definite, and in the latter case the function v is not semi-definite of the sign opposite to $w(x)$ on N.

Then the equilibrium state $x = 0$ of the system (84) is unstable. ∎

The well-known Krasovskii's theorem on instability also admits a generalization on the basis of the matrix Liapunov function.

Let G^* be a set in N, which does not contain the entire trajectories of the system (84) except the point $x = 0$. ∎

Theorem 31. Let the system (84) be such that :

1) the conditions 1)-2) of Theorem 25 are fulfilled;
2) the function v (94) is not negative semi-definite on N;
3) the derivative D_+v (95) of function v (94) is positive semi-definite on G^* and positive outside G^*.

Then the equilibrium state $x = 0$ of the system (84) is unstable. ∎

II.4.3.4. Applications

Let us consider the known problem of stability in the product space, which leads to a consideration of the system of differential equations of the perturbed motion

$$\text{a)} \quad \frac{d\mathbf{y}}{dt} = Y(\mathbf{y}) + Y^*(\mathbf{y};\mathbf{z}) \; ;$$
$$\text{b)} \quad \frac{d\mathbf{z}}{dt} = Z(\mathbf{z}) + Z^*(\mathbf{y};\mathbf{z}) \; . \tag{98}$$

Here $\mathbf{y} \in R^p$, $\mathbf{z} \in R^q$, $Y : R^p \to R^p$, $Y^* : R^p \times R^q \to R^p$, $Z : R^q \to R^q$, $Z^* : R^p \times R^q \to R^q$; in addition, functions $Y, Z \; ; Y^*, Z^*$ are continuous on $R^p, R^q \; ; R^p \times R^q$ and vanish for $\mathbf{y} = \mathbf{z} = \mathbf{0}$. The problem is to determine in what manner stability of the equilibrium state $\mathbf{x} = \mathbf{y} = \mathbf{0}$ of the system (98) on $R^p \times R^q$ is connected with that of nonlinear approximation

$$\text{a)} \quad \frac{d\mathbf{y}}{dt} = Y(\mathbf{y}) \; ,$$
$$\text{b)} \quad \frac{d\mathbf{z}}{dt} = Z(\mathbf{z}) \; . \tag{99}$$

Here functions Y, Z are the same as in the system (98). Let us introduce some assumptions.

Assumption 9. Let there exist time-invariant neighbourhoods $N_y \subseteq R^p$ and $N_z \subseteq R^q$ of the equilibrium states $\mathbf{y} = \mathbf{0}$ and $\mathbf{z} = \mathbf{0}$ consequently, and the matrix-function

$$\mathbf{B}(\mathbf{y},\mathbf{z}) = \begin{bmatrix} v_{11}(\mathbf{y}) & v_{12}(\mathbf{y},\mathbf{z}) \\ v_{21}(\mathbf{y},\mathbf{z}) & v_{22}(\mathbf{z}) \end{bmatrix} , \tag{100}$$

with elements v_{ij} ; $i,j = 1,2$, which satisfy estimations characteristic to the quadratic forms

$$\begin{aligned} v_{11}(\mathbf{y}) &\ge c_{11}\|\mathbf{y}\|^2 && \forall(\mathbf{y} \ne \mathbf{0}) \in N_y \; ; \\ v_{22}(\mathbf{z}) &\ge c_{22}\|\mathbf{z}\|^2 && \forall(\mathbf{z} \ne \mathbf{0}) \in N_z \; ; \\ v_{12}(\mathbf{y},\mathbf{z}) &\ge c_{12}\|\mathbf{y}\|\|\mathbf{z}\| && \forall(\mathbf{y} \ne \mathbf{0} , \mathbf{z} \ne \mathbf{0}) \in N_y \times N_z \; ; \\ v_{21}(\mathbf{y},\mathbf{z}) &= v_{12}(\mathbf{y},\mathbf{z}) && \forall(\mathbf{y} \ne \mathbf{0} , \mathbf{z} \ne \mathbf{0}) \in N_y \times N_z \; . \end{aligned} \tag{101}$$

Assumption 10. Let there exist constants ρ_{ij} , $i = 1,2$; $j = 1,2,\dots,8$ such that on N_y, N_z or on $N_y \times N_z$:

$$\begin{aligned} &\left(\frac{\partial v_{11}}{\partial \mathbf{y}} , Y\right) \le \rho_{11}\|\mathbf{y}\|^2 \; ; \quad \left(\frac{\partial v_{11}}{\partial \mathbf{y}} , Y^*\right) \le \rho_{12}\|\mathbf{y}\|^2 + \rho_{13}\|\mathbf{y}\|\|\mathbf{z}\| \; ; \\ &\left(\frac{\partial v_{22}}{\partial \mathbf{z}} , Z\right) \le \rho_{21}\|\mathbf{z}\|^2 \; ; \quad \left(\frac{\partial v_{22}}{\partial \mathbf{z}} , Z^*\right) \le \rho_{22}\|\mathbf{z}\|^2 + \rho_{23}\|\mathbf{y}\|\|\mathbf{z}\| \; ; \\ &\left(\frac{\partial v_{12}}{\partial \mathbf{y}} , Y\right) \le \rho_{14}\|\mathbf{y}\|^2 + \rho_{15}\|\mathbf{y}\|\|\mathbf{z}\| \; ; \\ &\left(\frac{\partial v_{12}}{\partial \mathbf{y}} , Y^*\right) \le \rho_{16}\|\mathbf{y}\|^2 + \rho_{17}\|\mathbf{y}\|\|\mathbf{z}\| + \rho_{18}\|\mathbf{z}\|^2 \; ; \end{aligned} \tag{102}$$

$$(\frac{\partial v_{12}}{\partial z}, Z) \le \rho_{24}\|z\|^2 + \rho_{25}\|y\|\,\|z\| \;;$$

$$(\frac{\partial v_{12}}{\partial z}, Z^*) \le \rho_{26}\|y\|^2 + \rho_{27}\|y\|\,\|z\| + \rho_{28}\|z\|^2 \;.$$

Theorem 32. Let for the system (98) :

1) all conditions of Assumptions 9, 10 be fulfilled;

2) the matrix

$$G = \begin{bmatrix} c_{11} & c_{12} \\ c_{21} & c_{22} \end{bmatrix}, \quad c_{12} = c_{21},$$

be positive definite;

3) the matrix

$$S = \begin{bmatrix} \sigma_{11} & \sigma_{12} \\ \sigma_{21} & \sigma_{22} \end{bmatrix}, \quad \sigma_{12} = \sigma_{21},$$

be negative definite, where (Djordjević, 1983)

$$\sigma_{11} = \eta_1^2(\rho_{11}+\rho_{12}) + 2\eta_1\eta_2(\rho_{14}+\rho_{16}+\rho_{26}),$$

$$\sigma_{22} = \eta_2^2(\rho_{21}+\rho_{22}) + 2\eta_1\eta_2(\rho_{18}+\rho_{24}+\rho_{28}),$$

$$\sigma_{12} = \frac{1}{2}(\eta_1^2\rho_{13} + \rho_{23}\eta_2^2) + \eta_1\eta_2(\rho_{15}+\rho_{25}+\rho_{17}+\rho_{27}), \quad \eta_1,\eta_2 > 0 \qquad (103)$$

Then the equilibrium state $\mathbf{y} = \mathbf{z} = \mathbf{0}$ of system (98) is asymptotically stable.

If the conditions of Assumptions 9, 10 are fulfilled for $N_y = R^p$, $N_z = R^q$ and the conditions 2)-3) of the theorem hold, then the equilibrium state $\mathbf{y} = \mathbf{z} = \mathbf{0}$ of the system (98) is asymptotically stable in the whole. ∎

The proof is similar to that of (Djordjević, 1983). On the basis of estimations (101) it is not difficult to illustrate that the estimation from below

$$v \ge u^T C^T G C u, \qquad (104)$$

where

$$u^T = (\|y\|, \|z\|) \quad ; \quad C = \begin{bmatrix} \eta_1 & 0 \\ 0 & \eta_2 \end{bmatrix},$$

holds for the function $v(y,z) = \eta^T B(y,z)\eta$, $\eta = (\eta_1\ \eta_2)^T$.

On the basis of estimations (101) for a derivative $\dot{v}(\mathbf{y},\mathbf{z})$ of the function v defined by the expression

$$\dot{v}(y,z) = \eta^T \dot{B}(y,z)\eta$$

it is not difficult to find the estimation

$$\dot{v}(y,z) \le u^T S u \ . \tag{105}$$

In view of the conditions 2)-3) of Theorem 32 and estimations (104), (105) all conditions of the Theorem 27 are fulfilled by the function v and its derivative.

If in estimations (101) we replace the sign of inequality by the opposite one, then by using the method similar to the one given above, we can obtain an estimation

$$\dot{v}(y,z) \ge u^T S' u \ ,$$

which allows us to state conditions of instability of the equilibrium state $y = z = 0$ of the system (84) on the basis of the Theorem 29.

Theorem 32 illustrates that the asymptotic stability of the equilibrium state $y = z = 0$ of the system (98) can hold even in case when the equilibrium state $y = 0$ of the system (99)-a) and $z = 0$ of the system (99)-b) has no property of asymptotic quasi-stability of Theorem 6.

COMMENTS ON REFERENCES

1. Fundamentals of the theory of differential and integral inequalities are contained in the works by Chaplygin (1919), Gronwall (1919), Ważewski (1950), Azbelev (1956), Azbelev, Tsalyuk (1958, 1962, 1964), Alekseyev (1965), Zubov (1957), Luzin (1951), Matrosov (1967), Melnikov (1975), Antosiewicz (1958,1965), Yoshizawa (1966), Conti (1956), Hahn (1967), et al. Numerous results are summarized in monographs by : Walter (1964), Bekenbach, Bellman (1965), Šzarski (1967), Lakshmikantham, Leela (1969), Mitrinović (1970), Beesack (1975), Rabczuk (1976), Filatov, Sharova (1976), Martynyuk, Gutowski (1979).

2. Definitions in 1.3 are stated by taking into account some results of the works given in 3 of Notes and References to Chapter I. The vector Liapunov functions were a topic of analysis in a number of works [Bellman (1962), Matrosov (1962), Deo (1971), Lakshmikantham (1965, 1963), LaSalle (1975), et al.], and they found a wide application while analyzing stability of large-scale systems (see 1 of Notes and References to Chapter III).

3. The Liapunov matrix functions originally have arisen in consideration of the derivatives higher than of the first order of the vector Liapunov functions. As far as the authors know, for the first time

they were described by Martynyuk in the book [Martynyuk, Gutowski (1979); see also Martynyuk (1984a)].

4. The comparison principle in the motion stability theory is developed on the basis of both scalar Liapunov functions [see *An Outline* by Martynyuk (1985)] and vector functions [see monographs by Grujić (1972), Martynyuk (1975), Michel, Miller (1977), Šiljak (1978), Matrosov, Anapolskii, Vasilyev (1980)]. Let us also mention papers by Bitsoris (1976), Bitsoris, Burgat (1976), Gentina, Borne, Burgat, Bernussou, Grujić (1979), Grujić, Gentina, Borne, Burgat, Bernussou (1978), in which the comparison principle is also defined via vector norms.

5. Integral and differential inequalities, considered in 2.2, were the topic of analysis in the works by Gutowski, Radziszewski (1970,1971), Martynyuk, Gutowski (1979), Tikhonov (1965,1969).

6. In the Theorem 6 Persidskii's concept of quasi-stability, which was applied by Matrosov (1963) while formulating his theorems of stability on the basis of the comparison method, is not utilized here. Approach, developed in the present book, is based on the idea of an initial system extension by the auxiliary scalar or vector equation [Hatvani (1975), Martynyuk (1975,1984b)].

7. Statements 14.a) and b) are proved in papers by Sevastyanov (1951), Kotelyanskii (1952); 14.c) in the papers by Fiedler and Ptak (1962) and Persidskii (1969); 14.d) in that by Araki (1975) and also LaSalle (1976).

8. The Theorem 11 is proved independently in both papers by Martynyuk, Obolensky (1978) and Burgat, Bernussou, Grujić, Gentina, Borne (1978).

9. Without the monotonicity property of the right part of the non-stationary system, criteria of autonomous comparison systems stability have not a direct generalization to non-autonomous ones.

10. In 4 the results of papers by Martynyuk (1984,1985) are presented, in which ideas of the Liapunov matrix-functions application to the stability theory is developed. Illustrative results are derived jointly with Shegai (see Martynyuk, Shegai, 1986).

Alekseyev, V.M. (1965), Theorem on an integral inequality and several of its applications. *Mat. Sbornik, 68*, No.2, 251-273 (in Russian).

Aminov, M.Sh. (1959), Some problems of motion and stability of a rigid body of a variable mass. *Tr. Kazan. aviats. in-ta. Ser. Matematika i Mekhanika, X.*

Antosiewicz, H.A. (1958), A survey of Lyapunov's second method. In *Contributions to the Theory of Nonlinear Oscillations*. Princeton Univ. Press, 141-166.

Antosiewicz, H.A. (1965), Recent contributions to Lyapunov's second method. *Colloq. Int. CNRS, 148*, No.2, 29-37.

Araki, M. (1975), Application of M-matrices to the stability problems of composite dynamical systems. *J. Math. Anal. and Appl., 52*, No.2, 309-321.

Azbelev, N.V. (1956), On boundaries of Chaplygin's theorem on differential inequalities applicability. *Mat. Sbornik, 39*, No.2, 161-178 (in Russian).

Azbelev, N.V., and Z.B. Tsalyuk (1958), On Chaplygin's problem. *Ukr. Mat. Zhurnal, 10*, No.1, 3-12 (in Russian).

Azbelev, N.V., and Z.B. Tsalyuk (1962), On integral inequalities, I. *Mat. Sbornik, 58*, No.3, 325-342 (in Russian).

Azbelev, N.V., and Z.B. Tsalyuk (1964), On differential and integral inequalities. *Proc. of the 4th Math. Congress, Moscow, 2*, 384-391 (in Russian).

Barbashin, Ye.A. (1970), *Liapunov's Function*. Nauka, Moscow (in Russian).

Beckenbach, E.F., and R. Bellman (1965), *Inequalities*, second revised printing, Springer Verlag, Berlin.

Beesack, P.R. (1975), *Gronwhall Inequalities*. Cavleton Math. Lecture Notes, No.11.

Bellman, R. (1954), *Theory of Stability of Ordinary Differential Equations Solutions*. Izd-vo Inostr. Lit., Moscow (in Russian).

Bellman, R. (1962), Vector Lyapunov functions. *J. SIAM Control*, Ser.A, *1*, No.1, 32-34.

Bitsoris, G. (1976), Sur la stabilité asymptotique des systèmes interconnectés. *C.R. Acad. Sci. A, 282*, 301.

Bitsoris, G., and C. Burgat (1976), Stability conditions and estimates of stability region of complex systems. *Int. J. Systems Sci., 7*, No.8, 128-135.

Burgat, C., J. Bernussou, Lj.T. Grujić, P. Borne, and J.C. Gentina (1978), Les perturbations structurelles arbitraires et périodiques. *RAIRO Automat./Syst. Anal. and Contr., 12*, No.3, 245-267.

Cesari, L. (1964), *Asymptotic Behaviour and Stability of Ordinary Differential Equations Solutions*. Mir, Moscow (in Russian).

Chaplygin, S.A. (1919), New method of integration of a general differential equation of the train motion. *Byul. Nauch.-Eksperim. in-ta Putej Soobshchen, 1a*, No.9, 308-334 (in Russian).

Chaplygin, S.A. (1976), New method of approximate integration of differential equations. *Selected Works*, Nauka, Moscow, 307-360 (in Russian).

Conti, R. (1956), Sulla prolungabilita della soluzioni di un sistema di equazioni differenziali ordinarie. *Boll. Unione Mat. Ital.*, Ser. III, *Anno XI*, No.4, 510-514.

Corduneanu, K. (1960), On application of differential inequalities to the theory of stability. *Analele Stint. ale Univ. "Al. i Cusa" din Tasi*, sec.1, *6*, No.1, 47-58.

Demidović, B.P. (1967), *Lectures on Mathematical Theory of Stability*. Nauka, Moscow (in Russian)..

Deo, S.G. (1971), On vector Lyapunov functions. *Proc. Amer. Math. Soc., 31*, No.3, 575-580.

Djordjević, M.Z. (1983), *Stability Analysis of Large-Scale Systems whose Subsystems may be Unstable*. EHT, Zurich, Report No. 83-01.

Erugin, N.P. (1946), Reduced systems. *Tr. Mat. in-ta AN SSSR, 13*, No.3, 3-140 (in Russian).

Fiedler, M., and V. Ptak (1962), On matrices with non-positive off-diagonal elements and positive principal minors. *Czech. Nat. J., 12*, No.87, 382-400.

Filatov, A.N., and L.V. Sharova (1976), *Integral Inequalities and the Theory of Nonlinear Oscillations*. Nauka, Moscow (in Russian).

Gentina, J.C., P. Borne, C. Burgat, J. Bernussou, and Lj.T. Grujić (1979), Sur la stabilité des systèmes de grande dimension, normes vectorielles. *RAIRO Automat./Syst. Anal. and Control, 13*, No.1, 57-75.

Goncharenko, V.I., L.G. Lobas, and N.V. Nikitina (1981), On a single statement of the problem on shimmi of castoring wheels. *Prikl. Mekhanika, 17*, No.8, 82-88 (in Russian).
Grebenikov, Ye.A., and Yu.A. Ryabov (1979), *Constructive Methods of Nonlinear Systems Analysis*. Nauka, Moscow (in Russian).
Gronwall, T.H. (1919), A note on the derivatives with respect to a parameter of the solutions of a system of differential equations. *Ann. Math., 20*, No.2, 292-296.
Grujić, Lj.T. (1972), *Large-Scale Systems Stability*. Ph.D. dissertation, Faculty of Mechanical Engineering, Belgrade (in Serbo-Croatian). (Published in 1974)
Gutowski, R. (1976), *Rownania Rozniczkowe Zwycajne*. Wyd-wo Naukowo-Techniczne, Warsawa.
Gutowski, R., and B. Radziszewski (1970), Asymptotic behaviour and properties of solutions of a system of non-linear second order ordinary differential equations describing motion of mechanical systems. *Arch. Mech. Stosow, 22*, No.6, 675-694.
Gutowski, R., and B. Radziszewski (1971), Asymptotic behaviour and properties of solutions of nonlinear ordinary differential equations of first order describing the motion of a mechanical system. *Ibid., 23*, No.1, 17-25.
Hahn, W. (1967), *Stability of Motion*. Springer Verlag, New York.
Hatvani, L. (1975), On application of differential inequalities to the theory of stability. *Vestn. Mosk. un-ta. Ser. Matem., Mekhanika*, No.3, 83-89 (in Russian).
Hicks, J.R. (1939), *Value and Capital*. Oxford University Press, Oxford, England.
Keldysh, M.V. (1945), Shimmi of the driving wheel of three-wheeled shassi. *Tr. Tzentr. Aerogidro-Dinam. in-ta*, No.564, 1-33 (in Russian).
Kotelyanskii, I.N. (1952), On some properties of matrices with positive elements. *Mat. Sbornik*, No.31, 497-506 (in Russian).
Krasnoselskii, M.A. (1962), *Positive Solutions of Operator Equations*. Fizmatgiz, Moscow (in Russian).
Krasovskii, N.N. (1959), *Certain Problems of the Motion Stability Theory*. Fizmatgiz, Moscow (in Russian); English translation (1963), *Stability of Motion*, Stanford Univ. Press, Stanford.
Lakshmikantham, V. (1963), Differential equations in Banach spaces extension of Lyapunov's method. *Boc. Cambridge Philos. Soc., 59*, No.2, 373-381.
Lakshmikantham, V. (1964), Functional differential systems and extension of Lyapunov's method. *J. Math. Anal. Appl., 8*, 392-405.
Lakshmikantham, V. (1965), Vector Lyapunov functions and conditional stability. *J. Math. Anal. Appl., 10*, No.2, 368-377.
Lakshmikantham, V., and S. Leela (1969), *Diffenrential and Integral Inequalities, Vol.I.* Acad. Press, New York.
LaSalle, J.P. (1975), Vector Lyapunov functions. *Bull. Inst. Math. Acad. Sinica, 3*, No.1, 139-150.
LaSalle, J.P. (1976), *The Stability of Dynamical Systems*. SIAM, Philadelphia.
Lazaryan, V.A., M.L. Korotenko, Yu.V. Demin, and N.A. Radchenko (1969), Stability of a coach with the elastic coupling of the pairs of wheels with the bogie frame. *Tr. Dnepropetr. in-ta Inzh. Transp.*, No.84, 3-8 (in Russian).
Lefschetz, S. (1957), *Differential Equations : Geometric Theory*. Interscience Publishers.
Liapunov, A.M. (1935), *General Problem of Stability of Motion*. ONTI, Moscow-Leningrad (in Russian) (The original was published 1892).
Lopatinskii, Ya.B. (1980), *Introduction to the Modern Theory of Differential Equations in Partial Derivatives*. Nauk. Dumka, Kiev (in Russian).
Luzin, N.N. (1951), On the method of acad. S.A. Chaplygin's approximate integration. *Uspekhi Mat. Nauk., 6*, No.6, 3-27 (in Russian).
Martynyuk, A.A. (1973), *Technical Stability in Dynamics*. Tekhnika, Kiev (in Russian).
Martynyuk, A.A. (1975), *Stability of Motion of Complex Systems*. Nauk. Dumka, Kiev (in Russian).
Martynuuk, A.A. (1984a), The Lyapunov matrix-function. *Nonlinear Analysis : Theory, Methods and Applications, 8*, No.10, 1223-1226.

Martynyuk, A.A. (1984b), Extension of the space state of dynamic systems and the problem of stability. *Colloquium on Qualitative Theory of Differential Equations, Szeged, August 27-31*, 58-59.

Martynyuk, A.A. (1985), Scalar equations of comparison in the stability of motion theory. *Prikl. Mekhanika, 21*, No.8, 3-19 (in Russian).

Martynyuk, A.A. (1986), On application of the Lyapunov matrix-functions in the theory of stability. *J. Nonlinear Analysis : Theory, Methods and Appl., 9*, No.12, 1495-1501.

Martynyuk, A.A., and R. Gutowski (1979), *Integral Inequalities and Stability of Motion*. Nauk. Dumak, Kiev (in Russian).

Martynyuk, A.A., and N.V. Nikitina (1981), On nonlinear systems of comparison in stability problems of large-scale systems. *Prikl. Mekhanika, 17*, No.12, 97-102 (in Russian).

Martynyuk, A.A., and N.V. Nikitina (1982), The comparison method in the problem of shimmi. *Ibid, 18*, No.11, 102-107 (in Russian).

Martynyuk, A.A., and A.Yu. Obolenskii (1978), *Stability of Autonomous Systems of Comparison Analysis*. Izd-vo in-ta Matemat., Kiev, An Ukr. SSR, Preprint 78.28 (in Russian).

Martynyuk, A.A., and A.Yu. Obolenskii (1980), On stability of autonomous Ważewskii's systems solutions. *Differents-Uravnenia, 16*, No.8, 1392-1407 (in Russian).

Martynyuk, A.A., and V.V. Segai (1986), To the theory of stability of the autonomous systems. *Prikl. Mekhanika, 22*, No.4, 97-102 (in Russian).

Martynyuk, A.A., and V.G. Verbitskii (1982), The domain of attraction estimation for nonlinear systems of the definite form. *Prikl. Mekhanika, 18*, No.10, 102-107 (in Russian).

Matrosov, V.M. (1962), To the theory of stability of motion. *Prikl. Matem. i Mekhanika, 26*, No.6, 992-1002 (in Russian).

Matrosov, V.M. (1963), To the theory of stability of Motion. II. *Tr. Kazan. aviats. in-ta. Ser. Matematika i Mekhanika*, No.80, 22-33 (in Russian).

Matrosov, V.M. (1967a), On differential equations and inequalities with discontinuous right parts. I. *Differents - Uravnenia, 3*, No.3, 395-409 (in Russian).

Matroxov, V.M. (1967b), On differential equations and inequalities with discontinuous right parts. II. *Differents - Uravnenia, 3*, No.5, 839-848 (in Russian).

Matrosov, V.M., L.Yu. Anapolskii, and S.N. Vasilyev (1980), *Method of Comparison in Mathematical Theory of Systems*. Nauka, Novosibirsk (in Russian).

Melnikov, G.I. (1956), Some problems of the direct Liapunov method. *Dokl. AN SSSR, 110*, No.3, 326-329 (in Russian).

Melnikov, G.I. (1975), *Dynamics of Nonlinear Mechanical and Electro-Mechanical Systems*. Mashinostrojenije, Leningrad (in Russian).

Michel, A.N., and R.K. Miller (1977), *Qualitative Analysis of Large-Scale Dynamical Systems*. Academic Press, New York.

Mitrinović, D.S. (1970), *Analytic Inequalities*. Acad. Press, Berlin.

Nemytskii, V.V. (1955), Estimation of the domain of nonlinear systems asymptotic stability. *Dokl. AN SSSR, 101*, No.5, 805-80[illegible] (in Russian).

Nemytskii, V.V., and V.V. Stepanov (1949), *Qualitative Theory of Differential Equations*. GITTL, Moscow-Leningrad (in Russian).

Optial, Z. (1960), Sur l'allure asymptotique des solutions de certaines équations différentielles de la mécanique non linéaire. *Ann. pol. math., 8*, No.2, 105-124.

Persidskii, K.P. (1951), Some critical cases of denumerable systems. *Izv. AN Kaz. SSR. Ser. Mat. i Mekh.*, No.5, 3-24 (in Russian).

Persidskii, S.K. (1969), On some theorems of the second Liapunov method. *Differents - Uravnenia, 5*, No.4, 678-687.

Piontkovskii, A.A., and L.D. Rutkovskaya (1967), Investigations of certain problems of the stability theory via the vector Liapunov function method. *Avt. i Telem., 28*, No.10, 23-31 (in Russian).

Rabczuk, R. (1976), *Elementy Nierownosci Rozniczkowych*. Panstw-wyd-wo Naukowe, Warsawa.

Rouch, N., P. Habets, and M. Laloy (1980), *Stability Theory by Liapunov's Direct method*. MIR, Moscow (in Russian).

Sevastyanov, B.A. (1951), Theory of braching stochastic process. *Uspekhi Matem. Nauk, 6*, 47-49 (in Russian).
Šiljak, D.D. (1978), *Large-Scale Dynamic Systems : Stability and Structure*. North-Holland, New York.
Szarski, J. (1967), *Differential Inequalities*. Panstw. Wyd-Wo Naukowe, Warsawa.
Tikhonov, A.A. (1965), On stability of motion under the constantly effecting perturbations. *Vestn. Leningr. un-tan* No.1, 95-101 (in Russian).
Tikhonov, A.A. (1969), To the problem of stability of motion under the constantly effecting perturbations. *Vestn. Leningr. un-ta*, No.19, 116-122 (in Russian).
Tikhonov, A.A. (1974), *On Some Nonlinear Problems of Dynamics*. Dissertation, Lengosuniversitet (in Russian).
Vedrov, V.S. (1938), *Dynamic Stability of the Aeroplane*. Oborongiz, Moscow (in Russian).
Walter, W. (1964), *Differential and Integral Inequalities*. Springer-Verlag, Berlin.
Ważewski, T. (1950), Systèmes des équations et des inégalités différentielles ordinaires aux deuxièmes membres monotones et leurs applications. *Ann. Soc. Pol. Math.*, *23*, 112-166.
Yoshizawa, T. (1966), *Stability Theory by Liapunov's Second Method*. Math. Soc. Japan, Tokyo.
Zubov, V.I. (1957), *A.M. Liapunov's Methods and Their Application*. Izd-vo Leningr. un-ta, Leningrad (in Russian).

CHAPTER III
LARGE-SCALE SYSTEMS IN GENERAL

III.1. INTRODUCTION

A system can be considered complex for the great number of its nonlinearities and/or their form, for its nonstationarity and for its rich structure. A system can be considered large-scale due to several intrinsic features among which we quote the following (see also Kukhtenko, 1966; Martynyuk, 1975) :

1) high-dimensionality;
2) manifold of the system structure (networks, trees, hierarchical structures, etc.);
3) multiple connections of the system elements (sub-systems interconnection in one level and between different levels of hierarchy);
4) manifold of the elements nature (machines, automata, robots, people-operators);
5) recurrence of change of the system composition and state (variability of the system's structure, connections, and composition);
6) multiple criteria of the system (difference between local criteria for sub-systems and global criterion for a system in the whole, their inconsistency).

A direct analysis of a dynamical property (such as stability, controllability, observability, optimality, robustness) of a large-scale system can be cumbersome or even yet impossible. In the framework of the stability analysis, this means that the (second) Liapunov method, the most general and powerful method for the stability analysis, cannot be effectively directly applied to large-scale systems. The cause is the lack of an algorithm for constructing a system Liapunov function.

A trial can be made to test indirectly a stability property of a large-scale system by solving one of the following problems :

Problem A. Decompose the whole system into interconnected subsystems and then find an aggregation form of the system yielding conditions under which its Liapunov stability can be deduced from stability of its disconnected subsystems and from qualitative properties of their interactions.

Problem B. Decompose the whole system into interconnected subsystems and find an aggregation form of a large-scale system which enables simultaneous reduction of the number and complexity of the conditions under which a Liapunov stability property of the whole system is inferred from the properties of interconnected subsystems without using information about stability properties of disconnected subsystems.

Bailey (1966, 1968) and Aoki (1968) initiated effective aggregation of large-scale systems.

New mathematical tools for solving these problems are the concept of vector Liapunov functions discovered by Bellman (1962) and Matrosov (1962, 1963, 1968), the concept of vector norms discovered by Robert (1964) and the concept of matrix Liapunov functions by Martynyuk (II). However, application of these concepts to the stability analysis relies on the Liapunov method (at least at the last stage of the analysis) and in certain cases on the comparison principle by Kamke (1932) and Ważewski (1950), which was further developed in the second Chapter.

Lakshmikantham (1965) and Kayande and Lakshmikantham (1966) used vector Liapunov functions to study conditional stability and conditionally invariant sets, respectively.

Bailey (1966, 1968) was the first to apply effectively the concept of vector Liapunov functions to the Liapunov stability analysis of large-scale systems by solving the Problem A. This problem was later on studied (in the framework of continuous-time systems) extensively for large-scale systems with

- exponentially stable subsystems by Matrosov (1963,1972), Piontkovskii and Rutkovskaya (1967), Michel and Porter (1970), Porter and Michel (1970,1971), Šiljak (1971-1973 ,1975 ,1977 ,1978), Thompson (1972), Araki and Kondo (1972), Thompson and Koenig (1972), Grujić and Šiljak (1972,1973), Grujić (1972 ,1974b,c ,1976b), Michel (1973,1974), Weissenberger (1973), Vakhonina, Zemlyakov and Matrosov (1973), Šiljak and Vukčević (1974), Martynyuk (1975), Michel and Miller (1977), Araki (1978), Grujić, Gentina, Borne, Burgat and Bernussou (1978), Saeki, Araki and Kondo (1980) and Ikeda and Šiljak (1980);

- asymptotically stable subsystems by Michel and Porter (1970), Grujić and Šiljak (1972, 1973), Šiljak (1972a, 1978), Grujić (1972, 1974b,c, 1976b), Michel (1973, 1974), Araki (1975, 1978), Blight and McClamroch (1975), Grujić, Gentina and Borne (1976), Bitsoris and Burgat (1976), Michel and Miller (1977), Araki (1978) and Sinha (1980);
- unstable subsystems by Thompson (1972), Grujić (1972, 1976b), Grujić and Šiljak (1972, 1973), Michel (1973, 1974), Michel and Miller (1977), Morari, Stephanopoulos and Aris (1977), Araki (1978) and Šiljak (1978);
- equi-ultimately bounded subsystems by Kloeden (1975);
- partially stable subsystems by Bondi, Fergola and Gambardella (1979).

Matrosov (1962, 1968) was the first to treat the Problem B, which was further studied by Gentina, Borne and Laurent (1972, 1973), Grujić (1974a,c ,1977a), Grujić, Gentina and Borne (1976), Borne and Benrejeb (1977), Borne, Benrejeb and Cocquerelle (1977), Grujić, Gentina, Borne, Burgat and Bernussou (1978), Gentina, Borne, Burgat, Bernussou and Grujić (1979), Grujić and Burgat (1979, 1980), Burgat and Grujić (1979) and Vidyasagar (1980) in the framework of time-continuous large-scale systems.

Different stability domains of time-continuous large-scale systems were studied at first by Weissenberger (1973) and later on by Bitsoris and Burgat 1976), Morari, Stephanopoulos and Aris (1977), Šiljak (1978) and Grujić and Ribbens-Pavella (1978, 1979).

The surveys together with authors' original results were given by Šiljak (1972a, 1978), Martynyuk (1975), Michel and Miller (1977), Araki (1978), Grujić, Gentina, Borne, Burgat and Bernussou (1978), Burgat, Bernussou, Grujić, Gentina and Borne (1978), Sandel, Varaiya, Athans and Safonov (1978) and Gentina, Borne, Burgat, Bernussou and Grujić (1979).

III.2. DESCRIPTION AND DECOMPOSITION OF LARGE-SCALE SYSTEMS

A large-scale system S of dimension n is governed by

$$\frac{d\mathbf{x}}{dt} = \mathbf{f}(t,\mathbf{x},\mathbf{P}) \ , \qquad (1)$$

where $\mathbf{f}$ is a member of functional family F, $\mathbf{f} \in F$,

$$F = \{\mathbf{f}^1,\mathbf{f}^2,\dots,\mathbf{f}^N\} \ , \ \mathbf{f}^k \in C(T\times R^n\times R^{s\times q},R^n) \ , \ \forall k = 1,2,\dots,N \ , \qquad (2)$$

N is a natural number. The matrix $\mathbf{P} = [\mathbf{p}_1^T, \mathbf{p}_2^T, \dots, \mathbf{p}_s^T]^T \in R^{s \times q}$ reflects internal (e.g. parameter) and/or external perturbations. The class of all admissible matrices $\mathbf{P}$ is denoted by P,

$$P = \{\mathbf{P} : \mathbf{P}_1 \le \mathbf{P}(t) \le \mathbf{P}_2 \ , \forall t \in T\} \ .$$

The matrices $\mathbf{P}_1$ and $\mathbf{P}_2$ are completely defined. The set P may be singleton $\{\mathbf{0}\}$. The number s denotes the number of the subsystems S_i of S. The i-th interconnected subsystem S_i is described by

$$\frac{d\mathbf{x}_i}{dt} = \mathbf{f}_i(t, \mathbf{x}, \mathbf{p}_i) \ , \ \forall i = 1, 2, \dots, s \ . \tag{3}$$

The vectors $\mathbf{x}$ and $\mathbf{f}$ are partitioned as $\mathbf{x} = [\mathbf{x}_1^T, \mathbf{x}_2^T, \dots, \mathbf{x}_s^T]^T$ and $\mathbf{f} = [\mathbf{f}_1^T, \mathbf{f}_2^T, \dots, \mathbf{f}_s^T]^T$ so that $\mathbf{f}_i \in F_i$,

$$F_i = \{\mathbf{f}_i^1, \mathbf{f}_i^2, \dots, \mathbf{f}_i^N\} \ , \ \ \mathbf{f}_i^k \in C(T \times R \times R^{1 \times q}, R^{n_i}) \ , \ \forall k = 1, 2, \dots, N \ , \tag{4}$$

and $n = n_1 + n_2 + \dots + n_s$.

The number N, the families F and F_i $(\forall i = 1, 2, \dots, s)$ and (possibly arbitrary) variations of the superscript $k = k(t)$ over the set

$$N = \{1, 2, \dots, N\} \ , \ k(t) \in N \ , \ \forall t \in R \ ,$$

describe (possibly arbitrary) structural variations of the whole system (Šiljak, 1971-1973, 1975, 1977, 1978; Grujić, 1972; Burgat, Bernussou, Grujić, Gentina and Borne, 1978). The whole system is structurally invariant iff $k(t)$ is constant, $k(t) \equiv K$, i.e. the set N is singleton, $N = \{K\}$. The number N is the number of all possible structures of the whole system.

The function $\mathbf{f}_i$ satisfies the condition $\mathbf{f}_i(t, \mathbf{x}, \mathbf{p}_i) = \mathbf{0}$ for all $t \in R$ iff $\mathbf{x} = \mathbf{0}$ and $\mathbf{p}_i = \mathbf{0}$. The notation $\mathbf{x}^i = [\mathbf{0}^T, \mathbf{0}^T, \dots, \mathbf{x}_i^T, \mathbf{0}^T, \dots, \mathbf{0}^T]^T$ and $\mathbf{g}_i : T \times R^{n_i} \to R^{n_i}$ defined by $\mathbf{g}_i(t, \mathbf{x}_i) = \mathbf{f}_i(t, \mathbf{x}^i, \mathbf{0})$ describe the i-th disconnected subsystem $\hat{S}_i$ (5),

$$\frac{d\mathbf{x}_i}{dt} = \mathbf{g}_i(t, \mathbf{x}_i) \ . \tag{5}$$

The vector $\mathbf{x}_i$ is the state vector of the i-th disconnected subsystem $\hat{S}_i$, $\mathbf{x}_i \in R^{n_i}$. Evidently the function $\mathbf{h}_i$ defined by

$$\mathbf{h}_i(t, \mathbf{x}, \mathbf{p}_i) = \mathbf{f}_i(t, \mathbf{x}, \mathbf{p}_i) - \mathbf{g}_i(t, \mathbf{x}_i) \ , \ \forall i = 1, 2, \dots, s \ , \tag{6}$$

describes the action of the whole system S on its i-th interconnected subsystem S_i (3) that can be now described by

$$\frac{d\mathbf{x}_i}{dt} = \mathbf{g}_i(t, \mathbf{x}_i) + \mathbf{h}_i(t, \mathbf{x}, \mathbf{p}_i) \ , \ \forall i = 1, 2, \dots, s \ . \tag{7}$$

Let H_i denote the set of all possible h_i ,

$$H_i = \{h_i^1, h_i^2, \ldots, h_i^N\} \ , \ h_i^j(t,x,p_i) = f_i^j(t,x,p_i) - g_i(t,x_i) \ ,$$
$$\forall j = 1,2,\ldots,N \ , \ \forall i = 1,2,\ldots,s \ . \tag{8}$$

Evidently, $f_i^j \in F_i$ implies $h_i^j \in H_i$, and vice versa, $\forall i = 1,2,\ldots,s$.

In order to describe more concisely the structurally variable large-scale system S let the following notation be introduced. The *structural parameter* $s_{ij}: T \to \{0,1\}$ is binary valued function of t , or $s_{ij} : T \to [0,1]$, and represents (i,j)-th element of the *structural matrix* $S_i : R \to R^{n_i \times Nn_i}$ *of the* i-*th interconnected subsystem* S_i ,

$$S_i = [s_{i1} I_i, s_{i2} I_i, \ldots, s_{iN} I_i] \ , \ I_i = \mathrm{diag}\{1,1,\ldots,1\} \in R^{n_i \times n_i} \ .$$

Notice that it may be, but need not be, required that $s_{ij}(t) = 1$ implies $s_{ik}(t) = 0$ for all $k \neq j$.

The function $h_i : R \times R^n \times R^{1\times q} \to R^{Nn_i}$,

$$h_i = \begin{bmatrix} h_i^1 \\ h_i^2 \\ \vdots \\ h_i^N \end{bmatrix} \ , \ \forall i = 1,2,\ldots,s \ ,$$

defines all possible interactions of the i-th interconnected subsystem S_i (7) that can be now described by

$$\frac{dx_i}{dt} = g_i(t,x_i) + S_i(t)\, h_i(t,x,p_i) \ , \ \forall i = 1,2,\ldots,s \ , \tag{9}$$

provided all $s_{ij} : T \to \{0,1\}$ are binary valued functions.

Let $S_i : T \to R^{n \times Nn}$ be defined by

$$S = \begin{bmatrix} S_1 & 0_{12} & 0_{13} & \cdots & 0_{1s} \\ 0_{21} & S_2 & 0_{23} & \cdots & 0_{2s} \\ - & - & - & & - \\ 0_{s1} & 0_{s2} & 0_{s3} & \cdots & S_s \end{bmatrix} \ , \ 0_{ij} \in R^{n_i \times n_j} \ .$$

The matrix $S(t)$ describes all structural variations of the system S and will be called *the structural matrix of the system* S . The set of all possible $S(t)$ will be denoted by S_S and referred to as *the structural set of the system*,

$$S_s = \left\{ S : S = \begin{bmatrix} S_1 & O_{12} & \cdots & O_{1s} \\ O_{21} & S_2 & \cdots & O_{2s} \\ - & - & & - \\ O_{s1} & O_{s2} & \cdots & S_s \end{bmatrix} , \; S_i = [s_{i1}I_i, s_{i2}I_i, \ldots, s_{iN}I_i] , \; s_{ij} \in \{0,1\} \right\} \qquad (10)$$

In (10) there $s_{ij} \in [0,1]$ is permissible in cases (9) is obtained rather directly than via (1), (5)-(8).

If we set $g = [g_1^T, g_2^T, \ldots, g_s^T]^T$ and $h = [h_1^T, h_2^T, \ldots, h_s^T]^T$ then the over-all system S can be described by

$$\frac{dx}{dt} = g(t,x) + S(t)\, h(t,x,P) \; , \; P \in P \; , \; S(t) \in S_s \; , \; \forall t \in R \; . \qquad (11)$$

Notice that structural variations can be random representing unpredictable structural perturbations or can be intentionally designed so as in structurally variable control systems.

In the sequel, if it is not otherwise defined, D^+v and v^* will be taken along motions of the system (11).

III.3. STRUCTURAL STABILITY PROPERTIES OF LARGE-SCALE SYSTEMS

Many structural stability properties can be defined in the sense of Liapunov, but only those of them which will be used in the sequel are defined.

Definition 1. The equilibrium state $x = 0$ of the system (11) is

- *structurally asymptotically stable (in the whole) over* $P \times S_s$ iff it is asymptotically stable (in the whole) for every $(P,S) \in P \times S_s$, respectively,
- *structurally uniformly asymptotically stable (in the whole) over* $P \times S_s$ iff it is uniformly asymptotically stable (in the whole) for every $(P,S) \in P \times S_s$, respectively,
- *structurally exponentially stable (in the whole) over* $P \times S_s$ iff it is exponentially stable (in the whole) for every $(P,S) \in P \times S_s$, respectively. ■

The structural stability concept generalizes and involves the significant connective stability concept by Šiljak (1971-1973, 1978).

We denote a vector $[x^T, u^T]^T \in R^{n+s}$ by z and consider a system

$$\frac{dx}{dt} = g_1(t,x) + S(t)\, h(t,x,P) \; ,$$

$$\frac{du}{dt} = g_2(t,u,x,P,S^*) \ , \quad S^* \in S_s \ , \tag{12}$$

where $g_2 : R \times R^s \times R^n \times P \times S_s \to R^s$; $g_2(t,0,0,P,S^*) = 0$, $\forall (t,P,S^*) \in R \times P \times S_s$, and, besides, functions g_1 , h , g_2 possess smoothness, ensuring existence of the unique solution of equation (12) :

$$[x^T(t;t_o,x_o) \ , u^T(t;t_o,u_o)]^T = z^T(t;t_o,z_o) \ ,$$

through each point of $R \times N \times R^s$. Let Greek letters denote numbers.

Definition 2. The state $[x^T,u^T]^T = 0$ of system (12) is :

a) u-*stable with respect to* T_i , iff for every $t_o \in T_i$ and every $\epsilon \in \mathring{R}_+$ there exists $\delta(t_o,\epsilon) > 0$ such that for $\|z_o\| < \delta(t_o,\epsilon)$ an inequality $\|u(t;t_o,u_o)\| < \epsilon$, $\forall t \in T_o$ is fulfilled;

b) *uniformly* u-*stable with respect to* T_i , iff conditions of the definition a) are fulfilled and for every $\epsilon \in R_+$ the consequent maximal value δ_M satisfies the condition

$$\inf [\delta_M(t,\epsilon) : t \in T_i] > 0 \ ;$$

c) u-*stable in the whole with respect to* T_i iff conditions of the definition a) are fulfilled and $\delta_M(t,\epsilon) \to +\infty$ for $\epsilon \to +\infty$ and $t \in T_i$;

d) *uniformly* u-*stable in the whole with respect to* T_i , iff conditions of the definitions b) and c) are fulfilled. ∎

An expression "*with respect to* T_i " in definitions a)-d) can be omitted, iff $T_i = R$.

Definition 3. The state $[x^T,u^T]^T = 0$ of system (12) is :

a) u-*attractive with respect to* T_i iff for any $t_o \in T_i$ and $\zeta \in]0,+\infty[$ there exist $\Delta(t_o) > 0$ and $\tau(t_o,z_o,\zeta) \in [0,+\infty[$ such that for $\|z_o\| < \Delta(t_o)$ an inequality $\|u(t;t_o,u_o)\| < \zeta$, $\forall t \in]t_o + \tau(t_o,z_o,\zeta)$, $+\infty[$ holds;

b) u_o-*uniformly* u-*attractive with respect to* T_i iff conditions of the Definition 3.a) are fulfilled and for any $t_o \in T_i$ and any $\zeta \in]0,+\infty[$ there exist $\Delta(t_o) > 0$ and $\tau_u(t_o,\Delta(t_o),\zeta) \in [0,+\infty[$ such that

$$\sup [\tau_m(t_o,z_o,\zeta) : z_o \in B_\Delta(t_o)] = \tau_u(t_o,\Delta(t_o),\zeta) \ ;$$

c) t_o-*uniformly* u-*attractive with respect to* T_i , iff conditions of the Definition 3.a) are fulfilled and for any $\zeta \in]0,+\infty[$ there exist $\Delta > 0$ and $\tau_u(z_o,\zeta) \in [0,+\infty]$ such that

$$\sup [\tau_m(t_o,z_o,\zeta) : t_o \in T_i] = \tau_u(z_o,\zeta) \ ;$$

d) *uniformly* u-*attractive with respect to* T_i , iff conditions of the

Definition 3.a),c) are fulfilled, i.e. conditions of the Definition 3.a) hold and for every $\zeta \in]0,+\infty[$ there exist $\Delta > 0$ and $\tau_u(\Delta,\zeta) \in [0,+\infty[$ such that

$$\sup\,[\tau_m(t_o,z_o,\zeta) : (t_o,z_o) \in T_i \times B_\Delta] = \tau_u(\Delta,\zeta) \,.$$

e) Properties 3.a)-d) are fulfilled in the whole, iff conditions of the Definition 3.a) are valid for any $\Delta(t_o) \in]0,+\infty[$ and $t_o \in T_i$. ∎

An expression "*with respect to* T_i " in definitions 3.a)-e) can be omitted, iff $T_i = R$.

Definition 4. The state $[x^T,u^T]^T = 0$ of system (12) is :

a) *asymptotically* u-*stable with respect to* T_i , iff it is u-stable with respect to T_i and u-attractive with respect to T_i ;
b) *equi-asymptotically* u-*stable with respect to* T_i , iff it is u-stable with respect to T_i and u_o-uniformly u-attractive with respect to T_i ;
c) *quasi-uniformly asymptotically* u-*stable with respect to* T_i , iff it is uniformly u-stable with respect to T_i and t_o-uniformly u-attractive with respect to T_i ;
d) *uniformly asymptotically* u-*stable with respect to* T_i , iff it is uniformly u-stable with respect to T_i and uniformly u-attractive with respect to T_i ;
e) *exponentially* u-*stable with respect to* T_i iff there exist $\Delta > 0$ and constant values $\alpha \geq 1$ and $\beta > 0$ such that for $\|z_o\| < \Delta$ an inequality

$$\|u(t;t_o,u_o)\| \leq \alpha\|z_o\| \exp\,[-\beta(t-t_o)] \;, \quad \forall t \in T_o \;, \; t_o \in T_i \;,$$

is fulfilled;
f) *exponentially* u-*stable in the whole with respect to* T_i iff conditions of the Definition 4.e) are fulfilled for $\Delta = +\infty$.
g) It possesses properties 4.a)-d) *in the whole*, iff the corresponding types of u-stability and u-attraction hold in the whole. ∎

In Definitions 4.a)-g) an expression "*with respect to* T_i " can be omitted, iff $T_i = R$ [25] .

---

[25] See 2) of Comments on References to Ch. III.

III.4. AGGREGATION FORMS OF LARGE-SCALE SYSTEMS AND CONDITIONS OF STRUCTURAL STABILITY

The aggregation form of a large-scale system has an essential significance for obtaining effective stability conditions. The essence of the aggregation method consists in a substitution of the initial large-scale system by a system of smaller dimensionality so that solutions of the initial one and of obtained systems could be found in one-to-one manner. At first it may seem that such a substitution is not correct, for a system of the n-th order has n characteristic motions, each of which can exist for a definite realization of structural perturbations. Further on we illustrate that correctness of the aggregation form is closely connected with the characteristic features of the initial system S.

III.4.1. Aggregation forms and solutions for the Problem A

III.4.1.1. The first aggregation form and stability criteria

For the subsequent analysis it is useful to introduce the following notation (Grujić, 1972) :

$$D_t^+ v_i(t,\mathbf{x}_i) = \limsup \left\{ \frac{v_i(t+\theta,\mathbf{x}_i) - v_i(t,\mathbf{x}_i)}{\theta} : \theta \to 0^+ \right\} ,$$

$$\delta_{ij} = \begin{cases} 0 , & i \neq j \\ 1 , & i = j \end{cases} , \quad \delta_{ij} \text{ is the Kronecker symbol } ,$$

$$\mathbf{I}_{ij} = \operatorname{diag}\{(1-\delta_{1j}),(1-\delta_{2j}),\dots,(1-\delta_{jj}),0,\dots,0\} \in R^{n_i \times n_i} ,$$

$$\mathbf{x}_i(t+\theta;t,\mathbf{x}_i) = \mathbf{x}_i + \Delta\mathbf{x}_i ,$$

$$D_{x_{ij}}^+ v_i(t,\mathbf{x}_i) = \limsup \left\{ \frac{v_i(t+\theta , \mathbf{x}_i + \mathbf{I}_{i,j+1}\Delta\mathbf{x}_i) - v_i(t+\theta , \mathbf{x}_i + \mathbf{I}_{ij}\Delta\mathbf{x}_i)}{\Delta x_{ij}} : \begin{matrix} \theta \to 0^+ \\ \|\Delta\mathbf{x}_i\| \to 0 \end{matrix} \right\}$$

and

$$D_{\mathbf{x}_i}^+ v_i(t,\mathbf{x}_i) = (D_{x_{i1}}^+ v_i(t,\mathbf{x}_i) , D_{x_{i2}}^+ v_i(t,\mathbf{x}_i) , \dots , D_{x_{in_i}}^+ v_i(t,\mathbf{x}_i))^T .$$

It is easy to verify that (Grujić, 1972; Grujić, Gentina, Borne, Burgat and Bernussou, 1978) :

$$D^+ v_i(t,\mathbf{x}_i) \le D_t^+ v_i(t,\mathbf{x}_i) + [D_{\mathbf{x}_i}^+ v_i(t,\mathbf{x}_i)]^T \frac{d\mathbf{x}_i}{dt} .$$

This inequality should be in mind throughout this chapter.

Now, the first aggregation form can be introduced by

Assumption 1. There exist an open connected neighbourhood N_i of $\mathbf{x}_i = \mathbf{0}$, functions $v_i : R \times R^{n_i} \to R_+$ and $\Omega_i : R \times R^{n_i} \to R_+$ and real numbers

α_{ii} and $\alpha_{ij}(S_i)$ such that :

1) v_i and Ω_i are positive definite on N_i , $\forall i = 1,2,\dots,s$,
2) $\alpha_{ij}(S_i) \geq 0$, $\forall i,j = 1,2,\dots,s$, $i \neq j$, $\forall S \in S_s$,
3) $\alpha_{ii} < 0$, $\forall i = 1,2,\dots,s$,
4) $D_t^+ v_i(t,x_i) + [D_{x_i}^+ v_i(t,x_i)]^T g_i(t,x_i) \leq \alpha_{ii}\Omega_i(t,x_i)$, $\forall(t,x_i) \in R\times N_i$, $\forall i = 1,2,\dots,s$,
5) $[D_{x_i}^+ v_i(t,x_i)]^T S_i h_i(t,x,p_i) \leq \sum_{j=1}^{s} \alpha_{ij}(S_i)\,\Omega_j(t,x_j)$, $\forall(t,x) \in R\times N$, $\forall(P,S) \in P\times S_s$, $\forall i = 1,2,\dots,s$,

where $N \subseteq N_1\times N_2\times\dots\times N_s$, N is a connected neighbourhood of $x = 0$.

This aggregation form slightly generalizes that of Grujić and Šiljak (1972, 1973).

Let $a_{ij}(S_i)$ be the (i,j)-th element of the aggregation matrix $A(S)$,

$$A(S) = (a_{ij}(S_i)) \ , \quad a_{ij}(S_i) = \alpha_{ii}\delta_{ij} + \alpha_{ij}(S_i) \ , \quad \forall i,j = 1,2,\dots,s \ , \ \forall S \in S_s \ . \tag{13}$$

Evidently, A is a Metzler matrix (Chapter II.3.4) due to 2) of Assumption 1.

The condition 4 of Assumption 1 is a requirement imposed on the disconnected subsystem S_i (5), and the condition 5 of the same assumption is imposed on the interaction h_i in order to establish a solution for the Problem A.

Theorem 1. (I) *It is sufficient for structural asymptotic stability of* $x = 0$ *of the system (11) over* $P\times S_s$ *that*

(a) *Assumption 1 holds,*

(b) *there is positive real number* ξ_i *or* $\xi_i = +\infty$ *such that the set* $V_{i\zeta}(t)$ *is asymptotically contractive for every* $\zeta \in]0,\xi_i[$ *and every* $i = 1,2,\dots,s$,

(c) *there is* $A \in R^{s\times s}$ *such that* $A(S)$ *(13) obeys* $A(S) \leq A$, $\forall S \in S_s$, *elementwise, where* $A = A(S^*)$ *for some* $S^* \in S_s$ *is admissible,*

and

(d) *the aggregation matrix* A *is stable, or equivalently,*

$$(-1)^i \begin{vmatrix} a_{11} & a_{12} & \cdots & a_{1i} \\ a_{21} & a_{22} & \cdots & a_{2i} \\ - & - & & - \\ a_{i1} & a_{i2} & \cdots & a_{ii} \end{vmatrix} > 0 \ , \quad \forall i = 1,2,\dots,s \ .$$

(II) *If, in addition to (I),* $N_i = R^{ni}$ *,* v_i *is radially unbounded and* $\xi_i = +\infty$ *for every* $i = 1,2,\dots,s$ *, then* $x = 0$ *of the system is structurally asymptotically stable in the whole over* $P \times S_s$ *.*

Proof. (I) Let a vector $c \in \check{R}_+^s$ be elementwise positive vector that can be arbitrarily chosen and $b \in R^s$ be determined from $b = -(A^T)^{-1}c$. The vector b is elementwise positive due to 2) and 3) of Assumption 1 and (d) [Fiedler and Ptak (1962)]. The condition (b) guarantees that $V_{1\zeta_1}(t) \times V_{2\zeta_2}(t) \times \dots \times V_{s\zeta_s}(t)$ is asymptotically contractive for every $\zeta_i \in]0,\xi_i[$, $\forall i = 1,2,\dots,s$. Let $v(t,x) = b^T v(t,x)$, where $v = [v_1,v_2,\dots,v_s]^T$, and $V_\zeta(t)$ be the largest connected neighbourhood of $x = 0$ such that $v(t,x) < \zeta$, $\forall x \in V_\zeta(t)$, $\forall t \in R$. Hence $V_\zeta(t)$ is asymptotically contractive for every $\zeta \in]0,\xi[$ where $\xi = \min\{b_i\xi_i : i = 1,2,\dots,s\}$ and b_i is the i-th element of $b = [b_1,b_2,\dots,b_s]^T$. Further, the hypothesis and the conditions 4) and 5) of Assumption 1 imply existence of a connected neighbourhood N of $x = 0$, $N \subseteq N_1 \times N_2 \times \dots \times N_s$ such that

$$D^+v(t,x) \le b^T A(S)\, w(t,x) \ , \quad \forall (t,x,P,S) \in R \times N \times P \times S_s \ , \tag{14}$$

where

$$w = [\Omega_1,\Omega_2,\dots,\Omega_s]^T \ .$$

Positive definiteness of Ω_i on N_i , $\forall i = 1,2,\dots,s$, the definition of N , positiveness of all b_i of b , the condition (c) of Theorem 1 and (14) yield

$$D^+v(t,x) \le -c^T w(t,x) \ , \quad \forall (t,x,P,S) \in R \times N \times P \times S_s \ . \tag{15}$$

The function v satisfies all conditions of Theorem 13 of Section I.3.2.5 for every $(P,S) \in P \times S_s$. Hence, $x = 0$ of (11) is structurally asymptotically stable over $P \times S_s$.

(II) $N_i = R^{ni}$ together with the hypothesis of the Assumption 1 guarantee positive definiteness of v_i and Ω_i , $\forall i = 1,2,\dots,s$, hence, positive definiteness of v . Radial unboundedness of all v_i implies radial unboundedness of v . All conditions of Corollary 2 of Section I.2.3.5 are satisfied on $P \times S_s$ which completes the proof. ∎

Notice that v_i functions used in Theorem 1 are not decrescent.

The structural asymptotic stability guaranteed by Theorem 1 is not uniform. If the structural uniform asymptotic stability is desirable then another criterion should be used.

Theorem 2. (I) *It is sufficient for structural uniform asymptotic stability of* $x = 0$ *of the system (11) over* $P \times S_s$ *that*

(a) Assumption 1 holds;

(b) all v_i *are decrescent functions;*

(c) there is $A \in R^{s \times s}$ *such that* $A(S)$ *(13) obeys* $A(S) \le A$, $\forall S \in S_s$, *elementwise, where* $A = A(S^*)$ *for some* $S^* \in S_s$ *is admissible;*

and

(d) the aggregation matrix A *is stable.*

(II) *If, in addition to (I),* $N_i = R^{n_i}$ *and* v_i *is radially unbounded for every* $i = 1,2,\ldots,s$ *, then* $x = 0$ *of the system is structurally uniformly asymptotically stable in the whole over* $P \times S_s$.

Proof. Let b and v be determined as in the proof of Theorem 1. The function v is decrescent due to the condition I.b). Now, the assertion under (I) is true due to Theorem 7 of Section I.3.2.4 and the proof of Theorem 1, and the assertion under (II) holds due to Theorem 8 of Section I.3.2.4 and the proof of Theorem 1. ∎

Theorem 2 guarantees structural exponential stability of $x = 0$ of the system (11) under special features of all v_i and Ω_i .

Theorem 3. (I) *It is sufficient for structural exponential stability of* $x = 0$ *of the system (11) over* $P \times S_s$ *that*

(a) Assumption 1 holds;

(b) there exist positive numbers η_{i1} , η_{i2} *and* η_{i3} *such that both* $\eta_{i1}\|x_i\| \le v_i(t,x_i) \le \eta_{i2}\|x_i\|$ *and* $\Omega_i(t,x_i) = \eta_{i3}\|x_i\|$ *hold for all* $(t,x_i) \in R \times N_i$ *and* $i = 1,2,\ldots,s$;

(c) there is $A \in R^{s \times s}$ *such that* $A(S) = (a_{ij}(S_i))$ *obeys* $A(S) \le A$, $\forall S \in S_s$, *for* $a_{ij}(S_i) = [\alpha_{ij}\delta_{ij} + \alpha_{ij}(S_i)]\eta_{j3}$, *where* $A = A(S^*)$ *for some* $S^* \in S_s$ *is admissible;*

(d) A *is stable.*

(II) *If, in addition to (I),* $N_i \in R^{n_i}$ *for every* $i = 1,2,\ldots,s$ *then* $x = 0$ *of the system is structurally exponentially stable in the whole over* $P \times S_s$.

Proof. Let b and v be determined as in the proof of Theorem 1. Then the condition (b) and the proof of Theorem 1 imply existence of positive numbers η_1 , η_2 and η_3 such that both $\eta_1\|x\| \le v(t,x) \le \eta_2\|x\|$ and $D^+v(t,x) \le -\eta_3\|x\|$ are valid for all $(t,x) \in R \times N$. Hence, the assertion under (I) is true in view of Theorem 9 of the Section I.3.2.4 and that under (II) is true in view of Theorem 10 of the same section. ∎

Example 1. Let the fourth order system composed of two second order subsystems be considered,

$$\frac{d\mathbf{x}_1}{dt} = \frac{1}{1+t^2}\left\{-(1+2t)\,\mathbf{x}_1 + s_{11}(t)\begin{bmatrix} -\operatorname{sat} 0.1\,x_{11} \\ 0 \end{bmatrix} + s_{12}(t)\begin{bmatrix} 0.5\,x_{22} \\ 0 \end{bmatrix} + s_{13}(t)\begin{bmatrix} 0 \\ 0.1\,x_{21} \end{bmatrix}\right\}$$

$$\frac{d\mathbf{x}_2}{dt} = \frac{1}{1+t^2}\left\{-2(1+t)\,\mathbf{x}_2 + s_{21}(t)\begin{bmatrix} 0.4\,x_{11} \\ 0 \end{bmatrix} + s_{22}(t)\begin{bmatrix} 0 \\ 0.4\,x_{12} \end{bmatrix} + s_{23}(t)\begin{bmatrix} 0 \\ \operatorname{sat} 0.2\,x_{22} \end{bmatrix}\right\}$$

where $\operatorname{sat}\xi = \xi$ for $|\xi| \le 1$ and $\operatorname{sat}\xi = \operatorname{sign}\xi$ for $|\xi| \ge 1$. In this example, $P = \{\mathbf{0}\}$ and the structural matrices $\mathbf{S}_i(t)$ have the form

$$\mathbf{S}_i(t) = \begin{bmatrix} s_{i1}(t) & 0 & s_{i2}(t) & 0 & s_{i3}(t) & 0 \\ 0 & s_{i1}(t) & 0 & s_{i2}(t) & 0 & s_{i3}(t) \end{bmatrix}, \quad \forall i=1,2$$

and

$$\mathbf{S}(t) = \begin{bmatrix} \mathbf{S}_1(t) & 0 \\ 0 & \mathbf{S}_2(t) \end{bmatrix} .$$

The structural set of the system is found as

$$S_s = \left\{\mathbf{S}(t) : \mathbf{S}(t) = \begin{bmatrix} \mathbf{S}_1(t) & 0 \\ 0 & \mathbf{S}_2(t) \end{bmatrix}, \right.$$

$$\mathbf{S}_i(t) = [s_{i1}(t)\,\mathbf{I}_2\,,\, s_{i2}(t)\,\mathbf{I}_2\,,\, s_{i3}(t)\,\mathbf{I}_2]\,,$$

$$\left. s_{ij}(t) \in \{0,1\}\,,\ \forall t \in R\,,\ \forall i=1,2\,,\ \forall j=1,2,3 \right\} .$$

Notice that the system structural variations described by the structural set S_s are not admissible in the framework of the connective stability concept and that $s_{ij}(t) = s_{mn}(t)$ is admissible for all $i,m = 1,2$ and $j,n = 1,2,3$. All possible system interactions are described by $\mathbf{S}_1(t)\,\mathbf{h}_1(t,\mathbf{x})$ and $\mathbf{S}_2(t)\,\mathbf{h}_2(t,\mathbf{x})$ where

$$h_1(t,\mathbf{x}) = \begin{bmatrix} -\operatorname{sat} 0.1 x_{11} \\ 0 \\ 0.5 x_{22} \\ 0 \\ 0 \\ 0.1 x_{21} \end{bmatrix}, \quad h_2(t,\mathbf{x}) = \begin{bmatrix} 0.4 x_{11} \\ 0 \\ 0 \\ 0.4 x_{12} \\ 0 \\ \operatorname{sat} 0.2 x_{22} \end{bmatrix}$$

The form of disconnected subsystems

$$\frac{d\mathbf{x}_1}{dt} = -\frac{1+t}{1+t^2}\,\mathbf{x}_1 \;, \quad \frac{d\mathbf{x}_2}{dt} = -\frac{2+t}{1+t^2}\,\mathbf{x}_2 \;,$$

suggests the form of v_i ,

$$v_i(t,\mathbf{x}_i) = (1+t^2)(|x_{i1}| + |x_{i2}|) \;, \; i=1,2 \;.$$

The sets $\mathcal{V}_{i\zeta}(t)$ are now determined by

$$\mathcal{V}_{i\zeta}(t) = \{\mathbf{x}_i : |x_{i1}| + |x_{i2}| < \zeta(1+t^2)^{-1}\} \;, \; \forall i=1,2 \;,$$

which shows that they are asymptotically contractive for every $\zeta \in \,]0,+\infty[$. It is further obtained that $\Omega_i(t,\mathbf{x}_i) = |x_{i1}| + |x_{i2}|$, $i=1,2$, and $\alpha_{11} = -1$, $\alpha_{11}(S_1) = 0.1\,s_{11}$, $\alpha_{12}(S_1) = 0.5\,s_{12} + 0.1\,s_{13}$, $\alpha_{21}(S_2) = 0.4\,(s_{21} + s_{22})$, $\alpha_{22} = -2$, $\alpha_{22}(S_2) = 0.2\,s_{23}$ obey all conditions of Assumption 1. Hence,

$$\mathbf{A}(S) = \begin{bmatrix} -1 + 0.1\,s_{11} & , \; 0.5\,s_{12} + 0.1\,s_{13} \\ 0.4\,(s_{21} + s_{22}) & , \; -2 + 0.2\,s_{23} \end{bmatrix}$$

Evidently,

$$\mathbf{A}(S) \le \begin{bmatrix} -0.9 & , \; 0.6 \\ 0.8 & , \; -1.8 \end{bmatrix}, \; \forall S \in \mathcal{S}_s \;,$$

which yields

$$\mathbf{A} = \begin{bmatrix} -0.9 & , \; 0.6 \\ 0.8 & , \; -1.8 \end{bmatrix}.$$

The matrix $\mathbf{A}$ obeys the condition (d) of Theorem 1. Since all conditions of (II) of this theorem are satisfied then it follows that $\mathbf{x}=\mathbf{0}$ of the system is structurally asymptotically stable in the whole over $\mathcal{S}_s$.

III.4.1.2. The second aggregation form and stability criteria

The second aggregation form is defined by

Assumption 2. There exist an open connected neighbourhood N_i of $x_i = 0$, functions v_i and Ω_i and real numbers α_{ii} and $\alpha_{ij}(S)$ such that

1) v_i and Ω_i are positive definite on N_i, $\forall i = 1,2,\dots,s$,
2) $\alpha_{ii} < 0$, $\forall i = 1,2,\dots,s$,
3) $\alpha_{ij}(S_i) \geq 0$, $\forall i,j = 1,2,\dots,s$, $i \neq j$, $\forall S \in S_s$,
4) $D_t^+ v_i(t,x_i) + [D_{x_i}^+ v_i(t,x_i)]^T g_i(t,x_i) \leq \alpha_{ii}\Omega_i^2(t,x_i)$, $\forall (t,x_i) \in R \times N_i$, $\forall i = 1,2,\dots,s$,

and

5) $[D_{x_i}^+ v_i(t,x_i)]^T S_i h_i(t,x,p_i) \leq \Omega_i(t,x_i) \sum_{j=1}^{s} \alpha_{ij}(S_i)\Omega_j(t,x_j)$,

$\forall (P,S) \in P \times S_s$, $\forall (t,x) \in R \times N$, $\forall i = 1,2,\dots,s$. ■

This aggregation form slightly generalizes that of Michel (1973, 1974).

Theorem 4. (I) *For structural asymptotic stability of* $x = 0$ *of the system (11) over* $P \times S_s$ *it is sufficient that*

(a) Assumption 2 holds,

(b) there is positive number ξ_i *or* $\xi_i = +\infty$ *such that the set* $V_{i\zeta}(t)$ *is asymptotically contractive for every* $\zeta \in]0,\xi_i[$ *and every* $i = 1,2,\dots,s$,

(c) there is $A \in R^{s\times s}$ *such that* $A(S)$ *(13) obeys* $A(S) \leq A$, $\forall S \in S_s$, *where* $A = A(S^*)$ *for some* $S^* \in S_s$ *is admissible,*

and

(d) the aggregation matrix A *is stable.*

(II) *If, in addition to (I),* $N_i = R^{ni}$, v_i *is radially unbounded and* $\xi_i = +\infty$ *for every* $i = 1,2,\dots,s$ *then* $x = 0$ *of the system is structurally asymptotically stable in the whole over* $P \times S_s$.

Proof. (I) The conditions 2) and 3) of the Assumption 2, (a),(c) and (d) of Theorem 4 guarantee (LaSalle, 1976) the existence of a positive diagonal $D = \mathrm{diag}\{d_1,d_2,\dots,d_s\}$ such that $A^T D + DA$ is negative definite. Let for such D the function v be defined by $v(t,x) = d^T v(t,x)$ for the vector $d = [d_1,d_2,\dots,d_s]^T$. Elementwise positiveness of the vector d, positive definiteness of v_i on N_i for all $i = 1,2,\dots,s$, and the properties of all N_i imply existence of a neighbourhood N, $N \subseteq N_1 \times N_2 \times \dots \times N_s$, of $x = 0$ on which v is positive definite. The condition (b) guarantees existence of a positive number ξ, or $\xi = +\infty$, such that the largest connected neighbourhood $\mathcal{V}_\zeta(t)$ of $x=0$ obeying $v(t,x) < \zeta$, $\forall(t,x) \in R\times\mathcal{V}_\zeta(t)$ is asymptotically contractive for every $\zeta \in]0,\xi[$. In view of the conditions 4) and 5) of the Assumption 2 and (a) and (c) of Theorem 4, it follows

$$D^+v(t,x) \le w^T(t,x)(A^TD + DA)\,w(t,x) \quad , \quad \forall(t,x) \in R\times\mathcal{N} \quad , \quad \forall(P,S) \in \mathcal{P}\times\mathcal{S}_s \; . \tag{16}$$

Positive definiteness of Ω_i on $\mathcal{N}_i$, $\forall i = 1,2,\dots,s$, $\mathcal{N} \subseteq \mathcal{N}_1\times\mathcal{N}_2\times\dots\times\mathcal{N}_s$ and negative definiteness of $A^TD + DA$ imply negative definiteness of the right-hand side of (16) on $\mathcal{N}$ for every $(P,S) \in \mathcal{P}\times\mathcal{S}_s$. All conditions of Theorem 13 of the Section I.3.2.5 are satisfied for every $(P,S) \in \mathcal{P}\times\mathcal{S}_s$. Hence, $x=0$ of (11) is structurally asymptotically stable over $\mathcal{P}\times\mathcal{S}_s$.

(II) Since $\mathcal{N}_i = R^{n_i}$ and v_i is radially unbounded, then v is also radially unbounded and positive definite. In view of all $\xi_i = +\infty$ all conditions of Corollary 2 of the Section I.3.2.5 are satisfied for every $(P,S) \in \mathcal{P}\times\mathcal{S}_s$, which completes the proof. ∎

The structural asymptotic stability guaranteed by Theorem 4 is not uniform, for which the next criterion is established.

Theorem 5. (I) *For structural uniform asymptotic stability of* $x=0$ *of the system (11) over* $\mathcal{P}\times\mathcal{S}_s$ *it is sufficient that*

(a) Assumption 2 holds,

(b) the function v_i *is decrescent on* $\mathcal{N}_i$, $\forall i = 1,2,\dots,s$,

(c) there is $A \in R^{s\times s}$ *such that* $A(S)$ *(13) obeys* $A(S) \le A$, $\forall S \in \mathcal{S}_s$, *where* $A = A(S^*)$ *for some* $S^* \in \mathcal{S}_s$ *is admissible*

and

(d) the aggregation matrix A *is stable.*

(II) *If, in addition to (I),* $\mathcal{N}_i = R^{n_i}$ *and* v_i *is radially unbounded for every* $i = 1,2,\dots,s$, *then* $x=0$ *of the system (11) is structurally uniformly asymptotically stable in the whole over* $\mathcal{P}\times\mathcal{S}_s$.

Proof. Let D and v be defined such as in the proof of Theorem 4. In view of the condition (b), the function v is decrescent on the appropriate neighbourhood $\mathcal{N}$ of $x=0$. In the proof of Theorem 4 it was shown that v is positive definite on $\mathcal{N}$ and that (16) holds. Hence, all conditions of Theorem 7, the Section I.3.2.3, are satisfied which proves the assertion under (I). When all $\mathcal{N}_i = R^{n_i}$ and v_i are radially unbounded then v is radially unbounded and positive definite. (16) is still valid. All conditions of Theorem 8, the Section I.3.2.3, are fulfilled, which completes the proof. ∎

Structural exponential stability of $x = 0$ of the large-scale system (11) is guaranteed by

Theorem 6. (I) *For structural exponential stability of* $x = 0$ *of the system (11) over* $P \times S_s$ *it is sufficient that*

(a) Assumption 2 holds,

(b) there exist positive numbers η_{i1}, η_{i2} *and* η_{i3} *such that both* $\eta_{i1}\|x_i\|^2 \leq v_i(t,x_i) \leq \eta_{i2}\|x_i\|^2$ *and* $\Omega_i(t,x_i) = \eta_{i3}\|x_i\|$ *hold for all* $(t,x_i) \in R \times N_i$ *and* $i = 1,2,\dots,s$ *,*

(c) there is $A \in R^{s \times s}$ *such that* $A(S) = [a_{ij}(S_i)]$ *obeys* $A(S) \leq A$ *,* $\forall S \in S_s$ *, for* $a_{ij}(S_i) = [\alpha_{ii}\delta_{ij} + \alpha_{ij}(S_i)]\eta_{i3}\eta_{j3}$ *, where* $A = A(S^*)$ *for some* $S^* \in S_s$ *is admissible,*

(d) A *is stable.*

(II) *If, in addition to (I),* $N_i = R^{n_i}$ *for every* $i = 1,2,\dots,s$ *then* $x = 0$ *of the system (11) is structurally exponentially stable in the whole over* $P \times S_s$.

Proof. Let D , d and v be defined such as in the proof of Theorem 4 and let λ_M be the maximal eigenvalue of $A^T D + DA$. In view of negative definiteness of $A^T D + DA$, the eigenvalue λ_M is negative. The conditions (b) and (16) imply existence of positive numbers η_1 , η_2 and η_3 , in fact $\eta_3 = -\lambda_M$, such that both $\eta_1\|x\|^2 \leq v(t,x) \leq \eta_2\|x\|^2$ and $D^+v(t,x) \leq -\eta_3\|x\|^2$ hold for all $(t,x,P,S) \in R \times N \times P \times S_s$, which proves assertion under (I) due to Theorem 9 of the Section I.3.2.3. When $N_i = R^{n_i}$ for all $i = 1,2,\dots,s$ then $N = R^n$ and all conditions of Theorem 10 of the same section are satisfied for all $(P,S) \in P \times S_s$, which completes the proof. ■

Example 2. A simplified mathematical model of an orbiting astronomical observatory control system can be given in the Lur'e like form (Grujić, 1972),

$$\frac{dx_i}{dt} = A_{i1}x_1 + A_{i2}x_2 + A_{i3}x_3 + \nu Bf(\Sigma) \ , \ \Sigma = Cx \ , \ \forall i=1,2,3 \ ,$$

where $n_1 = n_2 = n_3 = 3$,

$$A_{11} = \begin{bmatrix} 0 & , -a_1 & , & 0 \\ a_2 & , -a_3 & , & a_4 \\ -a_5 & , \ 0 & , & -a_5 \end{bmatrix} , \quad A_{12} = -A_{13} = \begin{bmatrix} 0 & , & 0 & , & 0 \\ -a_6 & , & 0 & , & 0 \\ -a_7 & , & 0 & , & 0 \end{bmatrix} ,$$

$$A_{ij} = 0 \ , \ i = 2,3 \ , \ j = 1,2,3 \ , \ j \neq i \ ;$$

$$A_{22} = A_{23} = \begin{bmatrix} 0 , & -a_1 , & 0 \\ 2a_2 , & -a_3 , & a_4 \\ -2a_5 , & 0 , & -a_5 \end{bmatrix} , \quad B = \begin{bmatrix} 0 , & 0 , & 0 \\ \delta_{i1} , & \delta_{i2} , & \delta_{i3} \\ 0 , & 0 , & 0 \end{bmatrix} , \quad \delta_{ij} = \begin{bmatrix} 0 , \; i \neq j \\ 1 , \; i = j \end{bmatrix}$$

$$C = \begin{bmatrix} r_{11}^T , & r_{12}^T , & r_{13}^T \\ 0 , & r_{22}^T , & 0 \\ 0 , & 0 , & r_{33}^T \end{bmatrix} ,$$

$$r_{1i}^T = [\rho_{1i}^1 , \rho_{1i}^2 , \rho_{1i}^3] , \quad r_{jj}^T = [\rho_{j1} , \rho_{j2} , \rho_{j3}] , \quad \begin{bmatrix} i = 1,2,3 \\ j = 2,3 \end{bmatrix}$$

$$f(\Sigma) = [\psi_1(\sigma_1), \psi_2(\sigma_2), \psi_3(\sigma_3)]^T , \quad \Sigma = (\sigma_1, \sigma_2, \sigma_3)^T ,$$

$$\frac{\psi_i(\sigma_i)}{\sigma_i} \in [0,1] , \quad \forall \sigma_i \in R \; ; \quad \psi_i(\sigma_i) \in C(R,R) , \quad \forall i = 1,2,3 \; .$$

The state $\mathbf{x} = 0$ is the unique equilibrium state of the system. The elements of all matrices A_{ij} , B and C as well as ν are known real numbers. The form

$$\frac{d\mathbf{x}_i}{dt} = A_{ii}\mathbf{x}_i , \quad \forall i = 1,2,3 ,$$

of the disconnected subsystems suggests the choice of v_i functions in the form $v_i(\mathbf{x}_i) = \mathbf{x}_i^T H_i \mathbf{x}_i$, $\forall i = 1,2,3$, where H_i is the matrix solution of the Liapunov matrix equation $A_{ii}^T H_i + H_i A_{ii} = -G_i$, $\forall i = 1,2,3$, for arbitrary positive definite symmetric matrix $G_i \in R^{3\times 3}$. Since A_{ii} is stable, then H_i is also positive definite symmetric matrix. For such a choice of all v_i the following matrix A is obtained in view of Assumption 2 :

$$A = \begin{bmatrix} -\lambda_m(G_1) + 2\nu\|H_1\| \cdot \|r_{11}\| , & 2\|H_1\|(\|A_{12}\| + \nu\|r_{12}\|) , & 2\|H_1\|(\|A_{13}\| + \nu\|r_{13}\|) \\ 0 , & -\lambda_m(G_2) + 2\nu\|H_2\| \cdot \|r_{22}\| , & 0 \\ 0 , & 0 , & -\lambda_m(G_3) + 2\nu\|H_3\| \cdot \|r_{33}\| \end{bmatrix} .$$

$\lambda_m(G_i)$ is the minimal eigenvalue of the matrix G_i , and $\lambda_M(G_i)$ is its maximal eigenvalue.

If

$$\|r_{ii}\| < \frac{\lambda_m(G_i)}{2\nu\|H_i\|} , \quad \forall i = 1,2,3 ,$$

then all conditions of (II) of Theorem 6 are satisfied for $\eta_{i1} = \lambda_m(H_i)$, $\eta_{i2} = \lambda_M(H_i)$, $\eta_{i3} = 1$, $\forall i = 1,2,3$.

Hence, $x=0$ of the system is exponentially stable in the whole.

III.4.1.3. The third aggregation form

In contrast to the previous two aggregation forms, we consider such a form for which an aggregate system (a comparison system) is nonlinear.

Assumption 3. There exist open connected neighbourhoods N_i, $N_i \subseteq R^{ni}$ of the states $x_i = 0$, functions v_i, Ω_i and real numbers $\alpha_i \leq 0$, $\forall i = 1,2,\dots,s$, such that

1) v_i are positive definite on N_i, $\forall i = 1,2,\dots,s$;
2) Ω_i satisfy a W°-condition, ensure existence of a local solution of the Cauchy problem for a corresponding system of differential equations, and they are continuous;
3) there exists a neighbourhood U of the point $u=0$ such that for all $u \in \bar{U}$, $u \neq 0$, $\Omega(t,P,S,u) \neq 0$; $\Omega(t,P,S,0) = 0$;
4) $D_t^+ v_i(t,x_i) + [D_{x_i}^+ v_i(t,x_i)]^T g_i(t,x_i) \leq \alpha_i v_i(t,x_i)$, $\forall (t,x_i) \in R \times N_i$, $\forall i = 1,2,\dots,s$;
5) $[D_{x_i}^+ v_i(t,x_i)]^T S_i h_i(t,x,p_i) \leq \Omega_i(t,P,S,v_1,\dots,v_s)$, $\forall (t,x) \in R \times N$, $\forall (P,S) \in P \times S_s$, $\forall i = 1,2,\dots,s$. ■

This form of aggregation allows us to state the following.

Theorem 7. *Let*

(a) conditions of the Assumption 3 be fulfilled;

(b) there exist positive values ξ_i or $\xi_i = +\infty$ such that the sets $V_{i\zeta}(t)$ are asymptotically contractive for any $\zeta \in]0,\xi[$ and all $i = 1,2,\dots,s$;

(c) there exist $S^* \in S_i$ such that for functions $\Omega_i = \Omega_i(t,P,S,u)$ an estimation $\Omega_i(t,P,S,u) \leq \Omega_i(t,P,S^*,u)$, $\forall (t,P,S,u) \in R \times P \times S_s \times U$, $\forall i = 1,2,\dots,s$, is fulfilled;

(d) the state $u=0$ of the comparison system

$$\frac{du_i}{dt} = \alpha_i u_i + \Omega_i(t,P,S^*,u_1,\dots,u_s), \quad \forall i = 1,2,\dots,s,$$

be asymptotically stable.

Then the state $x=0$ of the system (11) is structurally asymptotically stable on $P \times S_s$.

If conditions (a)-(d) are fulfilled for $N_i = R^{ni}$, $U = R_+^s$ and functions v_i which are radially unbounded for $\xi_i = +\infty$, $i = 1,2,\dots,s$, then the state $x=0$ of system (11) is structurally asymptotically stable in the whole on $P \times S_s$.

Proof. Relying on the conditions (a) of Theorem 7 and 1) of Assumption 3, we conclude that the set $V_{1\zeta_1}(t) \times V_{2\zeta_2}(t) \times \dots \times V_{s\zeta_s}(t)$ is asymptotically contractive for every $\zeta_i \in]0,\xi_i[$, $\forall i = 1,2,\dots,s$. Fulfillment of conditions 4), 5) of Assumption 3 ensures the following estimation along the motions of the system (11)

$$D^+ v_i(t,x_i) \le \alpha v_i + \Omega_i(t,P,S,v_i,\dots,v_s) ,$$
$$\forall (t,x,P,S) \in R\times N\times P\times S_s , \quad \forall i = 1,2,\dots,s .$$

The corresponding comparison system

$$\frac{du_i}{dt} = \alpha_i u_i + \Omega_i(t,P,S^*,u_1,\dots,u_s) , \quad \forall i = 1,2,\dots,s ,$$

for $u_{i_0} \in V_{i\zeta_i}(t_0)$, $i = 1,2,\dots,s$, in view of conditions 2), 3) of the Assumption 3 and Proposition 10, Ch. II, allows us to obtain estimations

$$v_i(t,x_i) \le u_i(t;t_0, u_0, P,S^*) , \quad \forall t \in R , \quad i = 1,2,\dots,s .$$

Conditions (a), (d) of the Theorem 7 and the obtained estimation ensures structural asymptotic stability on $P\times S_s$ of the state $x = 0$ of the system (11).

If conditions of the Theorem 7 are fulfilled for $N_i = R^{ni}$, $\xi_i = +\infty$, $U = R^s_+$, then the radial unboundedness and positive definiteness in the whole of functions v_i together with conditions (b)-(d) of the theorem ensure asymptotic stability in the whole of the zero solution of the comparison system. This proves the second statement of the Theorem 7. ∎

Theorem 8. *Let*

(a) conditions of the Assumption 3 be fulfilled;

(b) all functions v_i *be decrescent on* N_i , $\forall i = 1,2,\dots,s$;

(c) there exist $S^* \in S_s$ *such that for functions* $\Omega_i(t,P,S,u)$ *an estimation* $\Omega_i(t,P,S,u) \le \Omega_i(t,P,S^*,u)$, $\forall (t,P,S,u) \in R\times P\times S_s\times U$ *holds;*

(d) the state $u = 0$ *of the comparison system*

$$\frac{du_i}{dt} = \alpha_i u_i + \Omega_i(t,P,S^*,u_1,\dots,u_s) , \quad i = 1,2,\dots,s$$

be uniformly asymptotically stable.

Then the state $x = 0$ *of the system (11) is structurally uniformly asymptotically stable on* $P\times S_s$.

If conditions (a)-(d) are fulfilled for $N_i = R^{ni}$ *and functions* v_i are *radially unbounded for all* $i = 1,2,\dots,s$ *, then the state* $x=0$ *of the system (11) is structurally uniformly asymptotically stable in the whole on* $P\times S_s$.

Proof. We obtain estimations

$$v_i(t,x_i) \le u_i(t;t_o, u_o, P, S^*) \ , \quad \forall t\in R \ , \ i = 1,2,\dots,s$$

similarly as in the Theorem 7. It follows from the condition that the state $u=0$ is uniformly asymptotically stable and from decreasing of functions v_i that the state $x=0$ of the system (11) is structurally asymptotically stable on $P\times S_s$.

If conditions (a)-(d) of the Theorem 8 are fulfilled for $N_i = R^{ni}$, $U = R_+^s$, radially unbounded functions v_i , that are positive definite in the whole and decrescent for all $i = 1,2,\dots,s$, then the estimations of functions v_i and uniform asymptotic stability in the whole of the state $u=0$ ensure fulfillment of the second statement of the Theorem 8. ∎

Theorem 9. *Let*

(a) conditions of the Assumption 3 with constants $\alpha_i < 0$ *,* $\forall i = 1,2,\dots,s$ *be fulfilled;*

(b) there exist positive values η_{i1}, η_{i2} *such that*

$$\eta_{i1}\|x_i\| \le v_i(t,x_i) \le \eta_{i2}\|x_i\|$$

for all $(t,x_i)\in R\times N_i$ *,* $i = 1,2,\dots,s$ *;*

(c) there exist $S^*\in S_s$ *such that for functions* $\Omega_i(t,P,S,u)$ *an estimation*

$$\Omega_i(t,P,S,u) \le \Omega_i(t,P,S^*,u) \ , \ i = 1,2,\dots,s \ , \ \forall (t,P,S,u)\in R\times P\times S_s\times U$$

holds;

(d) the state $u=0$ *of the comparison system*

$$\frac{du_i}{dt} = \alpha_i u_i + \Omega_i(t,P,S^*,u_1,\dots,u_s) \ , \ i = 1,2,\dots,s$$

be exponentially stable.

Then the state $x=0$ *of the system (11) is structurally exponentially stable on* $P\times S_s$.

If conditions (a)-(d) are fulfilled for $N_i = R^{ni}$ *,* $U = R_+^s$ *,* $\forall i\in[1,s]$ *, then the state* $x=0$ *of the system (11) is structurally exponentially stable in the whole on* $P\times S_s$.

Proof. Due to conditions (a)-(c) of the Theorem 9 we determine estimations

$$v_i(t,x_i) \le u_i(t;t_o, u_o, P, S^*) , \forall t \in R , i = 1,2,\dots,s ,$$

where $$u_{io} \ge \eta_{i2} \| x_{io} \| , i = 1,2,\dots,s .$$

As far as the state $u=0$ of the comparison system is exponentially stable, we can find such $\Delta>0$ and positive values $\gamma \ge 1$ and $\beta>0$ that it follows from condition $u_{io}<\Delta$ that

$$\sum_{i=1}^{s} u_i(t;t_o, u_o, P, S^*) \le \gamma \sum_{i=1}^{s} u_{io} \exp[-\beta(t-t_o)] .$$

From conditions (a) of the theorem and 1) of the Assumption 3, we get

$$\eta_1 \|x\| \le v(t,x) = \sum_{i=1}^{s} v_i(t,x_i) \le \gamma \sum_{i=1}^{s} u_{io} \exp[-\beta(t-t_o)] .$$

Hence

$$\|x(t;t_o,x_o)\| \le \gamma^* \sum_{i=1}^{s} u_{io} \exp[-\beta(t-t_o)] ,$$

where $\gamma^* = \eta_1^{-1}\gamma$.

This proves the first statement of the Theorem 9.

If conditions (a)-(d) of the Theorem 9 are fulfilled for $N_i = R^{n_i}$, $U = R_+^s$, $i \in [1,s]$, then the value Δ can be chosen equal to $+\infty$ in estimations and, hence, an exponential estimation of the solution of the system (11) will be fulfilled in the whole. The second statement of the Theorem 9 follows from all this. ■

III.4.1.4. Conclusion on solutions for the Problem A

Solutions for the Problem A require decomposition of the large-scale systems into interconnected subsystems (S_i) so that the form of the corresponding disconnected subsystems $(\hat{S}_i)$ is suitable for effective construction of their Liapunov functions. Then, exact information about stability properties of all disconnected subsystems $(\hat{S}_i)$ is needed for the final stability test. Furthermore, the aggregation conditions (the conditions 5 of the Assumptions 1 and 2) imposed on the interactions are expressed in terms of the functions used as comparison functions (Ω_i) for Dini derivatives of the Liapunov functions (v_i) taken along motions of the disconnected subsystems $(\hat{S}_i)$.

The number of required stability tests is for one greater than the number of subsystems (i.e. $s+1$).

III.4.2. Aggregation forms and solutions for the Problem B

Unlike the former aggregation forms, the next aggregation forms are related to interconnected subsystems S_i . Let $v_i \in C(R \times R^{n_i}, R)$ and

$$v_i^*(t,x,p_i,S_i) = \begin{cases} D_t^+ v_i(t,x_i) + [D_{x_i}^+ v_i(t,x_i)]^T [g_i(t,x_i) + S_i h_i(t,x,p_i)] , \\ \text{or} \\ D^+ v_i(t,x) \quad , \quad i = 1,2,\dots,s . \end{cases}$$

The function v_i is the i-th component of the vector (tentative Liapunov) function v, $v = [v_1,v_2,\dots,v_s]^T$.

III.4.2.1. The fourth aggregation form and stability criteria

The fourth aggregation form is defined by

Assumption 4. There exist connected neighbourhoods N_i , $N_i \subseteq R^{ni}$, of $x_i = 0$, functions v_i and $\zeta_i : R\times R \to R$ and real numbers $a_{ij}(S_i)$, $b_{ij}(S_i)$ and c_{ij} such that

1) $a_{ij}(S_i) \geq 0$, $b_{ij}(S_i) \geq 0$, $c_{ij} \geq 0$, $\forall i,j = 1,2,\dots,s$, $i \neq j$, $\forall S \in S_s$,

2) v_i is positive definite on N_i , $\forall i = 1,2,\dots,s$,

3) $\zeta_i(0) = 0$, $\zeta_i(\rho_i') \leq \zeta_i(\rho_i'')$, $\forall \rho_i', \rho_i'' \in R$, $\rho_i' \leq \rho_i''$, $\forall i = 1,2,\dots,s$,

4) $0 \leq \zeta_i(\rho_i) / \rho_i < \kappa_i$, $\forall \rho_i \in R$, $\rho_i \neq 0$, $0 < \kappa_i \leq +\infty$, where

$$\rho_i = \sum_{j=1}^{s} c_{ij} v_j , \quad \forall i = 1,2,\dots,s ,$$

5) $v_i^*(t,x,p_i,S_i) \leq \sum_{j=1}^{s} a_{ij}(S_i)\, v_j(t,x_j) + \sum_{j=1}^{s} b_{ij}(S_i)\, \zeta_j(\rho_j)$,

$\forall i = 1,2,\dots,s$, $\forall (t,x,P,S) \in R\times N\times P\times S_s$,

and

6) there are $a_{ij} \in R$, $b_{ij} \in R$, $a_{ij} = a_{ij}(S^*)$ and $b_{ij} = b_{ij}(S^*)$ for some $S^* \in S_s$ are admissible, $a_{ij}(S_i) \leq a_{ij}$, $b_{ij}(S_i) \leq b_{ij}$, $\forall i,j = 1,2,\dots,s$, $\forall S \in S_s$. ■

This aggregation form introduces comparison matrices $A = (a_{ij})$, $B = (b_{ij})$ and $C = (c_{ij})$, where $A,B,C \in R^{s\times s}$, and the comparison vector functions $z = [\zeta_1,\zeta_2,\dots,\zeta_s]^T$ and $r = [\rho_1,\rho_2,\dots,\rho_s]^T$. The conditions 4)-6) of the Assumption 4 imply the comparison vector differential inequality of the Lur'e type

$$v^*(t,x,P,S) \leq Av(t,x) + Bz(r) \;,\; r = Cv(t,x) \;,\; \forall (t,x,P,S) \in R\times N\times P\times S_s , \tag{17a}$$

or simply

$$v^* \leq Av + Bz(r) \;,\; r = Cv . \tag{17b}$$

Let $K = \mathrm{diag}\{\kappa_1,\kappa_2,\dots,\kappa_s\}$.

Theorem 10. (I) *For structural asymptotic stability of* $x = 0$ *of the system (11) over* $P\times S_s$ *it is sufficient that*

(a) Assumption 4 holds,

(b) there is positive number ξ_i *or* $\xi_i = +\infty$ *such that the set* $V_{i\zeta}(t)$ *is asymptotically contractive for every* $\zeta \in]0,\xi_i[$ *and every* $i = 1,2,\dots,s$,

(c) either the matrix $B = (b_{ij})$ *or* $C = (c_{ij})$ *is positive diagonal,*

(d) the pair (A,B) *is controllable,* (A,C) *is observable and the matrix* A *is stable,*

(e) there is a non-negative diagonal $s \times s$ *matrix* Θ *such that the matrix* $Q(\omega^2)$,

$$Q(\omega^2) = K^{-1} + \mathrm{He}(I + j\omega\Theta)\, C(A - j\omega I)^{-1} B \,, \tag{18}$$

is positive definite for every $\omega \in [0,+\infty]$.

(II) *If, in addition to (I),* $N_i = R^{n_i}$, $\xi_i = +\infty$ *and* v_i *is radially unbounded for every* $i = 1,2,\dots,s$ *then* $x = 0$ *of the system (11) is structurally asymptotically stable in the whole over* $P \times S_s$.

Proof. (I) Let $u \in R^s$ be described by the comparison differential equation of the Lur'e type

$$\dot{u} = Au + Bz(p) \quad , \quad p = Cu \,. \tag{19}$$

The conditions (c)-(e) guarantee absolute stability of $u = 0$ on $[0,K[$. This result, the comparison principle (Kamke, 1932, Ważevski, 1950; Matrosov, 1968, see also Chapter II) that is valid over $P \times S_s$ due to the conditions 1,3-6 of the Assumption 4, positive definiteness of all v_i that is guaranteed by 2 of the Assumption 4 and the conditions (a) and (b) prove structural asymptotic stability of $x = 0$ of (11) over $P \times S_s$. Then, the assertion under (II) is evidently true because all $N_i = R^{n_i}$, $\xi_i = +\infty$ and all v_i are radially unbounded. ∎

Uniformity of the structural asymptotic stability, which is not guaranteed by Theorem 10, is ensured by

Theorem 11. (I) *For structural uniform asymptotic stability of* $x = 0$ *of the system (11) over* $P \times S_s$ *it is sufficient that*

(a) Assumption 4 holds,

(b) the function v_i *is decrescent on* N_i , $\forall i = 1,2,\dots,s$,

(c) either the matrix $B = (b_{ij})$ *or* $C = (c_{ij})$ *is positive diagonal,*

(d) the pair (A,B) *is controllable,* (A,C) *is observable and the matrix* A *is stable,*

(e) there is a non-negative diagonal $s \times s$ *matrix* Θ *such that the matrix* $Q(\omega^2)$ *(18) is positive definite for every* $\omega \in [0,+\infty]$.

(II) *If, in addition to (I),* $N_i = R^{n_i}$ *and* v_i *is radially unbounded for every* $i = 1,2,\dots,s$ *then* $x = 0$ *of the system (11) is structurally uniformly asymptotically stable in the whole over* $P \times S_s$.

Proof. (I) The conditions (c)-(e) guarantee absolute stability of $u=0$ of the comparison system (19) on $[0,K[$. This result, the comparison principle (Kamke, 1932; Ważewski, 1950; Matrosov, 1968; see also Chapter II) that is valid over $P \times S_s$ due to the conditions 1, 3-6 of the Assumption 4, positive definiteness of all v_i that is guaranteed by 2 of the same assumption and the conditions (a) and (b) prove structural uniform asymptotic stability of $x=0$ of (11) over $P \times S_s$. If, in addition, all $N_i = R^{n_i}$ and v_i are radially unbounded then $x=0$ of (11) is structurally uniformly asymptotically stable in the whole over $P \times S_s$. ■

Notice that the fourth aggregation form is useful only if there is $i \in \{1,2,\dots,s\}$ for which $\kappa_i = +\infty$. Otherwise all $\kappa_i < +\infty$ and hence, due to $\kappa_i > 0$ and $b_{ij} \geq 0$ (1 of the Assumption 4), the comparison differential inequality (17) takes the form

$$v^* \leq (A + BKC)\, v \qquad \text{or} \qquad v^* \leq \hat{A} v \ , \ \hat{A} = A + BKC \ ,$$

which is a special case of the first aggregation form. In the case there is $\kappa_i = +\infty$ then absolute stability of $u=0$ of the comparison system (19) is not proved to be exponential, and therefore, structural exponential stability of $x=0$ of (11) over $P \times S_s$ cannot be guaranteed via the comparison system stability in such a case.

Example 3. The sixth order system is described by $(n_i = 2 \ , \ i = 1,2,3 \ ; \ s = 3)$

$$\frac{dx_1}{dt} = x_1 + \begin{pmatrix} 0 \\ 1 \end{pmatrix} \phi_1(\sigma_1) + \begin{bmatrix} -\dfrac{6+12t^2}{1+t^2}\, x_{11} + (|x_{21}| - 0.3|x_{31}| - 3|x_{11}| - 2|x_{12}|)^3 \operatorname{sign} x_{11} \\ -\dfrac{5+15t^2}{1+2t^2}\, x_{12} + (0.7|x_{22}| - 2|x_{11}| - 5|x_{12}|)^3 \operatorname{sign} x_{12} \end{bmatrix}$$

$$\frac{dx_2}{dt} = -6x_2 + \begin{pmatrix} 1 \\ 1 \end{pmatrix} \phi_2(\sigma_2) + \underbrace{\left[\begin{pmatrix} s_{21} & 0 \\ 0 & s_{21} \end{pmatrix} \begin{pmatrix} s_{22} & 0 \\ 0 & s_{22} \end{pmatrix} \begin{pmatrix} s_{23} & 0 \\ 0 & s_{23} \end{pmatrix} \begin{pmatrix} s_{24} & 0 \\ 0 & s_{24} \end{pmatrix} \right]}_{S_2} \underbrace{\begin{bmatrix} h_2^1 \\ h_2^2 \\ h_2^3 \\ h_2^4 \end{bmatrix}}_{h_2}$$

where

$$\mathbf{h}_2^1 = \begin{bmatrix} -(4+p_{21})x_{21} + (-2|x_{21}| - 6|x_{22}| + 0.9|x_{31}|)^5 \operatorname{sign} x_{21} \\ -(8+p_{22})x_{22} + (-3|x_{21}| - 2|x_{22}| + |x_{31}| + 0.4|x_{32}|)^5 \operatorname{sign} x_{22} \end{bmatrix},$$

$$\mathbf{h}_2^2 = \mathbf{h}_2^1 - \begin{bmatrix} x_{21} \\ 0 \end{bmatrix},$$

$$\mathbf{h}_2^3 = \begin{bmatrix} -(4+p_{21})x_{21} + (-2|x_{21}| - 3|x_{22}|)^5 \operatorname{sign} x_{21} \\ -(8+p_{22})x_{22} + (-2|x_{21}| - 7|x_{22}| + |x_{31}|)^5 \operatorname{sign} x_{22} \end{bmatrix},$$

$$\mathbf{h}_2^4 = \mathbf{h}_2^3 - \begin{bmatrix} x_{21} \\ 0 \end{bmatrix},$$

and

$$\mathbf{h}_2 = \begin{bmatrix} \mathbf{h}_2^1 \\ \mathbf{h}_2^2 \\ \mathbf{h}_2^3 \\ \mathbf{h}_2^4 \end{bmatrix},$$

$$\frac{d\mathbf{x}_3}{dt} = 2\mathbf{x}_3 + \begin{bmatrix} 0.2 \\ 0.2 \end{bmatrix} \phi_3(\sigma_3) +$$

$$+ \begin{bmatrix} -\frac{10(1+t)}{1+0.5t} x_{31} + (|x_{11}| + 0.6|x_{12}| - 2.2|x_{31}| - 2|x_{32}|)^{1/3} \operatorname{sign} x_{31} \\ -\frac{10(1+2t)}{1+t} x_{32} + (0.4|x_{11}| - 3|x_{31}| - 8|x_{32}|)^{1/3} \operatorname{sign} x_{32} \end{bmatrix},$$

where

$$\sigma_1 = x_{11} + x_{12}, \quad \sigma_2 = x_{21} + 2x_{22}, \quad \sigma_3 = 2x_{31} + 3x_{32},$$

$$\frac{\phi_1(\sigma_1)}{\sigma_1} \in [0,2], \quad \frac{\phi_2(\sigma_2)}{\sigma_2} \in [0,1], \quad \frac{\phi_3(\sigma_3)}{\sigma_3} \in [0,5],$$

$$\phi_i(\sigma_i) \in C(R,R), \quad i = 1,2,3,$$

$$P = \left\{ \mathbf{P} : \begin{bmatrix} 0 & 0 & 0 \\ -4 & -6 & 0 \\ 0 & 0 & 0 \end{bmatrix} \le \mathbf{P} \le \begin{bmatrix} 0 & 0 & 0 \\ 1 & 4 & 0 \\ 0 & 0 & 0 \end{bmatrix} \right\},$$

$$\mathbf{S} = \mathrm{diag}\{\mathbf{S}_1,\mathbf{S}_2,\mathbf{S}_3\} \ ,$$

$$\mathcal{S}_s = \left\{ \mathbf{S} : \begin{bmatrix} \mathbf{I}_2 & 0 & 0 \\ 0 & \mathbf{0} & 0 \\ 0 & 0 & \mathbf{I}_2 \end{bmatrix} \le \begin{bmatrix} \mathbf{S}_1 & 0 & 0 \\ 0 & \mathbf{S}_2 & 0 \\ 0 & 0 & \mathbf{S}_3 \end{bmatrix} \le \begin{bmatrix} \mathbf{I}_2 & 0 & 0 \\ 0 & \binom{10101010}{01010101} & 0 \\ 0 & 0 & \mathbf{I}_2 \end{bmatrix} \right\} .$$

These disconnected subsystems, which can be described by

$$\frac{d\mathbf{x}_i}{dt} = \alpha_i \mathbf{x}_i + \mathbf{b}_i \phi_i(\sigma_i) \ , \quad \sigma_i = \mathbf{c}_i^T \mathbf{x}_i \ , \quad i = 1,2,3 \ ,$$

are of the Lur'e-Postnikov type, the first and third of which are unstable.

The form of the interactions suggests the form of v_i ,

$$v_i(\mathbf{x}_i) = |x_{i1}| + |x_{i2}| \ , \ \forall i = 1,2,3 \ .$$

It now results

$$\mathbf{v}^* \le -2\mathbf{v} + 2\mathbf{z}(\mathbf{r}) \ , \quad \mathbf{r} = \mathbf{C}\mathbf{v}$$

together with

$\mathbf{A} = -2\mathbf{I}$, $\mathbf{B} = 2\mathbf{I}$, $\mathbf{r} = (\rho_1,\rho_2,\rho_3)^T$, $\mathbf{z}(\mathbf{r}) = (\rho_1^3,\rho_2^5,\rho_3^{1/3})^T$, $\kappa_i = \infty$, $i = 1,2,3$

$$\mathbf{C} = \begin{bmatrix} -2 & 1 & 0 \\ 0 & -2 & 1 \\ 1 & 0 & -2 \end{bmatrix}$$

Hence,

$$\zeta_1(\rho_1) = \rho_1^3 \ , \quad \zeta_1(\rho_1') \le \zeta_1(\rho_1'') \quad \text{for} \quad \rho_1' \le \rho_1'' \ ,$$
$$\zeta_2(\rho_2) = \rho_2^5 \ , \quad \zeta_2(\rho_2') \le \zeta_2(\rho_2'') \quad \text{for} \quad \rho_2' \le \rho_2'' \ ,$$
$$\zeta_3(\rho_3) = \rho_3^{1/3} \ , \quad \zeta_3(\rho_3') \le \zeta_3(\rho_3'') \quad \text{for} \quad \rho_3' \le \rho_3'' \ .$$

All conditions of the Assumption 4 have been verified. The pair $(\mathbf{A},\mathbf{B})$ is controllable, $(\mathbf{A},\mathbf{C})$ is observable and the matrix $\mathbf{A}$ is stable. Let $\Theta = \frac{1}{2}\mathbf{I}$ so that

$$\mathbf{K}^{-1} + \mathrm{He}(\mathbf{I} + j\omega\Theta)\,\mathbf{C}(\mathbf{A} - j\omega\mathbf{I})^{-1}\,\mathbf{B} = \frac{1}{2}\begin{bmatrix} 4 & -1 & -1 \\ -1 & 4 & -1 \\ -1 & -1 & 4 \end{bmatrix} > 0 \ , \quad \forall\omega \in [0,+\infty] \ .$$

All conditions of (II) of Theorem 11 are satisfied. The equilibrium state $\mathbf{x} = \mathbf{0}$ of the whole system is structurally uniformly asymptotically stable in the whole over $P \times \mathcal{S}_s$ for all admissible nonlinearities ϕ_i in the sector $[0,+\infty[$.

In this example the stability test has been carried out via the third order comparison system despite the whole system has been of the sixth order. The vector Liapunov function used in the example has had the form

$$v(x) = \begin{bmatrix} |x_{11}| + |x_{12}| \\ |x_{21}| + |x_{22}| \\ |x_{31}| + |x_{32}| \end{bmatrix} .$$

III.4.2.2. The fifth aggregation form

Although the fifth aggregation form can be sometimes considered as a special case obtained for $B = 0$ (hence, $C = 0$) from the fourth aggregation form, it is in general another aggregation form defined by

Assumption 5. There are connected neighbourhoods N_i, $N_i \subseteq R^{ni}$, of $x_i = 0$, functions v_i and Ω_i and real numbers $a_{ij}(S_i)$ such that

1) $a_{ij}(S_i) \geq 0$, $i \neq j$, $\forall i,j = 1,2,\dots,s$, $\forall S \in S_s$,
2) v_i and Ω_i are both positive definite on N_i, $\forall i = 1,2,\dots,s$,
3) $v_i^*(t,x,p_i,S_i) \leq \sum_{j=1}^{s} a_{ij}(S_i)\, \Omega_j(t,x_j)$, $\forall i = 1,2,\dots,s$,

   $\forall (t,x,P,S) \in R \times N \times P \times S_s$,

and

4) there are real numbers a_{ij}, b_{ij}, $a_{ij} = a_{ij}(S_i^*)$ and $b_{ij} = b_{ij}(S_i^*)$ for some $S^* \in S_s$ are admissible, such that $a_{ij}(S_i) \leq a_{ij}$, $\forall i,j = 1,2,\dots,s$, $\forall S \in S_s$.

This form introduces only one $(s \times s)$ aggregation matrix $A = (a_{ij})$ and the comparison vector differential inequality

$$v^*(t,x,P,S) \leq A(S)\, w(t,x) \leq Aw(t,x) \quad , \forall (t,x,P,S) \in R \times N \times P \times S_s . \tag{20}$$

Theorem 12. (I) *For structural asymptotic stability of* $x = 0$ *of the system (11) over* $P \times S_s$ *it is sufficient that*

(a) Assumption 5 holds,

(b) there is positive number ξ_i *or* $\xi_i = +\infty$ *such that the set* $V_{i\zeta}(t)$ *is asymptotically contractive for every* $\zeta \in]0,\xi_i[$ *and every* $i = 1,2,\dots,s$,

and

(c) the matrix A *is stable.*

(II) *If, in addition to (I),* $N_i = R^{ni}$, $\xi_i = +\infty$ *and* v_i *is radially unbounded for every* $i = 1,2,\dots,s$, *then* $x = 0$ *of the system (11) is structurally asymptotically stable in the whole over* $P \times S_s$.

Proof. (I). In view of the condition (a) and the conditions 1) and 4) of the Assumption 5, it follows that all off diagonal elements of the matrix $\mathbf{A}$ are non-negative. Hence, the condition (c) guarantees existence of s-vector $\mathbf{b}$ elementwise positive such that for arbitrary s-vector $\mathbf{c}$ elementwise positive the following holds (Fiedler and Ptak, 1962) :

$$\mathbf{A}^T\mathbf{b} = -\mathbf{c} . \tag{21}$$

For such a vector $\mathbf{b}$ a tentative Liapunov function v is defined by $v(t,\mathbf{x}) = \mathbf{b}^T\mathbf{v}(t,\mathbf{x})$.

Now, the conditions (a) of the theorem, 1),3) and 4) of the Assumption 5 and (21) yield

$$v^*(t,\mathbf{x},\mathbf{P},\mathbf{S}) \le -\mathbf{c}^T\mathbf{w}(t,\mathbf{x}) , \ \forall(t,\mathbf{x},\mathbf{P},\mathbf{S}) \in R\times N\times P\times S_s .$$

This result, positive definiteness of both v_i and Ω_i on N_i and asymptotic contraction of $V_{i\zeta}(t)$ for all $i = 1,2,\dots,s$, prove that v obeys all conditions of Theorem 13 of the Section I.3.2.5 over $P\times S_s$. Hence, $\mathbf{x} = 0$ of (11) is structurally asymptotically stable over $P\times S_s$.

(II) Now, $N_i = R^{ni}$, $\xi_i = +\infty$ and radial unboundedness of v_i for every $i = 1,2,\dots,s$, show that v obeys all conditions of Corollary 2 of the Section I.3.2.5 over $P\times S_s$, which completes the proof. ∎

If it is desirable that the structural asymptotic stability be uniform then

Theorem 13. (I) *For structural uniform asymptotic stability of* $\mathbf{x} = 0$ *of the system (11) over* $P\times S_s$ *it is sufficient that*

(a) Assumption 5 holds,

(b) the function v_i *is decrescent on* N_i , $\forall i = 1,2,\dots,s$,

and

(c) the matrix $\mathbf{A}$ *is stable.*

(II) *If, in addition to (I),* $N_i = R^{ni}$ *and* v_i *is radially unbounded then* $\mathbf{x} = 0$ *of the system (11) is structurally uniformly asymptotically stable in the whole over* $P\times S_s$.

Proof. (I) The condition (a) and the conditions 1) and 4) of the Assumption 5, as in the proof of Theorem 12, guarantee that for arbitrary s-vector $\mathbf{c} > 0$ there is s-vector $\mathbf{b} > 0$ satisfying (21). For such a vector $\mathbf{b}$ the function $v = \mathbf{b}^T\mathbf{v}$ obeys

$$v^*(t,\mathbf{x},\mathbf{P},\mathbf{S}) \le -\mathbf{c}^T\mathbf{w}(t,\mathbf{x}) , \ \forall(t,\mathbf{x},\mathbf{P},\mathbf{S}) \in R\times N\times P\times S_s . \tag{22}$$

This result, positive definiteness and decrescency of v_i on N_i and positive definiteness of Ω_i on N_i for every $i = 1,2,\dots,s$, prove that the function v obeys all conditions of Theorem 7 of the Section I.3.2.4 over $P\times S_s$. Hence, $x=0$ of (11) is structurally uniformly asymptotically stable over $P\times S_s$.

(II) When, in addition to (I), $N_i = R^{n_i}$ and v_i is radially unbounded for every $i = 1,2,\dots,s$, then the function v obeys all conditions of Theorem 8 of the Section I.3.2.4 over $P\times S_s$, which completes the proof. ■

The structural exponential stability requires specific properties of the functions v_i and Ω_i as shown by

Theorem 14. (I) *For structural exponential stability of* $x=0$ *of the system (11) over* $P\times S_s$ *it is sufficient that*

(a) Assumption 5 holds,

(b) there are positive numbers η_{i1}, η_{i2} *and* η_{i3} *obeying* $\eta_{i1}\|x_i\| \le v_i(t,x_i) \le \eta_{i2}\|x_i\|$, *and* $\Omega_i(t,x_i) = \eta_{i3}\|x_i\|$ *for every* $(t,x_i)\in R\times N_i$ *and* $i = 1,2,\dots,s$,

and

(c) the matrix A *is stable.*

(II) *If, in addition to (I),* $N_i = R^{n_i}$ *for every* $i = 1,2,\dots,s$ *then* $x=0$ *of the system (11) is structurally exponentially stable in the whole over* $P\times S_s$.

Proof. (I) Let the vector $b>0$ and the function v be defined so as in the proof of Theorem 13. Then (22) holds. In view of the condition (b), it now follows that there are positive numbers η_1, η_2 and η_3 and a connected neighbourhood N of $x=0$ such that both

$$\eta_1\|x\| \le v(t,x) \le \eta_2\|x\|,$$

and

$$v^*(t,x,P,S) \le -\eta_3\|x\|, \quad \forall (t,x,P,S)\in R\times N\times P\times S_s,$$

which shows that all conditions of Theorem 9 of the Section I.3.2.4 are satisfied over $P\times S_s$. Hence, $x=0$ of (11) is structurally exponentially stable over $P\times S_s$.

(II) $N_i = R^{n_i}$, $\forall i = 1,2,\dots,s$, implies $N = R^n$, which together with (I) shows that all conditions of Theorem 10 of the Section I.3.2.4 are satisfied over $P\times S_s$. Hence, then $x=0$ of (11) is structurally exponentially stable in the whole over $P\times S_s$. ■

Example 4. A sixth order Lur'e system is composed of three second order interconnected subsystems (S_i) described by

$$\frac{d\mathbf{x}_i}{dt} = \begin{bmatrix} -10 & 1 \\ 4 & -10 \end{bmatrix} \mathbf{x}_i + \begin{bmatrix} 0 \\ s_{i1} \end{bmatrix} \phi_i(\sigma_i) + \begin{bmatrix} 25s_{i2} \\ 25s_{i2} \end{bmatrix} \psi_i(\hat{\sigma}_i) \quad , \quad \forall i = 1,2,3 \; ,$$

$$\sigma_i = 2x_{i1} + x_{i2} \; , \; \phi_i(\sigma_i)\,\sigma_i^{-1} \in [0,2] \; , \; \psi_i(\hat{\sigma}_i)\,\hat{\sigma}_i^{-1} \in [1,2] \; , \; \forall i = 1,2,3 \; ,$$

$$\phi_i(\sigma_i) \in C(R,R) \; , \; \psi_i(\hat{\sigma}_i) \in C(R,R) \; ,$$

and

$$\hat{\sigma}_1 = -2(x_{11} + x_{12}) + 0.005(x_{21} + x_{22}) \; ,$$

$$\hat{\sigma}_2 = -2(x_{21} + x_{22}) + 0.005(x_{31} + x_{32}) \; ,$$

$$\hat{\sigma}_3 = 0.005(x_{21} + x_{22}) - 2(x_{31} + x_{32}) \; .$$

The structural matrices $\mathbf{S}_i$ and $\mathbf{S}$ have the form

$$\mathbf{S}_i = \begin{bmatrix} 0 & s_{i2} \\ s_{i1} & s_{i2} \end{bmatrix} \quad \text{for} \quad \mathbf{h}_i = \begin{bmatrix} \phi_i \\ 25\psi_i \end{bmatrix} \; , \quad \forall i = 1,2,3 \; ,$$

and

$$\mathbf{S} = \text{diag}\{\mathbf{S}_1, \mathbf{S}_2, \mathbf{S}_3\} \; ,$$

where $s_{ij}(t) = s_{ik}(t)$ is admissible for $j \neq k$, $\forall i,j,k = 1,2,3$. The structural set $\mathcal{S}_s$ is determined now by

$$\mathcal{S}_s = \left\{ \mathbf{S} : \mathbf{0} \le \mathbf{S} \le \text{diag}\left\{ \begin{bmatrix} 0 & 1 \\ 1 & 1 \end{bmatrix}, \begin{bmatrix} 0 & 1 \\ 1 & 1 \end{bmatrix}, \begin{bmatrix} 0 & 1 \\ 1 & 1 \end{bmatrix} \right\} \right\}$$

Let $v_i(\mathbf{x}_i) = |x_{i1}| + |x_{i2}|$, $\forall i = 1,2,3$. Then, $\Omega_i(\mathbf{x}_i) = v_i(\mathbf{x}_i)$, $N_i = R^2$, $\forall i = 1,2,3$,

$$\begin{aligned} a_{11}(\mathbf{S}_1) &= -6 + 4s_{11} \; , & a_{12}(\mathbf{S}_1) &= 0.5s_{12} \; , & a_{13}(\mathbf{S}_1) &= 0 \; , \\ a_{21}(\mathbf{S}_2) &= 0 \; , & a_{22}(\mathbf{S}_2) &= -6 + 4s_{21} \; , & a_{23}(\mathbf{S}_2) &= 0.5s_{22} \; , \\ a_{31}(\mathbf{S}_3) &= 0.5s_{32} \; , & a_{32}(\mathbf{S}_3) &= 0 \; , & a_{33}(\mathbf{S}_3) &= -6 + 4s_{31} \; , \end{aligned}$$

and

$$\mathbf{S}^* = \text{diag}\left\{ \begin{bmatrix} 0 & 1 \\ 1 & 1 \end{bmatrix}, \begin{bmatrix} 0 & 1 \\ 0 & 1 \end{bmatrix}, \begin{bmatrix} 0 & 1 \\ 1 & 1 \end{bmatrix} \right\}$$

obey all conditions of Assumption 5. In this case

$$\mathbf{A} = \begin{bmatrix} -2 & 0.5 & 0 \\ 0 & -2 & 0.5 \\ 0.5 & 0 & -2 \end{bmatrix}$$

obeys $\mathbf{A} \ge \mathbf{A}(\mathbf{S})$, $\forall \mathbf{S} \in \mathcal{S}_s$, because

$$A(S) = \begin{bmatrix} -6+4s_{11} & 0.5\,s_{12} & 0 \\ 0 & -6+4s_{21} & 0.5\,s_{22} \\ 0.5\,s_{32} & 0 & -6+4s_{31} \end{bmatrix}.$$

The function v_i is positive definite in the whole, decrescent and radially unbounded for every $i = 1,2,3$. Hence, all conditions of (II) of Theorem 13 are satisfied. The equilibrium state $x = 0$ is structurally asymptotically stable in the whole over S_s.

The fifth aggregation form can be considered as an adaptation of the first aggregation form for the Problem B.

III.4.2.3. The sixth aggregation form

The sixth aggregation form is defined by

Assumption 6. There exist a connected neighbourhood N_i of $x_i = 0$, functions v_i and Ω_i and real numbers $a_{ij}(S_i)$ such that

1) $a_{ij}(S_i) \geq 0$, $i \neq j$, $\forall i,j = 1,2,\dots,s$,
2) v_i and Ω_i are both positive definite on N_i, $\forall i = 1,2,\dots,s$,
3) $v_i^*(t,x,p_i,S_i) \leq \Omega_i(t,x_i) \sum_{j=1}^{s} a_{ij}(S_i)\,\Omega_j(t,x_j)$, $\forall i = 1,2,\dots,s$, $\forall (t,x,P,S) \in R\times N\times P\times S_s$,

and

4) there is $S^* \in S_s$ such that for $a_{ij} = a_{ij}(S_i^*)$, $a_{ij}(S_i) \leq a_{ij}$, $\forall i,j = 1,2,\dots,s$, $\forall S \in S_s$. ∎

Let $A = (a_{ij})$ be determined by the Assumption 6.

Theorem 15. (I) *For structural asymptotic stability of* $x = 0$ *of the system (11) over* $P\times S_s$ *it is sufficient that*

(a) Assumption 6 holds,
(b) there is positive number ξ_i *or* $\xi_i = +\infty$ *such that the set* $V_{i\zeta}(t)$ *is asymptotically contractive for every* $\zeta \in]0,\xi_i[$ *and every* $i = 1,2,\dots,s$,

and

(c) the aggregation matrix A *is stable.*

(II) *If, in addition to (I),* $N_i = R^{n_i}$, $\xi_i = +\infty$ *and* v_i *is radially unbounded then* $x = 0$ *of the system (11) is structurally asymptotically stable in the whole over* $P\times S_s$.

This theorem is proved in the same way as Theorem 4. ∎

Uniformity of the structural asymptotic stability is guaranteed by

Theorem 16. (I) *For structural uniform asymptotic stability of* $x=0$ *of the system (11) over* $P \times S_s$ *it is sufficient that*

(a) Assumption 6 holds,

(b) the function v_i *is decrescent on* N_i , $\forall i=1,2,\dots,s$,

and

(c) the aggregation matrix A *is stable.*

(II) *If, in addition to (I),* $N_i = R^{ni}$ *and* v_i *is radially unbounded for every* $i=1,2,\dots,s$ *then* $x=0$ *of the system (11) is structurally uniformly asymptotically stable in the whole over* $P \times S_s$

Proof of this theorem is essentially the same as that of Theorem 5. ∎

Refined properties of the functions v_i and Ω_i lead tc the exponential character of the structural asymptotic stability, which is precisely expressed by

Theorem 17. (I) *For structural exponential stability of* $x=0$ *of the system (11) over* $P \times S_s$ *it is sufficient that*

(a) Assumption 6 holds,

(b) there exist positive numbers η_{i1}, η_{i2} *and* η_{i3} *such that both* $\eta_{i1}\|x_i\|^2 \le v_i(t,x_i) \le \eta_{i2}\|x_i\|^2$ *and* $\Omega_i(t,x_i) = \eta_{i3}\|x_i\|$ *hold for every* $(t,x_i) \in R \times N_i$ *and* $i=1,2,\dots,s$,

and

(c) the aggregation matrix A *is stable.*

(II) *If, in addition to (I),* $N_i = R^{ni}$ *for every* $i=1,2,\dots,s$ *then* $x=0$ *of the system (11) is structurally exponentially stable in the whole over* $P \times S_s$.

This theorem is essentially proved in the same way as Theorem 6. ∎

Example 5. A sixth order Lur'e system is composed of three second order interconnected subsystems (S_i) with unstable linear parts,

$$\frac{dx_i}{dt} = \begin{bmatrix} -1 & 1 \\ 2 & -1 \end{bmatrix} x_i + \begin{bmatrix} 10 & s_{11} \\ s_{12} & 10 \end{bmatrix} \begin{bmatrix} \phi_{i1}(\sigma_{i1}) \\ \phi_{i2}(\sigma_{i2}) \end{bmatrix} + \begin{bmatrix} s_{13} \\ s_{14} \end{bmatrix} \psi_i(\sigma)$$

$$\sigma_{i1} = -x_{i1} \ , \ \sigma_{i2} = -x_{i2} \ , \ \sigma = 0.5(x_{11} + x_{21} + x_{31}) \ ,$$

$$\phi_{ij}(\sigma_{ij})\,\sigma_{ij}^{-1} \in [1,2] \ , \ \psi_i(\sigma)\,\sigma^{-1} \in [0,1] \ , \ \forall i=1,2,3 \ , \ \forall j=1,2 \ ,$$

$$\phi_{ij}(\sigma_{ij}) \in C(R,R) \ , \ \psi_i(\sigma) \in C(R,R) \ , \ \forall i=1,2,3 \ , \ \forall j=1,2 \ .$$

The structural matrices S_i and S have the form

$$S_i = \begin{bmatrix} 1 & 0 & s_{i1} & 0 & s_{i3} \\ 0 & s_{i2} & 0 & 1 & s_{i4} \end{bmatrix}$$

$$\text{for} \quad h_i = (10\phi_{i1}, \phi_{i1}, \phi_{i2}, 10\phi_{i2}, \psi_i)^T \;, \quad \forall i = 1,2,3 \;,$$

$$S = \text{diag}\{S_1, S_2, S_3\}$$

The structural set S_s is determined by

$$S_s = \left\{ S : \text{diag}\left\{ \begin{bmatrix} 1 & 0 & 0 & 0 & 0 \\ 0 & 0 & 0 & 1 & 0 \end{bmatrix}, \begin{bmatrix} 1 & 0 & 0 & 0 & 0 \\ 0 & 0 & 0 & 1 & 0 \end{bmatrix}, \begin{bmatrix} 1 & 0 & 0 & 0 & 0 \\ 0 & 0 & 0 & 1 & 0 \end{bmatrix} \right\} \le S \le \right.$$

$$\left. \le \text{diag}\left\{ \begin{bmatrix} 1 & 0 & 1 & 0 & 1 \\ 0 & 1 & 0 & 1 & 1 \end{bmatrix}, \begin{bmatrix} 1 & 0 & 1 & 0 & 1 \\ 0 & 1 & 0 & 1 & 1 \end{bmatrix}, \begin{bmatrix} 1 & 0 & 1 & 0 & 1 \\ 0 & 1 & 0 & 1 & 1 \end{bmatrix} \right\} \right\} .$$

The functions v_i are accepted in the form $v_i(x_i) = \|x_i\|^2$, $\forall i = 1,2,3$. They are positive definite in the whole, decrescent and radially unbounded. Now,

$$\Omega_i(x_i) = \|x_i\| \;, \quad N_i = R^2 \;, \quad \forall i = 1,2,3 \;,$$

$$a_{11}(S_1) = -9.5 + s_{11} + s_{12} + 0.5(s_{13} + s_{14}) \;, \; a_{12}(S_1) = 0.5(s_{13} + s_{14}) \;, \; a_{13}(S_1) = 0.5(s_{13} + s_{14}) \;,$$

$$a_{21}(S_2) = 0.5(s_{23} + s_{24}) \;, \; a_{22}(S_2) = -9.5 + s_{21} + s_{22} + 0.5(s_{23} + s_{24}) \;, \; a_{23}(S_2) = 0.5(s_{23} + s_{24}) \;,$$

$$a_{31}(S_3) = 0.5(s_{33} + s_{34}) \;, \; a_{32}(S_3) = 0.5(s_{33} + s_{34}) \;, \; a_{33}(S_3) = -9.5 + s_{31} + s_{32} + 0.5(s_{33} + s_{34}) \;,$$

and

$$a_{11} = -6.5 \;, \quad a_{12} = 1 \;, \quad a_{13} = 1$$
$$a_{21} = 1 \;, \quad a_{22} = -6.5 \;, \quad a_{23} = 1$$
$$a_{31} = 1 \;, \quad a_{32} = 1 \;, \quad a_{33} = -6.5$$

obey all conditions of Assumption 6 and (II) of Theorem 17. Hence, $x = 0$ is structurally exponentially stable in the whole over S_s.

It is interesting to note that the sixth aggregation form admits sign indefiniteness of the comparison functions.

Assumption 7. There are a connected neighbourhood N_i of $x_i = 0$, functions v_i and Ω_i and real numbers $a_{ij}(p_i, S_i)$ such that

1) v_i is positive definite on N_i, $\forall i = 1,2,\dots,s$,
2) there is $\xi_i > 0$ such that the set $V_{i\zeta}(t)$ is asymptotically contractive for every $\zeta \in \,]0, \xi_i[$, $\forall i = 1,2,\dots,s$,

and

3) $v_i^*(t,x,p_i,S_i) \le \Omega_i(t,x_i) \sum_{j=1}^{s} a_{ij}(p_i,S_i)\, \Omega_i(t,x_i)$, $\forall i = 1,2,\dots,s$,
$\forall(t,x,P,S) \in R\times N\times P\times S_s$. ∎

Now, the aggregation matrix $A(P,S)$ depends on (P,S) , $A(P,S) = [a_{ij}(p_i,S_i)]$.

Theorem 18. (I) *For structural asymptotic stability of* $x=0$ *of the system (11) over* $P\times S_s$ *it is sufficient that both*

(a) Assumption 7 holds,

and

(b) the aggregation matrix $A(P,S)$ *is negative semi-definite for every* $(P,S) \in P\times S_s$.

(II) *If, in addition to (I),* $N_i = R^{n_i}$, $\xi_i = +\infty$ *and* v_i *is radially unbounded for every* $i = 1,2,\dots,s$ *then* $x=0$ *of the system (11) is structurally asymptotically stable in the whole over* $P\times S_s$.

Proof. (I) Let $v(t,x) = v_1(t,x_1) + v_2(t,x_2) + \dots + v_s(t,x_s)$. The condition (a) and 1) of Assumption 7 prove positive definiteness of v on N . The condition 2) of Assumption 7 and the definition of v guarantee existence of $\xi > 0$ such that the largest neighbourhood $V_\zeta(t)$ of $x=0$ on which $v(t,x) < \zeta$ for all $t \in R$ is asymptotically contractive. The condition 3) of Assumption 7 implies along motions of (11)

$$D^+v(t,x,P,S) \le w^T(t,x)\, A(P,S)\, w(t,x) \ , \ \forall(t,x,P,S) \in R\times N\times P\times S_s \ .$$

This result and the condition (b) prove

$$D^+v(t,x,P,S) \le 0 \ , \ \forall(t,x,P,S) \in R\times N\times P\times S_s \ .$$

All conditions of Theorem 13 of the Section I.3.2.5 are satisfied over $P\times S_s$. Hence, $x=0$ of (11) is structurally asymptotically stable over $P\times S_s$.

When, in addition to (I), $N_i = R^{n_i}$, $\xi_i = +\infty$ and v_i is radially unbounded, then $N = R^n$, v is positive definite in the whole and radially unbounded, $V_\zeta(t)$ is asymptotically contractive for every $\zeta \in\,]0,+\infty[$ and $D^+v(t,x,P,S) \le 0$, $\forall(t,x,P,S) \in R\times R^n\times P\times S$.

All conditions of Corollary 2 of Section I.3.2.5 are satisfied over $P\times S_s$, which completes the proof. ∎

Example 6. The fourth order system composed of two second order subsystems is described by

$$\dot{\mathbf{x}}_1 = \begin{bmatrix} 0 & , & -1 \\ -2 & , & -2 \end{bmatrix} \mathbf{x}_1 + \begin{bmatrix} -\dfrac{9+t+5t^2}{1+t^2} x_{11} + 2p_{11}s_{11}x_{21} + p_{11}s_{11}x_{22} \\ -\dfrac{t}{1+t^2} x_{12} + \frac{3}{2}p_{11}s_{11}x_{21} + p_{11}s_{11}x_{22} \end{bmatrix},$$

$$\dot{\mathbf{x}}_2 = \begin{bmatrix} 0 & , & -1 \\ -1 & , & -1 \end{bmatrix} \mathbf{x}_2 + \begin{bmatrix} 6p_{21}s_{21}x_{11} + 2p_{21}s_{21}x_{12} - \dfrac{4+t+4t^2}{1+t} x_{21} \\ 3p_{21}s_{21}x_{11} + p_{21}s_{21}x_{12} - \dfrac{t}{1+t^2} x_{22} \end{bmatrix}.$$

The system structural matrix $\mathbf{S}$ is found as

$$\mathbf{S} = \begin{bmatrix} s_{11}, & 0, & 0, & 0 \\ 0, & s_{11}, & 0, & 0 \\ 0, & 0, & s_{21}, & 0 \\ 0, & 0, & 0, & s_{21} \end{bmatrix} = \begin{bmatrix} s_{11}\begin{bmatrix} 1,0 \\ 0,1 \end{bmatrix} & 0 \\ 0 & s_{21}\begin{bmatrix} 1,0 \\ 0,1 \end{bmatrix} \end{bmatrix} = \begin{bmatrix} \mathbf{S}_1 & 0 \\ 0 & \mathbf{S}_2 \end{bmatrix},$$

and

$$\mathcal{S}_s = \{\mathbf{S} : \mathbf{0} \le \mathbf{S} \le \mathbf{I}_4\}.$$

The parameter perturbation matrix $\mathbf{P}$ has the form

$$\mathbf{P} = \begin{bmatrix} p_{11} \\ p_{21} \end{bmatrix},$$

and the set $\mathcal{P}$ of all admissible $\mathbf{P}$ is

$$\mathcal{P} = \left\{ \mathbf{P} : \begin{bmatrix} -0.75 \\ -0.25 \end{bmatrix} \le \mathbf{P} \le \begin{bmatrix} 0.75 \\ 0.25 \end{bmatrix} \right\}.$$

Let $v_i = (1+t^2)\|\mathbf{x}_i\|^2$, $i = 1,2$, be accepted. Then,

$$\dot{v}_1(t,\mathbf{x},\mathbf{p}_1,\mathbf{S}_1) \le$$

$$\le \sqrt{1+t^2}\,(3x_{11}+x_{12})\left[-\sqrt{1+t^2}\,(3x_{11}+x_{12}) + \tfrac{3}{2}p_{11}s_{11}\sqrt{1+t^2}\,(2x_{21}+x_{22})\right]$$

and

$$\dot{v}_2(t,\mathbf{x},\mathbf{p}_2,\mathbf{S}_2) \le$$

$$\le \sqrt{1+t^2}\,(2x_{21}+x_{22})\left[-\sqrt{1+t^2}\,(x_{21}+x_{22}) + 2p_{21}s_{11}(3x_{11}+x_{12})\right],$$

$\forall(t,\mathbf{x},\mathbf{P},\mathbf{S}) \in \mathcal{R}\times\mathcal{R}^4\times\mathcal{P}\times\mathcal{S}_s$, which yields

$$\Omega_1(t,x_1) = \sqrt{1+t^2}\;(3x_{11}+x_{12}) \quad , \quad \Omega_2(t,x_2) = \sqrt{1+t^2}\;(2x_{21}+x_{22}) \; .$$

The functions Ω_1 and Ω_2 are sign indefinite and linear in x_1 and x_2, respectively. The aggregation matrix $\mathbf{A}(\mathbf{P},\mathbf{S})$ now results in the form

$$\mathbf{A}(\mathbf{P},\mathbf{S}) = \begin{bmatrix} -1 & , \frac{2}{3}p_{11}s_{11} \\ 2p_{21}s_{21} \; , & -1 \end{bmatrix} .$$

It is negative definite on $P\times S_s$.

The set $V_{i\zeta}(t)$,

$$V_{i\zeta}(t) = \left\{ x_i \; : \; x_{i1}^2 + x_{i2}^2 < \frac{\zeta}{1+t^2} \right\} , \; i = 1,2 \; ,$$

is asymptotically contractive for all $\zeta \in \,]0,+\infty[$. Hence, $\xi_i = +\infty$, $\forall i = 1,2$. All conditions of (II) of Theorem 18 are verified. The equilibrium state $\mathbf{x}=\mathbf{0}$ of the system is structurally asymptotically stable in the whole over $P\times S_s$.

When the functions $\mathbf{g}_i$ and $\mathbf{h}_i$ in (9), hence $\mathbf{g}$ and $\mathbf{h}$ in (11), are time-invariant, then the system (11) is described by

$$\frac{d\mathbf{x}}{dt} = \mathbf{g}(\mathbf{x}) + \mathbf{S}(t)\,\mathbf{h}(\mathbf{x},\mathbf{P}) \tag{23}$$

and the functions v_i , hence v, should be chosen independent of t .

Theorem 18 cannot be applied in such cases, for which the following can be useful.

Assumption 8. There are a connected neighbourhood N_i of $\mathbf{x}_i=\mathbf{0}$, functions v_i and Ω_i and real numbers $a_{ij}(\mathbf{p}_i,\mathbf{S}_i)$ such that

1) v_i is positive definite on N_i and time-invariant, $\forall i = 1,2,\dots,s$,
2) along motions of the system (23)

$$v_i^*(\mathbf{x},\mathbf{p}_i,\mathbf{S}_i) \le \Omega_i(\mathbf{x}_i) \sum_{j=1}^{s} a_{ij}(\mathbf{p}_i,\mathbf{S}_i)\,\Omega_j(\mathbf{x}_j) \; ,$$

$\forall i = 1,2,\dots,s$, $\forall(\mathbf{x},\mathbf{P},\mathbf{S}) \in N\times P\times S_s$,

and

3) the singleton $\{\mathbf{0}\}$ is the largest invariant set of (23) in the set Z , $Z = \{\mathbf{x} : \Omega_1^2(\mathbf{x}_1) + \Omega_2^2(\mathbf{x}_2) + \dots + \Omega_s^2(\mathbf{x}_s) = 0\}$. ∎

Theorem 19. (I) *For structural asymptotic stability of* $\mathbf{x}=\mathbf{0}$ *of the system (23) over* $P\times S_s$ *it is sufficient that both*

(a) Assumption 8 holds

and

(b) the aggregation matrix $A(P,S)$ *is negative definite for every* $(P,S) \in P\times S_s$.

(II) *If, in addition to (I),* $N_i = R^{n_i}$ *and* v_i *is radially unbounded for every* $i = 1,2,\dots,s$ *then* $x = 0$ *of the system (23) is structurally asymptotically stable in the whole over* $P\times S_s$.

Proof. (I) Let $v(x) = v_1(x_1) + v_2(x_2) + \dots + v_s(x_s)$. In view of the condition (a) and 1) of Assumption 8 the function v is positive definite on N. The condition 2) of Assumption 8 now yields $v^*(x,P,S) \le \le \lambda_M(P,S)\|w(x)\|^2$, $\forall(x,P,S) \in N\times P\times S_s$, for $w = (\Omega_1,\Omega_2,\dots,\Omega_s)^T$.

The condition (b) ensures $\lambda_M(P,S) < 0$, $\forall(P,S) \in P\times S_s$. Hence,

$$v^*(x,P,S) \begin{cases} < 0 , & \forall x \in N/Z , \\ \le 0 , & \forall x \in Z , \end{cases} \quad \forall(P,S) \in P\times S_s .$$

This result and positive definiteness of v on N ensure precompactness of solutions of (23) relative to N. The condition 3) of Assumption 8 and the invariance principle (LaSalle, 1976) prove now that $x = 0$ of the system (23) is asymptotically stable for every $(P,S) \in P\times S_s$.

(II) When, in addition to (I), $N_i = R^{n_i}$ and v_i is radially unbounded then $N = R^n$, all solutions of (23) are precompact relatively to R^n for every $(P,S) \in P\times S_s$ and the invariance principle is satisfied on R^n for every $(P,S) \in P\times S_s$. Hence, $x = 0$ of (11) is structurally asymptotically stable in the whole over $P\times S_s$. ■

Example 7. The fourth order stationary system is described by

$$\frac{dx_1}{dt} = \begin{bmatrix} -6x_{11} + 2x_{12} + \text{sat}\,\sigma_1 + 4\sin\sigma_2 \\ -8x_{12} + 2x_{11} + \text{sat}\,\sigma_1 + 4\sin\sigma_2 \end{bmatrix} , \quad \sigma_i = x_{i1} + x_{i2} , \ \forall i = 1,2 ,$$

$$\frac{dx_2}{dt} = \begin{bmatrix} \text{sat}\,\sigma_1 - 2x_{21} + s_{21}x_{22} - 3\,\text{sat}\,\sigma_2 - 2\,\text{sign}\,\sigma_2 \\ \text{sat}\,\sigma_1 - 3x_{22} + s_{22}x_{21} - 3\,\text{sat}\,\sigma_2 - 2\,\text{sign}\,\sigma_2 \end{bmatrix} .$$

The structural matrices S_1 , S_2 and S are given by

$$S_1 = \begin{bmatrix} 1 & 0 \\ 0 & 1 \end{bmatrix} , \quad S_2 = \begin{bmatrix} s_{21} & 1 & 0 & 0 \\ 0 & 0 & s_{22} & 1 \end{bmatrix} , \quad S = \text{diag}\{S_1 \ S_2\}$$

for

$$h_1 = \begin{bmatrix} -6x_{11} + 2x_{12} + \text{sat}\ \sigma_1 + 4\sin\sigma_2 \\ -8x_{12} + 2x_{11} + \text{sat}\ \sigma_1 + 4\sin\sigma_2 \end{bmatrix}, \quad h_2 = [h_2^1, h_2^2, h_2^3, h_2^4]^T ,$$

where

$$h_2^1(x) = x_{22}, \quad h_2^2(x) = \text{sat}\ \sigma_1 - 2x_{21} - 0.1\,\text{sat}\ \sigma_2 - 2\,\text{sign}\,\sigma_2 ,$$
$$h_2^3(x) = x_{21}, \quad h_2^4(x) = \text{sat}\ \sigma_1 - 3x_{22} - 0.1\,\text{sat}\ \sigma_2 - 2\,\text{sign}\,\sigma_2 .$$

The structural set S_s takes the form of

$$S_s = \left\{ S : \text{diag}\left\{ \begin{bmatrix} 1 & 0 \\ 0 & 1 \end{bmatrix}, \begin{bmatrix} 0 & 1 & 0 & 0 \\ 0 & 0 & 0 & 1 \end{bmatrix} \right\} \le \mathbf{S} \le \text{diag}\left\{ \begin{bmatrix} 1 & 0 \\ 0 & 1 \end{bmatrix}, \begin{bmatrix} 1 & 1 & 0 & 0 \\ 0 & 0 & 1 & 1 \end{bmatrix} \right\} \right\} .$$

Let $v_i(\mathbf{x}_i) = \|\mathbf{x}_i\|^2$, $\forall i = 1,2,3$. Evidently, every v_i is positive definite, decrescent and radially unbounded. The following parameters obey the condition 2) of Assumption 8,

$$a_{11}(\mathbf{S}_1) = -2, \quad a_{12}(\mathbf{S}_1) = 4 ,$$
$$a_{21}(\mathbf{S}_2) = 0, \quad a_{22}(\mathbf{S}_2) = -3 ,$$

for $N_1 = N_2 = R$, $\Omega_1(\mathbf{x}_1) = |x_{11} + x_{12}|$ and $\Omega_2(\mathbf{x}_2) = |\text{sat}\ \sigma_2|$.

The set Z is determined in this case by

$$Z = \{\mathbf{x} : (x_{11} + x_{12})^2 + [\text{sat}\,(x_{21} + x_{22})]^2 = 0\} .$$

Hence, $\mathbf{x} \in Z$ iff both $x_{11} = -x_{12}$ and $x_{21} = -x_{22}$, which show that there is an invariant set in Z of the system if $\dot{x}_{11} = -\dot{x}_{12}$ and $\dot{x}_{21} = -\dot{x}_{22}$. However, for $\mathbf{x} \in Z$ it results from the state equations

$$\dot{\mathbf{x}}_1 = \begin{bmatrix} -8x_{11} \\ -10x_{11} \end{bmatrix} \quad \text{and} \quad \dot{\mathbf{x}}_2 = \begin{bmatrix} (s_{21} - 1)\,x_{21} \\ (s_{22} + 3)\,x_{21} \end{bmatrix},$$

i.e. that $\dot{x}_{i1} \ne \dot{x}_{i2}$ for every $i = 1,2$, as soon as $x_{i1} \ne 0$, $\forall i = 1,2$. Hence, the largest invariant set of the system in Z is the singleton $\{0\}$. Since the matrix $\mathbf{A}(\mathbf{S})$,

$$\mathbf{A}(\mathbf{S}) = \begin{bmatrix} -2 & 4 \\ 0 & -3 \end{bmatrix}$$

is negative definite (because its symmetric part is negative definite), then all conditions of (II) of Theorem 19 are satisfied. Therefore, $\mathbf{x} = \mathbf{0}$ of the system is structurally asymptotically stable in the whole over S_s.

Notice that Ω_1 and Ω_2 are not positive definite functions.

III.4.2.4. *The seventh aggregation form and stability criteria*

The seventh aggregation form is defined by

Assumption 9. There are a connected neighbourhood N_i of $x_i = 0$, functions v_i, Ω_i and ζ_i and real numbers a_{ij} and b_{ij} such that

1) v_i is differentiable in $(t, x_i) \in R \times N_i$ and positive definite on N_i, $\forall i = 1,2,\dots,s$,
2) Ω_i is positive definite on R_+, $\forall i = 1,2,\dots,s$,
3) $\zeta_i(0) = 0$, $0 \le \zeta_i(v_i)/\Omega_i(v_i) < \kappa_i \le +\infty$, $\forall v_i \in \check{R}_+$, $\forall i = 1,2,\dots,s$,
4) $\displaystyle\int_0^{v_i} \frac{\zeta_i(\xi)}{\Omega_i(\xi)} d\xi$ is defined and continuous in v_i for all $v_i \in R_+$ and tends to $+\infty$ if and only if $v_i \to +\infty$,
5) there is $\xi_i > 0$ such that the set $V_{i\zeta}(t)$ is asymptotically contractive for all $\zeta \in]0, \xi_i[$, $\forall i = 1,2,\dots,s$,

and

6) $\dot{v}_i(t, x, p_i, S_i) \le \Omega_i[v_i(t, x_i)] \sum_{j=1}^{s} \{a_{ij}\Omega_j[v_j(t, x_j)] + b_{ij}\zeta_j[v_j(t, x_j)]\}$,
$\forall i = 1,2,\dots,s$, $\forall (t, x, P, S) \in R \times N \times P \times S_s$. ■

This aggregation form determines two aggregation matrices A and B, and matrix K, $A = (a_{ij})$, $B = (b_{ij})$, $K = \text{diag}\{\kappa_1, \kappa_2, \dots, \kappa_s\}$, and vector non-linearity $z = [\zeta_1, \zeta_2, \dots, \zeta_s]^T$.

Theorem 20. (I) *For structural asymptotic stability of* $x = 0$ *of the system (11) over* $P \times S_s$ *it is sufficient that*

(a) Assumption 9 holds,
(b) A *is stable,*
(c) there is a non-negative diagonal $s \times s$ *matrix* Θ *such that both*
1) $2K^{-1} - \Theta B - B^T\Theta$
and
2) $-[A^T + A + (A^T\Theta + I + B)(2K^{-1} - \Theta B - B^T\Theta)^{-1}(A^T\Theta + I + B)^T]$
are positive definite.

(II) *If, in addition to (I),* $N_i = R^{n_i}$, $\xi_i = +\infty$ *and* v_i *is radially unbounded then* $x = 0$ *of the system (11) is structurally asymptotically stable in the whole over* $P \times S_s$.

Proof. (I) Let v be defined for $\Theta = \text{diag}\{\theta_1, \theta_2, \dots, \theta_s\}$ by

$$v(t, x) = \sum_{i=1}^{s} \left[v_i(t, x_i) + \theta_i \int_0^{v_i(t, x_i)} \frac{\zeta_i(\xi)}{\Omega_i(\xi)} d\xi\right] \tag{24}$$

The conditions (a) of the theorem and 1), 3) and 4) of Assumption 9 imply positive definiteness of v on N. The condition 5) of the same assumption guarantees existence of $\xi > 0$ such that the set $V_\zeta(t)$ is asymptotically contractive for every $\zeta \in]0,\xi[$. Let

$$H = \frac{1}{2} I \ , \quad D = \frac{1}{2}(A^T\Theta + I) \ , \quad \Gamma = K^{-1} - \frac{1}{2}(\Theta B + B^T\Theta) \ .$$

Then

$$A^T + A + (A^T\Theta + I + B)(2K^{-1} - \Theta B - B^T\Theta)^{-1}(A^T\Theta + I + B)^T = A^T H + HA + (D + HB)\Gamma^{-1}(D + HB)^T .$$

The requirements 1) and 2) of the condition (c) guarantee (due to Grujić, 1977) existence of X , Y and positive definite $s \times s$ matrix Q such that

$$Y^T Y = \Gamma \tag{25a}$$

$$A^T H + HA + XX^T = -Q \tag{25b}$$

$$XY + HB = -D \ . \tag{25c}$$

This result and the condition 6) of Assumption 9 yield

$$\begin{aligned}
&\dot{v}(t,x,P,S) \le \\
&\le w^T(t,x)\,Aw(t,x) + w^T(t,x)\,Bz(t,x) + z^T(t,x)\,\Theta Aw(t,x) + z^T(t,x)\,\Theta Bz(t,x) \\
&= w^T(t,x)(A^T H + HA + XX^T)\,w(t,x) - [X^T w(t,x) + Yz(t,x)]^T[X^T w(t,x) + Yz(t,x)] \\
&\qquad - [w^T(t,x) - z^T(t,x)\,K^{-1}]\,z(t,x) \\
&\le -w^T(t,x)\,Qw^T(t,x) \quad , \quad \forall (t,x,P,S) \in R \times N \times P \times S_s \ .
\end{aligned}$$

Hence, all conditions of Theorem 13 of the Section I.3.2.5 are satisfied over $P \times S_s$ and the assertion under (I) is true.

(II) When, in addition to (I), $N_i = R^{n_i}$, $\xi_i = +\infty$ and v_i is radially unbounded, $\forall i = 1,2,\dots,s$, then v is positive definite in the whole and radially unbounded, and the set $V_\zeta(t)$ is asymptotically contractive for every $\zeta \in]0,+\infty[$. All conditions of Corollary 2 of the Section I.3.2.5 are satisfied over $P \times S_s$, which completes the proof. ∎

When the condition 5 of Assumption 9 is not satisfied, which happens in the case the system (11) is time-invariant, then the sixth aggregation form can be relied on

Assumption 10. There are a connected neighbourhood N_i of $x_i = 0$, functions v_i , Ω_i and ζ_i and real numbers a_{ij} and b_{ij} such that

1) v_i is differentiable in $(t,x_i) \in R \times N_i$, positive definite on N_i and decrescent on N_i , $\forall i = 1,2,\dots,s$,
2) Ω_i is positive definite on R_+ , $\forall i = 1,2,\dots,s$,
3) $\zeta_i(0) = 0$, $0 \le \zeta_i(v_i)/\Omega_i(v_i) < \kappa_i \le +\infty$, $\forall v_i \in \check{R}_+$, $\forall i = 1,2,\dots,s$,

4) $\displaystyle\int_0^{v_i} \frac{\zeta_i(\xi)}{\Omega_i(\xi)}\,d\xi$ is defined and continuous in $v_i \in \mathbb{R}_+$ and tends to $+\infty$ if and only if v_i tends to $+\infty$,

and

5) $\dot v_i(t,x,p_i,S_i) \le \Omega_i[v_i(t,x_i)] \sum_{j=1}^{s} \{a_{ij}\Omega_j[v_j(t,x_j)] + b_{ij}\zeta_j[v_j(t,x_j)]\}$

$\forall i = 1,2,\dots,s$, $\forall (t,x,P,S) \in \mathbb{R}\times N\times P\times S_s$. ■

Theorem 21. (I) *For structural uniform asymptotic stability of* $x=0$ *of the system (11) over* $P\times S_s$ *it is sufficient that*

(a) Assumption 10 holds,

(b) A *is stable,*

and

(c) there is a non-negative diagonal $s\times s$ *matrix* Θ *such that both*

1) $2K^{-1} - \Theta B - B^T\Theta$

and

2) $-[A^T + A + (A^T\Theta + I + B)(2K^{-1} - \Theta B - B^T\Theta)^{-1}(A^T\Theta + I + B)^T]$

are positive definite.

(II) *If, in addition to (I),* $N_i = \mathbb{R}^{n_i}$ *and* v_i *is radially unbounded for every* $i=1,2,\dots,s$ *then* $x=0$ *of the system (11) is structurally uniformly asymptotically stable in the whole over* $P\times S_s$.

Proof. (I) Let the function v be defined by (24). The condition (a) and the conditions 1), 3) and 4) of Assumption 10 prove that v is positive definite on N. The well defined form of v for all ζ_i and Ω_i, the fact that v_i is decrescent on N_i, which are implied by the hypothesis of the theorem and conditions 1) and 3) of Assumption 10, and the condition 4) of the same assumption guarantee that v is decrescent on N. Following now the part of proof of Theorem 20, related to $\dot v$ so that

$$\dot v(t,x,P,S) \le -w^T(t,x)\,Qw(t,x) \quad , \ \forall (t,x,P,S) \in \mathbb{R}\times N\times P\times S_s \ , \qquad (26)$$

in view of the conditions (c-1) and (c-2), it follows that $\dot v$ is majorized by a negative definite function $(-w^TQw)$ on N because Q is positive definite. Hence, all conditions of Theorem 7 of the Section I.3.2.4 are satisfied on $P\times S_s$, which proves the assertion under (I).

(II) When, in addition to (I), $N_i = \mathbb{R}^{n_i}$ and v_i is radially unbounded for every $i=1,2,\dots,s$, then v is also radially unbounded and obeys all conditions of Theorem 8 of the Section I.3.2.4, which completes the proof. ■

The requirement that all Ω_i be positive definite can be omitted for the system (23) as shown in what follows.

Assumption 11. There are a connected neighbourhood N_i of $x_i = 0$, functions v_i, Ω_i and ζ_i all independent of time t, and real numbers a_{ij} and b_{ij} such that

1) v_i is differentiable in $x_i \in N_i$ and positive definite on N_i, $\forall i = 1,2,\ldots,s$,
2) $\zeta_i(0) = 0$, $0 \le \zeta_i(v_i)/\Omega_i(v_i) < \kappa_i \le +\infty$, $\forall v_i \in \breve{R}_+$, $\forall i = 1,2,\ldots,s$,
3) $\int_0^{v_i} \frac{\zeta_i(\xi)}{\Omega_i(\xi)} d\xi$ is defined and continuous in $v_i \in \breve{R}_+$ and tends to $+\infty$ if and only if v_i tends to $+\infty$,
4) along the motions of the system (23)
$$\dot{v}_i(x, p_i, S_i) \le \Omega_i[v_i(x_i)] \sum_{j=1}^{s} \{a_{ij}\Omega_j[v_j(x_j)] + b_{ij}\zeta_j[v_j(x_j)]\},$$
$\forall i = 1,2,\ldots,s$, $\forall (x, P, S) \in N \times P \times S$,

and

5) the singleton $\{0\}$ is the largest invariant set of the system (23) in the set $Z = \{x : \Omega_1^2(x_1) + \Omega_2^2(x_2) + \ldots + \Omega_s^2(x_s) = 0\}$. ∎

Theorem 22. (I) *For structural asymptotic stability of* $x = 0$ *of the system (23) over* $P \times S_s$ *it is sufficient that*

(a) Assumption 11 holds,
(b) A *is stable,*
and
(c) there is a non-negative diagonal $s \times s$ *matrix* Θ *such that both*
1) $2K^{-1} - \Theta B - B^T\Theta$
and
2) $-[A^T + A + (A^T\Theta + I + B)(2K^{-1} - \Theta B - B^T\Theta)^{-1}(A^T\Theta + I + B)^T]$
are positive definite.

(II) *If, in addition to (I),* $N_i = R^{n_i}$ *and* v_i *is radially unbounded for every* $i = 1,2,\ldots,s$ *then* $x = 0$ *of the system (23) is structurally asymptotically stable in the whole over* $P \times S_s$.

Proof. (I) Let the function v be defined by (24). Following the proofs of Theorem 20 and Theorem 21 it follows that v is positive definite on N, and along solutions of the system (23) the inequality (26) is valid for Q symmetric positive definite matrix obeying (25). Hence, for $\lambda_m(Q)$ being the minimal eigenvalue of Q, $\lambda_m(Q) > 0$,

$$\dot{v}(\mathbf{x},P,S) \le -\lambda_m(Q)\|\mathbf{w}(\mathbf{x})\|^2 . \tag{27}$$

Since in the set $Z = \{\mathbf{x} : \|\mathbf{w}(\mathbf{x})\|^2 = 0\}$ the singleton $\{\mathbf{0}\}$ is the largest invariant set of (23) due to 5) of Assumption 11, (27) holds, and since all solutions of (23) are precompact relative to the largest set V_ζ in N due to positive definiteness of v on N and (27), then all conditions of the invariance principle (LaSalle, 1976) are satisfied over $P\times S_s$. Hence, $\mathbf{x}=\mathbf{0}$ of the system (23) is structurally asymptotically stable over $P\times S_s$.

(II) When, in addition to (I), $N_i = R^{ni}$ and v_i is radially unbounded, $\forall i = 1,2,\ldots,s$, then $N = R^n$, v is also radially unbounded and solutions of (23) are precompact relative to R^n, which prove that $\mathbf{x}=\mathbf{0}$ of (23) is structurally asymptotically stable in the whole over $P\times S_s$. ■

Example 8. The fourth order system composed of two second order subsystems is described by

$$\frac{d\mathbf{x}_1}{dt} = \begin{bmatrix} -11(1+t+2t^2)(x_{11}^2+x_{12}^2) - \frac{1}{2}(x_{11}^2+5x_{12}^2)^{1/5}x_{11} \\ -12(1+t^2)(x_{11}^2+x_{12}^2) - 2(4x_{11}^2+x_{12}^2)^{1/5}x_{12} \end{bmatrix} + s_{11}\begin{bmatrix} \frac{1}{4}x_{22}^3x_{11} \\ 0 \end{bmatrix} + s_{12}\begin{bmatrix} 0 \\ -0.1(1+\sin t)x_{21}^3x_{12} \end{bmatrix},$$

$$S_1 = \begin{bmatrix} s_{11}, & 0, & s_{12}, & 0 \\ 0, & s_{11}, & 0, & s_{12} \end{bmatrix},$$

$$\frac{d\mathbf{x}_2}{dt} = \begin{bmatrix} -13x_{21} - 0.5(x_{21}^2+4x_{22}^2)x_{21} \\ -10x_{22} - 2(x_{21}^2+x_{22}^2)x_{22} \end{bmatrix} + s_{21}\begin{bmatrix} 0.05(1.5+\cos t)x_{11}x_{21} \\ 0 \end{bmatrix} + s_{22}\begin{bmatrix} 0 \\ 0.10\,x_{12}x_{22} \end{bmatrix},$$

$$S_2 = \begin{bmatrix} s_{21}, & 0, & s_{22}, & 0 \\ 0, & s_{21}, & 0, & s_{22} \end{bmatrix}.$$

The form of nonlinearities suggests $v_i(\mathbf{x}_i) = \|\mathbf{x}_i\|^2$, which is positive definite, decrescent and radially unbounded. Now, along motions of the system,

$$\dot{v}_1(t,\mathbf{x},S_1) \le v_1(\mathbf{x}_1)[-20v_1(\mathbf{x}_1) - v_1(\mathbf{x}_1) - v_1^{1/5}(\mathbf{x}_1) + 0.5v_2^{3/2}(\mathbf{x}_2)] ,$$

$$\dot{v}_2(t,\mathbf{x},S_2) \le$$
$$\le v_2^{1/2}(\mathbf{x}_2)[0.5v_1(\mathbf{x}_1) + 0.5v_1(\mathbf{x}_1) + 0.5v_1^{1/5}(\mathbf{x}_1) - 20v_2(\mathbf{x}_2) - v_2^{3/2}(\mathbf{x}_2)] ,$$

$$\forall(t,\mathbf{x},S) \in R\times R^n\times S_s$$

where

$$S_s = \left\{ S : 0 \le \begin{bmatrix} S_1 & 0 \\ 0 & S_2 \end{bmatrix} = \begin{bmatrix} s_{11} & 0 & s_{12} & 0 & & & & \\ 0 & s_{11} & 0 & s_{12} & & O & & \\ & & & & s_{21} & 0 & s_{22} & 0 \\ & O & & & 0 & s_{21} & 0 & s_{22} \end{bmatrix} \le \right.$$

$$\left. \le \begin{bmatrix} 1 & 0 & 1 & 0 & & & & \\ 0 & 1 & 0 & 1 & & O & & \\ & & & & 1 & 0 & 1 & 0 \\ & O & & & 0 & 1 & 0 & 1 \end{bmatrix} \right\} .$$

It follows that $N = R^4$,

$$\Omega_1(v_1) = v_1 \ , \ \Omega_2(v_2) = v_2^{1/2} \ , \ \zeta_1(v_1) = v_1 + v_1^{1/5} \ , \ \zeta_2(v_2) = v_2^{3/2} \ ,$$

and

$$\dot{v}_1 \le \Omega_1(-20\Omega_1 - \zeta_1 + 0.5\zeta_2) \ , \ \dot{v}_2 \le \Omega_2(0.5\Omega_1 - 20\Omega_2 + 0.5\zeta_1 - \zeta_2) \ ,$$

so that
$$\frac{\zeta_1(v_1)}{\Omega_1(v_1)} \in [0,+\infty[\ , \ \frac{\zeta_2(v_2)}{\Omega_2(v_2)} \in [0,+\infty[\ , \ \forall v_1,v_2 \in \breve{R}_+ .$$

Hence, $\mathbf{K}^{-1} = \mathbf{0}$ and

$$\mathbf{A} = \begin{bmatrix} -20 \ , & 0 \\ 0.5 \ , & -20 \end{bmatrix} \ , \ \mathbf{B} = \begin{bmatrix} -1 \ , & 0.5 \\ 0.5 \ , & -1 \end{bmatrix} \ , \ \mathbf{z}(\mathbf{v}) = \begin{bmatrix} \zeta_1(v_1) \\ \zeta_2(v_2) \end{bmatrix} .$$

The matrix $\mathbf{A}$ is stable. Let $\Theta = 20^{-1}I_2$ which implies that both

$$2\mathbf{K}^{-1} - (\Theta\mathbf{B} + \mathbf{B}^T\Theta) = \frac{1}{10}\begin{bmatrix} 1 & -0.5 \\ -0.5 & 1 \end{bmatrix}$$

and

$$-[A^T + A + (A^T\Theta + I + B)(2K^{-1} - \Theta B - B^T\Theta)^{-1}(A^T\Theta + I + B)^T] = \begin{bmatrix} 29.992 & , & 4.75 \\ 4.75 & , & 30 \end{bmatrix}$$

are positive definite. All conditions of (II) of Theorem 21 are satisfied. Hence, the equilibrium state $x=0$ of the system is structurally uniformly asymptotically stable in the whole over S_s .

III.4.2.5. The eighth aggregation form and stability criteria

The eighth aggregation form is the most general. It is defined by

Assumption 12. There are a connected neighbourhood N_i of $x_i = 0$ and functions v_i , Ω_i and μ_{ij} such that

1) v_i and Ω_i are positive definite on N_i , $\forall i = 1,2,\dots,s$,
2) there is $\xi_i > 0$ such that the set $V_{i\zeta}(t)$ is asymptotically contractive for every $\zeta \in]0,\xi_i[$ and every $i = 1,2,\dots,s$,
3) $\sum\limits_{i=1}^{s} v_i^*(t,x,p_i,S_i) \le \sum\limits_{i,j=1}^{s} \mu_{ij}(t,x,P,S)\, \Omega_i(t,x_i)\, \Omega_j(t,x_j)$,

   $\forall (t,x,P,S) \in R\times N\times P\times S_s$,

and

4) there are $\underline{\mu}_{ij} \in]-\infty,+\infty[$ and $\bar{\mu}_{ij} \in]-\infty,+\infty]$, $\underline{\mu}_{ij} \le \bar{\mu}_{ij}$, such that

$$\underline{\mu}_{ij} \le \mu_{ij}(t,x,P,S) \begin{cases} \le \bar{\mu}_{ij} \ , & \bar{\mu}_{ij} < +\infty \\ < \bar{\mu}_{ij} \ , & \bar{\mu}_{ij} = +\infty \end{cases} ,$$

   $\forall i,j = 1,2,\dots,s$, $\forall (t,x,P,S) \in R\times N\times P\times S$. ■

This aggregation form introduces in general a functional matrix N , $N(t,x,P,S) = [\mu_{ij}(t,x,P,S)]$. Let $L \in R^{s\times s}$, $L = (\ell_{ij})$, and
$L = \{L : \underline{\mu}_{ij} \le \ell_{ij} \le \bar{\mu}_{ij}$ if $\bar{\mu}_{ij} < +\infty$, $\underline{\mu}_{ij} \le \ell_{ij} < \bar{\mu}_{ij}$ if $\bar{\mu}_{ij} = +\infty$, $\forall i,j = 1,2,\dots,s\}$.
The set L is (bounded or unbounded) hyperparallelepiped. The set of all its vertices is denoted by L_v ,

$$L_v = \left\{ L : \ell_{ij} = \underline{\mu}_{ij} \text{ or } \ell_{ij} \begin{cases} = \bar{\mu}_{ij} \ , & \bar{\mu}_{ij} < +\infty \\ \to \bar{\mu}_{ij} \ , & \bar{\mu}_{ij} = +\infty \end{cases} , \ \forall i,j = 1,2,\dots,s \right\} .$$

Let $M(L,\epsilon) = L + \epsilon I$, where $\epsilon > 0$ iff there is $\bar{\mu}_{ij} = +\infty$ and $\epsilon = 0$ iff all $\bar{\mu}_{ij} < +\infty$. The matrices M and L are introduced to reduce stability condition from a stability test of a functional matrix N on $R\times N\times P\times S_s$ to a stability test of the matrix $M(L,\epsilon)$ for every $L \in L_v$.

Theorem 23. (I) *For structural asymptotic stability of* $x=0$ *of the system (11) over* $P\times S_s$ *it is sufficient that both*

(a) Assumption 12 holds
and
(b) the matrix $M(L,\epsilon)$ *is negative definite for every* $L \in L_V$.

(II) *If, in addition to (I),* $N_i = R^{n_i}$, $\xi_i = +\infty$ *and* v_i *is radially unbounded for every* $i = 1,2,\ldots,s$, *then* $x = 0$ *of the system is structurally asymptotically stable in the whole over* $P \times S_S$.

Proof. (I) In view of the condition (a) and 1) of Assumption 12, it follows that $v = v_1 + v_2 + \ldots + v_s$ is positive definite on N. The condition 2) of the same assumption guarantees existence of $\xi > 0$ such that the set $V_\zeta(t)$ is asymptotically contractive for every $\zeta \in]0,\xi[$. The condition 3) of Assumption 12 yields

$$v^*(t,x,P,S) \le w^T(t,x)\, N(t,x,P,S)\, w(t,x) \ , \ \forall (t,x,P,S) \in R \times N \times P \times S_S \ .$$

The condition 4) of Assumption 12 and the definition of L , M and ϵ prove that for every $(t,x,P,S) \in R \times N \times P \times S_S$ there is $L \in L$ such that $N(t,x,P,S) = L$. Hence, for such L ,

$$w^T(t,x)\, N(t,x,P,S)\, w(t,x) = w^T(t,x)\, L w(t,x)$$
$$= w^T(t,x)\, M(L,\epsilon)\, w(t,x) - \epsilon \| w(t,x) \|^2 \ .$$

Since $M(L,\epsilon)$ is negative definite for every $L \in L_V$ according to the condition (b), then it is negative definite for every $L \in L$ (Grujić, 1974c-1976b) which, together with positive definiteness of Ω_i on N_i for every $i = 1,2,\ldots,s$, proves negative definiteness of

$$w^T(t,x)\, N(t,x,P,S)\, w(t,x)$$

on N over $P \times S_S$. All conditions of Theorem 13 of the Section I.3.2.5 have been verified over $P \times S_S$. Hence, $x = 0$ of (11) is structurally asymptotically stable over $P \times S_S$.

(II) If, in addition to (I), $N_i = R^{n_i}$, $\xi_i = +\infty$ and v_i is radially unbounded then $N = R^n$, $\xi = +\infty$ and v is radially unbounded. All conditions of Corollary 2 of the Section I.3.2.5 are satisfied over $P \times S_S$, which completes the proof. ■

For uniformity of the structural asymptotic stability the next assumption will be accepted.

Assumption 13. There are a connected neighbourhood N_i of $x_i = 0$ and functions v_i , Ω_i and μ_{ij} such that

1) v_i is positive definite on N_i and decrescent on N_i , $\forall i = 1,2,\ldots,s$,
2) Ω_i is positive definite on N_i , $\forall i = 1,2,\ldots,s$,

3) $\sum_{i=1}^{s} v_i^*(t,x,p_i,S_i) \le \sum_{i,j=1}^{s} \mu_{ij}(t,x,P,S)\, \Omega_i(t,x_i)\, \Omega_j(t,x_j)$,

$\forall(t,x,P,S) \in R\times N\times P\times S_s$,

and

4) there are $\underline{\mu}_{ij} \in]-\infty,+\infty[$ and $\bar{\mu}_{ij} \in]-\infty,+\infty]$, $\underline{\mu}_{ij} \le \bar{\mu}_{ij}$, such that

$$\underline{\mu}_{ij} \le \mu_{ij}(t,x,P,S) \begin{cases} \le \bar{\mu}_{ij} , & \bar{\mu}_{ij} < +\infty \\ < \bar{\mu}_{ij} , & \bar{\mu}_{ij} = +\infty \end{cases} ,$$

$\forall i,j = 1,2,\dots,s$, $\forall(t,x,P,S) \in R\times N\times P\times S_s$. ∎

Theorem 24. (I) *For structural uniform asymptotic stability of* $x=0$ *of the system (11) over* $P\times S_s$ *it is sufficient that*

(a) Assumption 13 holds

and

(b) the matrix $M(L,\epsilon)$ *is negative definite for every* $L \in L_v$.

(II) *If, in addition to (I),* $N_i = R^{n_i}$, $\xi_i = +\infty$ *and* v_i *is radially unbounded for every* $i = 1,2,\dots,s$ *then* $x=0$ *of the system (11) is structurally uniformly asymptotically stable in the whole over* $P\times S_s$.

Proof. (I) The condition (a) and 1) of Assumption 13 prove positive definiteness of $v = v_1 + v_2 + \dots + v_s$ on N as well as that it is decrescent on N. The conditions 3) and 4) of Assumption 13 and the definition of ϵ guarantee existence of $L \in L$ for every $(t,x,P,S) \in R\times N\times P\times S_s$ such that $v^*(t,x) \le w^T(t,x)\, M(L,\epsilon)\, w(t,x) - \epsilon\|w(t,x)\|^2$. The condition (b) and 2) of Assumption 13 show now that all requirements of Theorem 7 of the Section I.3.2.4 are satisfied over $P\times S_s$. Hence, $x=0$ of (11) is structurally uniformly asymptotically stable over $P\times S_s$.

(II) When, in addition to (I), $N_i = R^{n_i}$ and v_i is radially unbounded, $\forall i = 1,2,\dots,s$, then $N = R^n$ and v is also radially unbounded. All conditions of Theorem 8 of the Section I.3.2.4 are satisfied over $P\times S_s$, which completes the proof. ∎

Exponential character of the structural asymptotic stability requires special features of all v_i and Ω_i .

Assumption 14. There are a connected neighbourhood N_i of $x_i = 0$, functions v_i and μ_{ij} and positive numbers η_{i1} and η_{i2} such that

1) $\eta_{i1}\|x_i\|^2 \le v_i(t,x_i) \le \eta_{i2}\|x_i\|^2$, $\forall(t,x_i) \in R\times N_i$, $\forall i = 1,2,\dots,s$,

2) $\sum_{i=1}^{s} v_i^*(t,x,p_i,S_i) \le \sum_{i,j=1}^{s} \mu_{ij}(t,x,P,S)\|x_i\|\cdot\|x_j\|$,

$\forall(t,x,P,S) \in R\times N\times P\times S_s$,

and

3) there are $\underline{\mu}_{ij} \in]-\infty,+\infty[$ and $\bar{\mu}_{ij} \in]-\infty,+\infty]$, $\underline{\mu}_{ij} \le \bar{\mu}_{ij}$, such that

$$\underline{\mu}_{ij} \le \mu_{ij}(t,x,P,S) \begin{cases} \le \bar{\mu}_{ij} \ , & \bar{\mu}_{ij} < +\infty \\ < \bar{\mu}_{ij} \ , & \bar{\mu}_{ij} = +\infty \end{cases} ,$$

$\forall i,j = 1,2,\ldots,s$, $\forall(t,x,P,S) \in R\times N\times P\times S_s$. ■

Theorem 25. (I) *For structural exponential stability of* $x=0$ *of the system (11) over* $P\times S_s$ *it is sufficient that*

(a) Assumption 14 holds,

and

(b) the matrix $M(L,\epsilon)$ *is negative definite for every* $L\in L_v$.

(II) *If, in addition to (I),* $N_i = R^{n_i}$ *then* $x=0$ *of the system (11) is structurally exponentially stable in the whole over* $P\times S_s$.

Proof. (I) The condition (a) and 1) of Assumption 14 guarantee existence of positive numbers η_1 and η_2 such that the function $v = v_1 + v_2 + \ldots + v_s$ obeys $\eta_1\|x\|^2 \le v(t,x) \le \eta_2\|x\|^2$, $\forall(t,x) \in R\times N$.
The condition 2) of Assumption 14 implies

$$v^*(t,x,P,S) \le w^T(t,x)\, N(t,x,P,S)\, w(t,x) \ , \ \forall(t,x,P,S) \in R\times N\times P\times S_s \ ,$$

which together with the definition of ϵ and the condition 3) of Assumption 14 guarantee that for every $(t,x,P,S) \in R\times N\times P\times S_s$ there exists $L\in L$ such that $N(t,x,P,S) = L$ and

$$v^*(t,x,P,S) \le w^T(t,x)\, M(L,\epsilon)\, w(t,x) - \epsilon\|w(t,x)\|^2 \ , \ \forall(t,x,P,S) \in R\times N\times P\times S_s \ ,$$

where now $w = (\|x_1\|,\|x_2\|,\ldots,\|x_s\|)^T$. When all $\bar{\mu}_{ij} < +\infty$ then $\epsilon = 0$ and $M(L,0) = L$. Since both L and $M(L,\epsilon)$ are negative definite, $\forall L\in L$, because it is negative definite for every $L\in L_v$ (Grujić, 1974c-1976b), then there is positive number η_3 such that

$$v^*(t,x,P,S) \le -\eta_3\|x\|^2 \ , \ \forall(t,x,P,S) \in R\times N\times P\times S_s \ .$$

All conditions of Theorem 9 of the Section I.3.2.4 have been verified over $P\times S_s$. Hence, $x=0$ of (11) is structurally exponentially stable over $P\times S_s$.

(II) When, in addition to (I), $N_i = R^{n_i}$, $\forall i = 1,2,\ldots,s$, then $N = R^n$ and all conditions of Theorem 10 of the Section I.3.2.4 are satisfied over $P \times S_s$, which completes the proof. ∎

In the case of the system (23) the condition on definiteness of the functions Ω_i in Assumption 13 can be avoided as shown in what follows.

Assumption 15. There are a connected neighbourhood N_i of $x_i = 0$ and functions v_i , Ω_i and μ_{ij} such that

1) v_i is time invariant and positive definite on N_i , $\forall i = 1,2,\dots,s$,
2) along motions of the system (23)

$$\sum_{i=1}^{s} v_i^*(x,p_i,S_i) \le \sum_{i,j=1}^{s} \mu_{ij}(x,P,S)\, \Omega_i(x_i)\, \Omega_j(x_j) \ , \ \forall (x,P,S) \in N \times P \times S_s \ ,$$

3) the singleton $\{0\}$ is the largest invariant set of the system (23) in the set $Z = \{x : \Omega_1^2(x_1) + \Omega_2^2(x_2) + \dots + \Omega_s^2(x_s) = 0\}$,

and

4) there are $\underline{\mu}_{ij} \in]-\infty,+\infty[$ and $\bar{\mu}_{ij} \in]-\infty,+\infty]$, $\underline{\mu}_{ij} \le \bar{\mu}_{ij}$, such that

$$\underline{\mu}_{ij} \le \mu_{ij}(x,P,S) \begin{cases} \le \bar{\mu}_{ij} \ , & \bar{\mu}_{ij} < +\infty \\ < \bar{\mu}_{ij} \ , & \bar{\mu}_{ij} = +\infty \end{cases} ,$$

$\forall i,j = 1,2,\dots,s$, $\forall (x,P,S) \in N \times P \times S_s$. ∎

Theorem 26. (I) *For structural asymptotic stability of* $x = 0$ *of the system (23) over* $P \times S_s$ *it is sufficient that both*

(a) Assumption 15 holds

and

(b) the matrix $M(L,\epsilon)$ *is negative definite for every* $L \in L_v$.

(II) *If, in addition to (I), the set* $N_i = R^{n_i}$ *and* v_i *is radially unbounded for every* $i = 1,2,\dots,s$ *then* $x = 0$ *of the system (23) is structurally asymptotically stable in the whole over* $P \times S_s$.

Proof. (I) The condition (b) of the theorem guarantees (Grujić, 1974c-1976b) negative definiteness of $M(L,\epsilon)$ for every $L \in L$. The condition (a) and 2) and 4) of Assumption 15 guarantee along motions of (23)

$$D^+ v(x,P,S) < 0 \ , \ \forall x \in N/Z \ , \ \forall (P,S) \in P \times S_s \ ,$$

and

$$D^+ v(x,P,S) \le 0 \ , \ \forall x \in Z \cap N \ , \ \forall (P,S) \in P \times S_s$$

where $v = v_1 + v_2 + \dots + v_s$.

These results and positive definiteness of v on N that is implied by 1) of Assumption 15 guarantee precompactness of solutions of (23) relative to N over $P \times S_s$. All conditions of the invariance principle are satisfied (LaSalle, 1976) over $P \times S_s$. Hence, $x = 0$ of (23) is structurally asymptotically stable over $P \times S_s$.

(II) When, in addition to (I), $N_i = R^{n_i}$ and v_i is radially un-unbounded then $N = R^n$ and v is also radially unbounded. Hence, $x = 0$ is structurally asymptotically stable in the whole over $P \times S_s$. ∎

Example 9. The fourth order system composed of two second order subsystems is described by

$$\frac{dx_i}{dt} = -\alpha x_i + s_{i1} \begin{bmatrix} 4 \\ 0.5 \end{bmatrix} \phi(\sigma) \ , \ \sigma = [-16,-1,-8,-2]\, x \ , \ i = 1,2 \ ,$$

together with

$$\frac{\phi(\sigma)}{\sigma} \in \,]0,+\infty[\ , \ \forall \sigma \in R \ , \ \sigma \neq 0 \ , \ \alpha \in \,]0,+\infty[\ , \ \alpha = \text{const.} \ ,$$

and

$$S_s = \left\{ S : S = \begin{bmatrix} S_1 & 0 \\ 0 & S_2 \end{bmatrix} = \begin{bmatrix} s_{11} & 0 & 0 & 0 \\ 0 & s_{11} & 0 & 0 \\ 0 & 0 & s_{21} & 0 \\ 0 & 0 & 0 & s_{21} \end{bmatrix} , \ s_{11} = s_{21} \in \{0,1\} \right\} .$$

Let $v_1(x_1) = 2x_{11}^2 + x_{12}^2$, $v_2(x_2) = x_{21}^2 + 2x_{22}^2$ and $v = v_1 + v_2$. It is now derived

$$\begin{aligned} \dot{v}(x,S) &= -2\alpha(2x_{11}^2 + x_{12}^2 + x_{21}^2 + 2x_{22}^2) - s_{11}\phi(\sigma)\,\sigma \\ &\leq -2\alpha(\Omega_1^2 + \Omega_2^2) \ , \quad \forall (x,S) \in R^4 \times S_s \ , \end{aligned}$$

where

$$\Omega_1 = \sqrt{2x_{11}^2 + x_{12}^2} \quad \text{and} \quad \Omega_2 = \sqrt{x_{21}^2 + 2x_{22}^2} \ .$$

Hence, $\underline{\mu}_{11} = \bar{\mu}_{11} = -2\alpha$, $\underline{\mu}_{12} = \bar{\mu}_{12} = 0$, $\underline{\mu}_{21} = \bar{\mu}_{21} = 0$, $\underline{\mu}_{22} = \bar{\mu}_{22} = -2\alpha$, and $M(L,\epsilon) = -2\alpha I$, $L = \{L : L = -2\alpha I\}$, $\epsilon = 0$.

Since α is positive valued parameter then $M(L,\epsilon)$ is negative definite $\forall L \in L$. The functions v_1 and v_2 are positive definite in the whole and radially unbounded, $N_1 = R^2$, $N_2 = R^2$, and Ω_1 and Ω_2 are positive definite. All conditions of (II) of Theorem 24 are satisfied. Hence, $x = 0$ of the system is structurally uniformly asymptotically stable in the whole over S_s .

Example 10. The fourth-order system is described by

$$\frac{dx_i}{dt} = (-1)^i x_j + \begin{bmatrix} -\xi x_{i1} + \zeta s_{i1} \,\text{sat}\, x_{j2} \\ -\xi x_{i2} + \zeta s_{i2} \,\text{sat}\, x_{j1} \end{bmatrix} ; \ \xi, \zeta \in \,]0,+\infty[\ , \ \forall i,j = 1,2 \ , \ i \neq j \ .$$

In this case $s_{i1} \neq s_{i2}$ is admissible and

$$h_i^1(x) = (-1)^i x_j - \xi x_i \ , \ h_i^2(x) = \zeta \begin{bmatrix} \text{sat}\, x_{j2} \\ 0 \end{bmatrix} , \ h_i^3(x) = \zeta \begin{bmatrix} 0 \\ \text{sat}\, x_{j1} \end{bmatrix} ,$$

$$h_i = [h_i^1, h_i^2, h_i^3]^T \ , \ \forall i,j = 1,2 \ , \ i \neq j \ ,$$

so that

$$S_i = \begin{bmatrix} 1 & 0 & s_{i1} & 0 & 0 & 0 \\ 0 & 1 & 0 & 0 & 0 & s_{i2} \end{bmatrix} , \ \forall i = 1,2 \ , \ S = \text{diag}\{S_1, S_2\} \ .$$

The structural set S_s is defined by

$$S_s = \Big\{ S : \text{diag}\{[I_2, 0] \ [I_2, 0]\} \leq$$
$$\leq S \leq \text{diag}\left\{ \begin{bmatrix} 1 & 0 & 1 & 0 & 0 & 0 \\ 0 & 1 & 0 & 0 & 0 & 1 \end{bmatrix}, \begin{bmatrix} 1 & 0 & 1 & 0 & 0 & 0 \\ 0 & 1 & 0 & 0 & 0 & 1 \end{bmatrix} \right\} \Big\}$$

Let $v_i(x_i) = \|x_i\|^2$, $\forall i = 1,2$. The function v_i obeys all conditions of (II) of Theorem 26, $\forall i = 1,2$. Now, it can be easily verified that $\mu_{11} = -\xi$, $\mu_{12} = \zeta(s_{11} + s_{12}) \in [0, 2\zeta]$, $\mu_{21} = \zeta(s_{21} + s_{22}) \in [0, 2\zeta]$, $\mu_{22} = -\xi$, $N_i = R^2$ and $\Omega_i(x_i) = \|x_i\|$, $\forall i = 1,2$, obey all conditions of Assumption 15. Evidently $\epsilon = 0$ in $M(L, \epsilon)$, where now

$$M(L,0) = \begin{bmatrix} \ell_{11} & \ell_{12} \\ \ell_{21} & \ell_{22} \end{bmatrix} = L \ ,$$

$$L = \left\{ L : \begin{bmatrix} -\xi & 0 \\ 0 & -\xi \end{bmatrix} \leq L \leq \begin{bmatrix} -\xi & 2\zeta \\ 2\zeta & -\xi \end{bmatrix} \right\} ,$$

and

$$L_v = \left\{ \begin{bmatrix} -\xi & 0 \\ 0 & -\xi \end{bmatrix}, \begin{bmatrix} -\xi & 2\zeta \\ 0 & -\xi \end{bmatrix}, \begin{bmatrix} -\xi & 0 \\ 2\zeta & -\xi \end{bmatrix}, \begin{bmatrix} -\xi & 2\zeta \\ 2\zeta & -\xi \end{bmatrix} \right\} .$$

Hence, $\xi > 2\zeta$ guarantees negative definiteness of $M(L) = L$ for all $L \in L_v$. All conditions of (II) of Theorem 26 are satisfied and, therefore, $x = 0$ of the system is structurally asymptotically stable in the whole over S_s .

Notice that application of the sixth aggregation form, which is based on separate majorization of each $\dot{v}_i$, yields

$$\mu_{11} = -\xi \ , \ \mu_{12} = 2 + \zeta(s_{11}+s_{12}) \in [2, 2+2\zeta] \ ,$$

$$\mu_{21} = 2 + \zeta(s_{21}+s_{22}) \in [2, 2+2\zeta] \ , \ \mu_{22} = -\xi \ ,$$

$$\mathcal{L} = \left\{ L : \begin{bmatrix} -\xi & 2 \\ 2 & -\xi \end{bmatrix} \le L \le \begin{bmatrix} -\xi & 2+2\zeta \\ 2+2\zeta & -\xi \end{bmatrix} \right\}$$

and

$$\mathcal{L}_v = \left\{ \begin{bmatrix} -\xi & 2 \\ 2 & -\xi \end{bmatrix}, \begin{bmatrix} -\xi & 2+2\zeta \\ 2 & -\xi \end{bmatrix}, \begin{bmatrix} -\xi & 2 \\ 2+2\zeta & -\xi \end{bmatrix}, \begin{bmatrix} -\xi & 2+2\zeta \\ 2+2\zeta & -\xi \end{bmatrix} \right\} .$$

Hence, in this case $\xi > 2+2\zeta$ guarantees the stability property, which is more restrictive then the former condition $\xi > 2\zeta$.

III.4.2.6. The ninth aggregation form

The next aggregation form is a development of the third one for a solution of Problem B. A more complete result is obtained here for a class of autonomous large-scale systems with precise estimations.

Assumption 16. There exist open connected neighbourhoods N_i , $N_i \subseteq R^{ni}$ of the states $x_i = 0$, functions v_i and Ω_i such that :

1) functions v_i are time-invariant, positive definite on N_i , $i = 1,2,\dots,s$,
2) along motions of the system (23) the following estimations are fulfilled :

$$v_i^*(x, p_i, S_i) \le \Omega_i(P, S, v_1, \dots, v_s) \ , \ \forall i = 1,2,\dots,s \ , \ \forall (x,P,S) \in N \times P \times S_s \ ,$$

   where $N = N_1 \times \dots \times N_s$ is a connected neighbourhood of $x = 0$,
3) functions Ω_i are continuous; they satisfy the W° - condition and ensure existence of a continuous solution of the system

$$\frac{du_i}{dt} = \Omega_i(P, S, u_1, \dots, u_s) \ , \ \forall i = 1,2,\dots,s \ , \ \forall (P,S,u) \in P \times S_s \times U \ ,$$

4) there exists a neighbourhood U of the state $u = 0$ such that for all $u \in U$, $u \ne 0$, $\Omega_i(P,S,u_1,\dots,u_s) \ne 0$ and $\forall (P,S) \in P \times S_s$, $\Omega_i(P,S,0,\dots,0) = 0$, $\forall i = 1,2,\dots,s$. ∎

Theorem 27. *Let*

(a) conditions of the Assumption 16 be fulfilled;

(b) there exist $S^* \in S_s$ *such that for functions* $\Omega_i(P,S,u_1,\dots,u_s)$ *an estimation* $\Omega_i(P,S,u_1,\dots,u_s) \le \Omega_i(P,S^*,u_1,\dots,u_s)$, $\forall i = 1,2,\dots,s$, $\forall (P,S,u) \in P \times S_s \times U$ *is valid;*

(c) there exist a vector $u^\circ \in U$ *such that the system of inequalities* $\Omega_i(P,S^*,u_1^\circ,\dots,u_s^\circ) < 0$, $\forall i = 1,2,\dots,s$ *is valid.*

Then the state $x=0$ *of the system (23) is structurally asymptotically stable on* $P\times S_s$.

If conditions (a)-(c) are fulfilled for $N_i = R^{n_i}$, $U = R_+^s$ *and functions* v_i *are radially unbounded for every* $i = 1,2,\dots,s$ *, then the state* $x=0$ *of the system (23) is structurally asymptotically stable in the whole on* $P\times S_s$.

Proof. In view of the conditions 2), 3) of the Assumption 16 and (a), (b) of the theorem, for the system (23) we get a comparison system

$$\frac{du_i}{dt} = \Omega_i(P,S^*,u_1,\dots,u_s) \;, \quad \forall i \in [1,s] \;, \quad \forall (P,S^*,u) \in P\times S_s\times U \;. \qquad (28)$$

The solution $u_i(t;t_o, u_o, P,S^*)$ of this system will be considered for $u_{io} \ge v_i(x_{io})$, $\forall i \in [1,s]$. For functions v_i the condition $v_i[x_i(t)] \le u_i(t;t_o,u_o,P,S^*)$, $\forall i \in [1,s]$, is fulfilled along the motions of the system (23) in view of the comparison principle (see Ch. II). Conditions (a) of the theorem and 1) of the Assumption 16 ensure positive definiteness of the function v ,

$$v(x) = v_1(x_1) + v_2(x_2) + \dots + v_s(x_s) \;,$$

and, according to the Assumption 10, Ch. I , there exists a function $\varphi_1 \in K_{[0,\alpha[}$, where $\alpha = \sup\{\|x\| : x\in N\}$ such that

$$\varphi_1(\|x(t;t_o,x_o)\|) \le v(x(t;t_o,x_o)) = \sum_{i=1}^{s} v_i(x_i(t;t_o,x_{io})) \le$$

$$\le \sum_{i=1}^{s} u_i(t;t_o, u_o) \;, \quad \forall t\in T_o \;, \; \forall x_o \in N \;.$$

Condition (c) of the theorem and Theorem 11, Ch. II ensure asymptotic stability of the state $u=0$ of the system (28). In this connection for any $\epsilon>0$, $t_o\in R$ we can find $\eta^*>0$ and $\sigma(\epsilon)>0$ such that

$$\sum_{i=1}^{s} u_i(t;t_o, u_o) < \varphi_1(\epsilon) \;, \quad \forall t \ge t_o+\sigma \;,$$

follows from the inequality $u_{io} < \eta^*$, $\forall i \in [1,s]$.

Further on, due to continuity we can find $\eta>0$ such that $v_i(x_{io}) < \eta^*$ follows from the condition $\|x_{io}\| < \eta$. Hence,

$$\|x(t;t_o,x_o)\| \le \varphi_1^{-1}(v(x(t;t_o,x_o))) \le \varphi_1^{-1}\Big(\sum_{i=1}^{s} u_i(t;t_o, u_o)\Big)$$

$$< \varphi_1^{-1}(\varphi_1(\epsilon)) = \epsilon \;, \quad \forall t \ge t_o+\sigma \;,$$

and the state $x=0$ of the system (23) is structurally asymptotically stable on $P\times S_s$.

When conditions (a)-(c) of the Theorem 27 are fulfilled for $N_i = R^{n_i}$, $U = R_+^s$ and radially unbounded functions v_i , $\forall i = 1,2,\dots,s$, then the function v in x is radially unbounded and condition (c) is fulfilled for $u_{io} \to +\infty$, but $u_{io} \neq +\infty$, $\forall i = 1,2,\dots,s$. The state $u=0$ of the system (28) is asymptotically stable in the whole. Hence, the state $x=0$ of the system (23) is structurally asymptotically stable in the whole on $P\times S_s$. ■

Let us note that if in the Assumption 16 and Theorem 27 estimations in conditions 2) and (b) are fulfilled with the sign of equality, respectively, then conditions of the Theorem 27 are the necessary and sufficient ones for structural asymptotic stability (in the whole) on $P\times S_s$ of the state $x=0$ of system (23).

Assumption 17. There exist open connected neighbourhoods N_i , $N_i \subseteq R^{n_i}$ of the states $x_i = 0$, functions v_i , Ω_i and constants $\alpha_i < 0$, $\forall i = 1,2,\dots,s$, and also the positive numbers η_{i1}, η_{i2} such that :

1) for functions v_i estimates $\eta_{i1}\|x_i\| \le v_i(x_i) \le \eta_{i2}\|x_i\|$, $\forall x_i \in N_i$, $\forall i \in [1,s]$, are valid;
2) along motions of the system (23) the next estimates are fulfilled : $v_i^*(x,p_i,S_i) \le \alpha_i v_i + \Omega_i(P,S,v_1,\dots,v_s)$, $\forall i = 1,2,\dots,s$, $\forall(x,P,S,u) \in$ $\in N\times P\times S_s\times U$;
3) functions $\Omega_i(P,S,v_1,\dots,v_s)$ satisfy the W^o - condition and ensure existence of a continuous solution of the Cauchy problem for a system of differential equations

$$\frac{du_i}{dt} = \alpha u_i + \Omega_i(P,S,u_1,\dots,u_s) \ , \ \forall i = 1,2,\dots,s \ , \ \forall(P,S,u) \in P\times S_s\times U \ ,$$

   for the corresponding initial conditions;
4) there exists a neighbourhood U of the point $u=0$ such that for $u \in \bar{U}$, $u \neq 0$, $\Omega_i(P,S,u) \neq 0$, $\Omega_i(P,S,0) = 0$, $\forall(P,S) \in P\times S_s$. ■

Theorem 28. *Let :*

(a) conditions of the Assumption 17 be fulfilled;

b) there exist $S^* \in S_s$ *such that for functions* $\Omega_i(P,S,u_1,\dots,u_s)$ *an estimation* $\Omega_i(P,S,u_1,\dots,u_s) \le \Omega_i(P,S^*,u_1,\dots,u_s)$, $\forall i = 1,2,\dots,s$, $\forall(P,S,u) \in P\times S_s\times U$ *, is fulfilled;*

c) the state $u=0$ *of the system*

$$\frac{du_i}{dt} = \alpha_i u_i + \Omega_i(P,S^*,u_1,\dots,u_s) \ , \ \forall i = 1,2,\dots,s \ ,$$

be exponentially stable.

Then the state $x = 0$ *of the system (23) is structurally exponentially stable on* $P \times S_s$.

If conditions (a)-(c) are fulfilled for $N_i = R^{ni}$, $U = R_+^s$, *then the state* $x = 0$ *of the system (23) is structurally exponentially stable in the whole on* $P \times S_s$.

The proof of the theorem is similar to that of the Theorem 9. ∎

III.4.2.7. The tenth aggregation form

In the framework of the Problem B, the next aggregation form is based on the concept of a generalized comparison system [27].

Assumption 18. There exist open, connected neighbourhoods N_i , $N_i \subseteq R^{ni}$ of the states $x_i = 0$, functions v_i and r_i such that :

1) functions v_i are positive definite on N_i , $\forall i = 1,2,\dots,s$;
2) $r_i = r_i(t,x,P,S,u)$, $\forall i = 1,2,\dots,s$, satisfy the W^o - condition with respect to u and $\forall (t,x,P,S,u) \in R \times N \times P \times S_s \times U$, $r_i(t,0,P,S,0) = 0$, $\forall i \in [1,s]$;
3) there exist constants $\xi_i > 0$ such that the set $V_{i\zeta}(t)$ is asymptotically contractive for every $\zeta \in]0,\xi_i[$, $\forall i = 1,2,\dots,s$;
4) along the motions of the system (11) the next estimates : $v_i^*(t,x,p_i,S_i) \le r_i(t,x,P,S,v_1,\dots,v_s)$, $\forall i \in [1,s]$, $\forall (t,x,P,S,u) \in$ $\in R \times N \times P \times S_s \times U$ are fulfilled;
5) there exists a continuous solution of the system

$$\frac{du}{dt} = r(t,x,P,S,u) \ , \ \forall (t,x,P,S,u) \in R \times N \times P \times S_s \times U \ . \ \blacksquare$$

This aggregation form leads to the consideration of an extended system

$$\begin{aligned} \frac{dx}{dt} &= g(t,x) + S(t)\, h(t,x,P) \ ; \\ \frac{du}{dt} &= r(t,x,P,S^*,u) \ , \quad S^* \in S_s \ . \end{aligned} \tag{29}$$

Theorem 29. *Let :*

(a) conditions of the Assumption 18 be fulfilled;

(b) there exist $S^* \in S_s$ *such that for functions* $r_i(t,x,P,S,u)$ *an estimation*

$$r_i(t,x,P,S,u) \le r_i(t,x,P,S^*,u) \ , \ \forall (t,x,P,S,u) \in R \times N \times P \times S_s \times U \ ,$$

is fulfilled;

---

[27] See 4) of Comments on References to Ch. III.

(c) *the state* $(x^T,u^T)^T=0$ *of the system (29) be asymptotically* u*-stable.*

Then the state $x=0$ *of the system (11) is structurally asymptotically stable on* $P\times S_s$.

If conditions (a)-(c) are fulfilled for $N_i=R^{ni}$, $U=R_+^s$, $\xi_i=+\infty$, *radially unbounded functions* v_i , *then the state* $x=0$ *of the system (11) is structurally asymptotically stable in the whole on* $P\times S_s$.

Proof. It follows from conditions (a) of the theorem and 1) of Assumption 18 that the function $v=v_1+v_2+\dots+v_s$ is positive definite on N . Condition 3) of the Assumption 18 ensures existence of $\zeta>0$ such that the set $V_\zeta(t)$ is asymptotically contractive for every $\zeta\in]0,\xi[$. Due to conditions (b) of the theorem and 4) of the Assumption 18 we obtain an aggregated system

$$\frac{du_i}{dt}=r_i(t,x,P,S,u_1,\dots,u_s)\ ,\ i=1,2,\dots,s\ ,\ \forall(t,x,P,S,u)\in R\times N\times P\times S_s\times U\ ,$$

which for $u_{io}\geq v_i(t_o,x_o)$ has a solution $u_i(t;t_o,u_o;p_i,S_i)$ for all $(t,x,P,S)\in R\times N\times P\times S_s$. The first statement of the Theorem 29 follows from asymptotic u-stability of the state $(x^T,u^T)^T=0$ of the system (29), estimates

$$v_i(t,x_i(t;t_o,x_{io};p_i,S_i))\leq u_i(t;t_o,u_o;p_i,S_i)\ ,\ \forall t\in T_o\ ,\ \forall i\in[1,s]\ , \tag{30}$$

and from the Theorem 6, Chapter II.

If conditions (a)-(c) of the Theorem 29 are fulfilled for $N_i=R^{ni}$, $U=R_+^s$, $\xi_i=+\infty$, $i\in[1,s]$ and for functions v_i being radially unbounded for all $i=1,2,\dots,s$, then $N=R^n$, $U=R^s$, $\zeta=+\infty$ and the function v is also radially unbounded.

It follows from all that and also from asymptotic u-stability in the whole of the state $(x^T,u^T)^T=0$ of the system (29) and estimates (30) that the state $x=0$ of the system (11) is structurally asymptotically stable in the whole on $P\times S_s$. ■

In order to obtain conditions for the uniform structural asymptotic stability of the state $x=0$ of the system (11) in the framework of the given aggregation form, we must formulate a number of conditions.

Assumption 19. There exist open connected neighbourhoods N_i , $N_i\subseteq R^{ni}$, of the states $x_i=0$ and functions v_i and r_i such that :

1) functions v_i are positive definite on N_i and decrescent on N_i , $\forall i=1,2,\dots,s$;
2) $r_i=r_i(t,x,P,S,u)$, $\forall i=1,2,\dots,s$, satisfy the W^o-condition with

respect to u, and $\forall(t,x,P,S,u) \in R\times N\times P\times S_s\times U$, $r_i(t,0,P,S,0) = 0$, $\forall i \in [1,s]$;

3) along motions of the system (11) the estimates

$$v_i^*(t,x,p_i,S_i) \le r_i(t,x,P,S,v_1,\dots,v_s) ,$$
$$\forall(t,x,P,S,u) \in R\times N\times P\times S_s\times U , \quad \forall i \in [1,s] ,$$

are fulfilled,

4) there exists a continuous solution of the system

$$\frac{du_i}{dt} = r_i(t,x,P,S,u_1,\dots,u_s) , \quad i = 1,2,\dots,s , \quad \forall(t,x,P,S,u) \in R\times N\times P\times S_s\times U . \blacksquare$$

Theorem 30. *Let :*

(a) conditions of the Assumption 19 be fulfilled;

(b) there exist $S^* \in S_s$ *such that for a function* $r_i(t,x,P,S,u)$ *an estimation*

$$r_i(t,x,P,S,u) \le r_i(t,x,P,S^*,u) ,$$
$$\forall i \in [1,s] , \quad \forall(t,x,P,S,u) \in R\times N\times P\times S_s\times U ,$$

(c) the state $(x^T,u^T)^T = 0$ *of the system (29) be uniformly asymptotically* u *-stable.*

Then the state $x=0$ *of the system (11) is structurally uniformly asymptotically stable on* $P\times S_s$.

If conditions (a)-(c) are fulfilled for $N_i = R^{n_i}$, $U = R_+^s$, *and functions* v_i *are radially unbounded for all* $i = 1,2,\dots,s$, *then the state* $x=0$ *of the system (11) is structurally uniformly asymptotically stable in the whole on* $P\times S_s$.

Proof. The conditions (a) of the theorem and 1) of the Assumption 19 show that the function $v = v_1 + v_2 + \dots + v_s$ is positive definite and decrescent on N. Using the conditions 2)-4) of the Assumption 19, (b) of the Theorem 30, and the comparison principle we find estimates $v_i(t;x_i(t;t_o,x_{io})) \le u_i(t;t_o,u_o)$. Hence, similarly to the proof of the Theorem 27 we claim that the state $x=0$ of the system (11) is structurally uniformly asymptotically stable on $P\times S_s$.

If conditions (a)-(c) of the Theorem 30 are fulfilled for $N_i = R^{n_i}$, $U = R^s$ and radially unbounded functions v_i , then the function v is also radially unbounded and decrescent on $N = R^n$. As a result of this and of the comparison principle the state $x=0$ of the system (11) is structurally uniformly asymptotically stable in the whole on $P\times S_s$. $\blacksquare$

Conditions for exponential stability impose the special requirements on functions v_i .

Assumption 20. There exist open connected neighbourhoods N_i, $N_i \subseteq R^{n_i}$ of the states $x_i = 0$, functions v_i, r_i and positive numbers η_{i1}, η_{i2} such that

1) $\eta_{i1}\|x_i\| \le v_i(t,x_i) \le \eta_{i2}\|x_i\|$, $\forall(t,x_i) \in R\times N_i$, $\forall i \in [1,s]$;
2) $r_i = r_i(t,x,P,S,u)$, $\forall i \in [1,s]$, satisfies the W^o - condition with respect to u, and $\forall(t,x,P,S,u) \in R\times N\times P\times S_s\times U$, $r_i(t,0,P,S,0) = 0$, $\forall i = 1,2,\dots,s$;
3) along motions of the system (11) the following estimates are fulfilled :
$$v_i^*(t,x,p_i,S_i) \le r_i(t,x,P,S,v_1,\dots,v_s)\ ,$$
$$\forall i \in [1,s]\ ,\ \forall(t,x,P,S,u) \in R\times N\times P\times S_s\times U\ ;$$
4) there exists a continuous solution of the system
$$\frac{du_i}{dt} = r_i(t,x,P,S,u_1,\dots,u_s)\ ,\ i = 1,2,\dots,s\ ,\ \forall(t,x,P,S,u) \in R\times N\times P\times S_s\times U\ . \blacksquare$$

Theorem 31. *Let*

(a) conditions of the Assumption 20 be fulfilled;
(b) there exist $S^* \in S_s$ *such that for functions* $r_i(t,x,P,S,u)$ *the estimates*
$$r_i(t,x,P,S,u) \le r_i(t,x,P,S^*,u)\ ,$$
$$\forall(t,x,P,S,u) \in R\times N\times P\times S_s\times U\ ,$$
are fulfilled;
(c) the state $(x^T,u^T)^T = 0$ *of the system (29) be exponentially* u*-stable.*

Then the state $x = 0$ *of the system (11) is structurally exponentially stable on* $P\times S_s$.

If conditions (a)-(c) are fulfilled for $N_i = R^{n_i}$, $U = R_+^s$, *then the state* $x = 0$ *of the system (11) is structurally exponentially stable in the whole on* $P\times S_s$.

Proof. Conditions (a) of the theorem and 1) of the Assumption 20 ensure existence of the positive numbers η_1 and η_2 such that the function $v = v_1 + v_2 + \dots + v_s$ satisfies an estimation $\eta_1\|x\| \le v(t,x) \le \eta_2\|x\|$, $\forall(t,x) \in R\times N$.

Together with the comparison principle conditions (b) of the theorem and 2)-4) of the Assumption 20 lead to estimations
$$v_i(t,x_i(t;t_o,x_o;p_i,S_i)) \le u_i(t;t_o,u_o;p_i,S_i)\ ,\ \forall i \in [1,s]\ ,$$
$$v(t,x(t;t_o,x_o;P,S)) \le \sum_{i=1}^{s} u_i(t;t_o,u_o;P,S)\ ,$$
respectively. Hence, in view of the condition (c) of the theorem it is

easy to find out that for a solution $\chi(t;t_0,x_0;P,S)$ of the system (11) an exponential estimation is fulfilled on $P\times S_s$, which proves the first statement of the theorem.

If conditions (a)-(c) of the Theorem 31 are fulfilled for $N_i = R^{n_i}$, $U = R^s_+$, $\forall i = 1,2,\dots,s$, then $N = R^n$ and as a consequence of the comparison principle and estimates of the funciton v we determine that the state $x = 0$ of the system (11) is exponentially stable in the whole on $P\times S_s$. ■

III.4.2.8. Conclusion on solutions for the Problem B

Solutions for the problem B have been based on the system decomposition into interconnected subsystems. They do not require either stability or stability test of disconnected subsystems. However, they enable one shot stability test of the whole system. Seven different aggregation forms have been developed. They have been used for establishing criteria for different stability properties. Besides, they enable influence of the form of interactions on the choice of the form of a tentative Liapunov function.

III.4.3. The structural stability analysis of a large-scale system with non-asymptotically stable subsystems

In examples 3, 5 of the present chapter we consider a case, when indepedent subsystems of a large-scale system have the unstable states $x_i = 0$. Here we shall analyze a case of the stability theory of large-scale systems, when the equilibria of subsystems are stable ("neutrally stable"). The obtained comparison systems (aggregated systems) do not possess, as a rule, a linear part, which makes their analysis more difficult.

III.4.3.1. Eleventh aggregation form

The present aggregation form is realized in the framework of a solution of the Problem A and is connected with the third form.

Assumption 21. There exist open connected neighbourhoods N_i, $N_i \subseteq R^{n_i}$, of the states $x_i = 0$ and functions v_i and q_i such that :

1) functions v_i are positive definite on N_i, $\forall i = 1,2,\dots,s$;
2) functions q_i are locally Lipschitzian and they satisfy the W^o - condition;
3) along motions of the system (11) an estimate

$$D_t^+ v_i(t,x_i) + [D_{x_i}^+ v_i(t,x_i)]^T g_i(t,x_i) \le 0 ,$$

$$\forall i = 1,2,\dots,s , \ \forall (t,x_i) \in R\times N_i$$

is fulfilled;

4) $[D_{x_i}^+ v_i(t,x_i)]^T S_i h_i(t,x,p_i) \le q_i(t,P,S,v_1,\dots,v_s)$, $\forall i \in [1,s]$, $\forall(t,x,P,S) \in R\times N\times P\times S_s$, and $q_i(t,P,S,0,\dots,0) = 0$, $\forall i \in [1,s]$, $\forall t \in R$. ■

Theorem 32. *Let*

(a) conditions of the Assumption 21 be fulfilled;

(b) there exist positive numbers ξ_i *(it is possible that* $\xi_i = +\infty$*) such that the sets* $V_{i\zeta}(t)$ *are asymptotically contractive for any* $\zeta \in]0,\xi_i[$ *and every* $i = 1,2,\dots,s$ *;*

(c) there exist $S^* \in S_s$ *such that an estimation*

$$q_i(t,P,S,u) \le q_i(t,P,S^*,u) , \ i \in [1,s]$$

is fulfilled $\forall(t,P,S,u) \in R\times P\times S_s\times U$ *;*

(d) the state $u = 0$ *of the system*

$$\frac{du_i}{dt} = q_i(t,P,S^*,u_1,\dots,u_s) , \ i = 1,2,\dots,s$$

be asymptotically stable.

Then the state $x = 0$ *of the system (11) is structurally asymptotically stable on* $P\times S_s$.

If conditions (a)-(d) are fulfilled for $N_i = R^{n_i}$, $U = R_+^s$, *functions* v_i *are radially unbounded and* $\xi_i = +\infty$, $\forall i \in [1,s]$ *, then asymptotic stability in the whole of the state* $u = 0$ *of the system ensures structural asymptotic stability in the whole of the state* $x = 0$ *of the system (11).*

The proof of the theorem is similar to that of the Theorem 7 of this chapter. ■

Theorem 33. *Let :*

(a) conditions of the Assumption 21 be fulfilled;

(b) functions v_i *,* $i = 1,2,\dots,s$ *, be decrescent;*

(c) there exist $S^* \in S_s$ *such that estimation*

$$q_i(t,P,S,u) \le q_i(t,P,S^*,u) , \ \forall i \in [1,s] ,$$

is fulfilled $\forall(t,P,S,u) \in R\times P\times S_s\times U$ *;*

(d) the state $u = 0$ *of the system*

$$\frac{du_i}{dt} = q_i(t,P,S^*,u_1,\dots,u_s) , \ i = 1,2,\dots,s ,$$

be uniformly asymptotically stable.

Then the state $x=0$ *of the system (11) is structurally uniformly asymptotically stable on* $P \times S_s$.

If conditions (a)-(d) are fulfilled for $N_i = R^{ni}$, $U = R_+^s$, *functions* v_i *are radially unbounded* $\forall i \in [1,s]$, *functions* q_i *satisfy the Lipschitz conditions and that of Ważewski for* $u \in R_+^s$, *then the state* $x=0$ *of the system (11) is structurally uniformly asymptotically stable in the whole on* $P \times S_s$.

The proof is similar to that of the Theorem 8. ∎

Theorem 34. *Let :*

(a) conditions of the Assumption 21 be fulfilled;

(b) there exist positive numbers η_{i1}, η_{i2} *such that*

$$\eta_{i1}\|x_i\| \le v_i(t,x_i) \le \eta_{i2}\|x_i\| \ , \ \forall (t,x_i) \in R \times N_i \ , \ i = 1,2,\dots,s \ ;$$

(c) there exist $S^* \in S_s$ *such that the estimate*

$$q_i(t,P,S,u) \le q_i(t,P,S^*,u) \ , \ \forall i \in [1,s] \ , \ \forall (t,P,S,u) \in R \times P \times S_s \times U \ ,$$

is fulfilled;

(d) the state $u=0$ *of the system*

$$\frac{du_i}{dt} = q_i(t,P,S^*,u) \ , \ i = 1,2,\dots,s \ ,$$

be exponentially stable.

Then the state $x=0$ *of the system (11) is structurally exponentially stable on* $P \times S_s$.

If conditions (a)-(d) are fulfilled for $N_i = R^{ni}$, $U = R_+^s$, *and functions* q_i *satisfy the Lipschitz condition and that of Ważewski for* $u \in R_+^s$, *then the state* $x=0$ *of the system (11) is structurally exponentially stable in the whole on* $P \times S_s$.

The proof is similar to that of the Theorem 9. ∎

III.4.3.2. The twelfth aggregation form

We consider the system (11) in the presence of a small positive parameter in the interconnections of subsystems [28]

$$\frac{dx}{dt} = g(t,x) + \mu \, S(t) \, h(t,x,P) \tag{31}$$

under the invariant structure, i.e. $S(t) \equiv I$. We make the following assumptions on the independent subsystems (32),

[28] See 5) of Comments on References to Ch. III.

$$\frac{dx_i}{dt} = g_i(t, x_i) \quad , \quad i = 1,2,\dots,s \quad , \tag{32}$$

Vector-functions g_i are defined, continuous and they satisfy the Lipschitz condition with respect to x_i in the domain

$$\Gamma = \left\{ (t,x) : t \in [0,+\infty[\quad , \quad \sum_{j=1}^{n_i} |x_j| < k \right\} \quad , \quad i = 1,2,\dots,s \quad , \quad k \in \breve{R}_+ \quad . \tag{33}$$

The equilibrium state $x_i = 0$, $\forall i = 1,2,\dots,s$, is t_o - uniformly stable. Functions $h(t,x,P)$ are supposed to be such that there exists the unique solution of the Cauchy problem for the system (31), and $h(t,0,P) = 0$, $\forall (t,P) \in R_+ \times P$.

Let vectors x_i be ordered in a sequence α_i , $i = 1,2,\dots,s$. In the sequence $\{\alpha_i\}$ there is $\alpha_1 = 1$ and each sequence element α_k $(k = 2, \dots, s)$ is equal to the previous one or to index k . Let us denote

$$\varphi_i^{\alpha_1}(t, x_1, \dots, x_s, p) = [D_{x_i} v_i(t, x_i)]^T h_i(t, x, P) \quad ,$$

$$\varphi_i^{\alpha_i}(t, x_{\alpha_i}, \dots, x_s, p) = [D_{x_i} v_i(t, x_i)]^T h_i(t, 0, \dots, 0, x_{\alpha_i}, \dots, x_s, p) \quad ,$$

$$\bar{\varphi}_i^{\alpha_i}(t_o, x_{\alpha_i o}, \dots, x_{so}, p) = \mu_t \{ \varphi_i^{\alpha_i}(t, \bar{x}_{\alpha_i}(t), \dots, \bar{x}_s(t), p) \} \quad ,$$

where μ_t is an averaging operator with respect to an explicitly contained time t , $\bar{x}_k(t) = \bar{x}_k(t; t_o, x_{ko})$ (for t_o, x_{ko} from the domain Γ (33)) , $k = 1,2,\dots,s$, is a solution of the system (32).

We determine a vector-function $\hat{v} = (\hat{v}_1(x_1), \dots, \hat{v}_s(x_s))^T$, components of which - $\hat{v}_i$ are continuous and defined in the domain $\Gamma_i = \{ x_i : \|x_i\| < k \}$.

We denote

$$E(\hat{v}_i = 0) \triangleq \{ x_i \in \Gamma_i : \hat{v}_i(x_i) = 0 \} \quad ,$$

$$\rho(x_i, E(\hat{v}_i = 0)) = \inf [\|x_i - x_i'\| \; , \; x_i' \in E(\hat{v}_i = 0)] \quad .$$

Definition 5. We shall say that $\bar{\varphi}_i^{\alpha_i}$ is smaller than zero in the set $E(\hat{v}_i = 0)$ iff $\forall \eta_i, \epsilon_i$ $(0 < \eta_i < \epsilon_i < k)$

$$\exists r_i(\eta_i, \epsilon_i) > 0 \;\&\; \delta(\eta_i, \epsilon_i) > 0 \qquad \bar{\varphi}_i^{\alpha_i}(t_o, x_{\alpha_i o}, \dots, x_{so}, p) < -\delta_i$$

for

$$\eta_i \le \|x_{io}\| \le \epsilon_i \quad , \quad \rho(x_{io}, E(\hat{v}_i = 0)) < r_i \quad , \quad \forall t_o \in [0, +\infty[\quad . \quad \blacksquare$$

Assumption 22. There exist open connected neighbourhoods N_i , $N_i \subseteq R^{n_i}$ of the states $x_i = 0$, functions v_i and $\hat{v}_i$ such that the following conditions are fulfilled :

1) functions v_i are positive definite, continuous, differentiable and decrescent on N_i , $\forall i \in [1,s]$;
2) functions $\hat{v}_i$ are continuous and non-positive on N_i , $\forall i \in [1,s]$;

3) along motions of subsystems (32) :

$$D_t v_i(t,x_i) + [D_{x_i} v_i(t,x_i)]^T g_i(t,x_i) \le \hat{v}_i(x_i) \le 0 ,$$
$$\forall (t,x_i) \in R\times N_i , \quad \forall i \in [1,s] ;$$

4) functions $\varphi_i^{\alpha i}$, $\forall i \in [1,s]$, satisfy the Lipschitz condition on N with the constant L;

5) functions h_i , $\forall i = 1,2,\dots,s$, are bounded on N . ∎

Theorem 35. *Let :*

(a) conditions of the Assumption 22 be fulfilled;

(b) there exist sequence of $\bar{\varphi}_i^{\alpha i}$, $\forall i \in [1,s]$, *uniformly with respect to* $t_o, x_{\alpha_i o}, \dots, x_{so}$ *;*

(c) $\bar{\varphi}_i^{\alpha i}$ *be defined in the set* $E(\hat{v}_i = 0)$, $i \in [1,s]$, *smaller than zero.*

Then the state $x=0$ *of the system (31) is* t_o *-uniformly stable.*

Proof. Let there be given $\epsilon \in]0,k[$ and $t_o \in]0,+\infty[$. The theorem will be proved if it appears that for every solution $x_i(t) = x_i(t;t_o,x_o)$ of the system (31) an estimate

$$\sum_{i=1}^{s} \|x_i(t;t_o,x_o)\| < \epsilon \quad \text{for} \quad \sum_{i=1}^{s} \|x_{io}\| < \eta , \quad 0 < \eta < \mu_o ,$$

where $\eta(\epsilon) > 0$ and $\mu_o(\epsilon) > 0$ will not depend on t_o , will be fulfilled for all $t > t_o$.

Let us assume that $\epsilon_s = \epsilon/s$. As a consequence of the condition 1) of the Assumption 22 for the function v_s some constants $c_s > 0$ and $\eta_s > 0$ can be given such that any point x_s of the moving surface $v_s(t,x_s) = c_s$ for $t \ge 0$ will satisfy an inequality $\eta_s \le \|x_s\| \le \epsilon_s/2$.

According to the condition (c) of the theorem there exists a number $\delta_s(\eta_s,\epsilon_s) > 0$ for which $\bar{\varphi}_s^{\alpha_s} < -\delta_s(\eta_s,\epsilon_s)$ and $\epsilon_{s-1} = \min\{\delta_s/6L(\alpha_s-1), \epsilon_s\}$ can be defined. Continuing this process, we can choose constants c_{s-1} , η_{s-1} for the function $v_{s-1}(t,x_{s-1})$ and by means of $\epsilon_{s-1}, \eta_{s-1}$ we can determine the value δ_{s-1} . Thus, for $v_i(t,x_i)$, $i = s-2,\dots,1$ constants $\epsilon_i = \min\{\delta_{i+1}/6L(\alpha_{i+1}-1), \epsilon_{i+1}\}$, η_i and $\delta_i(\eta_i,\epsilon_i)$ can be defined. In addition, for the points x_i , satisfying the condition $v_i(t,x_i) = c_i$, an inequality $\eta_i \le \|x_i\| \le \epsilon_i/2$ will be fulfilled for any $t \ge 0$. Fulfilling all conditions of the Theorem 35 we can verify that for $t \ge 0$, $\forall i = 1,2,\dots,s$, $x_i(t)$ satisfies an estimate $\|x_i(t)\| < \epsilon$ for $t \ge 0$ and $\|x_{io}\| < \eta_i$, $i = 1,2,\dots,s$.

Estimates illustrated above are general for all $i \in [1,s]$. We consider a solution $x_i(t)$ leaving the domain determined by $\|x_i\| < \eta_i$ and at a moment $t = t'_{io}$ intersecting the hypersurface $v_i(t,x_i) = c_i$ at the point x'_{io}. For $t > t'_{io}$ behaviour of the function $v_i(t,x_i(t))$ can be estimated as follows :

$$v_i(t,x_i(t)) \leq c_i + \int_{t'_{io}}^{t} \hat{v}_i(x_i(\tau))\, d\tau + \mu \int_{t'_{io}}^{t} \varphi_i^{\alpha_i}(\tau, x_1(\tau),\dots,x_s(\tau))\, d\tau . \tag{34}$$

Estimating the third term in formula (34) we assume that for all $j \in [i+1,s]$ and $t \geq t_o$ solutions $x_j(t)$ remain in Γ_j. There are two possible cases.

1) Let there be $\rho(x'_{io}, E(\hat{v}_i = 0)) \geq r_i(\eta_i,\epsilon_i)$. We determine from $h_i(t,0,\dots,0,P) = 0$ and the condition 4) of the Assumption 22 that $|\varphi_i^{\alpha_1}(t,x_1,\dots,x_s,P)| \leq u_i$ for $(t,x) \in R_+ \times N$, $u_i \in \check{R}_+$, $i \in [1,s]$. Chosing $\mu < \mu' < \gamma_i/2u_i$, $(\gamma_i = \inf|\hat{v}_i(x_i)|)$,

$$\mathcal{D}_i = \{x_i : \rho(x_i, E(\hat{v}_i = 0)) > r_i/2 ,\ \eta_i \leq \|x_i\| \leq \epsilon_i\}$$

we obtain an estimate $v_i(t,x_i(t)) \leq c_i - \gamma_i(t - t'_{io})/2$ for all $t > t'_{io}$, for which $\|x_i(t)\| < \eta_i$, $\rho(x_i(t), E(\hat{v}_i = 0)) > r_i/2$. Hence, it is easy to obtain that the component $x_i(t)$ will not leave the domain given by $\|x_i\| < \epsilon_i$ and will not intersect the hypersurface $v_i(t,x_i(t)) = c_i$ as far as the estimate $\rho(x_i(t), E(\hat{v}_i = 0)) > r_i/2$ is correct.

2) Let there be $\rho(x'_{io}, E(\hat{v}_i = 0)) > r_i(\eta_i,\epsilon_i)$. The third term in the estimate (34) will be represented in the form :

$$\int_{t'_{io}}^{t} \varphi_i^{\alpha_1}(\tau, x_1(\tau),\dots,x_s(\tau))\, d\tau =$$

$$= \int_{t'_{io}}^{t} \left[\varphi_i^{\alpha_1}(\tau,x_1(\tau),\dots,x_s(\tau)) - \varphi_i^{\alpha_i}(\tau,\bar{x}_{\alpha_i}(\tau),\dots,\bar{x}_s(\tau))\right] d\tau + \tag{35}$$

$$+ \int_{t'_{io}}^{t} \varphi_i^{\alpha_i}(\tau,\bar{x}_{\alpha_i}(\tau),\dots,\bar{x}_s(\tau))\, d\tau ,$$

where $\bar{x}_k(t) = \bar{x}_k(t;t'_{ko},x'_{ko})$, $x'_{ko} = \bar{x}_k(t'_{ko};t'_{ko},x'_{ko})$, $k = \alpha_i,\dots,s$, is a solution of the system (32). According to the condition 4) of the Assumption 22

$$|\varphi_i^{\alpha_1}(t,x_1,\dots,x_s) - \varphi_i^{\alpha_i}(t,\bar{x}_{\alpha_i},\dots,\bar{x}_s)| \leq L\Big(\sum_{n=1}^{\alpha_i - 1} \|x_n\| + \sum_{k=\alpha_i}^{s} \|x_k - \bar{x}_k\|\Big) .$$

In the course of this consideration of $x_i(t)$ (index i is increasing) we determine $\|x_k\| < \epsilon_n$ $(n = 1,\dots,\alpha_i - 1)$ for $\mu < \mu_{\alpha_i - 1}$. Taking into account the choice of ϵ_n, $n = 1,2,\dots,\alpha_i - 1$ for $\mu < \mu_{\alpha_i - 1}$ we obtain

$$L \sum_{n=1}^{\alpha_i - 1} \|x_n\| < \frac{\delta_i}{6} .$$

Using the conditions (b) and (c) of the Theorem 35, we determine that at an interval $[t'_{io}, t'_{io} + 2\ell_i]$, where ℓ_i can be chosen sufficiently large, the last term in (35) will be smaller than $5\delta_i(t - t'_{io})/6$ for $t > t'_{io} + \ell$. As a consequence of the Lipschitz condition on $g_i(t,x)$ and the condition 5) of the Assumption 22, by means of the choice of μ''_i, we show that $\|x_i(t) - \bar{x}_i(t)\| < \epsilon_i/2$, i.e. the solution $x_i(t)$ will not leave the domain $\|x_i\| < \epsilon$ at this interval and

$$\sum_{k=\alpha_i}^{s} \|x_k - \bar{x}_k\| < \frac{\delta_i}{6L} .$$

Hence, for $t \in]t'_{io} + \ell_i, t'_{io} + 2\ell_i[$ and $\mu < \min\{\mu''_i, \mu_{\alpha_i - 1}\}$ an inequality

$$\int_{t'_{io}}^{t} \varphi_i^{\alpha_1}(\tau, x_1(\tau), \dots, x_s(\tau))\, d\tau < -\frac{\delta_i}{2}(t - t'_{io})$$

is valid. It means that the last term in the expression (34) will be negative for $t \geq t'_{io} + \ell_i$. That is why the solution $x_i(t)$ after leaving the hypersurface $v_i(t,x_i) = c_i$, remains in the domain defined by $v_i(t,x_i) \leq c_i$ and, moreover, for some $t_i^* \in]t'_{io}, t'_{io} + \ell_i[$ will come inside the hypersurface $v_i(t,x_i) = c_i$. From uniformity of estimates with respect to t'_{io}, x'_{io} for $\mu < \mu_i = \min\{\mu'_i, \mu''_i, \mu_{\alpha_i - 1}\}$ at every subsequent interval one of the two cases given above is valid. Hence, $x_i(t)$ will not leave the domain $\|x_i\| < \epsilon$ for any $t \geq t_o$.

If after $v_1(t, x_1(t))$ we estimate all $v_i(t, x_i(t))$ in sequence in terms of $i \in [2,s]$, then it is not difficult to find out that

$$\sum_{i=1}^{s} \|x_i(t)\| < \epsilon \quad \text{for} \quad t \geq t_o \geq 0 \quad \text{and} \quad \sum_{i=1}^{s} \|x_{io}\| < \eta \quad \text{for} \quad \eta < \mu_o .$$

Here $\mu_o = \min\{\mu_i\}$, $\eta(\epsilon) = \min\{\eta_i\}$, $i \in [1,s]$, do not depend on $t_o \in T_o$. The theorem is proved. ■

Example 11 (Kosolapov, 1979). We consider a system of equations of the perturbed motion

$$\frac{dx_1}{dt} = -x_1 + x_2^2 + \mu x_2 y_1 , \quad \frac{dx_2}{dt} = \mu(x_2 y_2 - x_1 x_2 + x_1 z + x_1^2 \sin t) ,$$

$$\frac{dy_1}{dt} = -y_1 + y_2 + \mu x_2^2 , \quad \frac{dy_2}{dt} = \mu(-x_2^2 y_2 z^2 - y_1 y_2^2 + (y_1 + y_2)\cos t) , \tag{36}$$

$$\frac{dz}{dt} = \mu(-z^3 + y_1 y_2 - x_1^2 z + z^2 \cos t) \ . \tag{36}$$

For $\mu = 0$ we get three independent subsystems

$$\frac{dx_1}{dt} = -x_1 + x_2^2 \ , \ \frac{dx_2}{dt} = 0 \ ; \ \frac{dy_1}{dt} = -y_1 + y_2 \ , \ \frac{dy_2}{dt} = 0 \ ; \ \frac{dz}{dt} = 0 \ . \tag{37}$$

Solutions of these subsystems have the form

$$x_1 = x_{2o}^2 + (x_{1o} - x_{2o}^2) \exp[-(t-t_o)] \ , \ x_2 = x_{2o} \ ,$$

$$y_1 = y_{1o} + (y_{1o} - y_{2o}) \exp[-(t-t_o)] \ , \ y_2 = y_{2o} \ ,$$

$$z = z_o \ .$$

Referring to subsystems (37), the Liapunov functions, their full derivatives, and all $\bar{\varphi}_i^{\alpha_i}$ can be defined as follows

$$v_1(\mathbf{y}) = y_2^2 + (y_1 - y_2)^2 \ , \ \frac{dv_1}{dt} = -2(y_1 - y_2)^2 = \hat{v}_1(\mathbf{y}) \le 0 \ ,$$

$$\bar{\varphi}_1^1(\mathbf{y}_o, z_o, \mathbf{x}_o) = -2y_{2o}^4 - 2y_{2o}^2 x_{2o}^2 z_o^2 \ ;$$

$$v_2(z) = z^2 \ , \ \frac{dv_2}{dt} = \hat{v}_2(z) \equiv 0 \ ,$$

$$\bar{\varphi}_2^2(z_o, \mathbf{x}_o) = -2z_o^4 - 2z_o^2 x_{2o}^4 \quad \text{for} \quad y_1 = 0 \ , \ y_2 = 0 \ ;$$

$$v_3(\mathbf{x}) = x_2^2 + (x_1 - x_2^2)^2 \ , \ \frac{dv_3}{dt} = -2(x_1 - x_2^2)^2 = \hat{v}_3 \le 0 \ ,$$

$$\bar{\varphi}_3^3(\mathbf{x}_o) = -2x_{2o}^4 \quad \text{for} \quad y_1 = 0 \ , \ y_2 = 0 \ , \ z = 0 \ .$$

All conditions of the Theorem 35 are fulfilled. Hence the state $x_1 = x_2 = y_1 = y_2 = z = 0$ of the system (36) is t_o - uniformly stable.

COMMENTS ON REFERENCES

Brief review of contributions to large-scale systems stability theory will be given only for stability properties in Liapunov's sense and only for continuous time systems (until 1982).

1. Bailey (1966,1968) and Aoki (1968) proposed the method of effective aggregation of a large-scale system. Bailey (1966,1968) was the first to apply the concept of a vector Liapunov function to analysis of the large-scale system stability by means of the Problem A solution. An essential part of all three illustrated approaches (the method of vector Liapunov functions, the method of vector norms, the minimax method) is the comparison principle (Chaplygin, 1919, Kamke, 1932, Ważewski, 1950), a development of which is given in Chapter II of the present

book. Lakshmikantham (1965) and Kayande and Lakshmikantham (1966) applied vector Liapunov functions to analysis of stability conditions and conditionally-invariant sets, respectively. Considerable development received ideas of Bailey (1966,1968) while investigating the large-scale systems in the framework of the Problem A with the following subsystems :

* exponentially stable [Matrosov (1963,1972); Piontkovskii and Rutkovskaya (1967); Zadorozhny (1969); Michel, Porter (1970); Porter, Michel (1970,1971); Šiljak (1971-1973 , 1975, 1977, 1978); Zemlyakov (1972); Thompson (1972); Araki, Kondo (1972); Thompson, Koenig (1972); Grujić, Šiljak (1972,1973); Grujić (1972 , 1974b,c , 1976b); Michel (1973,1974); Weissenberger (1973); Vakhonina, Zemlyakov, Matrosov (1973); Šiljak, Vukčević (1974); Martynyuk (1975); Michel, Miller (1977); Araki (1978); Grujić, Gentina, Borne, Burgat, Bernussou (1978); Saeki, Araki, Kondo (1980); Ikeda, Šiljak (1980)];
* asymptotically stable [Michel, Porter (1970); Grujić, Šiljak (1972, 1973); Šiljak (1972a,1978); Grujić (1972 , 1974b,c , 1976b); Michel (1973,1974); Araki (1975,1978); Blight, McClamrock (1975); Grujić, Gentina, Borne (1976); Bitsoris, Burgat (1976); Michel, Miller (1977); Araki (1978); Sinha (1980)];
* unstable [Thompson (1972); Grujić, Šiljak (1972,1973); Grujić (1972, 1976b) ; Martynyuk (1972,1973,1975); Michel (1973,1974); Goisa, Martynyuk (1974); Michel, Miller (1977); Morari, Stephanopoulos, Aris (1977); Araki (1978); Šiljak (1978)];
* equi-absolutely bounded [Kloeden (1975)];
* partially stable [Bondi, Fergola, Gambardella (1979)].

In the framework of the Problem B solution continuous large-scale systems were studied by Matrosov (1962,1968); Gentina, Borne, Laurent (1972,1973), Grujić (1974a,c , 1977a); Grujić, Gentina, Borne (1976); Borne, Benrejeb (1977); Grujić, Gentina, Borne, Burgat, Bernussou (1979); Grujić, Burgat (1979,1980); Burgat, Grujić (1979); Vidyasagar (1980).

Domains of attraction of large-scale systems were analyzed by Weissenberger (1973), and somewhat later, by Bitsoris, Burgat (1976); Morari, Stephanopoulos, Aris (1977); Šiljak (1978). Grujić and Ribbens-Pavella (1978,1979) studied the problem of asymptotic stability domain estimation in the framework of large-scale systems.

2. The concept of stability of motion with respect to a part of variables belongs to Liapunov (1893). The first theorems on partial stability on the basis of Liapunov's functions were obtained by Rumyantsev (1957). Let us mention here the works by Peiffer, Rouche (1969) and by

Rouche and Peiffer (1967) which greatly influenced the development of the partial stability theory. An outline of this concept development together with the original results is contained in the paper by Oziraner and Rumyantsev (1972) and in the book by Rouche, Habets and Laloy (1980).

3. The word "sat" is an abbreviation of the expression "saturation nonlinearity".

4. An idea of the analyzed system extension by means of a generalized comparison system [Matrosov (1963)] is utilized in the works by Hatvani (1975) and by Martynyuk (1975) and it is widely applied in the monograph by Martynyuk and Gutowski (1979). Application of this idea to analysis of the large-scale system structural stability is realized here for the first time.

5. Large-scale systems with weakly interconnected subsystems [Martynyuk (1972a,1975)] admit an effective application of the method of averaging, developed in nonlinear mechanics [Bogolyubov, Mitropolsky (1974)], together with the method of Liapunov's functions [Liapunov (1892)]. The given form of aggregation was developed by Kosolapov under the guidance of Martynyuk [Martynyuk, Kosolapov (1978); Kosolapov (1979)].

6. Šiljak pointed out to the authors that Persidskii used the notion of structural stability in a sense different than he defined connective stability and than structural stability defined in this book. The notion "structural stability" was used also by other authors, even they studied only influence of system parameters instead of system structural variations on stability. There is not any ambiguity herein because structural stability is used only in the sense defined in Chapter III.

Anashkin, O.V. (1978), On investigations of stability under persistently acting disturbances in the neutral case. *Different Uravn, 14*, No.6, 1124-1127 (in Russian).

Aoki, M. (1968), Control of large-scale dynamic systems by aggregation. *IEEE Trans. on Aut. Control, AC-13*, No.3, 246-253.

Araki, M. (1975), Application of M-matrices to the stability problems of composite dynamical systems. *J. Math. Analysis and Applications, 52*, 309-321.

Araki, M. (1978), Stability of large-scale non linear systems-quadratic-order theory of composite-system method using M-matrices. *IEEE Trans. on Aut. Control, AC-23*, No.2, 129-142.

Araki, M., and B. Kondo (1972), Stability and transient behaviour of composite systems. *IEEE Trans. on Aut. Control, AC-17*, No.4, 537-541.

Bailey, F.N. (1966), The application of Lyapunov's second method to interconnected systems. *J. SIAM Control*, Ser.A, *3*, No.3, 443-462.

Bailey, F.N. (1968), The concept of aggregation in system stability analysis. *2nd Asilomar Conf. on Circ. and Syst.*, 570-576.

Bellman, R. (1962), Vector Liapunov functions. *J. SIAM Control*, Ser.A, *1*, No.1, 32-34.

Bitsoris, G., and C. Burgat (1976), Stability conditions and estimates of the stability region of complex systems. *Int. J. Systems Science*, *7*, No.8, 911-928.

Blight, J.D., and N.H. McClamroch (1975), Graphical stability criteria for large-scale nonlinear multiloop systems. *Preprints of the 6th IFAC World Congress*, Part 44.5 : 1-6.

Bogoljubov, N.N., and Yu.A. Mitropolsky (1974), *Asymptotic Methods in the Theory of Non-Linear Oscillations*. Nauka, Moscow (in Russian).

Bondi, P., P. Fergola, and L. Gambardella (1979), Partial stability of large-scale systems. *IEEE Trans. on Aut. Control, AC-24*, No.1, 94-96.

Borne, P., and M. Benrejeb (1977a), Stability of non-linear composite systems. *1st World Conf. on Math. at the Service of Man*, Barcelona, Tech. Session 4/6, 1-12.

Borne, P., and M. Benrejeb (1977b), On the stability of a class of interconnected systems. Application to the forced working condi-dions. *4th IFAC Symp. on Multivariable Technological Systems*, Frederictown, 261-265.

Borne, P., M. Benrejeb, and J.L. Cocquerelle (1978), On the synthesis of a class of interconnected systems with structural variations - application to an electrical power plant. *MECO 78*, IASTED, Acta Press, 1-4.

Burgat, C., J. Bernussou, Lj.T. Grujić, J.C. Gentina, and P. Borne (1978), Sur la stabilité des systèmes de grande dimension. Les per-turbations structurelles arbitraires et périodiques. *R.A.I.R.O. Automatique/Systems Analysis and Control, 12*, No.3, 245-267.

Burgat, C., and Lj.T. Grujić (1979), Sur la stabilité asymtptotique d'un système de grande dimension admettant un système agrégé du type Lotka-Volterra. *C.R. Acad. Sc. Paris, 288*, Ser.A, 691-693.

Chaplygin, S.A. (1919), New method of integration of a general differential equation of the train motion. *Byul. Nauch.-Eksperim. in-ta Putej Soobshcheniya, vyp.1a*, No.9, 308-334 (in Russian).

Evans, F.J. (1978), Prospect for dynamic security monitoring in large-scale electric power systems. *Proc. 7th World IFAC Congress*, Helsinki, 1-14.

Fiedler, M., and V. Ptak (1962), On matrices with non-positive off-diagonal elements and positive principal minors. *Czech. Nat. J.*, *12*, No.87, 382-400.

Gentina, J.C., P. Borne, C. Burgat, J. Bernussou, and Lj.T. Grujić (1979), Sur la stability des systèmes de grande dimension. Normes vectorielles. *R.A.I.R.O. Automatique/Systems Analysis and Control*, *13*, No.1, 57-75.

Gentina, J.C., P. Borne, and F. Laurent (1972), Stabilité des systèmes continus non linéaires de grande dimension. *R.A.I.R.O.*, *6*, No.3,69-77.

Gentina, J.C., P. Borne, and F. Laurent (1973), On a study of stability and sensitivity of large continuous non-linear systems under uncer-tainty. *3rd IFAC Symp. on Sens. Adapt. and Optim.*, ISA, 412-417.

Goina, L.N., and A.A. Martynyuk (1974), Systems of oscillators analysis with weak interconnection in the neighbourhood of a special point. *Matem. Fizika, 15*, - (in Russian).

Grujić, Lj.T. (1972), *Large-scale systems stability*. Dissertation (published in 1974), Faculty of Mechanical Engineering, Belgrade (in Serbo-Coratian).

Grujić, Lj.T. (1974a), Stability analysis of large-scale systems with stable and unstable subsystems. *Int. J. Control, 20*, No.3, 453-463.

Grujić, Lj.T. (1974b), On multi-level absolute stability analysis of large-scale systems : the Popov method, the comparison principle and time-invariant systems. *Automatika*, Nos.1-2, 67-72.

Grujić, Lj.T. (1974c), On multi-level absolute stability analysis of large-scale systems : the Lyapunov method, the comparison principle and time varying systems. *Automatika*, Zagreb, Nos.4-5, 155-161.

Grujić, Lj.T. (1975), Stability of product sets. *Proc. 1975 Midwest Symposium on Circuits and Systems*, Concordia University, Montreal, 254-258.

Grujić, Lj.T. (1976a), Time-varying sets, aggregation and stability of large-scale systems. *IEEE Proc. on Int. Symp. Circ. and Systems*, Münich, 392-401.

Grujić, Lj.T. (1976b), General stability analysis of large-scale systems. *IFAC Symp. on Large-Scale Systems Theory and Applications*, Udine, 203-213.

Grujić, Lj.T. (1977a), Stability and instability of product sets. *Systems Science, 3*, No.1, 13-31.

Grujić, Lj.T. (1977b), Un lemme matriciel réciproque; application à la stabilité absolue. *C.R. Acad. Sci. Paris*, Ser.A, *284*, 1409-1412.

Grujić, Lj.T., and C. Burgat (1979), Estimations E_i du domaine de stabilité pour un système interconnecté de comparison du type Lotka-Volterra. *C.R. Acad. Sc. Paris*, Ser.A, *288*, 745-747.

Grujić, Lj.T., and C. Burgat (1980a), Lotka-Volterra-like approach to large-scale systems stability. *Int. J. Systems Sci., 11*, No.10, 1131-1144.

Grujić, Lj.T., and C. Burgat (1980b), Stability analysis of large-scale generalized Lotka-Volterra systems. *Ricerche di Automatica, 9*, No.2, 161-170.

Grujić, Lj.T., J.C. Gentina, and P. Borne (1976), General aggregation of large-scale systems by vector Lyapunov functions and vector norms. *Int. J. Control, 24*, No.4, 529-550.

Grujić, Lj.T., J.C. Gentina, P. Borne, C. Burgat, and J. Bernussou (1978), Sur la stabilité des systèmes de grande dimension. Fonctions de Lyapunov vectorielles. *R.A.I.R.O. Automatique/Systems Analysis and Control, 12*, No.4, 319-348.

Grujić, Lj.T., and M. Ribbens-Pavella (1978), Relaxed large-scale systems stability analysis applied to power systems. *IFAC VII World Congress*, 27-34.

Grujić, Lj.T., and M. Ribbens-Pavella (1979), Asymptotic stability of large-scale systems with application to power systems. I : domain estimation. *Electrical Power & Energy Systems, 1*, No.3, 151-157.

Grujić, Lj.T., and D.D. Šiljak (1972), Stability of large-scale systems with stable and unstable subsystems. *Proc. 1972 JACC*, Stanford University, Paper 17-3, 550-555.

Grujić, Lj.T., and D.D. Šiljak (1973), Asymptotic stability and instability of large-scale systems. *IEEE Trans. on Aut. Control, AC-18*, No.6, 636-645.

Hahn , W. (1967), *Stability of Motion*. Springer Verlag, Berlin.

Hatvani, L. (1975), On application of differential inequalities to the theory of stability. *Vestn. Mosk. un-ta. Ser. Matem. i Mekhanika*, No.3, 83-89 (in Russian).

Hatvani, L. (1980), On the continuation of solutions of differential equations by vector Lyapunov functions. *Proc. Amer. Math. Soc., 79*, No.1, 59-62.

Ikeda, M., and D.D. Šiljak (1980), Decentralized stabilization of linear time-varying systems. *IEEE Trans. on Aut. Control, AC-25*, No.1, 106-107.

Kamke, E. (1932), Zur Theorie der Systeme Gewöhnlicher Differential-Gleichungen II. *Acta Mathematica, 58*, 57-85.

Kayande, A.A., and V. Lakshmikantham (1966), Conditionally invariant sets and vector Lyapunov functions. *J. Math. Anal. and Applications, 14*, 285-293.

Khapayev, M.M. (1967), On a Liapunov like theorem. *Dokl. AN SSSR, 176*, No.6, 1262-1265 (in Russian).

Kloeden, P.E. (1975), Aggregation-decomposition and equi-ultimate boundedness. *J. Australian Math. Soc., XIX*, Series B, Part 2, 249-258.

Kosolapov, V.I. (1979), To complex systems stability. *Prykl. Mekhanika, 25*, No.7, 133-136 (in Russian).

Kuhtenko, A.I. (1968), Basic problems of the complex systems control theory. In *Complex Control Systems*, Inst. Kibernetiki AN SSSR, Kiev, 3-62 (in Russian).

Ladde, G.S., and D.D. Šiljak (1975), Stochastic stability and instability of model ecosystems. *6th IFAC World Congress*, Boston, 55.4, 1-7.

Lakshmikantham, V. (1975), Vector Lyapunov functions and conditional stability. *J. Math. Anal. and Applications, 10*, 368-377.

Liapunov, A.M. (1893), Investigations of a singular case of the problem of motion stability. *Mat. Sbor., 17*, No.2, 253-333 (in Russian).

Martynyuk, A.A. (1972a), On instability of the equilibrium position of a multidimensional system composed of "neutrally stable subsystems", *Prykl. Mekhanika, 8*, No.6, 77-82.

Martynyuk, A.A. (1972b), On a Liapunov like theorem on stability of a multidimensional system. *Ukr. Mat. Sb., 24*, No.4, 532-537 (in Russian).

Martynyuk, A.A. (1973), Qualitative investigations of behaviour of weakly coupled oscillators in a vicinity of the equilibrium position. *Prykl. Mekhanika, 9*, No.7, 122-126 (in Russian).

Martynyuk, A.A. (1975a), *Stability of Motion of Complex Systems*, Naukova dumka, Kiev (in Russian).

Martynyuk, A.A. (1975b), Decomposition and aggregation in the systems analysis. *Teoret. i Prikl Mekhanika*, No.1, 87-93 (in Russian).

Martynyuk, A.A. (1979), On development of the Liapunov functions in the theory of stability of complex systems. *Prykl. Mekhanika, 15*, No.10, 3-23 (in Russian).

Martynyuk, A.A., and R. Gutowski (1979), *Integral Inequalities and Stability of Motion*. Naukova Dumka, Kiev (in Russian).

Martynyuk, A.A., and V.I. Kosolapov (1978), The comparison principle and the averaging method in the problem of stability of not-asymptotically stable motions under persistent action of disturbances. *Preprint : AN Ukr. SSR, Inst. Matem.*, No.33, 1-24 (in Russian).

Martynyuk, A.A., and N.V. Nikitina (1981), On nonlinear systems of comparison in the problems of stability of large-scale systems. *Prikl. Mekhanika, 17*, No.12, 97-102 (in Russian).

Matrosov, V.M. (1962), On the theory of stability of motion. *Prikl. Matem. i Mekhanika, 26*, 92-100 (in Russian).

Matrosov, V.M. (1963), On the theory of stability of motion,II. *Trudy Kazan. Aviac. Inst., 80*, 22-33 (in Russian).

Matrosov, V.M. (1968a), Comparison principle and vector Liapunov functions, I. *Diff. Uravn., IV*, No.8, 1374-1386 (in Russian).

Matrosov, V.M. (1968b), Comparison principle and vector Liapunov functions, II. *Ibid., IV*, No.10, 1739-1752 (in Russian).

Matrosov, V.M. (1971), Vector Lyapunov functions in the analysis of nonlinear interco-nected systems. *Symp. Mathematica*, Bologna, *VI*, 209-242.

Matrosov, V.M. (1972), The method of vector Lyapunov functions in feedback systems. *Autom. i Telemekh.*, No.9, 63-75 (in Russian).

Michel, A.N. (1973), Stability analysis of interconnected systems. *Berichte der Mathematisch -Statistischen Sektion*, No.4, Graz., 1-107.

Michel, A.N. (1974), Stability analysis of interconnected systems. *J. SIAM Control, 12*, No.3, 554-579.

Michel, A.N., and R.K. Miller (1977), *Qualitative Analysis of Large-Scale Dynamical Systems*. Academic Press, New York.

Michel, A.N., and D.W. Porter (1970), Stability analysis of composite systems. *IEEE Trans. Aut. Control, AC-17*, No.2, 222-226.

Morari, M., G. Stephanopoulos, and R. Aris (1977), Finite stability regions for large-scale systems with stable and unstable subsystems. *Int. J. Control, 26*, No.5, 805-815.

Oziraner, A.S., and V.V. Rumyantsev (1972), Method of Liapunov functions in the problem of stability of motion concerning a part of variables. *Matem. i Mekhanika, 36*, No.2, 364-384 (in Russian).

Peiffer, K., and N. Rouche (1969), Liapunov's second method applied to partial stability. *J. de Mécanique, 8*, No.2, 323-334.

Piontkovskii, A.A., and L.D. Rutkovkaya (1967), Investigation of certain stability theory problems by vector Lyapunov function method. *Autom. i Telemekh.*, No.10, 23-31 (in Russian).

Porter, D.W., and A.N. Michel (1970), Stability of composite systems. *Proc. of 4th Asilomar Conf. on Circuits, Systems and Computers*, 634-638.

Porter, D.W., and A.N. Michel (1971), Stability analysis of composite systems with nonlinear interconnections. *Proc. of 1971 Midwest Symp.*, 6.6-1/6.6-10.

Robert, F. (1964), Normes vectorielles de vecteurs et de matrices. *R.F.T.I. -Chiffres, 17*, No.4, 261-299.

Rouche, N., P. Habets, and M. Laloy (1980), *Stability Theory via Direct Liapunov's Method*. MIR, Moscow (in Russian).

Rouche, N., and K. Peiffer (1967), Le théorème de Lagrange-Dirichlet et la deuxième méthode de Liapounoff. *Annales de la Société Scientifique de Bruxelles, 81*, 19-33.

Rumyantsev, V.V. (1957), To stability of motion with respect to a part of variables. *Vestn. Mosk. un-ta, Ser. Matem. i Mekhanika*, No.4, 9-16 (in Russian).

Saeki, M., M. Araki, and B. Kondo (1980), Local stability of composite systems-frequency-domain condition and estimate of the domain of attraction. *IEEE Trans. on Aut. Control, AC-25*, No.5, 936-940.

Sandell, N.R., P. Varaiya, M. Athans, and M.G. Safonov (1978), Survey of decentralized control methods for large scale systems. *IEEE Trans. on Aut. Control, AC-23*, No.2, 108-128.

Šiljak, D.D. (1971), On large-scale system stability. *Proc. 9th Ann. Allerton Conf. on Circ. and Syst. Theory*, University of Illinois, 731-741.

Šiljak, D.D. (1972a), Stability of large-scale systems. *Proc. 1972 IFAC 5th World Congress*, ISA, Pittsburgh, Paper C-32, 1-11.

Šiljak, D.D. (1972b), Stability of large-scale systems under structural perturbations. *IEEF Trans. SMC, SMC-2*, No.2, 657-663.

Šiljak, D.D. (1973a), Competitive economic systems : stability decomposition and aggregation. *IEEE Conf. Dec. and Control*, 265-275.

Šiljak, D.D. (1973b), On stability of large-scale systems under structural perturbations. *IEEE Trans. Syst. Man and Cybern., SMC-3*, 415-417.

Šiljak, D.D. (1975a), On stability of the arms race. *Report No. NGR; 05-017-010-7508*, University of Santa Clara, *NSF Conf. on Control Theory in Intern. Relations Research*, Indian University, 1-44.

Šiljak, D.D. (1975b), When is a complex ecosystem stable ? *Mathematical Biosciences*, 25, 25-50.

Šiljak, D.D. (1977a), On pure structure of dynamic systems. *Nonlinear Analysis; Theory, Methods and Applications*, 1, 397-413.

Šiljak, D.D. (1977b), On the stability of the arms race. Chapter 9 in *Mathematical Systems in International Relations Research*, J.V. Gillespie and D.A. Zinnes, Praeger Publ., N.Y.

Šiljak, D.D. (1978), *Large-Scale Dynamic Systems : Stability and Structure*. North-Holland, New York.

Šiljak, D.D., and M.B. Vukčević (1974), On hierarchic stabilization of large-scale linear systems. *8th Asilomar Conf. on Circuits, Systems and Computers*, Pacific Grove, 503-507.

Sinha, A.S.C. (1980), Lyapunov functions for a class of large-scale systems. *IEEE Trans. Aut. Cont., AC-25*, No.3, 558-560.

Thompson, W.E. (1972), Exponential stability of interconnected systems. *IEEE Trans. Aut. Control, AC-17*, No.2, 222-226.

Thompson, W.E., and H.E. Koening (1972), Stability of a class of interconnected systems. *Int. J. Control, 15*, No.4, 751-763.

Vakhonina, G.S., A.S. Zemlyakov, and V.M. Matrosov (1973), On techniques for construction of Lyapunov vector functions for linear systems. *Autom. i Telem.*, No.2, 5-16 (in Russian).

Vidyasagar, M. (1980), Decomposition techniques for large-scale systems with nonadditive interactions : stability and stabilizability. *IEEE Trans. Aut. Control, AC-25*, No.4, 773-778.

Weissenberger, S. (1973), Stability regions of large-scale systems. *Automatica, 9*, No.5, 653-663.

Wažewski, T. (1950), Systèmes des équations et des inégalités différentielles ordinaires aux deuxièmes membres monotones et leurs applications. *Annales de la Société Polonaire de Mathématiques, 23*, 112-166.

Zadorozhny, V.F. (1969), Asymptotic stability analysis of many-dimensional systems by means of the method of vector Liapunov functions. *Kibernetika i Vychislitel'naya Tekhnika*, No.1, 92-97 (in Russian).

Zadorozhny, V.F., and A.A. Martynyuk (1972), Estimation of influence of connections between sub-systems on stability of a linear non-stationary system. *Prikl. Mekhanika, 18*, No.9, 65-71 (in Russian).

Zadorozhny, V.F., and A.A. Martynyuk (1973a), On a general problem of aggregation solution as the problem of moments. *Matematicheskaya Fizika*, No.14, 49-54 (in Russian).

Zadorozhny, V.F., and A.A. Martynyuk (1973b), On solution of the general aggregation problem as the moments problem. *Mat. Physics*, 14, 49-55.

Zadorozhny, V.F., and A.A. Martynyuk (1975), The direct Liapunov method and L-problem of moments in the problems on stability of many-dimensional systems. In *Problems of Analytical Mechanics, Theories of Stability and Control*, Nauka, 151-154 (in Russian).

Zemlyakov, A.S. (1972), On a problem of the comparison system construction. *Tr. Kazan. Aviats. in-ta*, No. 144, 46-54 (in Russian).

CHAPTER IV

SINGULARLY PERTURBED LARGE-SCALE SYSTEMS

IV.1. INTRODUCTION

Singularly perturbed large-scale systems are usually characterized by existence of several small parameters μ_i , $\mu_i > 0$. Existence of these parameters often led to the dependence of the system aggregation matrix on them. In such cases it was difficult to estimate the upper allowable bounds of small parameters. Besides, the usual assumption was that $\mu_{i+1}/\mu_i \to 0$ as $\mu_i \to 0$.

From the engineering point of view, it is important to find a way to construct the aggregation matrix that will be independent of small parameters μ_i and whose order will be essentially lower than the order of the system. Also, it is important to estimate upper allowable values of small parameters μ_i . Solving these problems, we meet a problem of the overall system Liapunov function construction and the problem of the (complete, or up to a certain degree) mutual independence of small parameters. Such independence implies different time scales of subsystems, as explained in what follows.

Stability problems of singularly perturbed large-scale systems have been studied via vector Liapunov function method since 1976 (Lj.T. Grujić, 1976-1979, 1981).

IV.2. DESCRIPTION AND DECOMPOSITION OF SINGULARLY PERTURBED LARGE-SCALE SYSTEMS

A singularly perturbed structurally variable large-scale system is described by

$$\frac{dx_i}{dt} = f_i(t, \mathbf{x}, \mathbf{y}, \mathbf{p}_i, \mathbf{S}_i) \; , \quad i=1,2,\dots,q \tag{1a}$$

$$\mu_i \frac{dy_i}{dt} = g_i(t,x,y,M,p_{q+i},S_{q+i}) \;,\; i=1,2,\dots,r \;, \tag{1b}$$

where $x_i \in R^{n_i}$, $y_i \in R^{m_i}$, f_i and g_i are continuous functions of the appropriate order, μ_i is a positive parameter that can be arbitrarily small and, hence, $\mu_i \in]0,1]$, $M = \text{diag}\{\mu_1,\mu_2,\dots,\mu_r\}$. The upper admissible value of μ_i is denoted by μ_{im}. The set of all allowable M is denoted by $\mathbf{M}$,

$$\mathbf{M} = \{M: 0 \le M \le I\} \;,$$

and then,

$$\mathbf{M}_m = \{M: 0 < \mu_i < \mu_{im} \;,\; \forall i=1,2,\dots,r\} \;.$$

Here, $q+r=s$.

When the small parameters μ_i are mutually unrelated then the whole system has essentially r unrelated time scales t_i

$$t_i = \frac{t-t_0}{\mu_i} \;,\; i=1,2,\dots,r \;. \tag{2}$$

In such cases *time-scaling of the system is non-uniform*. However, time scales t_i can be mutually related (Lj.T. Grujić 1977, 1978, 1979a,b, 1981) via τ_i,

$$\frac{t_i}{t_1} = \tau_i \;,\; i=1,2,\dots,r \;, \tag{3}$$

in certain bounds,

$$\tau_i \in [\underline{\tau}_i, \bar{\tau}_i] \;,\; i=1,2,\dots,r \tag{4}$$

where $0 < \underline{\tau}_i \le \bar{\tau}_i < +\infty$, $\forall i=1,2,\dots,r$. In the cases when (3) and (4) hold then *time-scaling of the system is uniform* and from (2) and (3) it results

$$\underline{\tau}_i = \frac{\mu_1}{\mu_i} \;,\; i=1,2,\dots,r \;. \tag{5}$$

Evidently, $\underline{\tau}_1 = \tau_1 = \bar{\tau}_1 = 1$.

The i-th subsystem described by (6),

$$\frac{dx_i}{dt} = f_i(t,x,y,p_i,S_i) \;, \tag{6a}$$

$$\mu_i \frac{dy_i}{dt} = g_i(t,x,y,M,p_{q+i},S_{q+i}) \tag{6b}$$

will be referred to as *the i-th interconnected singularly perturbed subsystem* of the whole system (1), and

$$\frac{dx_i}{dt} = f_i(t,x^i,y^i,p_i,S_i) \;, \tag{7a}$$

$$\mu_i \frac{dy_i}{dt} = g_i(t,x^i,y^i,M,p_{q+i},S_{q+i}) \;, \tag{7b}$$

where $x^i = (0^T,\dots,0^T,x_i^T,0^T,\dots,0^T)^T \in R^n$, $n = n_1+n_2+\dots+n$, $x_i \in R^{n_i}$, $y^i = (0^T,\dots,0^T,y_i^T,0^T,\dots,0^T)^T \in R^m$, $m = m_1+m_2+\dots+m_r$, $y_i \in R^{m_i}$, will be referred to as *the i-th disconnected singularly perturbed subsystem*.

In the case $q=r$ it can be useful to consider

$$\frac{dx_i}{dt} = f_i(t, x^i, y^i, p_i, S_i) \ , \tag{8a}$$

$$0 = g_i(t, x^i, y^i, 0, p_{q+i}, S_i) \ , \tag{8b}$$

which will be called *the i-th disconnected degenerate subsystem* of the whole system, and

$$\frac{dy_i}{dt} = g_i(\alpha, b^i, y^i, 0, p_{q+i}, S_i) \tag{9}$$

that will be called *the i-th disconnected fast (boundary-layer) subsystem* of the whole system. In (9), $\alpha \in R$ and $b^i \in R^n$, $b^i = (0^T, \ldots, 0^T, b_i^T, 0^T, \ldots, 0^T)^T$, $b_i \in R^{n_i}$. When all μ_i are (formally) set equal to zero in (1) then

$$\frac{dx_i}{dt} = f_i(t, x, y, p_i, S_i) \ , \ i = 1, 2, \ldots, q \ , \tag{10a}$$

$$0 = g_i(t, x, y, 0, p_{q+i}, S_{q+i}) \ , \ i = 1, 2, \ldots, r \ , \tag{10b}$$

which will be called *the interconnected degenerate (reduced order) subsystem* of the whole system, and

$$\frac{dy_i}{dt_1} = \tau_i g_i(\alpha, b, y, 0, p_{q+i}, S_i) \ , \ i = 1, 2, \ldots, r \ , \tag{11}$$

will be called *the interconnected fast (boundary-layer) subsystem* of the whole system.

It is accepted that

$$0 = g_i(t, x, y, 0, p_{q+i}, S_{q+i}) \ , \ \forall (t, x, y) \in R \times N_x \times N_y \ ,$$

is true for every $(P, S) \in P \times S$ iff $y = 0$, and that

$$0 = g_i(t, x^i, y^i, 0, p_{q+i}, S_{q+i}) \ , \ \forall (t, x^i, y^i) \in R \times N_x \times N_y \ ,$$

is true for any $(P, S) \in P \times S_s$ iff $y^i = 0$. Hence, (8) is now equivalent to

$$\frac{dx_i}{dt} = f_i(t, x^i, 0, p_i, S_i) \ , \tag{12}$$

and (10) is equivalent to

$$\frac{dx_i}{dt} = f_i(t, x, 0, p_i, S_i) \ , \ i = 1, 2, \ldots, q \ . \tag{13}$$

IV.3. AGGREGATION AND STABILITY CRITERIA FOR SINGULARLY PERTURBED LARGE-SCALE SYSTEMS

IV.3.1. Introduction

The preceding analysis showed existence of different decomposition forms of singularly perturbed large-scale systems. When they are combined with different aggregation forms of large-scale systems in general, a great variety of aggregation forms results for singularly perturbed large-scale systems. However, the crucial point is whether different time scales t_i are mutually dependent, or not.

IV.3.2. Non-uniform time scaling

IV.3.2.1. General theory

In this section $q=r$ holds. The large-scale system is decomposed into q interconnected singularly perturbed subsystems (6) for which the following assumptions will be accepted. These assumptions define an aggregation form of the whole system (1).

Let $N_{ixo} = \{x_i : x_i \in N_{ix}, x_i \neq 0\}$ and $N_{iyo} = \{y_i : y_i \in N_{iy}, y_i \neq 0\}$.

Assumption 1. There exist connected neighbourhoods $N_{ix} \subseteq R^{n_i}$ of $0 \in R^{n_i}$ and $N_{iy} \subseteq R^{m_i}$ of $0 \in R^{m_i}$, functions ν_i , Ω_i and ω_i and non-negative numbers ζ_{i1} and ζ_{i2} , $\zeta_{i1} < 1$, so that for every $i = 1, 2, \ldots, r$:

1) $\nu_i(t, x_i) \in C^{(1,1)}(R \times N_{ixo}, R_+)$, it is positive definite on N_{ix} and there is $\psi_i > 0$ such that the set $V_{i\zeta}(t)$ is asymptotically contractive for every $\zeta \in]0, \psi_i[$,
2) Ω_i is positive definite on N_{ix} ,
3) ω_i is positive definite on N_{iy} ,
4) $\nu_{it}(t, x_i) + \nu^T_{ix_i}(t, x_i) f_i(t, x^i, 0, 0, 0) \leq -\Omega_i(t, x_i)$, $\forall (t, x) \in R \times N_x$, $x_i \neq 0$,
5) $\nu_{it}(t, 0) \leq 0$, $\forall t \in R$, and
6) $\nu^T_{ix_i}(t, x_i)[f_i(t, x^i, y^i, p_i, S_i) - f_i(t, x^i, 0, 0, 0)] \leq \zeta_{i1}\Omega_i(t, x_i) + \zeta_{i2}\omega_i(t, y_i)$, $x_i \neq 0$, $\forall (t, x, y, P, S) \in R \times N_x \times N_y \times P \times S_s$,

where $x^i = (0^T, \ldots, 0^T, x_i^T, 0^T, \ldots, 0^T)^T \in R^n$ and $y^i = (0^T, \ldots, 0^T, y_i^T, 0^T, \ldots, 0^T)^T \in R^m$. ■

Assumption 2. There exist functions v_i , non-negative numbers ξ_{i1} , ξ_{i2} , ξ_{i3} and ξ_{i4} and a positive integer ρ such that $\xi_{i1} < 1$, $\xi_{i2} < 1$ and for every $i = 1, 2, \ldots, r$:

1) $v_i(t, x_i, y_i) \in C^{(1,1,1)}(R \times N_{ixo} \times N_{iyo}, R_+)$ and v_i is positive definite on $N_{ix} \times N_{iy}$, or : $\{v_i(t, y_i) \in C^{(1,1)}(R \times N_{iyo}, R_+)$ and v_i is positive definite on $N_{iy}\}$,
2) there is $\Psi_i > 0$ such that the set $V_{i\zeta}(t)$ is asymptotically contractive for every $\zeta \in]0, \Psi_i[$,
3) $v^T_{iy_i}(t, x_i, y_i) g_i(t, x^i, y^i, 0, 0, 0) \leq -\omega_i(t, y_i)$, $y_i \neq 0$, $\forall (t, x, y) \in R \times N_x \times N_y$, and, respectively, $\{\forall (t, x, y) \in R \times N_x \times N_y\}$,
4) $v^T_{iy_i}(t, x_i, y_i)[g_i(t, x^i, y^i, M^i, p_{r+i}, S_{r+i}) - g_i(t, x^i, y^i, 0, 0, 0)] \leq \xi_{i1}\rho_i\Omega_i(t, x_i) + \xi_{i2}\omega_i(t, y_i)$, $y_i \neq 0$, $\forall (t, x, y, M^i, P, S) \in R \times N_x \times N_y \times M^i \times P \times S_s$,

and respectively, $\{\forall(t,x,y,M^i,P,S)\in R\times N_x\times N_y\times M^i\times P\times S_s\}$,
where $M^i=\mathrm{diag}\{0,\dots,0,\mu^i,0,\dots,0\}$, $M^i=\{M^i:0<\mu_i\le 1\}$,

5) $v_{it}(t,x_i,y_i)+v_{ix_i}^T(t,x_i,y_i)f_i(t,x^i,y^i,p_i,S_i)\le \xi_{i3}\Omega_i(t,x_i)$
$+\ \xi_{i4}\omega_i(t,y_i)$, $x_i\neq 0$,
and $v_{it}(t,0,y_i)\le \xi_{i4}\omega_i(t,y_i)$, $\forall(t,x,y,P,S)\in R\times N_x\times N_y\times P\times S_s$,
or respectively, $\{\forall(t,x,y,P,S)\in R\times N_x\times N_y\times P\times S_s\}$. ■

Assumption 3. There exist real number α_{ij} , $\alpha_{ij}\ge 0$ for $i\neq j$, such that for every $i=1,2,\dots,r$,

1) $\mu_i[\nu_i(t,x_i)+v_i(t,x_i,y_i)]_{x_i}^T[f_i(t,x,y,p_i,S_i)-f_i(t,x^i,y^i,p_i,S_i)]$
$+v_{iy_i}^T[g_i(t,x,y,M,p_{r+i},S_{r+i})-g_i(t,x^i,y^i,M^i,p_{r+i},S_{r+i})]$
$\le \mu_i\sum_{j=1}^{r}\alpha_{ij}\eta_j(t,x_j,y_j)$, $\forall(t,x,y,M,P,S)\in R\times N_x\times N_y\times M\times P\times S_x$, $x_i\neq 0$,
$y_i\neq 0$,

2) $v_{iy_i}^T(t,x_i,y_i)[g_i(t,x,y,M,p_{r+i},S_{r+i})-g_i(t,0,y^i,M^i,p_{r+i},S_{r+i})]$
$\le \mu_i\sum_{j=1}^{r}\alpha_{ij}\eta_j(t,x_j,y_j)$, $\forall(t,x,y,M,P,S)\in R\times N_x\times N_y\times M\times P\times S_s$, $x_i=0$,
$y_i\neq 0$, and

3) $[\nu_i(t,x_i)+v_i(t,x_i,y_i)]_{x_i}^T[f_i(t,x,y,p_i,S_i)-f_i(t,x^i,0,p_i,S_i)]$
$\le \sum_{j=1}^{r}\alpha_{ij}\eta_j(t,x_j,y_j)$, $\forall(t,x,y,M,P,S)\in R\times N_x\times N_y\times M\times P\times S_s$, $x_i\neq 0$, $y_i=0$,

where

$$\eta_i(t,x_i,y_i)=(1-\zeta_{i1}-\xi_{i1}\mu_i^{\rho-1}-\xi_{i3})\Omega_i(t,x_i)+(\zeta_{i2}+\xi_{i4})(\tilde{\mu}_i-\hat{\mu}_i)\omega_i(t,y_i)\ ,$$
$$\tilde{\mu}_i=(1-\xi_{i2})(\zeta_{i2}+\xi_{i4})^{-1}\ , \text{ and } \hat{\mu}_i\in\,]0,\tilde{\mu}_i[\ .\ ■$$

Notice that $\hat{\mu}_i$ is the required estimate of the upper bound μ_{im} of admissible μ_i so that $M_i=\{M:0<\mu_i<\hat{\mu}_i,\forall i=1,2,\dots,r\}$.

The preceding assumptions introduce the aggregation matrix A , $A\in R^{r\times r}$, $A=(a_{ij})$, $a_{ij}=-\delta_{ij}+\alpha_{ij}$.

The order r of the aggregation matrix is much less than the order $(m+n)$ of the whole system (1).

Theorem 1. *(I) For structural asymptotic stability of* $(x^T,y^T)^T=0$ *of the system (1) over* $\hat{M}\times P\times S_s$ *it is sufficient that both*
(a) Assumptions 1-3 hold; and
(b) the matrix A *is stable.*

(II) If, in addition to (I), $N_{ix}=R^{n_i}$, $N_{iy}=R^{m_i}$, $\psi_i=+\infty$, $\Psi_i=+\infty$, ν_i *and* v_i *are radially unbounded for every* $i=1,2,\dots,r$ *, then* $(x^T,y^T)^T=0$ *of the system (1) is structurally asymptotically stable in the whole over* $M\times P\times S_s$.

Proof. (I) Let $\mathbf{v} = (V_1,V_2,\ldots,V_r)^T$, $V_i = \nu_i + v_i$, $\forall i=1,2,\ldots,r$, and $v = \mathbf{b}^T\mathbf{v}$ where $\mathbf{b}\in R^r$ is determined for arbitrary $\mathbf{c}\in R^r$, $\mathbf{c}>0$ elementwise, from $\mathbf{b} = -(\mathbf{A}^T)^{-1}\mathbf{c}$.

Since all $a_{ij} \geq 0$ for $i\neq j$ due to the condition (a) and the hypothesis of Assumption 3, and since $\mathbf{A} = (a_{ij})$ is stable due to the condition (b) it results (M. Fielder and V. Ptak 1962, Chapter III.4.1.1) that the vector $\mathbf{b}$ is positive elementwise. This result, the condition 1 of Assumption 1 and the conditions 1 and 2 of Assumption 2 prove that the function v is positive definite on $N = N_x \times N_y$ and that there is $\psi>0$ such that the set $V_\zeta(t)$ is asymptotically contractive for every $\zeta\in]0,\psi[$. Along motions of (1) :

$$\begin{aligned}\dot V_i &= \nu_{it}(t,x_i) + \nu_{ix_i}^T(t,x_i)f_i(t,x,y,p_i,S_i) + v_{it}(t,x_i,y_i)\\&\quad + v_{ix_i}^T(t,x_i,y_i)f_i(t,x,y,p_i,S_i) + \frac{1}{\mu_i}\, v_{iy_i}^T(t,x_i,y_i)g_i(t,x,y,M,p_{r+i},S_{r+i})\\&= \nu_{it}(t,x_i) + \nu_{ix_i}^T(t,x_i)f_i(t,x^i,0,0,0) + \nu_{ix_i}^T(t,x_i)[f_i(t,x^i,y^i,p_i,S_i)\\&\quad - f_i(t,x^i,0,0,0)] + \frac{1}{\mu_i}\, v_{iy_i}^T(t,x_i,y_i)g_i(t,x^i,y^i,0,0,0)\\&\quad + \frac{1}{\mu_i}\, v_{iy_i}^T[g_i(t,x^i,y^i,M^i,p_{r+i},S_{r+i})-g_i(t,x^i,y^i,0,0,0)]\\&\quad + \frac{1}{\mu_i}\, v_{iy_i}^T[g_i(t,x,y,M,p_{r+i},S_{r+i})-g_i(t,x^i,y^i,M^i,p_{r+i},S_{r+i})]\\&\quad + v_{it}(t,x_i,y_i) + v_{ix_i}^T(t,x_i,y_i)f_i(t,x^i,y^i,p_i,S_i)\\&\quad + [\,\nu_i(t,x_i) + v_i(t,x_i,y_i)]_{x_i}^T[f_i(t,x,y,p_i,S_i)-f_i(t,x^i,y^i,p_i,S_i)]\ ,\end{aligned}$$

for $x_i\neq 0$ and $y_i\neq 0$, $\forall i=1,2,\ldots,r$, $\forall(t,x,y,M,P,S)\in R\times N_x\times N_y\times M\times P\times S_s$.

Now, 4 and 6 of Assumption 1, 3-5 of Assumption 2, and 1 of Assumption 3 imply

$$\begin{aligned}\dot V_i &\leq -\Omega_i + \zeta_{i1}\Omega_i + \zeta_{i2}\omega_i - \frac{1}{\mu_i}\,\omega_i + \xi_{i1}\mu_i^{\rho-1}\Omega_i + \frac{1}{\mu_i}\,\xi_{i2}\omega_i + \sum_{j=1}^{r}\alpha_{ij}\eta_j + \xi_{i3}\Omega_i\\&\quad + \xi_{i4}\omega_i\\&= -[(1-\zeta_{i1}-\xi_{i1}\mu_i^{\rho-1})\Omega_i + (\frac{1}{\mu_i}-\frac{1}{\mu_i}\,\xi_{i2}-\zeta_{i2}-\xi_{i4})\omega_i] + \sum_{j=1}^{r}\alpha_{ij}\eta_j\ ,\end{aligned}$$

$x_i\neq 0$, $y_i\neq 0$, $\forall i=1,2,\ldots,r$, $\forall(t,x,y,M,P,S)\in R\times N_x\times N_y\times M\times P\times S_s$.

This result, the definition of η_i , $\tilde\mu_i$, $\hat\mu_i$ (Assumption 3) and a_{ij} , and the fact that $0<\mu_i\leq 1$ yield for $x_i\neq 0$ and $y_i\neq 0$

$$\dot V_i \leq \sum_{j=1}^{r} a_{ij}\eta_j\ ,\ \forall i=1,2,\ldots,r\ ,\ \forall(t,x,y,M,P,S)\in R\times N_x\times N_y\times M\times P\times S_s\ ,$$

or for $\mathbf{e} = (\eta_1,\eta_2,\ldots,\eta_r)^T$,

$$\dot{\mathbf{v}}(t,x,y,M,P,S) \leq \mathbf{A}\mathbf{e}(t,x,y)\ ,\ \forall(t,x,y,M,P,S)\in R\times N_x\times N_y\times M\times P\times S_s\ ,\quad x_i\neq 0\ ,\ y_i\neq 0\ ,\ \forall i=1,2,\ldots,r\ . \tag{14}$$

When $x_i(t) = 0$ and $y_i(t) \neq 0$ for all $t \in [\sigma_{i1}, \sigma_{i1}+\epsilon_{i1}[$ and $x_i(\sigma_{i1}+\epsilon_{i1}) \neq 0$ iff $\epsilon_{i1} < +\infty$, $\epsilon_{i1} > 0$ then,

$$\begin{aligned}\dot{V}_i &= \nu_{it}(t,x_i) + v_{it}(t,x_i,y_i) + \frac{1}{\mu_i} v^T_{iy_i}(t,x_i,y_i) g_i(t,x,y,M,p_{r+i},S_{r+i}) \\ &= \nu_{it}(t,x_i) + v_{it}(t,x_i,y_i) + \frac{1}{\mu_i} v^T_{iy_i}(t,x_i,y_i) g_i(t,x^i,y^i,0,0,0) \\ &\quad + \frac{1}{\mu_i} v^T_{iy_i}(t,x_i,y_i)[g_i(t,x^i,y^i,M^i,p_{r+i},S_{r+i}) - g_i(t,x^i,y^i,0,0,0)] \\ &\quad + \frac{1}{\mu_i} v^T_{iy_i}(t,x_i,y_i)[g_i(t,x,y,M,p_{r+i},S_{r+i}) - g_i(t,x^i,y^i,M^i,p_{r+i},S_{r+i})]\end{aligned}$$

$\forall t \in [\sigma_{i1}, \sigma_{i1}+\epsilon_{i1}[$.

This result, 5 of Assumption 1 together with positive definiteness of ν_i on N_{ix}, 5 of Assumption 2, 2 of Assumption 3, the definition of η_i, $\tilde{\mu}_i$, $\hat{\mu}_i$ (Assumption 3) and $A = (a_{ij})$ imply

$$\dot{V}_i \leq \sum_{j=1}^{r} a_{ij}\eta_j \ , \quad \forall t \in [\sigma_{i1}, \sigma_{i1}+\epsilon_{i1}[\ .$$

Hence,

$$\begin{aligned}&\dot{v}(t,x,y,M,P,S) \leq Ae(t,x,y) \ , \ \forall t \in [\sigma_{i1}, \sigma_{i1}+\epsilon_{i1}[\ , \ \forall i=1,2,\dots,r \ , \\ &\qquad \forall (x,y,M,P,S) \in N_x \times N_y \times M \times P \times S_s \ , \ x = x(t) \ , \ y = y(t) \ .\end{aligned} \tag{15}$$

When $x_i(t) \neq 0$ and $y_i(t) = 0$ for all $t \in [\sigma_{i2}, \sigma_{i2}+\epsilon_{i2}[$ and $y_i(\sigma_{i2}+\epsilon_{i2}) \neq 0$ iff $\epsilon_{i2} < +\infty$, $\epsilon_{i2} > 0$, then

$$\begin{aligned}\dot{V}_i &= \nu_{it}(t,x_i) + \nu^T_{ix_i}(t,x_i,y_i) f_i(t,x,y,p_i,S_i) + v_{it}(t,x_i,y_i) \\ &\quad + v^T_{ix_i}(t,x_i,y_i) f_i(t,x,y,p_i,S_i) \\ &= \nu_{it}(t,x_i,y_i) + \nu^T_{ix_i}(t,x_i,y_i) f_i(t,x^i,0,0,0) \\ &\quad + \nu^T_{ix_i}(t,x_i)[f_i(t,x^i,0,p_i,S_i) - f_i(t,x^i,0,0,0)] + v_{it}(t,x_i,y_i) \\ &\quad + [\nu_i(t,x_i) + v_i(t,x_i,y_i)]^T_{x_i}[f_i(t,x,p_i,S_i) - f_i(t,x^i,0,p_i,S_i)] \\ &\quad + v^T_{ix_i}(t,x_i,y_i) f_i(t,x^i,0,p_i,S_i) \ , \quad \forall t \in [\sigma_{i2}, \sigma_{i2}+\epsilon_{i2}[\ .\end{aligned}$$

This result, 4 of Assumption 1, 5 of Assumption 2, $\xi_{i1} \geq 0$, $\mu_i > 0$, $\Omega_i \geq 0$, the definition of η_i, $\tilde{\mu}_i$, $\hat{\mu}_i$ (Assumption 3), $A = (a_{ij})$ and 3 of Assumption 3 yield

$$\dot{V}_i(t,x,y,M,P,S) \leq \sum_{j=1}^{r} a_{ij}\eta_j \ , \quad \forall t \in [\sigma_{i2}, \sigma_{i2}+\epsilon_{i2}[\ , \ \forall i=1,2,\dots,r \ .$$

Hence,

$$\begin{aligned}&\dot{v}(t,x,y,M,P,S) \leq Ae(t,x,y) \ , \ \forall t \in [\sigma_{i2}, \sigma_{i2}+\epsilon_{i2}[\ , \ \forall i=1,2,\dots,r \ , \\ &\qquad \forall (x,y,M,P,S) \in N_x \times N_y \times M \times P \times S_s \ , \ x = x(t) \ , \ y = y(t) \ .\end{aligned} \tag{16}$$

When $x_i(t) = 0$ and $y_i(t) = 0$, $\forall t \in [\sigma_{i3}, \sigma_{i3}+\epsilon_{i3}[$ and $0 < \epsilon_{i3} \leq +\infty$, then evidently,

$$V_i[t,x_i(t),y_i(t)] = \text{const.} \ , \ \forall t \in [\sigma_{i3},\sigma_{i3}+\epsilon_{i3}[\ , \ \forall i=1,2,\dots,r \ . \quad (17)$$

Now, (14)-(16) imply

$$\dot{v}(t,x,y,M,P,S) \le -c^T e(t,x,y) \ , \ \begin{cases} \forall(t,x,y,M,P,S) \in R\times N_x\times N_y\times M\times P\times S_s \\ \text{and} \quad x_i\neq 0 \ , \ y_i\neq 0 \ , \ \forall i=1,2,\dots,r \\ \qquad\qquad \text{or} \\ x=x(t) \ , \ y=y(t) \ , \ \forall t\in[\sigma_{ij},\sigma_{ij}+\epsilon_{ij}[\ , \\ \forall j=1,2,3 \ , \ \forall i=1,2,\dots,r \ , \\ \forall(x,y,M,P,S)\in N_x\times N_y\times\hat{M}\times P\times S_s \ . \end{cases} \quad (18)$$

Following now the proof of Theorem 13 of I-3.2.5, it results from (18) and the proved properties of v that $(x^T,y^T)^T = 0$ of (1) is structurally asymptotically stable over $\hat{M}\times P\times S_s$ (i.e. all conditions of Theorem 13 of I-3.2.5 are satisfied).

(II) When, in addition to (I), $N_{ix}=R^{n_i}$, $N_{iy}=R^{m_i}$, $\psi_i=+\infty$, $\Psi_i=+\infty$, ν_i and v_i are radially unbounded then $N_x=R^n$, $N_y=R^m$, $\psi=+\infty$ and v is radially unbounded, which completes the proof. ∎

Asymptotic contraction of the sets $V_{i\xi}(t)$ need not be requested as shown in what follows.

Assumption 4. There exist connected neighbourhoods $N_{ix}\subseteq R^{n_i}$ of $0\in R^{n_i}$ and $N_{iy}\subseteq R^{m_i}$ of $0\in R^{m_i}$, functions ν_i, Ω_i and ω_i and non-negative numbers ζ_{i1} and ζ_{i2} ; $\zeta_{i2}, \zeta_{i1}<1$, so that for every $i=1,2,\dots,r$,

1) $\nu_i(t,x_i)\in C^{(1,1)}(R\times N_{ix0},R_+)$, it is both positive definite on N_{ix} and decrescent on N_{ix},
2) Ω_i is positive definite on N_{ix},
3) ω_i is positive definite on N_{ix},
4) $\nu_{it}(t,x_i) + \nu^T_{ix_i}(t,x_i)f_i(t,x^i,0,0,0) \le -\Omega_i(t,x_i)$, $\forall(t,x_i)\in R\times N_{ix}$, $x_i\neq 0$,
5) $\nu_{it}(t,0)\le 0$, $\forall t\in R$,

and

6) $\nu^T_{ix_i}(t,x_i)[f_i(t,x^i,y^i,p_i,S_i) - f_i(t,x^i,0,0,0)] \le \zeta_{i1}\Omega_i(t,x_i) + \zeta_{i2}\omega_i(t,y_i)$, $x_i\neq 0$, $\forall(t,x,y,P,S)\in R\times N_x\times N_y\times P\times S_s$. ∎

Assumption 5. There exist functions v_i, non-negative numbers ξ_{i1}, ξ_{i2}, ξ_{i3} and ξ_{i4} and a positive integer ρ such that $\xi_{i1}<1$, $\xi_{i2}<1$

and for every $i=1,2,\dots,r$,

1) $v_i(t,x_i,y_i) \in C^{(1,1,1)}(R\times N_{ix0}\times N_{iy0},R_+)$ and v_i is both positive definite and decrescent on $N_{ix}\times N_{iy}$, or, $\{v_i(t,y_i) \in C^{(1,1)}(R\times N_{iy0},R_+)$ and v_i is both positive definite and decrescent on $N_{iy}\}$,

2) $v_{iy_i}^T(t,x_i,y_i)g_i(t,x^i,y^i,0,0,0) \le -\omega_i(t,y_i)$,

$\forall(t,x,y,P,S) \in R\times N_x\times N_y\times P\times S_s$,

3) $v_{iy_i}^T(t,x_i,y_i)[g_i(t,x^i,y^i,M^i,p_{r+i},S_{r+i})-g_i(t,x^i,y^i,0,0,0)]$

$\le \xi_{i1}\mu_i^{\rho}\Omega_i(t,x_i)+\xi_{i2}\omega_i(t,y_i)$, $y_i\ne 0$, $\forall(t,x,y,M^i,P,S)\in R\times N_x\times N_y\times M^i\times P\times S_s$,

and respectively, $\{\forall(t,x,y,M^i,P,S) \in R\times N_x\times N_y\times M^i\times P\times S_s\}$,

and

4) $v_{it}(t,x_i,y_i) + v_{ix_i}^T(t,x_i,y_i)f_i(t,x^i,y^i,p_i,S_i)$

$\le \xi_{i3}\Omega_i(t,x_i) + \xi_{i4}\omega_i(t,y_i)$, $x_i\ne 0$,

and $v_{it}(t,0,y_i) \le \xi_{i4}\omega_i(t,y_i)$, $\forall(t,x,y,P,S) \in R\times N_x\times N_y\times P\times S_s$,

and respectively, $\{\forall(t,x,y,P,S) \in R\times N_x\times N_y\times P\times S_s\}$. ■

Theorem 2. (I) *For structural uniform asymptotic stability of* $(x^T,y^T)^T=0$ *of the system (1) over* $\hat{M}\times P\times S_s$ *it is sufficient that both :*

(a) Assumptions 3-5 hold

and

(b) the matrix A *is stable.*

(II) *If, in addition to (I),* $N_{ix}=R^{n_i}$, $N_{iy}=R^{m_i}$, ν_i *and* v_i *are radially unbounded for every* $i=1,2,\dots,r$, *then* $(x^T,y^T)^T=0$ *of the system (1) is structurally uniformly asymptotically stable in the whole over* $\hat{M}\times P\times S_s$.

Proof. (I) In view of the condition (a), $\alpha_{ij}\ge 0$, $i\ne j$, (the hypothesis of Assumption 3), and hence, $a_{ij}\ge 0$ for $i\ne j$, and the condition (b) for arbitrary $c>0$, $c\in R^r$, the vector b , $b=-(A^T)^{-1}c$, is also elementwise positive (M. Fiedler and V. Ptak, 1962). For such a vector b , the function $v=b^T\mathbf{v}$ is both positive definite and decrescent on $N_x\times N_y$ due to 1 of Assumption 4 and 1 of Assumption 5. The conditions 2-6 of Assumption 4 are the same as conditions 2-6 of Assumption 1 and the conditions 2-4 of Assumptions 5 are the same as 3-5 of Assumption 2. Hence, by following the proof of Theorem 1 it results that (14)-(18) hold. The function v satisfies all conditions for uniform asymptotic stability of $(x^T,y^T)^T=0$ of (1) over $\hat{M}\times P\times S_s$ (Theorem 7 of the Section I-3.2.4), which proves the assertion under (I).

(II) When, in addition to (I), $N_{ix}=R^{n_i}$, $N_{iy}=R^{m_i}$, ν_i and v_i are radially unbounded, $\forall i=1,2,\dots,r$, then $N_x=R^n$, $N_y=R^m$ and v is also radially unbounded, which in view of Theorem 8 of the section I-3.2.4 completes the proof. ■

IV.3.2.2. Application to the structural absolute stability analysis

A singularly perturbed Lur'e like system is described by

$$\dot{x}_i = A_i x_i + q_{i1}\phi_{i1}(\sigma_{i1}) + \sum_{j=1}^{r} S^1_{ij}A_{ij}y_j \ , \quad \sigma_{i1} = c^T_{i1}x + c^T_{i2}y \ , \tag{19a}$$

$$\mu_i\dot{y}_i = \sum_{j=1}^{r} \mu_i S^1_{r+i,j}B_{ij}x_j + B_i y_i + q_{i2}\phi_{i2}(\sigma_{i2}) + q_{i3}\phi_{i3}(\sigma_{i3}) \ , \tag{19b}$$

$$\sigma_{i2} = \mu_i c^T_{i3}x_i + c^T_{i4}y_i \ , \quad \sigma_{i3} = \sum_{\substack{j=1\\ j\neq i}}^{r} \mu_i(c^T_{j5}S^2_{r+i,j}x_j + c^T_{j6}S^3_{r+i,j}y_j) \ , \quad \forall i=1,2,\dots,r \ , \tag{19c}$$

where for $\sigma_{ij}\neq 0$:

$$\sigma^{-1}_{ij}\phi_{ij}(\sigma_{ij}) \in [0,\kappa_{ij}] \ , \quad i=1,2,\dots,r \ , \quad j=1,2,3 \ . \tag{19d}$$

All matrices and vectors are of the appropriate order, and in particular S^1_{ij}, S^2_{ij} and S^3_{ij} are diagonal matrices. Let

$$S_i = \begin{bmatrix} S^1_{i,1}, & S^1_{i,2}, & \dots, & S^1_{i,i-1}, & I, & S^1_{i,i+1}, & \dots, & S^1_{i,r} \\ S^1_{r+i,1}, & S^1_{r+i,2}, & \dots, & S^1_{r+i,i-1}, & I, & S^1_{r+i,i+1}, & \dots, & S^1_{r+i,r} \\ S^2_{r+i,1}, & S^2_{r+i,2}, & \dots, & S^2_{r+i,i-1}, & 0, & S^2_{r+i,i+1}, & \dots, & S^2_{r+i,r} \\ S^3_{r+i,1}, & S^3_{r+i,2}, & \dots, & S^3_{r+i,i-1}, & 0, & S^3_{r+i,i+1}, & \dots, & S^3_{r+i,r} \end{bmatrix}$$

$S = \text{block diag}\ \{S_1,S_2,\dots,S_r\}$

and

$S_s = \{S: 0\le S^k_{ij}\le I \ , \ S^1_{ii}=I \ , \ S^1_{r+i,r}=I \ , \ S^m_{r+i,r}=0 \ , \ \forall m=2,3 \ , \forall i,j=1,2,\dots,r \ , \forall k=1,2,3\}$,

where I is, as usually, the identity matrix of the appropriate order. The disconnected singularly perturbed subsystems are obtained from (19) by setting x^i and y^i instead of x and y, respectively,

$$\dot{x}_i = A_i x_i + q_{i1}\phi_{i1}(\sigma_{i1}) + A_{ii}y_i \ , \quad \tilde{\sigma}_{i1} = c^T_{i1}x^i + c^T_{i2}y^i \ , \tag{20a}$$

$$\mu_i\dot{y}_i = \mu_i B_{ii}x_i + B_i y_i + q_{i2}\phi_{i2}(\sigma_{i2}) \ , \quad \forall i=1,2,\dots,r \ . \tag{20b}$$

Absolute stability of the subsystems (20) was studied in I.5.4. The results of that section will be used by supposing that all conditions of Theorem 20 of the section I.5.3 are satisfied for the i-th system (20). This means that both Assumption 4 and Assumption 5 are true and that

$$1 > \zeta_{i1}+\xi_{i1} \ , \quad \forall i=1,2,\dots,r \ , \tag{21a}$$

and $\hat{\mu}_i \in]0,\tilde{\mu}_i[$,

$$\tilde{\mu}_i = \frac{1-\xi_{i2}}{\zeta_{i2}} , \quad i=1,2,\dots,r . \tag{21b}$$

The comparison function η_i (Assumption 3) is obtained in the form

$$\eta_i(x_i,y_i) = \eta_{i3}(1-\zeta_{i1}-\xi_{i1})\|x_i\| + \zeta_{i2}(\tilde{\mu}_i-\hat{\mu}_i)\rho_{i3}\|y_i\| , \ \forall i=1,2,\dots,r . \tag{22}$$

Let $c_{ij}^k \in R^{n_k}$ be the k-th vector component of vector c_{ij} . Then, for $x_i \neq 0$ and $y_i \neq 0$,

$$\begin{aligned}
&\mu_i \nu_{ix_i}^T(x_i)[f_i(x,y,S_i)-f_i(x^i,y^i,S_i)] + v_{iy_i}^T[g_i(x,y,M,S_i)-g_i(x^i,y^i,M^i,S_i)] \\
&+\mu_i v_{ix_i}^T(y_i)[f_i(x,y,S_i)-f_i(x^i,y^i,S_i)] \\
&= \mu_i \frac{x_i^T}{\nu_i(x_i)}[H_{i1}+\frac{1}{2}\theta_{i1}\frac{\phi_{i1}(\sigma_{i1}^\circ)}{\sigma_{i1}^\circ}c_{i1}^i(c_{i1}^i)^T]\{q_{i1}[\phi_{i1}(\sigma_{i1})-\phi_{i1}(\tilde{\sigma}_{i1})] + \\
&+\sum_{\substack{j=1\\j\neq i}}^{r} S_{ij}^1 A_{ij}y_j\} + \frac{y_i^T}{v_i(y_i)}[\mu_i \sum_{\substack{j=1\\j\neq i}}^{r} S_{r+i,j}^1 B_{ij}x_j + q_{i3}\phi_{i3}(\sigma_{i3})] \\
&\leq \mu_i\{\eta_{i1}^{-1}\eta_{i2}[\sum_{\substack{j=1\\j\neq i}}^{r}\|A_{ij}\|\cdot\|y_j\| + \kappa_{i1}\|q_{i1}\| \sum_{j=1}^{r}(\|c_{i1}^j\|\cdot\|x_j\|+\|c_{i2}^j\|\cdot\|y_j\|)] + \\
&+\sum_{\substack{j=1\\j\neq i}}^{r}\|B_{ij}\|\cdot\|x_j\| + \|q_{i3}\|k_{i3}\sum_{\substack{j=1\\j\neq i}}^{r}(\|c_{j5}\|\cdot\|x_j\|+\|c_{j6}\|\cdot\|y_j\|)\} .
\end{aligned} \tag{23}$$

Let

$$\gamma_{ij} = k_{i1}\|q_{i1}\|\cdot\|c_{i1}^j\|\eta_{i1}^{-1}\eta_{i2} + (\|B_{ij}\| + \kappa_{i3}\|q_{i3}\|\cdot\|c_{j5}\|)(1-\delta_{ij}) , \tag{24a}$$

$$\epsilon_{ij} = (\|A_{ij}\| + \kappa_{i1}\|q_{i1}\|\cdot\|c_{i2}^j\|)\cdot\eta_{i1}^{-1}\eta_{i2} + \|q_{i3}\|\kappa_{i3}\|c_{j6}\|(1-\delta_{ij}) . \tag{24b}$$

Now, the following inequality will be used

$$\gamma_{ij}\|x_j\| + \epsilon_{ij}\|y_j\| \leq \alpha_{ij}\eta_j(x_j,y_j) \tag{25a}$$

provided that

$$\alpha_{ij} = \max\{\frac{\gamma_{ij}}{\eta_{j3}(1-\zeta_{j1}-\xi_{j1})}, \frac{\epsilon_{ij}}{\zeta_{j2}(\tilde{\mu}_j-\hat{\mu}_j)\rho_{j3}}\} , \quad i,j=1,2, ,r . \tag{25b}$$

(23)-(25) yield

$$\begin{aligned}
&\mu_i\nu_{ix_i}^T[f_i(x,y,S_i)-f_i(x^i,y^i,S_i)] + v_{iy_i}^T[g_i(x,y,M,S_i)-g_i(x^i,y^i,M^i,S_i)] \\
&+\mu_i v_{ix_i}^T[f_i(x,y,S_i)-f_i(x^i,y^i,S_i)] \leq \mu_i \sum_{j=1}^{r}\alpha_{ij}\eta_j(x_j,y_j) , \ x_i\neq 0 , \ y_i \neq 0 , \\
&\forall i=1,2,\dots,r , \ \forall(x,y,M,S) \in R^n\times R^m\times\hat{M}\times S_s .
\end{aligned} \tag{26a}$$

Repeating the preceding procedure it is obtained

$$\begin{aligned}
&v_{iy}^T[g_i(x,y,M,S_i)-g_i(0,y^i,M^i,S_i)] \leq \mu_i \sum_{j=1}^{r}\alpha_{ij}\eta_j(x_j,y_j) , \ x_i=0 , \ y_i\neq 0 , \\
&\forall i=1,2,\dots,r , \ \forall(x,y,M,S) \in R^n\times R^m\times\hat{M}\times S_s .
\end{aligned} \tag{26b}$$

and

$$[\nu_i(x_i)+v_i(y_i)]^T_{x_i}[f_i(x,y,S_i)-f_i(x^i,y^i,S_i)] \le \mu_i \sum_{j=1}^{r} \alpha_{ij}\eta_j(x_j,y_j) ,$$

$$x_i \ne 0 , \; y_i = 0 , \; \forall i=1,2,\dots,r , \; \forall (x,y,M,S) \in R^n \times R^m \times \hat{M} \times S_s . \qquad (26c)$$

(26a)-(26c) fulfil all conditions of Assumption 3. The aggregation matrix $A=(a_{ij})$, $a_{ij}=-\delta_{ij}+\alpha_{ij}$, is now determined by (24) and (25). If it is stable then the equilibrium state $(x^T,y^T)^T=0$ of the system (19) is structurally absolutely stable over $[0,K]\times\hat{M}\times P\times S_s$ in the sense of

Definition 1. *The equilibrium state* $(x^T,y^T)^T=0$ of the system (19) is *structurally absolutely stable over* $[0,K]\times\hat{M}\times P\times S_s$ iff it is absolutely stable on $[0\times K]$ for every $(M,P,S)\in M\times P\times S_s$. ■

Example 1. Let the system (19) be numerically defined as follows. It is the 12th order $(n=m=6)$ singularly perturbed Lur'e system decomposed into three $(q=r=3)$ interconnected singularly perturbed subsystems. They are determined by

$$A_i = \begin{bmatrix} 0 , & 1 \\ -1 , & -2 \end{bmatrix} , \; A_{ii} = I , \; A_{ij} = \gamma I , \; j \ne i , \; \gamma = \frac{1}{2000} , \; q_{i1} = \begin{bmatrix} 0 \\ 1/10 \end{bmatrix} ,$$

$$c^i_{i1} = \begin{bmatrix} -1/100 \\ 0 \end{bmatrix} , \; c^i_{i2} = \begin{bmatrix} 1 \\ 1 \end{bmatrix} , \; c^j_{i1} = \begin{bmatrix} \gamma \\ 0 \end{bmatrix} , \; c^j_{i2} = \begin{bmatrix} 0 \\ \gamma \end{bmatrix} , \; j \ne i , \; \kappa_{i1}=2 ,$$

$$B_i = \begin{bmatrix} -4 , & 1 \\ 1 , & -4 \end{bmatrix} , \; B_{ii} = 10^{-3} I , \; B_{ij} = \gamma I , \; j \ne i , \; q_{i2} = \begin{bmatrix} 1 \\ 1 \end{bmatrix} , \; q_{i3} = \begin{bmatrix} 0 \\ 1 \end{bmatrix} ,$$

$$c_{i3} = \begin{bmatrix} 10^{-3} \\ 0 \end{bmatrix} , \; c_{i4} = \begin{bmatrix} 1 \\ 0 \end{bmatrix} , \; c_{j5} = \begin{bmatrix} 0 \\ \gamma \end{bmatrix} , \; c_{j6} = \begin{bmatrix} \gamma \\ 0 \end{bmatrix} , \; \kappa_{i2}=\kappa_{i3}=1 .$$

The disconnected singularly perturbed subsystems (20) were considered in Example 13 of the section I.5.4. Using the results of that example and calculating a_{ij} according to (24) and (25) the aggregation matrix A is obtained as

$$A = \begin{bmatrix} -0.944 , & 0.012 , & 0.012 \\ 0.012 , & -0.094 , & 0.012 \\ 0.012 , & 0.012 , & -0.094 \end{bmatrix}$$

The stability property is tested by verifying the (Sevastyanov-Kotelyanskii-Hicks) conditions (III.4.1.1)

$$(-1)|a_{11}|>0\ ,\quad (-1)^2\begin{vmatrix} a_{11}\ , a_{12} \\ a_{21}\ , a_{22}\end{vmatrix}>0\ ,\quad (-1)^3\begin{vmatrix} a_{11}\ , a_{12}\ , a_{13} \\ a_{21}\ , a_{22}\ , a_{23} \\ a_{31}\ , a_{32}\ , \ 33\end{vmatrix}>0\ ,$$

which in this case take the form

$$(-1)(-0.944)>0\ ,\quad \begin{vmatrix} -0.944\ , & 0.012 \\ 0.012\ , & -0.944\end{vmatrix}>0\ ,\quad (-1)^3\begin{vmatrix} -0.944\ , & 0.012\ , & 0.012 \\ 0.012\ , & -0.944\ , & 0.012 \\ 0.012\ , & 0.012\ , & -0.944\end{vmatrix}>0\ .$$

Hence, all conditions of Theorem 2 have been verified. The equilibrium state $(\mathbf{x}^T,\mathbf{y}^T)^T=\mathbf{0}$ of the system is structurally absolutely stable over $[0,K]\times\hat{M}\times S$, where $K=\mathrm{diag}\{2,1,1,2,1,1,2,1,1\}$, $\hat{M}=\{M: M=\mathrm{diag}\{\mu_1,\mu_2,\mu_3\}\ ,\ \mu_i\in]0,0.447[\ ,\ \forall i=1,2,3\}$.

IV.3.2.3. Conclusion

Non-uniform time scaling of singularly perturbed large-scale systems is characterized by complete mutual independence of small parameters μ_i . The stability analysis of such systems was carried out on two hierarchical levels. On the first level stability of disconnected singularly perturbed subsystems was tested. Information about that and about qualitative properties of interactions was used on the second hierarchical level for construction of the aggregation matrix A . The order $q=r$ of the aggregation matrix is equal to the number of disconnected singularly perturbed subsystems and it is much less than the order $m+n$ of the whole system. The conceptual application of the general analysis was to the analysis of structural absolute stability of singularly perturbed large-scale Lur'e systems. The example illustrated the results, which were used for testing structural absolute stability of the twelfth order system. On the first level, absolute stability was verified for three disconnected singularly perturbed subsystems of the fourth order. On the second level, as a final test, was verification of stability of the aggregation third order matrix.

IV.3.3. Uniform time-scaling

Uniform time-scaling is defined by $\underline{\tau}$, $\bar{\tau}_i\in]0,+\infty[$, and τ_i such that both (3) and (4) hold. The stability analysis will be carried out by utilizing Lur'e like aggregation form and quadratic aggregation form.

IV.3.3.1. *Lur'e like aggregation form and stability criteria*

The Lur'e like aggregation form is defined by the next assumptions.

Assumption 6. There exist a connected neighbourhood N_{ix} of $x_i=0$, $N_{ix}\subseteq R^{ni}$, functions ν_i, Ω_i, ζ_i, α_{ij} and β_{ij} such that

1) ν_i is differentiable and positive definite on $R\times N_{ix}$ and there is $\rho_i>0$ such that the set $V_{i\zeta}(t)$ is asymptotically contractive for every $\zeta\in]0,\rho_i[$, $\forall i=1,2,\dots,q$,

2) Ω_i is positive definite on $R\times N_{ix}$, $\forall i=1,2,\dots,q$,

3) $\zeta_i\in C(R\times R,R)$, $\zeta_i(t,0)\equiv 0$ and for $\Omega_i\neq 0$:

$$\frac{\zeta_i(t,\Omega_i)}{\Omega_i}\in[0,\kappa_i[\ ,\ \forall(t,\Omega_i)\in R\times R\ ,\ \kappa_i\in]0,+\infty[\ ,\ \forall i=1,2,\dots,q\ ,$$

4) $\alpha_{ij},\beta_{ij}\in C(R,R)$, $\forall i,j=1,2,\dots,q$,

and

5) $$\nu_{it}(t,x_i)+\nu_{ix_i}^T(t,x_i)f_i(t,x,0,0,0)\le\Omega_i(t,x_i)\sum_{j=1}^{q}[\alpha_{ij}(t)\cdot\Omega_j(t,x_j)+$$
$$+\beta_{ij}(t)\zeta_j(t,\Omega_j)]\ ,\ \forall i=1,2,\dots,q\ ,\ \forall(t,x)\in R\times N_x\ .\ \blacksquare$$

Assumption 7. There exist a connected neighbourhood N_{iy} of $y_i=0$, $N_{iy}\subseteq R^{mi}$, functions v_i, ω_i, ξ_i, γ_{ij} and η_{ij} such that

1) v_i is differentiable on $R\times N_x\times N_{iy}$ and positive definite on N_{iy} for every $x\in N_x$, and there is $\Psi_i>0$ such that the set $V_{i\zeta}(t)$ is asymptotically contractive, $\forall\zeta\in]0,\Psi_i[$, $\forall i=1,2,\dots,p$,

2) ω_i is positive definite on N_{iy} for every $x\in N_x$, $\forall i=1,2,\dots,r$,

3) $\xi_i\in C(R\times R,R)$, $\xi_i(t,0)\equiv 0$ and for $\omega_i\neq 0$:

$$\frac{\xi_i(t,\omega_i)}{\omega_i}\in[0,\kappa_{q+i}[\ ,\ \forall(t,\omega_i)\in R\times R\ ,\ \kappa_{q+i}\in]0,+\infty]\ ,\ \forall i=1,2,\dots,r\ ,$$

4) $\gamma_{ij},\eta_{ij}\in C(R,R)$, $\forall i,j=1,2,\dots,r$,

and

5) $$\tau_i v_{iy_i}^T(t,x,y_i)g_i(t,x,y,M,p_{q+i},S_{q+i})$$
$$\le\omega_i(t,x,y_i)\sum_{j=1}^{p}[\gamma_{ij}(t)\omega_j(t,x,y_j)+\eta_{ij}(t)\xi_j(t,\omega_j)]\ ,\ \forall i=1,2,\dots,r\ ,$$
$$\forall\tau_i\in[\underline{\tau}_i,\bar{\tau}_i[\ ,\ \forall(t,x,y,M,P,S)\in R\times N_x\times N_y\times M\times P\times S_s\ .\ \blacksquare$$

Assumption 8. There exist $\rho_1,\rho_2\in]0,+\infty[$ such that both

1) $$\sum_{i=1}^{q}\|\nu_{ix_i}^T(t,x_i)[f_i(t,x,y,p_i,S_i)-f_i(t,x,0,0,0)]\|\le\rho_1\sum_{i=1}^{r}\omega_i^2(t,x,y_i)\ ,$$
$\forall(t,x,y,P,S)\in R\times N_x\times N_y\times P\times S_s$,

and

2) $\sum_{i=1}^{p} [v_{it}(t,x,y_i) + v_{ix}^T(t,x,y_i) f(t,x,y,P,S)] \le \rho_2 \sum_{i=1}^{p} \omega_i^2(t,x,y_i)$,

$\forall (t,x,y,P,S) \in R \times N_x \times N_y \times P \times S_s$,

where $f = (f_1^T, f_2^T, \dots, f_q^T)^T$. ∎

The aggregation of the system (1) defined by Assumptions 6 and 7 leads to the functional aggregation matrices A_i , B_i and C_i , i=1,2 , in the form

$$A_1(t) = (\alpha_{ij}(t)) \ , \ B_1(t) = (\beta_{ij}(t)) \ , \ C_1(t) \equiv I \ , \tag{27a}$$

$$A_2(t) = (\gamma_{ij}(t)) \ , \ B_2(t) = (\eta_{ij}(t)) \ , \ C_2(t) \equiv I \ . \tag{27b}$$

Let

$$\Theta_1 = \text{diag}\{\theta_1, \theta_2, \dots, \theta_q\} \ , \ \Theta_2 = \text{diag}\{\theta_{q+1}, \theta_{q+2}, \dots, \theta_{q+r}\} \ , \tag{28a}$$

$$K_1 = \text{diag}\{\kappa_1, \kappa_2, \dots, \kappa_q\} \ , \ K_2 = \text{diag}\{\kappa_{q+1}, \kappa_{q+2}, \dots, \kappa_{q+r}\} \ , \tag{28b}$$

$$L_i = 2K_i^{-1} - (\Theta_i B_i + B_i^T \Theta_i) \ , \ i=1,2 \ , \tag{29}$$

$$Q_i = -\frac{1}{2}[A_i^T + A_i + (A_i^T \Theta_i + I_i + B_i) L_i^{-1} (A_i^T \Theta_i + I_i + B_i)^T] \ , \ i=1,2 \ . \tag{30}$$

In the case Q_i is positive definite then there is $\epsilon_i \in]0,+\infty[$ obeying

$$Q_i(t) - \epsilon_i I_i > 0 \ , \ \forall t \in R \ , \ i=1,2 \ . \tag{31}$$

For such (preferably the greatest) ϵ_2 the estimate $\hat{\mu}_i$ of the upper bound μ_{im} of admissible μ_i will be determined by

$$\hat{\mu}_i = \min\{1 \ , \frac{\epsilon_2 \max}{\underline{\tau}_i \psi}\} \ , \ i=1,2,\dots,r \ , \tag{32}$$

where

$$\psi = \rho_1 \max\{(1+\theta_i \kappa_i) : i \in [1,q]\} + \rho_2 \max\{(1+\theta_i \kappa_i) : i \in [q+1,q+r]\} \ . \tag{33}$$

Theorem 3. (I) *For structural asymptotic stability over* $\hat{M} \times P \times S_s$ *of* $(x^T, y^T)^T = 0$ *of the system (1) it is sufficient that*

(a) Assumptions 6-8 hold,

(b) there exist non-negative diagonal matrices Θ_i , i=1,2 , *such that the functional matrices* L_i *and* Q_i *are positive definite for every* i=1,2 ,

(c) the functions Ω_i *and* ζ_i , i=1,2,…,q , *and* ω_i *and* ξ_i , i=1,2,…,r , *are all time independent,*

and

(d) $\hat{\mu}_i$, i=1,2,…,p *are determined by (31)-(33).*

(II) *If, in addition to (I),* $N_x \times N_y = R^{m+n}$, ν_i *is radially unbounded and* $\rho_i = +\infty$ *in Assumption 6 for every* i=1,2,…,q , v_i *is radially unbounded for every* $x \in R^n$ *and* $\Psi_i = +\infty$ *in Assumption 7 for every* i=1,2,…,r , *then* $(x^T, y^T)^T = 0$ *of the system (1) is structurally asymptotically stable in the whole over* $\hat{M} \times P \times S_s$.

(III) *For structural asymptotic stability over* $\hat{M} \times P \times S_s$ *of*

$(x^T,y^T)^T = 0$ *of the system (1) it is also sufficient that*

(e) Assumptions 6-8 hold,

(f) the functional matrices L_i *and* Q_i *are positive definite for every* $i=1,2$ *, and for* $\Theta_1=0$ *,* $\Theta_2=0$ *,*

and

(g) $\hat{\mu}_i$ *,* $i=1,2,\dots,r$ *are determined by (31)-(33).*

(IV) *If, in addition to (III),* $N_x \times N_y = R^{m+n}$ *,* ν_i *is radially unbounded and* $\rho_i=+\infty$ *in Assumption 6 for every* $i=1,2,\dots,q$ *,* v_i *is radially unbounded for every* $x \in R^n$ *and* $\Psi_i=+\infty$ *in Assumption 7 for every* $i=1,2,\dots,r$ *, then* $(x^T,y^T)^T = 0$ *of the system (1) is structurally asymptotically stable in the whole over* $\hat{M}\times P\times S_s$

Proof. (I) In view of (c) of the theorem, all ω_i , Ω_i , ζ_i and ξ_i are time independent. Let

$$v(t,x,y) = \sum_{i=1}^{q}\left\{\nu_i(t,x_i) + \theta_i \int_0^{\nu_i} \frac{\zeta_i[\Omega_i(\sigma)]}{\Omega_i(\sigma)}\, d\sigma\right\} + $$

$$+ \sum_{i=1}^{r}\left\{v_i(t,x,y_i) + \theta_{q+i}\int_0^{v_i} \frac{\xi_i[\omega_i(\sigma)]}{\omega_i(\sigma)}\, d\sigma\right\}$$

be a tentative Liapunov function of the whole system (1).

The conditions (a) and (b) of Theorem 3, the condition (1) of Assumption 6 and (1) of Assumption 7 prove both that v is positive definite decrescent on a connected neighbourhood N of $(x^T,y^T)^T = 0$ and that there is $\psi>0$ such that the set $V_\zeta(t)$ is asymptotically contractive $\forall\zeta\in]0,\psi[$. The Eulerian derivative of v along motions of (1) is found in the form

$$\dot{v} = \sum_{i=1}^{q}(\dot{\nu}_i + \theta_i\zeta_i\Omega_i^{-1}\dot{\nu}_i) + \sum_{i=1}^{r}(\dot{v}_r + \theta_{q+i}\xi_i\omega_i^{-1}\dot{v}_i) .$$

Now we use

$$\dot{\nu}_i(t,x_i) = \nu_{it}(t,x_i) + \nu_{ix_i}^T(t,x_i)f_i(t,x,0,0,0) + $$
$$+ \nu_{ix_i}^T[f_i(t,x,y,p_i,S_i) - f_i(t,x,0,0,0)] ,$$

$$\dot{v}_i(t,x,y_i) = v_{it}(t,x,y_i) + v_{ix_i}^T(t,x,y_i)f_i(t,x,y,P,S) + $$
$$+ \frac{1}{\mu_1}\tau_i v_{iy_i}^T(t,x,y_i)g_i(t,x,y,M,p_{q+i},S_{q+i}) ,$$

and Assumptions 6-8 in order to evaluate $\dot{v}$ from above,

$$\dot{v}(t,x,y,M,P,S) \le$$

$$\le \sum_{i,j=1}^{q}\{\Omega_i(x_i)[\alpha_{ij}(t)\Omega_j(x_j) + \beta_{ij}(t)\zeta_j(\Omega_j)][1 + \theta_i\zeta_i(\Omega_i)\Omega_i^{-1}(x_i)]\} +$$

$$+ \rho_1 \max\{(1+\theta_i\kappa_i): i\in[1,q]\} \sum_{i=1}^{r} \omega_i^2(x,y_i) +$$

$$+ \rho_2 \max\{(1+\theta_{q+i}\kappa_{q+i}): i\in[1,r]\} \sum_{i=1}^{r} \omega_i^2(x,y_i) +$$

$$+ \frac{1}{\mu_1} \sum_{i,j=1}^{r} \{\omega_i(x,y_i)[\gamma_{ij}(t)\omega_j(x,y_j) + \\ + \eta_{ij}(t)\xi_j(\omega_j)][1+\theta_{q+i}\xi_i(\omega_i)\omega_i^{-1}(x,y_i)]\} ,$$

$$\forall (t,x,y,M,P,S) \in R\times N_x\times N_y\times M\times P\times S_s .$$

After simple algebraic manipulations on the right-hand side of this inequality and using (27)-(33) it is obtained

$$\dot{v} \le w_1^T A_1 w_1 + w_1^T B_1 z_1 + z_1^T \Theta_1 A_1 w_1 + z_1^T \Theta_1 B_1 z_1 + \psi \| w_2 \|^2 + \\ + \frac{1}{\mu_1}(w_2^T A_2 w_2 + w_2^T B_2 z_2 + z_2^T \Theta_2 A_2 w_2 + z_2^T \Theta_2 B_2 z_2) , \tag{34}$$

where $w_1 = (\Omega_1,\Omega_2,\dots,\Omega_q)^T$, $z_1 = (\zeta_1,\zeta_2,\dots,\zeta_q)^T$, $w_2 = (\omega_1,\omega_2,\dots,\omega_r)^T$, $z_2 = (\xi_1,\xi_2,\dots,\xi_r)^T$. We now add to and subtract from the right-hand side of the last inequality both

$$\sigma_1 = (w_1^T - z_1^T K_1^{-1})z_1 \ge 0 \quad \text{and} \quad \sigma_2 = (w_2^T - z_2^T K_2^{-1})z_2 \ge 0 , \tag{35}$$

employ $C_i = I$ and define $H_i = \frac{1}{2} I$ together with

$$D_i = \frac{1}{2}(A_i^T \Theta_i + I) , \quad X_i Y_i = H_i B_i + D_i , \quad i=1,2 , \tag{36a}$$

$$\Gamma_i = K_i^{-1} - \frac{1}{2}(\Theta_i B_i + B_i^T \Theta_i) , \quad i=1,2 , \tag{36b}$$

$$-Q_i = A_i^T + A_i + X_i X_i^T , \quad i=1,2 . \tag{36c}$$

Under the condition (b) of the theorem it results that $Q_i = Q_i^T$ (30) is positive definite. Hence, there are $\epsilon_i \in]0,+\infty[$ such that (31) holds. As a result, it follows from (31)-(33), the condition (d) of the theorem and (34)-(36) that

$$\dot{v}(t,x,y,M,P,S) \le -\epsilon_1 \| w_1(x) \|^2 - (\frac{\epsilon_2}{\mu_1} - \psi) \| w_2(x,y) \|^2 , \\ \forall (t,x,y,M,P,S) \in R\times N_x\times N_y\times \hat{M}\times P\times S_s . \tag{37}$$

The right-hand side of this inequality is negative definite on $N_x \times N_y$ for every $\mu_1 \in]0,\hat{\mu}_1[$ due to the condition 2 of Assumption 6 and 7 and $\epsilon_1 > 0$. All conditions of Theorem 13 of the section I.3.2.5 have been verified on $\hat{M}\times P\times S_s$, which proves (I) of Theorem 3.

(II) In this case $N_x \times N_y = R^{m+n}$ and all ν_i and v_i are radially unbounded implying radial unboundedness of v . Since $\dot{v}$ is negative definite on $\hat{M}\times P\times S_s$ then $(x^T,y^T)^T = 0$ of (1) is structurally asymptotically stable in the whole over $\hat{M}\times P\times S_s$.

(III) In this case instead of (37) it is obtained for $\Theta_1=0$ and $\Theta_2=0$ that

$$\dot{v}(t,x,y,M,P,S) \le -\epsilon_1 \|w_1(t,x)\|^2 - (\frac{\epsilon_2}{\mu_1} - \psi)\|w_2(t,x,y)\|^2 ,$$
$$\forall(t,x,y,M,P,S) \in R\times N_x\times N_y\times \hat{M}\times P\times S_s \tag{38}$$

The right-hand side of this inequality is negative definite on $N_x\times N_y$ for every $\mu_1 \in]0,\hat{\mu}_1[$ due to the condition 2 of Assumption 6 and 7 and $\epsilon_1>0$. As it is showed under (I), v is positive definite on $N_x\times N_y$ and $V_\zeta(t)$ is asymptotically contractive, $\forall\zeta\in]0,\psi[$.
All conditions of Theorem 13 of the section I.3.2.5 are satisfied on $\hat{M}\times P\times S_s$, which proves that $(x^T,y^T)^T = 0$ of (1) is structurally asymptotically stable over $\hat{M}\times P\times S_s$.

(IV) Following the proof of (II) and (III) it is evident that the assertion under (IV) holds, which completes the proof. ∎

The proof of the Theorem 3 provides a new form of a system scalar Liapunov function v constructed on a higher hierarchical level by using the vector Liapunov function concept and new aggregation form defined by Assumptions 6 and 7. (Lj.T. Grujić, 1979b),

$$v = \sum_{i=1}^{q} \left\{\nu_i + \theta_i \int_0^{\nu_i} \frac{\zeta_i[\Omega_i(\sigma)]}{\Omega_i(\sigma)} d\sigma\right\} + \sum_{i=1}^{r} \left\{v_i + \theta_{q+i} \int_0^{v_i} \frac{\xi_i[\omega_i(\sigma)]}{\omega_i(\sigma)} d\sigma\right\}$$

This function can be also used to establish conditions for structural uniform asymptotic stability under the following assumptions :

Assumption 9. There exist a connected neighbourhood N_{ix} of $x_i=0$, $N_{ix}\subseteq R^{n_i}$, functions ν_i , Ω_i , α_{ij} and β_{ij} such that

1) ν_i is differentiable, positive definite and decrescent on $R\times N_{ix}$ $\forall i=1,2,\dots,q$,
2) Ω_i is positive definite on N_{ix} , $\forall i=1,2,\dots,q$,
3) $\zeta_i \in C(R\times R,R)$, $\zeta_i(t,0)\equiv 0$ and for $\Omega_i \neq 0$:
$$\frac{\zeta_i(t,\Omega_i)}{\Omega_i} \in [0,\kappa_i[\ , \ \forall(t,\Omega_i) \in R\times R \ , \ \kappa_i\in]0,+\infty] \ , \ \forall i=1,2,\dots,q ,$$
4) $\alpha_{ij},\beta_{ij} \in C(R,R)$, $\forall i,j=1,2,\dots,q$,

and

5) $\nu_{it}(t,x_i) + \nu_{ix_i}^T(t,x_i)f_i(t,x,0,0,0) \le$
$$\le \Omega_i(t,x_i) \sum_{j=1}^{q} [\alpha_{ij}(t)\Omega_j(t,x_j) + \beta_{ij}(t)\zeta_j(t,\Omega_j)] ,$$
$\forall i=1,2,\dots,q$, $\forall(t,x) \in R\times N_x$. ∎

Assumption 10. There exist a connected neighbourhood N_{iy} of $y_i=0$, $N_{iy}\subseteq R^{m_i}$, functions v_i, ω_i, ξ_i, γ_{ij} and η_{ij} such that

1) v_i is differentiable on $R\times N_x\times N_{iy}$ and both positive definite and decrescent on N_{iy} for every $x\in N_x$, $\forall i=1,2,\ldots,r$,

2) ω_i is positive definite on N_{iy} for every $x\in N_x$, $\forall i=1,2,\ldots,r$,

3) $\xi_i\in C(R\times R,R)$, $\xi_i(t,0)\equiv 0$ and for $\omega_i\neq 0$:

$$\frac{\xi_i(t,\omega_i)}{\omega_i}\in[0,\kappa_{q+i}[\ ,\ \forall(t,\omega_i)\in R\times R\ ,\ \kappa_{q+i}\in]0,+\infty[\ ,\ \forall i=1,2,\ldots,r\ ,$$

4) $\gamma_{ij},\eta_{ij}\in C(R,R)$, $\forall i,j=1,2,\ldots,r$,

and

5) $$\tau_i v_{iy_i}^T(t,x,y_i)g_i(t,x,y,M,p_{q+i},S_{q+i})\leq$$
$$\leq\omega_i(t,x,y_i)\cdot\sum_{j=1}^{p}[\gamma_{ij}(t)\omega_j(t,x,y_j)+\eta_{ij}(t)\xi_j(t,\omega_j)]\ ,\ \forall i=1,2,\ldots,r\ ,$$
$$\forall\tau_i\in[\underline{\tau}_i,\bar{\tau}_i]\ ,\ \forall(t,x,y,M,P,S)\in R\times N_x\times N_y\times M\times P\times S_s\ .\ \blacksquare$$

Theorem 4. (I) *For structural uniform asymptotic stability over* $\hat{M}\times P\times S_s$ *of* $(x^T,y^T)^T=0$ *of the system (1) it is sufficient that*

(a) Assumptions 8-10 hold,

(b) there exist non-negative diagonal matrices Θ_i, $i=1,2$, *such that the functional matrices* L_i *and* Q_i *are positive definite for every* $i=1,2$,

(c) the functions Ω_i *and* ζ_i, $i=1,2,\ldots,r$, *and* ω_i *and* ξ_i, $i=1,2,\ldots,r$, *are all time independent,*

and

(d) $\hat{\mu}_i$, $i=1,2,\ldots,r$, *are determined by (31)-(33).*

(II) *If, in addition to (I),* $N_x\times N_y\times R^{m+n}$, ν_i *is radially unbounded for every* $i=1,2,\ldots,q$, *and* v_i *is radially unbounded for every* $x\in R^n$ *and every* $i=1,2,\ldots,r$, *then* $(x^T,y^T)^T=0$ *of the system (1) is structurally uniformly asymptotically stable in the whole over* $\hat{M}\times P\times S_s$.

(III) *For structural uniform asymptotic stability of* $(x^T,y^T)^T=0$ *of the system (1) over* $\hat{M}\times P\times S_s$ *it is also sufficient that*

(e) Assumptions 8-10 hold,

(f) the functional matrices L_i *and* Q_i *are positive definite for every* $i=1,2$, *and for* $\Theta_1=0$, $\Theta_2=0$ *and*

(g) $\hat{\mu}_i$, $i=1,2,\ldots,p$, *are determined by (31)-(33).*

(IV) *If, in addition to (III),* $N_x\times N_y=R^{m+n}$, *functions* ν_i *are radially unbounded for every* $i=1,2,\ldots,q$, *and functions* v_i *are radially unbounded for every* $x\in R^n$ *and every* $i=1,2,\ldots,r$, *then*

$(\mathbf{x}^T,\mathbf{y}^T)^T=\mathbf{0}$ *of the system (1) is uniformly structurally asymptotically stable in the whole over* $\hat{M}\times\hat{P}\times S_s$.

Proof. Let v be defined by (39). The condition (1) of Assumptions 9 and 10 implies that v is differentiable, positive definite and decrescent on $N_x\times N_y$, and when $N_x\times N_y=R^{m+n}$ then it is also radially unbounded in view of (II) or (IV) of the theorem. Following the proof of the theorem 3 and using (3)-(5) of Assumptions 9 and 10 and Assumption 8 we get (37) for (I) and (II), and (38) for (III) and (IV). Hence, $\dot{v}$ is negative definite on $N_x\times N_y\times\hat{M}\times P\times S_s$ due to (2) of Assumptions 9 and 10, which completes the proof (Theorem 7 and Theorem 8 of I.3.2.4). ■

Example 2. An eight order non-linear non-stationary system is composed of two fourth order interconnected subsystems described by

$$\frac{d\mathbf{x}_1}{dt}=\begin{bmatrix}-11(1+t+2t^2)(x_{11}^2+x_{12}^2)-\frac{1}{2}(x_{11}^2+2x_{12}^2)^{1/5}+\frac{1}{4}s_{11}x_{12}^3 & , & 0\\ 0 \;,\; -12(1+t^2)(x_{11}^2+x_{12}^2)-2(3x_{11}^2+4x_{12}^2)^{1/5}-0.1\,(1+\sin t)\,s_{12}x_{21}^3\end{bmatrix}\mathbf{x}_1+$$

$$+\begin{bmatrix}-s_{13}y_{11} & , & 0\\ 0 & , & 2s_{14}y_{12}\end{bmatrix}\mathbf{y}_2=\mathbf{f}_1(t,\mathbf{x},\mathbf{y},S_1)\;,$$

$$\mu_1\frac{d\mathbf{y}_1}{dt}=\begin{bmatrix}-8+0.60\,\mu_1 s_{31}y_{22}^3-\frac{1}{2}(y_{11}^2+7y_{12}^2)^2+\mu_1 x_{21} & , & 0\\ 0 & , & -7-0.40\,s_{32}y_{21}^3-2(8y_{11}^2+5x_{12}^2)^2+\mu_2 x_{11}\end{bmatrix}\mathbf{y}_1=\mathbf{g}_1(\mathbf{x},\mathbf{y},\mathbf{M},S_3)\;,$$

$$\frac{d\mathbf{x}_2}{dt}=\begin{bmatrix}-13x_{21}+0.05\,(1.5+\cos t)s_{21}x_{11}-0.5\,(x_{21}^2+6x_{22}^2)x_{21}\\ -10x_{22}+0.10\,s_{22}x_{12}-2(7x_{12}^2+x_{22}^2)x_{22}\end{bmatrix}+$$

$$+\begin{bmatrix}s_{23}y_{22} & , & 0\\ 0 & , & s_{24}y_{21}\end{bmatrix}\mathbf{y}_1=\mathbf{f}_2(t,\mathbf{x},\mathbf{y},S_2)\;,$$

$$\mu_2\frac{d\mathbf{y}_2}{dt}=\begin{bmatrix}-17y_{21}+0.10\,\mu_2 s_{41}y_{11}-(y_{21}^2+6y_{22}^2)y_{21}\\ -10y_{22}+\frac{1}{5}s_{42}x_{12}y_{12}-2(4y_{21}^2+3y_{22}^2)y_{22}\end{bmatrix}=\mathbf{g}_2(\mathbf{x},\mathbf{y},\mathbf{M},S_4)\;,$$

where $\mathbf{M}=\mathrm{diag}\{\mu_1,\mu_2\}$, $M=\{\mathbf{M}: 0<\mu_i<1\;,\; i=1,2\}$, $\underline{\tau}_2=\frac{1}{2}$, $\bar{\tau}_2=1$ so that $\tau_2\in[\frac{1}{2},1]$, s_{ij} denotes $s_{ij}(t)$, $s_{13}\equiv s_{12}$, $s_{14}\equiv s_{11}$ and

$$S_1 = \begin{bmatrix} s_{11}, & 0, & s_{12}, & 0 \\ 0, & s_{11}, & 0, & s_{12} \end{bmatrix}, \quad S_2 = \begin{bmatrix} s_{21}, & 0, & s_{22}, & 0, & s_{23}, & 0, & s_{24}, & 0 \\ 0, & s_{21}, & 0, & s_{22}, & 0, & s_{23}, & 0, & s_{24} \end{bmatrix},$$

$$S_3 = \begin{bmatrix} s_{31}, & 0, & s_{32}, & 0 \\ 0, & s_{31}, & 0, & s_{32} \end{bmatrix}, \quad S_4 = \begin{bmatrix} s_{41}, & 0, & s_{42}, & 0 \\ 0, & s_{41}, & 0, & s_{42} \end{bmatrix},$$

so that

$$\mathbf{f}_1(t,\mathbf{x},\mathbf{y},\mathbf{S}_1) =$$

$$= \begin{bmatrix} [-11(1+t+2t^2)(x_{11}^2+x_{12}^2) - \frac{1}{2}(x_{11}^2+2x_{12}^2)^{1/5}]x_{11} \\ [-12(1+t^2)(x_{11}^2+x_{12}^2) - 2(3x_{11}^2+4x_{12}^2)^{1/5}]x_{12} \end{bmatrix} + \mathbf{S}_1 \cdot \begin{bmatrix} \frac{1}{4}x_{22}^3 \\ 2y_{12}y_{22} \\ -y_{11}y_{21} \\ -0.1(1+\sin t)\, x_{21}^3 \end{bmatrix}$$

$$\mathbf{f}_2(t,\mathbf{x},\mathbf{y},\mathbf{S}_2) = \begin{bmatrix} -13x_{21} - 0.5(x_{21}^2+6x_{22}^2)x_{21} \\ -10x_{22} - 2(7x_{12}^2+x_{22}^2)x_{22} \end{bmatrix} + \mathbf{S}_2 \cdot \begin{bmatrix} 0.05\,(1.5+\cos t)\,x_{11} \\ 0 \\ 0 \\ 0.10\,x_{12} \\ y_{11}y_{22} \\ 0 \\ 0 \\ y_{12}y_{21} \end{bmatrix}$$

$$\mathbf{g}_1(\mathbf{x},\mathbf{y},\mathbf{M},\mathbf{S}_3) = \begin{bmatrix} [-8 - \frac{1}{2}(y_{11}^2+7y_{12}^2)^2 + \mu_1 x_{21}]y_{11} \\ [-7 - 2(8y_{11}^2+5y_{12}^2)^2 + \mu_2 x_{11}]y_{12} \end{bmatrix} + \mathbf{S}_3 \cdot \begin{bmatrix} 0.60\,\mu_1 y_{22}^3 y_{11} \\ 0 \\ 0 \\ -0.40\,y_{21}^3 y_{12} \end{bmatrix}$$

$$\mathbf{g}_2(\mathbf{x},\mathbf{y},\mathbf{M},\mathbf{S}_4) = \begin{bmatrix} -17y_{21} - (y_{21}^2+6y_{22}^2)y_{21} \\ -10y_{22} - 2(4y_{21}^2+3y_{22}^2)y_{22} \end{bmatrix} + \mathbf{S}_4 \cdot \begin{bmatrix} 0.10\mu_2 y_{11} \\ 0 \\ 0 \\ \frac{1}{5}x_{12}y_{12} \end{bmatrix}$$

Let $N_{ix}=R^2$ and $N_{iy}=\{\mathbf{y}_i : \|\mathbf{y}_i\|<1\}$, $\forall i=1,2$. Hence, $N_{\mathbf{x}}=R^4$ and $N_{\mathbf{y}}=\{\mathbf{y}:\|\mathbf{y}\|<1\}$. All these sets are open, connected, $\mathbf{x}_i \in N_{ix}$ and $\mathbf{y}_i \in N_{iy}$, $\forall i=1,2$.

The form of the system nonlinearities suggests the following choice of the functions ν_i and v_i :

$$\nu_i(x_i) = \|x_i\|^2 \quad \text{and} \quad v_i(y_i) = \|y_i\|^2 \ , \ \forall i=1,2 \ .$$

Now, Assumptions 8-10 can be verified.

Verification of Assumption 9.

1) The function ν_i is differentiable, positive definite and decrescent on N_{ix} , $\forall i=1,2$.
2) The functions Ω_1 and Ω_2 , $\Omega_1=\nu_1$ and $\Omega_2=\nu_2^{1/2}$, are positive definite on N_{1x} and N_{2x} , respectively. There form is suggested by the analysis presented in the sequel under the item 5.
3) The functions ζ_1 and ζ_2 , $\zeta_1 = \Omega_1+\Omega_1^{1/5}$ and $\zeta_2 = \Omega_2^3$, obey $\zeta_i \in C(R,R)$, $\zeta_i(0)=0$, and for $\Omega_i \neq 0$:
$$\frac{\zeta_i(\Omega_i)}{\Omega_i} \in [0,+\infty[\ , \ \forall \Omega_i \in R \ , \ \forall i=1,2 \ .$$
The form of ζ_i is suggested by the analysis presented in the item 5.
4) In this example all α_{ij} and β_{ij} are real numbers due to the next analysis.
5) It now results that
$$\nu_{1t}(x_1) + \nu_{1x_1}^T(x_1)f_1(t,x,0,0,0) \le \Omega_1(-20\Omega_1 - \zeta_1 + \tfrac{1}{2}\zeta_2) \ , \ \forall x \in R^4$$
$$\nu_{2t}(x_1) + \nu_{2x_2}^T(x_2)f_2(t,x,0,0,0) \le \Omega_2(\tfrac{1}{2}\Omega_1 - 20\,\Omega_2 + \tfrac{1}{2}\zeta_1 - \zeta_2) \ , \ \forall x \in R^4 \ ,$$
which yield
$$A_1 = \begin{bmatrix} -20 \ , & 0 \\ \frac{1}{2} \ , & -20 \end{bmatrix} , \ B_1 = \begin{bmatrix} -1 \ , & \frac{1}{2} \\ \frac{1}{2} \ , & -1 \end{bmatrix} , \ K_1^{-1} = 0$$

Verification of Assumption 10.

1) The function v_i is differentiable on $R \times N_x \times N_{iy}$ and both positive definite and decrescent on N_{iy} for every $x \in N_x$, $\forall i=1,2$.
2) The functions $\omega_1=v_1^{1/2}$ and $\omega_2=v_2^{1/2}$, are positive definite on N_{1y} and N_{2y} , respectively, for every $x \in N_x$. Their form results from item 5.
3) The functions ξ_1 and ξ_2 , $\xi_1=\omega_1^5$ and $\xi_2=\omega_2^3$, obey $\xi_i \in C(R,R)$, $\xi_i(0)=0$ and for $\omega_i \neq 0$:
$$\frac{\xi_i(\omega_i)}{\omega_i} \in [0,+\infty[\ , \ \forall \omega_i \in R \ , \ \forall i=1,2 \ .$$
The analysis from item 5 suggests the preceding choice of ξ_i .
4) In this example all γ_{ij} and η_{ij} are real numbers due to the next analysis.
5) It is now obtained that (notice that $\underline{\tau}_1=\bar{\tau}_1=1$, i.e. $\tau_1=1$),
$$v_{1y_1}^T(y_1)g_1(x,y,M,S_3) \le \omega_1(-10\omega_1+2\xi_2-\xi_1) \ , \ \forall (x,y,M,S) \in R^4 \times N_y \times M \times S_s \ ,$$
$$\tau_2 v_{2y_2}^T(y_2)g_2(x,y,M,S_4) \le \omega_2(2\omega_1-10\omega_2-\xi_2) \ , \ \forall (x,y,M,S) \in R^4 \times N_y \times M \times S_s \ .$$

Hence,

$$A_2 = \begin{bmatrix} -10 , & 0 \\ 2 , & -10 \end{bmatrix} , \quad B_2 = \begin{bmatrix} -1 , & 2 \\ 0 , & -1 \end{bmatrix}$$

do not depend on small parameters μ_1 and μ_2 .

Verification of Assumption 8.

The Assumption 8 is verified as follows :

$$\sum_{i=1}^{2} \| [\text{grad}\, v_i(y_i)]^T [f_i(t,x,y,S_i) - f_i(t,x,0,0)] \| \le$$

$$\le \sum_{i=1}^{2} \| \text{grad}\, v_i(y_i) \| \| f_i(t,x,y,S_i) - f_i(t,x,0,0) \| \le 8 \{\omega_1^2 [v_1(y_1)] + \omega_2^2 [v_2(y_2)]\}$$

$\forall (t,x,y,S) \in R \times R^4 \times N_y \times S_s$.

Hence, $\rho_1 = 8$.

Assumptions 8-10 hold. The condition (a) of Theorem 4 is valid.

In order to test the condition (b) of Theorem 4 let $\Theta_1 = 20^{-1} I_2$. Then,

$$L_1 = 2K_1^{-1} - (\Theta_1 B_1 + B_1^T \Theta_1) = 10^{-1} \cdot \begin{bmatrix} 1 , & -\frac{1}{2} \\ -\frac{1}{2} , & 1 \end{bmatrix} > 0$$

and

$$Q_1 = -2^{-1} [A_1^T + A_1 + (A_1^T \Theta_1 + I_2 + B_1)(2K_1^{-1} - \Theta_1 B_1 - B_1^T \Theta_1)^{-1} (A_1^T \Theta_1 + I_2 + B_1)^T]$$

$$= \begin{bmatrix} 14.996 , & 2.375 \\ 2.375 , & 15 \end{bmatrix} > 0 .$$

Let $\Theta_2 = 0$ so that

$$L_2 = 2K_2^{-1} - (\Theta_2 B_2 + B_2^T \Theta_2) = 2I_2 > 0 ,$$

and

$$Q_2 = -2^{-1} [A_2^T + A_2 + (A_2^T \Theta_2 + I_2 + B_2)(2K^{-1} - \Theta_2 B_2 - B_2^T \Theta_2)^{-1} (A_2^T \Theta_2 + I_2 + B_2)^T]$$

$$= \begin{bmatrix} 9 , & -1 \\ -1 , & 9 \end{bmatrix} > 0 .$$

This result shows that $Q_2 - \epsilon_2 I_2$ is positive definite for every $\epsilon_2 \in [0,8[$.

Hence,

$$\hat{\mu}_1 = \min\{1 , \frac{\epsilon_{max}}{\psi}\} = 1 , \quad \hat{\mu}_2 = \min\{1 , \frac{\epsilon_{2max}}{\tau_2 \psi}\} = 1$$

which admit $\mu_i \in]0,1[$, $i=1,2$, together with $\mu_2 = \tau_2^{-1} \mu_1$, $\tau_2 \in [2^{-1}, 1]$. Hence, the conditions (b) and (d) have been verified. They are satisfied.

The functions Ω_i and ζ_i as well as ω_i and ξ_i , $i=1,2$, are all time independent. Hence, the condition (b) of Theorem 4 is satisfied. In view of I of Theorem 4, $(\mathbf{x}^T,\mathbf{y}^T)^T=\mathbf{0}$ of the system is structurally uniformly asymptotically stable over $M\times P\times S_s$ and $\mu_2=\tau^{-1}\mu_1$ together with $\tau_2\in[2^{-1},1]$.

Notice that the system considered in the preceding example is of the eight order. Its stability property was tested via positive definiteness of two second order matrices. This illustrates the great conceptual and numerical advantage of the simultaneous usage of and link between the vector Liapunov function concept and the singular perturbation approach.

IV.3.3.2. The quadratic aggregation form

The form of interactions often should be taken into account. It should often influence the form of the aggregation functions and/or comparison functions. In order to achieve this goal and to reduce simultaneously the number of stability tests, we shall propose another decomposition-aggregation form of the systems. Such a form will permit the reduction of the stability analysis to a single stability test that should be carried out on the highest hierarchical level.

Assumption 11. For every $i=1,2,\dots,q$ and every $j=1,2,\dots,r$ there exist connected neighbourhoods $N_{ix}\subseteq R^{n_i}$ of $\mathbf{0}\in R^{n_i}$ and $N_{jy}\subseteq R^{m_j}$ of $\mathbf{0}\in R^{m_j}$, functions v_i , ω_i and Ω_i , and real numbers β_{ij} , λ_{ij} and Λ_{ij} such that

1) $v_i(t,\mathbf{x}_i)\in C^{(1,1)}(R\times N_{ix},R_+)$, it is positive definite on N_{ix}, radially unbounded as soon as $N_{ix}=R^{n_i}$ and there is $\rho_i>0$ such that the set $\mathcal{V}_{i\zeta}(t)$ is asymptotically contractive for every $\zeta\in]0,\rho_i[$ and $i=1,2,\dots,q$,
2) $\omega_i(t,\mathbf{x}_i)=0$ iff $\mathbf{x}_i=\mathbf{0}$, $\forall i=1,2,\dots,q$,
3) $\Omega_i(t,\mathbf{y}_i)=0$ iff $\mathbf{y}_i=\mathbf{0}$, $\forall i=1,2,\dots,r$,
4) $v_{it}(t,\mathbf{x}_i)+v_{i\mathbf{x}_i}^T(t,\mathbf{x}_i)\mathbf{f}_i(t,\mathbf{x},\mathbf{0},\mathbf{0},\mathbf{0})\le\sum_{j=1}^{q}\beta_{ij}\omega_i(t,\mathbf{x}_i)\omega_j(t,\mathbf{x}_j)$, $\forall(t,\mathbf{x})\in R\times N_x$, $\forall i=1,2,\dots,q$,
5) $v_{i\mathbf{x}_i}^T(t,\mathbf{x}_i)[\mathbf{f}_i(t,\mathbf{x},\mathbf{y},\mathbf{p}_i,\mathbf{S}_i)-\mathbf{f}_i(t,\mathbf{x},\mathbf{0},\mathbf{0},\mathbf{0})]\le$
$\le\sum_{j=1}^{q}\lambda_{ij}\omega_i(t,\mathbf{x}_i)\omega_j(t,\mathbf{x}_j)+\sum_{j=1}^{r}\Lambda_{ij}\omega_i(t,\mathbf{x}_i)\Omega_j(t,\mathbf{y}_j)$,
$\forall(t,\mathbf{x},\mathbf{y},\mathbf{P},\mathbf{S})\in R\times N_x\times N_y\times P\times S_s$, $\forall i=1,2,\dots,q$. ■

Assumption 12. For every $i=1,2,\ldots,q$, and every $j=1,2,\ldots,r$ there exist connected neighbourhoods $N_{ix}\subseteq R^{n_i}$ of $0\in R^{n_i}$ and $N_{jy}\subseteq R^{m_j}$ of $0\in R^{m_j}$, functions ω_i , v_j and Ω_j and real numbers γ_{ijk} , ζ_{ijk} , ξ_{ij} and ρ_{ij} such that

1) $v_j(t,x,y_j)\in C^{(1,1,1)}(R\times N_x\times N_{jy},R_+)$, it is positive definite on $N_x\times N_{jy}$ (on N_{jy} as soon as v_j does not depend on x), radially unbounded uniformly in $x\in N_x$ as soon as $N_{jy}=R^{m_j}$ and there is $\Psi_j>0$ such that the set $V_{j\zeta}(t)$ is asymptotically contractive for every $\zeta\in]0,\Psi_j[$ and $j=1,2,\ldots,r$,

2) $v_{jt}(t,x,y_j)+v_{jx}^T(t,x,y_j)f(t,x,y,P,S)\le$
$$\le\sum_{i,k=1}^{q}\gamma_{jik}\omega_i(t,x_i)\omega_k(t,x_k)+\sum_{i,k=1}^{q,r}\zeta_{jik}\,\omega_i(t,x_i)\Omega_k(t,y_k)\ ,$$
$\forall(t,x,y,P,S)\in R\times N_x\times N_y\times P\times S_s$, where $f=(f_1^T,f_2^T,\ldots,f_q^T)$,

3) $\tau_j v_{jy_j}^T(t,x,y_j)g_j(t,x,y,M,p_{q+j},S_{q+j})\le$
$$\le\mu_1\sum_{i=1}^{q}\rho_{ji}\omega_i(t,x_i)\Omega_j(t,y_j)+\sum_{i=1}^{r}\xi_{ji}\Omega_i(t,y_i)\Omega_j(t,y_j)\ ,$$
$\forall(t,x,y,\mu_1,M,P,S)\in R\times N_x\times N_y\times]0,\mu_{1m}[\times M_m\times P\times S_s$. ■

Under Assumption 12 the function v_j may be the sum of two functions, one of which is positive definite in x and another one in y_j . In a particular case, v_j may be dependent only on y_j and then it should be positive definite in y_j . Besides, in the case that v_j does not depend on t and/or x then all $\gamma_{jik}=0$ and/or $\zeta_{jik}=0$, respectively.

Let the elements of aggregation matrices $B=(b_{ij})$, $C=(c_{ij})$ and $D=(d_{ij})$ be determined by

$$b_{ij}=2^{-1}[\beta_{ij}+\beta_{ji}+\lambda_{ij}+\lambda_{ji}+\sum_{k=1}^{r}(\gamma_{kij}+\gamma_{kji})]\ ,\ \forall i,j=1,2,\ldots,q\ ,\tag{40}$$

$$c_{ij}=\Lambda_{ij}+\rho_{ij}+\sum_{k=1}^{r}\zeta_{kij}\ ,\ \forall i=1,2,\ldots,q\ ,\ \forall j=1,2,\ldots,r\ ,\tag{41}$$

$$d_{ij}=2^{-1}(\xi_{ij}+\xi_{ji})\ ,\ \forall i,j=1,2,\ldots,r\ .\tag{42}$$

In order to estimate (by $\hat{\mu}_1$) the upper admissible bound of the small parameter μ_1 , we denote the maximal eigenvalues of the matrices B and D by $\Lambda(B)$ and $\Lambda(D)$, respectively and introduce

$$\tilde{\mu}_1=\frac{4\Lambda(B)\Lambda(D)}{\|C\|^2}\ ,\tag{43}$$

so that $\mu_1\in]0,\tilde{\mu}[$ may be selected. Evidently, $\Lambda(B)\Lambda(D)>0$ implies $\tilde{\mu}_1>0$. The required stability criterion can be now stated in terms of ν_i and v_j which are not decrescent in view of the property of the sets $\mathcal{V}_{i\zeta}(t)$ and $V_{j\zeta}(t)$ expressed in the condition 1 of Assumptions 11 and 12, respectively.

Theorem 5. (I) *For structural asymptotic stability over* $\hat{M}\times P\times S_s$ *of* $(x^T,y^T)^T=0$ *of the system (1) it is sufficient that*

(a) Assumptions 11 and 12 hold,

(b) the aggregation matrices $B=(b_{ij})$ *(40) and* $D=(d_{ij})$ *(42) are both negative definite,*

and

(c) $\tilde{\mu}_1$ *is determined by (43),* $\hat{\mu}_1\in]0,\tilde{\mu}[$, $\mu_1\in]0,\hat{\mu}_1[$ *and* $\hat{\mu}_i=\hat{\mu}_1\underline{\tau}_i^{-1}$ $\forall i=1,2,\dots,r$.

(II) *For structural asymptotic stability in the whole over* $\hat{M}\times P\times S$ *of* $(x^T,y^T)^T=0$ *of the system (1) it is sufficient that both*

(d) all the conditions under (I) hold for $N_x=R^n$ *and* $N_y=R^m$,

and

(e) that all $\rho_i=+\infty$ *and* $\Psi_j=+\infty$ *in Assumptions 11 and 12, respectively.*

Proof. (I) Let

$$v(t,x,y) = \sum_{i=1}^{q} \nu_i(t,x_i) + \sum_{j=1}^{r} v_j(t,x,y_j) .$$

This function is in $C^{(1,1,1)}(R\times N_x\times N_y)$ and positive definite, and there is $\psi>0$ such that the set $V_\zeta(t)$ is asymptotically contractive for every $\zeta\in]0,\psi[$ due to the condition 1 of Assumptions 11 and 12. Using the conditions 4 and 5 of Assumption 11 and the conditions 2 and 3 of Assumption 12 and (40)-(42) it results that

$$\dot{v}(t,x,y,M,P,S) \le \Lambda(B)\|w_1(t,x)\|^2+\|C\|\|w_1(t,x)\|\|w_2(t,y)\| + \frac{1}{\mu_1}\Lambda(D)\|w_2(t,y)\|$$

$$= [\|w_1(t,x)\|,\|w_2(t,y)\|]\begin{bmatrix}\Lambda(B) & \frac{1}{2}\|C\| \\ \frac{1}{2}\|C\| & \frac{1}{\mu_1}\Lambda(D)\end{bmatrix}\times\begin{bmatrix}\|w_1(t,x)\| \\ \|w_2(t,y)\|\end{bmatrix},$$

$$\forall\mu_1\in]0,\hat{\mu}_1[\ ,\ \hat{\mu}_1\in]0,\tilde{\mu}_1[\ ,\ \forall(t,x,y,M,P,S)\in R\times N_x\times N_y\times M_m\times P\times S_s\ , \tag{44}$$

where $w_1=(\omega_1,\omega_2,\dots,\omega_q)^T$ and $w_2=(\Omega_1,\Omega_2,\dots,\Omega_r)^T$. This result, the conditions 2 and 3 of Assumption 11 complete all requirements of Theorem 13 of the section I.3.2.5 on $\hat{M}\times P\times S_s$. Hence, the statement under (I) is true.

(II) When $N_x=R^n$ and $N_y=R^m$ then the function v is also radially unbounded and the set $V_\zeta(t)$ is asymptotically contractive for every $\zeta\in]0,+\infty[$, which together with the conditions under (I) prove structural asymptotic stability of $(x^T,y^T)^T=0$ of the system (1) over $\hat{M}\times P\times S_s$ (Corollary 2 of the section I.3.2.5 is satisfied over $\hat{M}\times P\times S_s$). ∎

In the case a uniform stability property is required then the conditions for asymptotic contraction of the sets $\mathcal{V}_{i\zeta}(t)$ and $\mathcal{V}_{j\zeta}(t)$ should be replaced by another one. This is precisely explained in Theorem 6 for which the following assumptions are needed.

Assumption 13. For every $i=1,2,\dots,q$ and every $j=1,2,\dots,p$ there exist connected neighbourhoods $N_{ix}\subseteq R^{n_i}$ of $0\in R^{n_i}$ and $N_{jy}\subseteq R^{m_j}$ of $0\in R^{m_j}$, functions ν_i, ω_i and Ω_j, and real numbers β_{ij}, λ_{ij} and Λ_{ij} such that

1) $\nu_i(t,x_i)\in C^{(1,1)}(R\times N_{ix},R_+)$, it is positive definite and decrescent on N_{ix}, and radially unbounded as soon as $N_{ix}=R^{n_i}$, $\forall i=1,2,\dots,q$,
2) ω_i is positive definite on N_{ix}, $\forall i=1,2,\dots,q$,
3) Ω_j is positive definite on N_{jy}, $\forall j=1,2,\dots,r$,
4) the conditions 4 and 5 of Assumption 11 hold. ∎

Assumption 14. For every $i=1,2,\dots,q$ and every $j=1,2,\dots,r$ there exist connected neighbourhoods $N_{ix}\subseteq R^{n_i}$ of $0\in R^{n_i}$ and $N_{jy}\subseteq R^{m_j}$ of $0\in R^{m_j}$, functions ω_i, v_j and Ω_j and real numbers γ_{ijk}, ζ_{ijk}, ξ_{ij} and ρ_{ij} such that

1) $v_j(t,x,y_j)\in C^{(1,1,1)}(R\times N_x\times N_{jy},R_+)$, it is positive definite and decrescent on $N_x\times N_{jy}$ (on N_{jy} only as soon as v_j does not depend on x) and radially unbounded uniformly in $x\in N_x$ as soon as $N_{jy}=R^{m_j}$, $\forall j=1,2,\dots,r$,
2) the conditions 2 and 3 of Assumption 12 are valid. ∎

Theorem 6. (I) *For structural uniform asymptotic stability over* $\hat{M}\times P\times S_s$ *of* $(x^T,y^T)^T=0$ *of the system (1) it is sufficient that*

(a) Assumptions 13 and 14 hold,

(b) the aggregation matrices $B=(b_{ij})$ (40) *and* $D=(d_{ij})$ (42) *are both negative definite,*

and

(c) $\tilde{\mu}_1$ *is determined by (40)-(43)* $\hat{\mu}_1\in]0,\tilde{\mu}_1[$, $\mu_1\in]0,\hat{\mu}_1[$ *and* $\hat{\mu}_i=\hat{\mu}_1\tau_i^{-1}$, $\forall i=1,2,\dots,r$.

(II) *For structural uniform asymptotic stability in the whole over* $\hat{M}\times P\times S_s$ *of* $(x^T,y^T)^T=0$ *of the system (1) it is sufficient that all the conditions under (I) hold for* $N_x=R^n$ *and* $N_y=R^m$.

Proof. (I) Let

$$v(t,x,y) = \sum_{i=1}^{q} \nu_i(t,x_i) + \sum_{j=1}^{r} v_j(t,x,y_j) .$$

This function is in $C^{(1,1,1)}(R\times N_x\times N_y,R_+)$, positive definite and decrescent due to the condition 1 of Assumptions 13 and 14. Using the conditions 4 of Assumption 13 and 2 of Assumption 14 it follows that (44) holds. The conditions 2 and 3 of Assumption 13, the properties of v and (44) prove structural uniform asymptotic stability of $(x^T, y^T)^T = 0$ of the system (1) over $\hat{M}\times P\times S_s$ because all conditions of Theorem 7 of the section I.3.2.4 are satisfied over $\hat{M}\times P\times S_s$.

(II) When $N_x = R^n$ and $N_y = R^m$ in addition to the conditions under (I) then all the conditions of Theorem 8 of the section I.3.2.4 are satisfied over $\hat{M}\times P\times S_s$. Hence, $(x^T, y^T)^T = 0$ of the system (1) is then structurally uniformly asymptotically stable in the whole over $\hat{M}\times P\times S_s$. ■

IV.3.3.2. Application to the structural absolute stability analysis

The preceding result can be applied to the absolute stability analysis of the Lur'e type singularly perturbed large-scale system

$$\dot{x}_i = A_i x_i + q_{i1}\phi_{i1}(\sigma_{i1}) + \sum_{j=1}^{p} S^1_{ij}A_{ij}y_j \ , \ \sigma_{i1} = \hat{c}^T_{i1}x_i + \hat{c}^T_{i2}y \ , \ \forall i=1,2,\dots,q \ ,$$

$$\mu_i\dot{y}_i = \sum_{j=1}^{q} \mu_i S^1_{q+i,j} B_{ij}x_j + B_i y_i + q_{i2}\phi_{i2}(\sigma_{i2}) + q_{i3}\phi_{i3}(\sigma_{i3}) \ , \ \forall i=1,2,\dots,r \ ,$$

$$\sigma_{i2} = \mu_i\hat{c}^T_{i3}x_i + \hat{c}^T_{i4}y_i \ , \ \sigma_{i3} = \sum_{j=1}^{q} \mu_i\hat{c}^T_{j5}S^2_{q+i,j}x_j + \sum_{j=1}^{p} \hat{c}^T_{j6}S^3_{q+i,j}y_j \ , \quad \forall i=1,2,\dots,r \ ,$$

$$\frac{\phi_{ij}(\sigma_{ij})}{\sigma_{ij}} \in [0,\kappa_{ij}] \subset R_+ \ , \ \begin{cases} i=1,2,\dots,q & \text{when } j=1 \\ i=1,2,\dots,r & \text{when } j=2,3 \end{cases} \ , \ \sigma_{ij} \neq 0 \ . \tag{45}$$

The structural matrices S^1_{ij} , S^2_{ij} , S^3_{ij} and S , as well the set S_s , are defined in the section IV.3.2.2.

It is assumed that the numbers $\underline{\tau}_i$ and $\bar{\tau}_i$ are given. The functions ν_i , ω_i , v_j and Ω_j and numbers β_{ij} , λ_{ij} , Λ_{ij} , γ_{ijk} , ζ_{ijk} , ξ_{ij} and ρ_{ij} are to be discovered so that Assumptions 13 and 14 are satisfied. A solution to be presented will be relied on a special form of ν_i and v_j .

Verification of Assumption 13.

It is supposed that A_i is a stable matrix, the pair (A_i, q_{i1}) is controllable, and that there exist numbers $\theta_i \in [0,+\infty[$ and $\epsilon_{i1} \in]0,+\infty[$ such that

$$\kappa^{-1}_{i1} + \mathrm{Re}(1+j\theta_i\omega)\hat{c}^T_{i1}(A_i - j\omega I_i)^{-1}q_{i1} - \epsilon_{i1}q^T_{i1}(A^T_i + j\omega I_i)^{-1}(A_i - j\omega I_i)^{-1}q_{i1} > 0 \ ,$$
$\forall\omega\in[0,+\infty]$.

Then,

$$\nu_i(x_i) = x^T_i H_i x_i + \theta_i \int_0^{\sigma^{\circ}_{i1}} \phi_{i1}(\sigma)d\sigma$$

is a Liapunov function of

$$\dot{x}_i = A_i x_i + q_{i1}\phi_{i1}(\sigma^{\circ}_{i1}) \ , \ \sigma^{\circ}_{i1} = \hat{c}^T_{i1}x_i \ ,$$

provided that H_i is the matrix solution of the Lur'e equations (46),

$$A_i^T H_i + H_i A_i + g_i g_i^T = -\epsilon_{i1} I_i \quad , \quad h_i + H_i q_i = -\sqrt{\gamma_i}\, g_i \tag{46}$$

for

$$\gamma_i = \kappa_{i1}^{-1} - \theta_i \hat{c}_{i1}^T q_i \quad , \quad h_i = 2^{-1}(\theta_i A_i^T \hat{c}_{i1} + \hat{c}_{i1}) \ .$$

In order to determine the numbers β_{ij}, we test first condition 4 of Assumption 13 which requires test of the conditions 4 and 5 of Assumption 11. Since ν_i is independent of t then $\nu_{it} \equiv 0$ and

$$\nu_{ix_i}^T(x_i) f_i(t,x,0,0,0) \le -\epsilon_{i1} \|x_i\|^2 \ , \quad \forall x_i \in R^{n_i} \ .$$

Hence, $\omega_i(x_i) = \|x_i\|$ and $\beta_{ij} = -\epsilon_{i1}\delta_{ij}$, $\forall i,j = 1,2,\ldots,q$.

The numbers λ_{ij} and Λ_{ij} are determined from condition 5 of Assumption 11,

$$\begin{aligned} &\nu_{ix_i}^T(x_i)[f_i(t,x,y,p_i,S_i) - f_i(t,x,0,0,0)] \le \\ &\le 2\|x_i\| \, \|H_i + \tfrac{1}{2}\theta_i \kappa_{i1} \hat{c}_{i1} \hat{c}_{i1}^T\| [2^{-1}\kappa_{i1} \Lambda(q_{i1}\hat{c}_{i1}^T + \hat{c}_{i1} q_{i1}^T)\|x_i\| + \\ &\quad + \sum_{j=1}^{r} (\kappa_{i1}\|q_{i1}\hat{c}_{i2}^T\| + \|A_{ij}\|)\|y_j\|] \ , \\ &\forall i = 1,2,\ldots,q \ , \ \forall (x,y,P,S) \in R^n \times R^m \times P \times S_s \ . \end{aligned}$$

The numbers λ_{ij} and Λ_{ij} are now determined by

$$\lambda_{ij} = \kappa_{i1}\|H_i + 2^{-1}\theta_i \kappa_{i1}\hat{c}_{i1}\hat{c}_{i1}^T\| \Lambda(q_{i1}\hat{c}_{i1}^T + \hat{c}_{i1} q_{i1}^T)\delta_{ij} \ , \ \forall i,j = 1,2,\ldots,q \ ,$$

$$\Lambda_{ij} = 2\|H_i + 2^{-1}\theta_i \kappa_{i1}\hat{c}_{i1}\hat{c}_{i1}^T\|(\kappa_{i1}\|q_{i1}\hat{c}_{i2}^T\| + \|A_{ij}\|) \ , \ \forall i = 1,2,\ldots,q \ , \ \forall j = 1,2,\ldots,r \ .$$

Verification of Assumption 14.

Let $v_i(y_i) = \|y_i\|^2$ be chosen. Then, $v_{it} \equiv 0$ and $v_{ix} \equiv 0$ so that all $\gamma_{ijk} = 0$ and $\zeta_{ijk} = 0$. Further, in order to verify the condition 2 of Assumption 14 it is necessary to test the condition 3 of Assumption 12,

$$\begin{aligned} &\tau_j v_{jy_j}(t,x,y_j) g_j(t,x,y,M,p_{q+i},S_{q+i}) \le \\ &\le \mu_1 \sum_{i=1}^{q} 2(\|B_{ji}\| + \kappa_{j2}\|q_{j2}\hat{c}_{j3}^T\|\delta_{ij} + \kappa_{j3}\|q_{j3}\hat{c}_{i5}^T\|)\|x_i\| \cdot \|y_j\| + \\ &\quad + \sum_{i=1}^{r} 2\{[2^{-1}\underline{\tau}_j \Lambda(B_j + B_j^T) + 2^{-1}\kappa_{j2}\bar{\tau}_j \Lambda(q_{j2}\hat{c}_{j4}^T + \hat{c}_{j4}^T q_{j2})]\,\delta_{ij} + \\ &\qquad + \kappa_{j3}\bar{\tau}_j \|q_{j3}\hat{c}_{i6}^T\|\} \ \|y_i\| \cdot \|y_j\| \ , \\ &\forall j = 1,2,\ldots,r \ , \ \forall (x,y,M,P,S) \in R^n \times R^m \times \hat{M} \times P \times S_s \ . \end{aligned}$$

It now results that $\Omega_i(y_i) = \|y_i\|$ and

$$\rho_{ij} = 2(\|B_{ij}\| + \kappa_{i2}\|q_{i2}\hat{c}_{i3}^T\|\delta_{ij} + \kappa_{i3}\|q_{i3}\hat{c}_{j5}^T\|) \ , \ \forall i = 1,2,\ldots,r \ , \ \forall j = 1,2,\ldots,q \ ,$$

$$\xi_{ij} = 2\{[2\underline{\tau}_i \Lambda(B_i + B_i^T) + \kappa_{i2}\bar{\tau}_i \Lambda(q_{i2}\hat{c}_{i4}^T + \hat{c}_{i4} q_{i2}^T)]\,\delta_{ij} + \kappa_{i3}\bar{\tau}_i \|q_{i3} c_{j6}^T\|\} \ ,$$

$\forall i,j = 1,2,\ldots,r$.

The elements b_{ij}, c_{ij} and d_{ij} of the aggregation matrices $\mathbf{B}$, $\mathbf{C}$ and $\mathbf{D}$ have been obtained via (40)-(42) in the form

$$b_{ij} = -(\epsilon_{i1}-\lambda_{ii})\delta_{ij} \ , \ \forall i,j=1,2,\dots,q \ ,$$

$$c_{ij} = \Lambda_{ij}+\rho_{ij} \ , \ \forall i=1,2,\dots,q \ , \ \forall j=1,2,\dots,r \ ,$$

$$d_{ij} = 2^{-1}(\xi_{ij}+\xi_{ji}) \ , \ \forall i,j=1,2,\dots,r \ .$$

The matrix $\mathbf{B}$ is diagonal. The necessary condition for the fulfilment of Theorem 6 is that all $b_{ii}<0$, i.e.

$$\epsilon_{i1} > \lambda_{ii} \ , \ \forall i=1,2,\dots,q \ .$$

If it is satisfied, then negative definiteness of $\mathbf{D}$ implies the structural absolute stability of $(\mathbf{x}^T,\mathbf{y}^T)^T = \mathbf{0}$ of the system (45) for every $\mathbf{M}\in\hat{\mathbf{M}}$, $\mu_1\in]0,\hat{\mu}_1[$, $\hat{\mu}_1\in]0,\tilde{\mu}_1[$, where $\tilde{\mu}_1$ is defined by (43), and for every $\mathbf{S}\in\mathbf{S}_s$. This result illustrates the essential reduction of the stability analysis. We should verify only simple inequalities (47) and negative definiteness of the symmetric matrix $\mathbf{D}$ of the order r . However, the consequence of such a strong order problem reduction can be a smaller estimate $\hat{\mu}_1$ of the upper allowable bound of μ_1 than that obtained via the multilevel hierarchical analysis.

IV.3.3.3. *Conclusion*

Uniform time scaling requires that all time scales be mutually related in certain bounds. In the case this requirement is satisfied the stability test is reduced to verification of simple algebraic conditions imposed on the system on the highest structural level only. The form of the conditions depend on the aggregation form, which was shown via Lur'e-like and quadratic-like aggregation forms with conceptual applications to the absolute stability analysis of large-scale singularly perturbed Lur'e systems.

IV.4. COMMENTS

The decomposition-aggregation approach to the stability analysis appears attractive when it is based on the vector Liapunov function. It can be more effective for singularly perturbed systems than for those which are not singularly perturbed. A greater order problem reduction can be achieved for the former.

The main problems for effective application of the method to larger class of the systems are the following :

- the form of aggregation functions ν_i and v_i is restricted essentially to few known applicable forms. This is related to the fundamental stability problem-construction of a system Liapunov function.
- If ν_i and v_j depend on states of the other subsystems different from the i-th and j-th subsystem, respectively, then their derivatives become very complex, and, hence, the sign test of the derivative of the resulting v function for the whole system appears cumbersome. However, if ν_i and v_j depend only on x_i and y_j, respectively, and possibly on t, then the form of the resulting v function is not influenced by the form of interactions. As a consequence of it, it appears difficult to assure the negative definiteness of the derivative of the function v of the whole system.
- Vector Liapunov function aggregation so far has led to various, and sometimes numerous, majorizations, moreover, often to rough majorizations. Stabilizing actions of interactions over subspaces are treated at best as neutral or destabilizing over the same subspaces. The benefit of this is simple stability test, and its drawback is too stringent requirement on disconnected subsystems.

The approach to singularly perturbed systems developed in Chapter I and in this Chapter is directly applicable to such systems with continuous but non-differentiable nonlinearities (Grujić, 1982a,b).

Fiedler, M., and V. Ptak (1962), On matrices with non-positive off diagonal elements and positive principal minors. *Czhech. Nat. J.*, *12*, No.87, 382-400.

Grujić, Lj.T. (1976a), Vector Liapunov functions and singularly perturbed large-scale systems. *Proc. 1976 JACC*, Lafayette, 408-416.

Grujić, Lj.T. (1976b), General stability analysis of large-scale systems. *IFAC Symp. on Large-scale systems theory and applications*, Udine, 203-213.

Grujić, Lj.T. (1977), Converse lemma and singularly perturbed large-scale systems. *1977 JACC*, 1107-1112.

Grujić, Lj.T. (1978), Singular perturbations, uniform asymptotic stability and large-scale systems . *Proc. 1978 JACC*, ISA, New York, 339-347.

Grujić, Lj.T. (1979a), Singular perturbations, large-scale systems and asymptotic stability of invariant sets. *Int. J. Systems Sci.*, *10*, No.12, 1323-1341.

Grujić, Lj.T. (1979b), Singular perturbations and large-scale systems. *Int. J. Control*, *29*, No.1, 159-169.

Grujić, Lj.T. (1979c), Sets and singularly perturbed systems. *Systems Science*, *5*, No.4, 327-338.

Grujić, Lj.T. (1981), Uniform asymptotic stability of non-linear singularly perturbed general and large-scale systems. *Int. J. Control*, *33*, No.3, 481-504.

Grujić, Lj.T. (1982a), Asymptotic stability conditions for singularly perturbed systems with non-differentiable non-linearities. *Automatika*, Zagreb, *23*, No 3-4, 83-84.

Grujić, Lj.T. (1982b), On asymptotic stability of large-scale singularly perturbed systems with non-differentiable nonlinearities. *Automatika*, Zagreb, *23*, No. 3-4, 85-86; *Proc. of IXth Int. Conf. on Non-linear Oscillations*, *2*, Naukova Dumka, Kiev (1984), 95-97.

CHAPTER V

LARGE-SCALE POWER SYSTEMS STABILITY

NOTATION

N.B. Besides the general notation used throughout the book, the following symbols are specific to this Chapter.

N	number of system's generators (or machines)
n , $n = N-1$	contrary to the previous chapters n does not denote here the dimension of system's state
$A_{ij} = E_i E_j Y_{ij}$	
D_i	mechanical damping coefficient of i-th generator
D_{ij}	electromagnetic damping coefficient between i-th and j-th generators
E_i	modulus of i-th generator's internal electromotive force (voltage)
$k_i = M_i^{-1}$	
M_i	inertia coefficient of i-th generator
P_{mi}	mechanical power delivered to the i-th generator from its turbine
P_{ei}	electrical power delivered by the i-th generator to the network
p_{mi}	variation of the mechanical power of i-th generator
P^o_{mi}	steady state value of P_{mi}
p^o_{mi}	steady state value of p_{mi}

$p_i = p_{mi} - p_{mi}^o$	
$p_{iN} = p_i - p_N$	
$\mathbf{Y}$	admittance matrix of the network reduced at the internal generator nodes
Y_{ij}	modulus of the ij-th element of $\mathbf{Y}$; $Y_{ij} = Y_{ji}$
$\alpha_i \mu_i^{-1}$	gain of the first order proportional regulator of i-th generator
$\beta_{ij} = A_{ij} M_i^{-1}$	
$\Gamma_i = \lambda_i + \lambda_{Ni} + \sum_{\substack{j=1 \\ j \neq 1}}^{n} \lambda_{ij}$	
δ_i	rotor angle of i-th generator relative to a reference
$\delta_{ij} = \delta_i - \delta_j$	
δ_i^o	the equilibrium under consideration of i-th generator
δ_{iN}^o	value of δ_{iN} at the equilibrium state
θ_{ij}	argument of the ij-th element of $\mathbf{Y}$; $\theta_{ij} = \theta_{ji}$
$\lambda_i = D_i M_i^{-1}$	if $\lambda_i = \lambda$, constant for $i = 1,2,\ldots,N$, one speaks of
λ	"uniform" (mechanical) damping
$\lambda_{ij} = D_{ij} M_i^{-1}$	
$\Lambda_{ij} = \lambda_{ij} - \lambda_{Nj} \quad \forall i = 1,2,\ldots,n$	
$\Lambda_{Ni} = \lambda_N - \lambda_i + \lambda_{iN}$	
μ_i^{-1}	time constant of the first order proportional speed regulator of i-th generator
$\sigma_{iN} = \delta_{iN} - \delta_{iN}^o$	
$\sigma_{ij} = \sigma_{iN} - \sigma_{jN}$	
Ω_i	rotor speed of i-th generator above the synchronous speed : $\Omega_i = \dot{\delta}_i$
Ω_i^o	the value of Ω_i at the steady state operation called "equilibrium state"
$\Omega_{ij} = \Omega_i - \Omega_j$	
$\omega_i = \Omega_i - \Omega_i^o$	

$\omega_{ij} = \omega_i - \omega_j$	
$\boldsymbol{\delta}$, $\boldsymbol{\sigma}$, $\boldsymbol{\Omega}$	without subscript these letters denote vectors
$(\boldsymbol{\delta},\boldsymbol{\Omega})$	state vector with variables δ_i , Ω_i , $i = 1,2,\ldots,n$
$(\boldsymbol{\delta}^s,0)$	value of the state vector at the SEP
$(\boldsymbol{\delta}^{ui},0)$	value of the state vector at the i-th UEP
CCT or t_c	Critical Clearing Time
COA	Center Of Angles
SEP	Stable Equilibrium Point of the system in its post-fault configuration
UEP	Unstable Equilibrium Point surrounding SEP
SDE	Stability Domain Estimate
PSDE	Practical SDE

V.1. INTRODUCTION

Modern electric power systems are a typical example of complex large-scale systems.

An electric power system is composed of several generation stations ("generators") which convert fuel, water or other types of energy into electricity, substations that distribute power to consumers ("loads"), and transmission lines that tie them together. The power system is requested to operate without interruption and under random energy supply requirements, despite electric power cannot be stored.

The nowadays continuously increasing complexity and interdependence of power systems make their stability assessment increasingly important and difficult. The difficulty increases even more with the economic requirement to operate them closer and closer to their capacity limits; at the same time, guaranteeing their stability becomes vital. Note that the concept of "stability" used here receives many interpretations. One of them corresponds to the so-called "transient stability", considered in this Chapter.

From a system theory point of view, and for the purpose of their transient stability analysis, electric systems belong to the class of large-scale, highly nonlinear systems operating under random perturbations and randomly varying parameters. Their transient stability assessment appears to be an extremely challenging problem, inasmuch as computational efficiency is of paramount importance. Indeed, their proper monitoring imposes stringent computing requirements with respect to core memory and execution time. Direct methods suitably adapted show particularly appealing and promising. This quite naturally explains the intensive research effort which has been devoted to this area for almost twenty-five years.

The present Chapter attempts to give an overview of how direct methods solve the above important practical problem, to bring out essential obtained results, to pinpoint the difficulties still existing, and to discuss some current and future research directions.

Two different methodologies are considered, namely the "scalar" and Belman's "vector" Liapunov approaches. The former is the first to having attracted researchers' attention. Its beginning, situated somewhere in the sixties, is marked by the construction of (scalar) Liapunov functions for power system dynamics. Several methods thus start being explored, such as : Zubov's method (Zubov, 1961), Popov's criterion (Popov, 1962), Moore and Anderson's theorem (Moore and Anderson, 1968), or simply calculation of a first integral of the equations of motion. Quite early, however, it appears that, whatever the method, constructing practical Liapunov functions imposes stringent simplifying assumptions for the system modelling. It also appears that in order to make direct methods practically attractive it is necessary to define "suitable" stability domain estimates, i.e. estimates combining accuracy and computational efficiency.

Thus, by the mid-seventies research efforts concentrate on the definition of Stability Domain Estimates, more workable and reliable than the "theoretical" one. Indeed, the latter imposes considerable computing burden, besides providing quite conservative practical results, which have moreover an uncertain degree of conservativeness. Section V.3 examines this question and shows that it has by now received satisfactory solutions. This is a crucial step towards the effective application of direct methods to power system stability analysis. Relaxing the direct methods as much as possible from simplifying assumptions is another important issue discussed in Section V.3. Promising approaches recently proposed are indicated.

Also by the mid-seventies, another research direction starts being investigated : Belman's approach and the derived "vector" Liapunov functions. This methodology aims essentially at tackling the problem of "how to relax the construction of Liapunov functions from severe simplifying assumptions". As will be shown in Section V.4, the "vector" approach has actually reached this objective but at the expense of an unacceptably high conservativeness; and despite nice theoretical results and improvements achieved, this drawback has not as yet been overcome.

N.B. This Chapter is not a survey. It does not intend to describe all contributions in the area and the references given at its end do not pretend do be complete, nor do they reflect the abondence of publications devoted to the subject. For a more comprehensive account, the reader may refer to existing survey publications (Ribbens-Pavella, 1971b; Fouad, 1975; Evans, 1978; Pai, 1981; Ribbens-Pavella and Evans, 1985a; Varaiya et al., 1985).

V.2. THE PHYSICAL PROBLEM AND ITS MATHEMATICAL MODELLING

N.B. The thrust of this Section is taken from Ribbens-Pavella and Evans (1985a).

V.2.1. Problem definition

This paragraph aims at giving the reader who is not familiar with power system dynamics the fundamental description of the physical system and of the conventional transient stability analysis. The power system engineer may go directly to paragraph V.2.2. The definitions given hereafter comply with the recommandations of CIGRE ("Conférence Internationale des Grands Réseaux Electriques", Barbier et al., 1978), and of IEEE Committee Report on Terms and Definitions (1982).

For the purposes of its transient stability analysis, an electric power system may be regarded as a set of electrical generating units (synchronous machines) and of loads interconnected through transmission lines (i.e. lines or transformers), including the associated equipment. Fig.1 gives a schematic representation of an N-machine system. The generators are characterized by their internal electromotive force "behind", i.e. in series with, their transient reactance; the electromotive force of say the i-th generator is represented by $\mathbf{E}_i = E_i\angle\delta_i$ where δ_i is its "rotor angle", i.e. the angle formed by an axis rotating with the i-th generator and an axis rotating at synchronous speed. All generators are interconnected through the network, which is characterized by its $\mathbf{Y}$ matrix, i.e. its admittance matrix reduced at

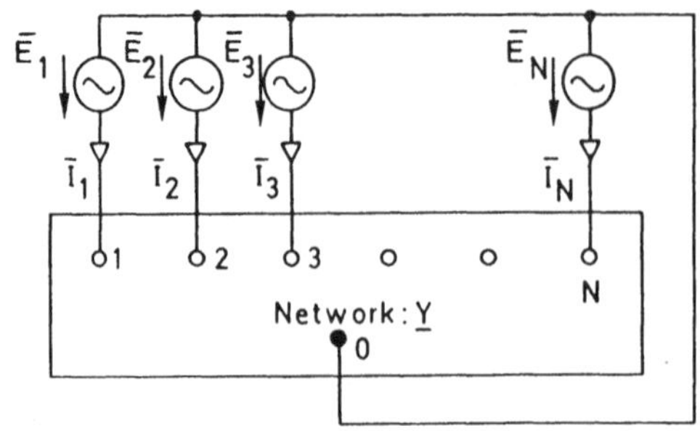

Figure 1

the internal generator nodes [1]. Note that Y accounts for the generator transient reactances, in addition to the line and transformer modelling and to the loads, modelled here as constant admittances.

For illustration purposes we have chosen a real power system comprising nine generating units. Fig. 2a describes this system along with its electric components and their values; Fig. 2b represents its corresponding network equivalent representation.

Definition 1. *A power system is said to be in a synchronous operation if all its interconnected synchronous machines are in synchronous operation with the network and with each other.*

All the system machines run then at the same electrical speed (product of its rotor angular velocity and the number of its pole pairs), the synchronous speed. During such a "steady-state" operation, the mechanical input power received by each generator equals, when neglecting its losses, the electrical output power delivered by it to the network. This equilibrium is upset when a large and sudden disturbance occurs.

Definition 2. *A large disturbance is a disturbance for which the equations that describe the dynamics of the power system cannot be linearized for the purpose of analysis.*

Such disturbances are : loss of loads, network significant changes such as short circuits, particularly three-phase short circuits at the generator busbars, and even more severe than that, loss of generators (or generators falling out of synchronism) which may be caused by one of the former large disturbances [2]. Occurrence of large disturbances imply significant changes of the Y matrix. It causes an imbalance between the mechanical input power to each generator and its electrical output power. The generator rotors start then "swinging" with respect to each other; the motion of, say, the i-th generator is then described by the Newton's law equation, referred in the literature as the "swing equation" (see Notation) :

---

[1] These are fictitious nodes taken behind the generator transient reactances.
[2] In the sequel the term "large" will be sometimes omitted.

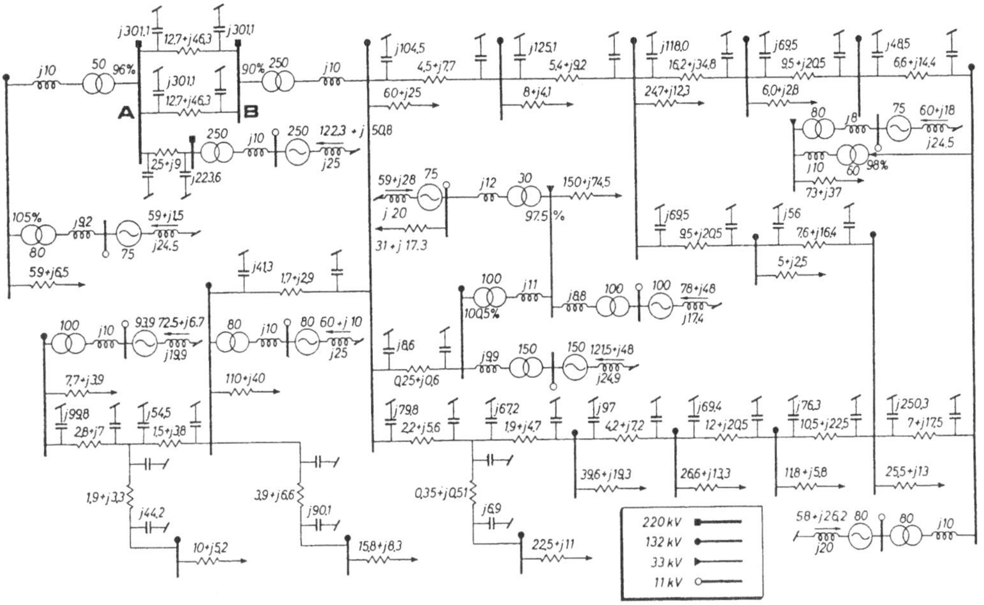

Figure 2.a

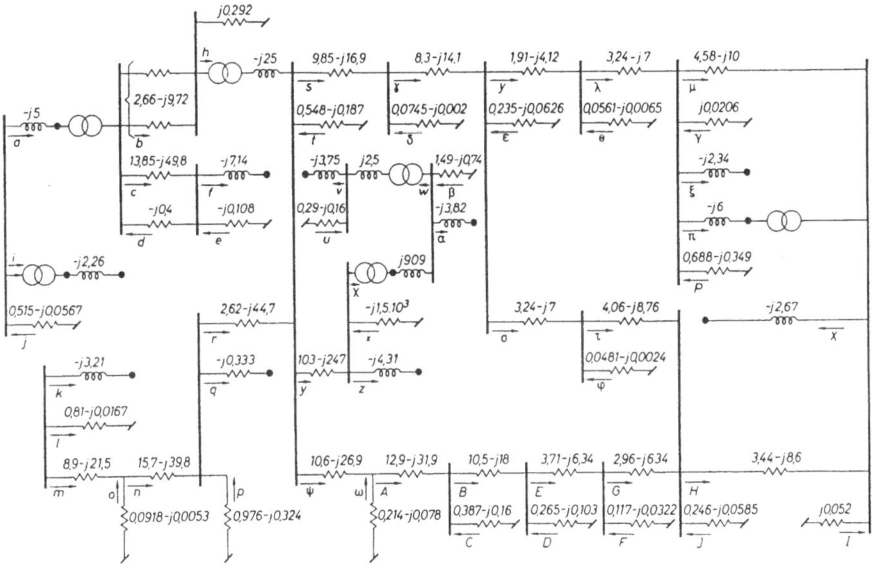

Figure 2.b

$$M_i\ddot{\delta}_i + D_i\dot{\delta}_i + \sum_{j=1}^{N} D_{ij}(\dot{\delta}_i - \dot{\delta}_j) = P_{mi} - P_{ei}(\boldsymbol{\delta}) \ . \tag{2.1}$$

Roughly speaking, the transient stability problem is the study of this dynamic system's capacity to recover, after a disturbance, a synchronous steady-state equilibrium condition. More precisely we give the following

Definition 3. *A power system is transiently stable for a particular (pre-fault) steady-state operating condition and for a particular disturbance if, following that disturbance, it reaches an acceptable steady-state operating condition.*

This definition suggests that the transient stability analysis implies consideration of three different operating phases (Figs.3) : the pre-fault (before occurrence of a disturbance), the fault-on (during disturbance action; one speaks of "faulted system"), and the post-fault one. Note also that the post-fault steady-state (synchronous) operating conditions that the system may reach long after clearing the disturbance do not necessarily (and generally) coincide with the pre-fault ones (see Fig. 3c).

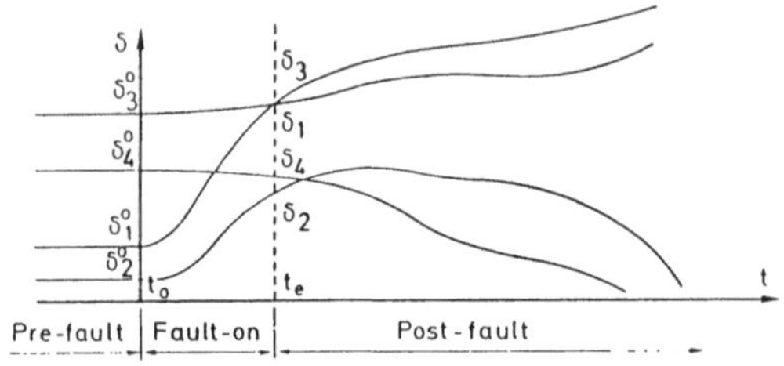

Figure 3.a

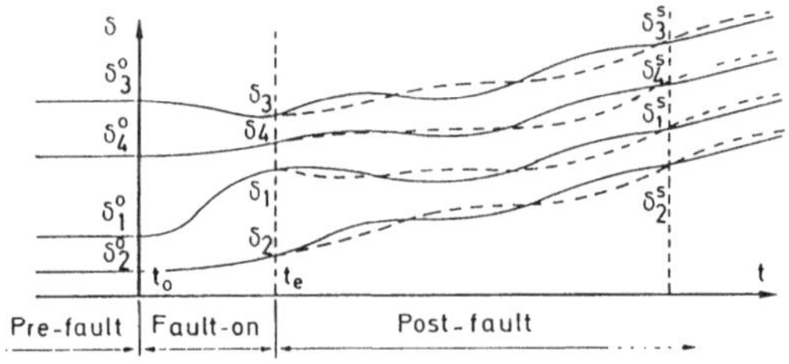

Figure 3.b

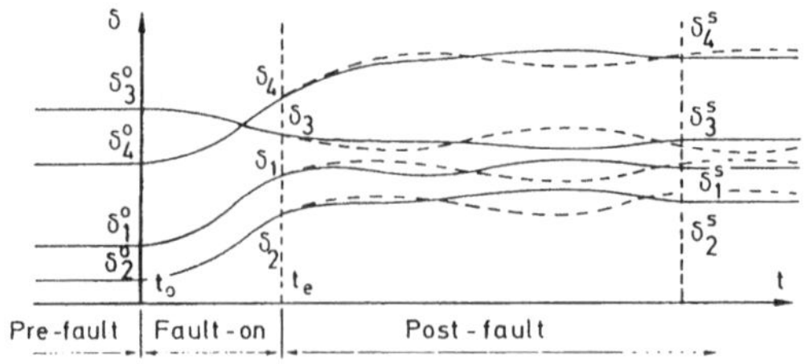

Figure 3.c

V.2.2. Conventional problem formulation

Transient stability analysis is very important in planning studies as well as in everyday system operation and monitoring. It is also an extremely complicated problem, for the rigorous solution of the above equations of motion would necessitate taking into account a large number of parameters which vary in time and are interdependent. As is usual in engineering problems, one then is led to make simplifying assumptions.

Whatever the approximations, the conventional transient stability problem is formulated in terms of Critical Clearing Time (CCT or t_c). More precisely, the standard exploration of a system's robustness from its transient stability point of view is performed through disturbance simulations (generally one at a time) and evaluation of the disturbance severity in terms of its CCT [1]. This conventional transient stability analysis may be formulated as follows : given a power system initially in a steady-state operation and assuming that a disturbance appears at t_o :

- is there a stable equilibrium position for the system after the disturbance is cleared ? and if yes,
- what is the critical clearing time ? i.e. the maximum time that the disturbance may remain without the system losing its *capability* to come back to a steady-state operation; and this time must lie within the capability of system protections and circuit breakers to clear the fault [1].

So, whatever the method, the conventional transient stability analysis has two steps (figs. 3) :

- *Step A* which studies the evolution of the (faulted) system from t_o to the clearing time t_e ;
- *Step B* which studies the evolution of the system in its post-fault (final) phase, from t_e onwards :
  - if the system is found to be unstable, t_e is greater than the critical clearing time t_c ,
  - if the system is found to be stable, $t_e < t_c$.

Both of the above two steps are investigated in the standard *step-by-step* method by numerically integrating the system swing equations of type (2.1) :[2]

[1] Admittedly this conventional "measure" of a system's robustness from its transient stability point of view is arguable for practical applications. It shows, however, extremely useful for the purpose of developing appropriate new theoretical methodologies; in a subsequent stage of the investigations, the new methods developed may be adapted to other, more appropriate stability "measures".

[2] Indeed, eq.(2.1) describes the generator motion for both the fault-on and post-fault periods, but with different parameters.

upon fixing a clearing time t_e , one runs the program first from t_o to t_e (step A), then one computes the *"swing curves"*, describing the rotor angles evolution with time from t_e onwards (step B). Determination of CCT requires several (generally 3 or more) trials of assumed values of t_e . Moreover, for a given t_e , computation of the swing curves of step B must be pursued for quite a long time before being able to decide whether the system shows a tendency towards stabilization or not. Obviously, the CCT computation by the standard step-by-step procedure appears to be quite a long and tedious process.

Remark 1. The kind of "stabilization" one should require of a power system depends upon the system modelling along with the related simplifying assumptions. Curves of Figs. 3b and 3c refer to four different cases. For example, with the simplified modelling described in Section 3.1 below and "uniform" mechanical damping, one can only expect from the system to possess the asymptotic stability of its *relative* rotor angles (see the solid line curves in Fig. 3b); if, moreover, all kinds of dampings are neglected, one cannot expect but the *simple* stability of the *relative* rotor motion (see the dotted line curves in Fig. 3b).

V.2.3. Definitions of stability domains and their estimates

For a while our attention will be focused on the difference between the notions "domain" and "region".

Referring to La Salle and Lefschetz (1961) a "region" is an open connected set. However, Santalo (1976) defined "domain" as an open and connected set, and "region" as the union of a domain with some, none, or all of its boundary points.

We want to emphasize that, for stability analysis of non-linear systems, only a neighbourhood (either open or closed or neither open nor closed) of the origin is of interest herein. Hahn (1967) used "domain" in this sense. The reason for using a neighbourhood that can be closed is that the domain of asymptotic stability of an equilibrium of a non-linear system can be closed (Grujić, 1976).

Let $\mathbf{x} \in R^M$, $\mathbf{f} : R^M \to R^M$ and

$$\dot{\mathbf{x}} = \mathbf{f}(\mathbf{x}) \tag{2.2}$$

Following Grujić (1975,1976), we accept :

Definition 4. A set $\mathcal{D}_s$, $\mathcal{D}_s \subseteq R^M$, is *the domain of the equilibrium state* $\mathbf{x}=0$ defined by

$$\mathcal{D}_s = \cup [\mathcal{D}_s(\epsilon) : \epsilon \in \mathring{R}_+] ,$$

where $\mathcal{D}_s(\epsilon)$ is such a neighbourhood of $x=0$ that $\|\chi(t;0,x_o)\| < \epsilon$ $\forall t \in R_+$, holds provided only that $x_o \in \mathcal{D}_s(\epsilon)$ for every $\epsilon \in \mathring{R}_+$.

The next definition has been commonly used (Krasovskii, 1963; Hahn, 1967; La Salle and Lefschetz, 1961).

Definition 5. A set $\mathcal{D}_a$, $\mathcal{D}_a \in R^M$, is *the domain of attraction of the equilibrium state* $x=0$ *of the system* (2.1) if and only if it is such a neighbourhood of $x=0$ that

$$\lim\ [\|\chi(t;0,x_o)\| : t \to +\infty] = 0$$

holds provided only that $x_o \in \mathcal{D}_a$. ∎

It is now natural to accept the definition of the domain of asymptotic stability of $x=0$ as proposed by Grujić (1975,1976,1981) :

Definition 6. A set $\mathcal{D}$, $\mathcal{D} \subseteq R^M$, is *the domain of asymptotic stability of* $x=0$ *of the system* (2.2) iff it is both a neighbourhood of $x=0$ and the intersection of its domain of stability and domain of attraction, that is, that $\mathcal{D} = \mathcal{D}_s \cap \mathcal{D}_a$ is a neighbourhood of $x=0$. ∎

The exact determination of the domain of asymptotic stability has great engineering and theoretical importance. Unfortunately, we can realize it only in special cases. For these reasons we investigate its estimate E defined as follows by referring to Grujić and Ribbens-Pavella (1977,1978), Grujić et al. (1979), Grujić (1981) :

Definition 7. A set E, $E \subseteq R^M$, is *an estimate set* (in brief, *estimate*) of the asymptotic stability domain $\mathcal{D}$ of $x=0$ of the system (2.2) if and only if

(i) E is a neighbourhood of $x=0$,

(ii) $E \subseteq \mathcal{D}$

and

(iii) E is positively invariant set of the system (2.2), that is, that $x_o \in E$ implies $\chi(t;0,x_o) \in E$ for every $t \in R_+$.

V.2.4. Liapunov's method applied to conventional transient stability analysis

The principle consists in replacing the computations implied by step B defined in Section V.2.2 *and their repetitions* by a time domain test. This in turn consists in comparing an initial Liapunov function value (Liapunov function value at the initial state) with its critical value (its value on the boundary of the stability domain). It thus requires construction of a Liapunov function and determination of the stability domain surrounding the Stable Equilibrium Point (SEP) of the system in its post-fault configuration. Stated differently, the Liapunov method applied to power system transient stability amounts to defining in the state space the domain of (asymptotic) stability for the post-fault SEP and to determining the intersection of system trajectory

To fix ideas, let (2.2) be the state model of the power system in its post-fault configuration, and let the state vector be composed of sub-vectors related to (some) rotor angles and to (some) rotor velocities

$$x = [\delta^T, \dot{\delta}^T]^T \ .$$

Further, let V be a Liapunov function used for (asymptotic) stability domain estimation and let V_ℓ be the value that it takes on the boundary of this domain. Then, evaluation of CCT amounts to computing the values of V along the system trajectory, i.e. for successive clearing times t_{e_i} , until reaching V_ℓ : the corresponding clearing time will be the CCT according to the Liapunov direct criterion. Fig. 4 describes this procedure; one sees that the CCT corresponds to t_{e2} . Note that for $t = t_{e3}$ the criterion does not guarantee system's stability anymore; but it does not guarantee instability either. One of the main sources of its conservative character lies in the well known fact that the stability theorems of the Liapunov direct method does not provide necessary and sufficient conditions for the form of the system Liapunov function [1].

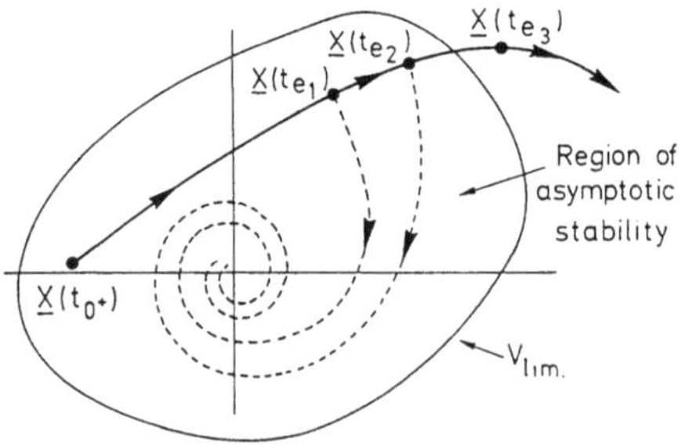

Figure 4

The practical procedure for computing CCT according to the Liapunov method is illustrated in Fig. 5 and summarized in the following three steps. For a given V function and disturbance assumed to start acting at t_o :

[1] Even though the existence of a Liapunov function is necessary and sufficient for uniform (asymptotic) stability (in the large) (Chapter I).

(i) compute the limit value, V_ℓ, that V takes on the boundary of its Stability Domain Estimate (SDE);

(ii) for successive clearing times t_{e_i} ($> t_o$), compute the corresponding values of the state variables; let $[\delta(t_{e_i}),\dot{\delta}(t_{e_i})]$ denote symbolically these latter;

(iii) compute the corresponding value $V[\delta(t_{e_i}),\dot{\delta}(t_{e_i})]$ at $t_{o+}, t_{e1}, \ldots$ until reaching V_ℓ :

$$t_{c\ell} \equiv CCT \quad \text{for} \quad V[\delta(t_{c\ell}),\dot{\delta}(t_{c\ell})] = V_\ell \,.$$

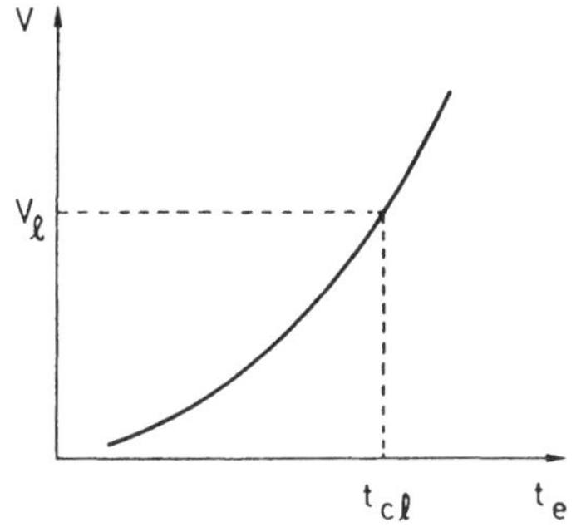

Figure 5

Remarks 2. a) It is obvious that this CCT is obtained with just one pass through integration of the faulted system equations.

b) It is also obvious that the success of the above approach depends essentially upon :

- the "quality" of the Liapunov function : it is important to construct the most "suitable" Liapunov function, V, for the physical system, i.e. that form of V which allows as refined as possible mathematical description while possessing the largest possible SDE;
- the speed of the numerical computation of this SDE (the time required for the computation of the V function being negligible in general).

V.2.5. System modelling

The modelling process has been dealt with extensively (Chorlton and Shackshaft, 1972; Dandeno, 1977). Models of (almost) any desired degree of precision can be specified for the generator and its controls. The equations of motion may be written in a number of reference frames - for example in terms of absolute angles, relative angles between machines, or relative to a "center of angle" for the system (Ribbens-Pavella, 1971b; Tavora and Smith, 1972; Evans, 1978; Athay et al., 1979a).

The swing equation of the i-th generator of an N-machine system is expressed by eq. (2.1). The expressions for P_{ei} are nonlinear functions of the generator angles and represent the internal transient behaviour of the synchronous machines, their mutual interactions, their damping effects (which are much more complex than velocity damping) and their response to excitation control action. Modelling with full accuracy becomes very complex.

It is certainly traditional for power system engineers to make "simplifying assumptions" according to the "need" of the problem and according to the time scale of interest. A good perspective on the time scale aspect is given in Quazza (1976). The effects of these assumptions have been well studied and compared with full scale tests (Chorlton and Shackshaft, 1972; Dandeno, 1977). For example, an indication of *"first swing" stability* [1] can still be got from the simplest assumptions (constant voltage behind transient reactance), mainly because the effects of initial acceleration and inertia tend to be dominant during this short period (up to about 1/2 second). However, the stability of subsequent swings ("resultant stability") is far more complex.

For everyday security monitoring purposes, simple assumptions become essential. Thus, e.g., for constant generator voltage behind transient reactance, P_{ei} is expressed by :

$$\begin{aligned} P_{ei}(\boldsymbol{\delta}) &= \mathrm{Re}\,[\bar{E}_i \bar{I}_i^*] = \mathrm{Re}\, \bar{E}_i \sum_{j=1}^{N} \bar{Y}_{ij}^* \bar{E}_j^* \\ &= E_i^2 Y_{ii} \cos\theta_{ii} + \sum_{\substack{j=1 \\ j\neq i}}^{N} E_i E_j Y_{ij} \cos(\delta_i - \delta_j - \theta_{ij}) \\ &= E_i^2 G_{ii} + \sum_{\substack{j=1 \\ j\neq i}}^{N} E_i E_j B_{ij} \sin(\delta_i - \delta_j) + E_i E_j G_{ij} \cos(\delta_i - \delta_j) \quad . \end{aligned} \tag{2.3}$$

Note that matrix $\mathbf{Y} = [\bar{Y}_{ij}] = [Y_{ij} \angle \theta_{ij}]$ changes with network topology and in particular during a fault, and for each different fault.

Another example is the consideration of first-order proportional governors; in this case P_{mi} present in eq. (2.1) is expressed by :

---

[1] The indication of "first swing" stability is available from examination of the swing curves of the most perturbed machines during say the first quarter cycle of their (pseudo-) sinusoidal oscillations. Note that the first swing stability is an essential consideration. In a survey conducted by the Power System Planning and Operation within the EPRI (Electric Power Research Institute) Gelopoulos and Laurent (1980) indicate that at least 85 % of the respondents look for first swing stability in more than 95 % of the cases they run.

$$\begin{aligned} P_{mi} &= P^{o}_{mi} + p_{mi} \\ \dot{p}_{mi} &= -\mu_i p_{mi} - \alpha_i \dot{\delta}_i \ . \end{aligned} \tag{2.4}$$

V.2.6. Mathematical formulation

Many state models may be built, depending upon the kind of assumptions made and the resulting physical modelling. Nevertheless, (some of the) salient intrinsic characteristics are present even in the less sophisticated state descriptions, as those considered below.

In eq. (2.1) describing the motion of the i-th generator, let us assume that

- P_{mi} and M_i are constant during the transients [1] ; constancy of P_{mi} implies that eqs. (2.4) degenerate to $P_{mi} = P^{o}_{mi}$;
- $D_{ij} = 0$;
- P_{ei} is expressed by eq. (2.3) with E_i , E_j constant during the transients [1] .

Deriving the appropriate state equations implies making the proper choice of state variables. This has extensively been explored in various ways by many authors; let us quote a few : energy considerations and physical system reduction (Di Caprio and Saccomanno, 1969, 1970); energetical and mathematical argumentation (Ribbens-Pavella, 1969, 1971b); controllability and observability concept (Sastry and Murthy, 1972a, 1972b); minimal realization aspects (Pai and Murthy, 1974). They all arrive at the conclusion that the state variables of concern here are relative rotor angles and absolute rotor speeds.

If, moreover, mechanical damping is either uniform or totally neglected, i.e. if

$$D_i = 0 \quad \text{or} \quad D_i M_i^{-1} = \lambda \ , \ \forall i = 1,2,\ldots,N \ , \tag{2.5}$$

then the state variables are relative rotor angles and relative rotor velocities as well. Thus, in both cases the motion of the overall system has to be described with respect to a reference frame. Such a possible reference is one of the system's generators; for example, one may choose arbitrarily the N-th generator to be this reference. Another possibility is to refer the system's motion to the so-called "Center Of Angle" (COA) defined by (Tavora and Smith, 1972; Athay et al., 1979a)

$$\delta_o = \left(\sum_{i=1}^{N} M_i \right)^{-1} \sum_{i=1}^{N} M_i \delta_i \ . \tag{2.6}$$

[1] i.e. constant during the fault-on and post-fault phases and equal to their corresponding pre-fault values.

From a theoretical viewpoint the choice between these above two frames of reference is immaterial; similarly, the choice among the N generators of the reference one is in principle indifferent. However, this may not be true from a practical viewpoint.

In the case where the mechanical dampings do not obey the particular conditions (2.5) and where N acts as the reference generator, eqs. (2.1) and (2.3) yield the following state equations :

$$\begin{aligned} \dot{\delta}_{iN} &= \Omega_i - \Omega_N , \qquad\qquad (i = 1,2,\dots,N) \\ \dot{\Omega}_i &= - D_i \Omega_i + M_i^{-1} \{P_{mi} - E_i^2 G_{ii} - \sum_{\substack{j=1 \\ j\neq i}}^{N} [E_i E_j B_{ij} \sin(\delta_{iN} - \delta_{jN}) \\ &\qquad + E_i E_j G_{ij} \cos(\delta_{iN} - \delta_{jN})]\} . \end{aligned} \tag{2.7}$$

The above system of equations may also be set in a matrix form. Indeed, denoting by $\mathbf{x}$ the (2N-1) state composite vector

$$\mathbf{x} = [\boldsymbol{\delta}^T, \boldsymbol{\omega}^T]^T$$

where

$$\boldsymbol{\delta} = [(\delta_{1N} - \delta_{1N}^s), \dots, (\delta_{nN} - \delta_{nN}^s)]^T$$

$$\boldsymbol{\Omega} = [\Omega_1, \Omega_2, \dots, \Omega_N]^T$$

and where δ_{iN}^s is the i-th component of the SEP (Stable Equilibrium Point), one arrives at the following formulation (e.g. see Henner, 1974) :

$$\begin{aligned} \dot{\mathbf{x}} &= \mathbf{A}\mathbf{x} = \mathbf{B}\tilde{\mathbf{f}}(\tilde{\boldsymbol{\sigma}}) \\ \boldsymbol{\sigma} &= \mathbf{C}\mathbf{x} \end{aligned} \tag{2.8}$$

In the above (Henner, 1974)

$$\tilde{\sigma}_k = \delta_{ij} - \delta_{ij}^s \quad , \quad k = (i-1)N - \frac{i(i+1)}{2} + j \quad , \quad i < j \quad ,$$

$\tilde{\mathbf{f}}(\tilde{\boldsymbol{\sigma}})$ is a vector-valued function with dimension $m = Nn/2$, whose k-th component is

$$f_k(\sigma_k) = E_i E_j [B_{ij} \sin(\sigma_k + \delta_{ij}^s - \sin\delta_{ij}^s) + G_{ij} \sin(\sigma_k - \delta_{ij}^s - \cos\delta_{ij}^s)] ,$$
$$(k = 1,2,\dots,m) , \tag{2.9}$$

$\mathbf{A}$, $\mathbf{B}$, $\mathbf{C}$ are appropriate constant matrices.

The above equivalent formulations (2.7) and (2.8) suggest two different types of Liapunov function constructions, namely : analytic calculation of first integrals of the eqs. (2.7), and Lur'e type Liapunov function for the system (2.8). But both face the same difficulty and circumvent it in the same way, viz. to construct a Liapunov function they require neglect of transfer conductances of the physical system.

Formally, this means that the coefficients G_{ij} present in (2.7) and (2.8) have to be set equal to zero. Indeed, as soon as $G_{ij} = 0$ in eq. (2.9), the function $\tilde{f}_k(\tilde{\sigma}_k)$ obeys the sector condition

$$\tilde{f}_k(\tilde{\sigma}_k)\,\tilde{\sigma}_k > 0 \qquad k = 1,2,\dots,m \tag{2.10}$$

for some values $\tilde{\sigma}_k$ about $\tilde{\sigma}_k = 0$:

$$\tilde{\sigma}_{k\,min} < \tilde{\sigma}_k < \tilde{\sigma}_{k\,max} \;;$$

Hence, eqs. (2.8) take on the standard Lur'e - Postnikov type form with multiple nonlinearities and satisfy the generalized Popov criterion of Moore and Anderson (1968). A Lur'e type Liapunov function derives then readily :

$$V(x) = x^T P x + \int_0^{\tilde{\sigma}} f(\tilde{\sigma})\, Q\, d\tilde{\sigma} \,. \tag{2.11}$$

A similar reasoning, developed below, shows that neglect of transfer conductances seems necessary for devising any energy type Liapunov function whatever the construction procedure used.

V.3. SCALAR LIAPUNOV APPROACH

V.3.1. Preliminaries

As has been mentioned, from the very beginning of their development, direct methods have exhibited attractive features but also serious difficulties. Their attractiveness lies mainly in the possibility to provide a synthetic answer to the stability problem without simulating the whole transient. The difficulties are of two types :

(i) The first is implied by the very construction of Liapunov functions which imposes quite simplified system description (at least in the multimachine case). The most stringent among them have been mentioned in Section V.2.5.

(ii) The second is related to the inability of the classical stability theory to provide reliable and computationally efficient SDEs; better means for evaluating SDEs are needed :

- in order to overcome the conservativeness that the *theoretical* SDE confers on the Liapunov criterion;
- in order to speed up significantly its computation which, otherwise counterbalances the computational advantage of the direct criterion over the numerical integration procedure.

Both of the above two types of difficulties along with possible means to alleviate them are examined in this Section. For reasons which will appear below, the so-called "energy type" V function is chosen as the basis of our reasoning and is studied first.

V.3.2. The "energy type" Liapunov function

This function is devised in the most simple case where E_i, P_{mi}, M_i are assumed constant, mechanical dampings obey eq. (2.5), and electromagnetic dampings are zero.

V.3.2.1. State model

Choosing arbitrarily the N-th generator as the reference, we get the state vector

$$\mathbf{x} = [\delta_{1N}, \delta_{2N}, \ldots, \delta_{nN}, \omega_{1N}, \omega_{2N}, \ldots, \omega_{nN}]^T = [\boldsymbol{\delta}^T \boldsymbol{\Omega}^T]^T \tag{3.1}$$

We shall symbolically denote it by $(\boldsymbol{\delta}, \boldsymbol{\Omega})$.

State equations derive then readily from (2.1) and (2.3) :

$$\begin{aligned} \dot{\delta}_{iN} &= \Omega_{iN} \\ \dot{\Omega}_{iN} &= -\lambda\Omega_{iN} + [P_{mi} - P_{ei}(\boldsymbol{\delta})]\, M_i^{-1} - [P_{mN} - P_{eN}(\boldsymbol{\delta})]\, M_N^{-1} \\ &= -\lambda\Omega_{iN} + f_i(\boldsymbol{\delta}) \qquad i = 1,2,\ldots,n \end{aligned} \tag{3.2}$$

The solutions of (3.2) are given by

$$\begin{aligned} \Omega^o_{iN} &= 0 && \text{(a)} \\ & && i = 1,2,\ldots,n \\ f_i(\boldsymbol{\delta}^o) &= 0 && \text{(b)} \end{aligned} \tag{3.3}$$

They are therefore obtained by solving the n nonlinear algebraic eqs. (3.3.b).

In what follows, we will be led to consider two types of solutions (equilibrium points) of the system in its post-fault configuration :

- the stable equilibrium point (SEP) of concern - if any; let $[\delta^s_{1N}, \delta^s_{2N}, \ldots, \delta^s_{nN}, 0, 0, \ldots, 0]^T$ be its components; they will be symbolically denoted by $(\boldsymbol{\delta}^s, \mathbf{0})$;
- unstable equilibrium points (UEP), surrounding the SEP; the i-th among these UEPs will be symbolically denoted by $(\boldsymbol{\delta}^{ui}, \mathbf{0})$.

V.3.2.2. Liapunov function

It is possible to construct a Liapunov function for system (3.2), by integrating the state equations (3.2), provided that conditions (3.4) below are satisfied

$$\frac{\partial f_i}{\partial \delta_\kappa} = \frac{\partial f_\kappa}{\partial \delta_i} \qquad i,\kappa = 1,2,\dots,n \tag{3.4}$$

This in turn implies neglecting (some of) the transfer conductances, i.e. setting equal to zero (some of) the expressions G_{ij} :

$$G_{ij} = Y_{ij} \cos \theta_{ij} = 0 \ . \tag{3.5}$$

Of course, conditions (3.4) are always verified when setting

$$\theta_{ij} = \frac{\pi}{2} \qquad (i \neq j) \ . \tag{3.6}$$

With this assumption, integration of eqs. (3.2) yields after some manipulations (e.g. see Ribbens-Pavella, 1969, 1971a) :

$$\begin{aligned} V(\boldsymbol{\delta},\boldsymbol{\Omega}) &= \sum_{i=1}^{n} \sum_{j=i+1}^{N} \{\tfrac{1}{2} M_i M_j \Omega_{ij}^2 - (P_i M_j - P_j M_i)(\delta_{ij} - \delta_{ij}^s) \\ &\qquad - (\sum_{i=1}^{N} M_i)\, E_i E_j B_{ij} (\cos \delta_{ij} - \cos \delta_{ij}^s)\} \\ &= V_\kappa(\boldsymbol{\Omega}) + V_p(\boldsymbol{\delta}) \end{aligned} \tag{3.7}$$

The derivative of $V(\boldsymbol{\delta},\boldsymbol{\Omega})$ is then expressed by :

$$\dot{V} = -\lambda \sum_{i=1}^{n} \sum_{j=i+1}^{N} M_i M_j \Omega_{ij}^2 \ .$$

One easily verifies that this V function is indeed a Liapunov function in a certain domain surrounding the SEP $(\boldsymbol{\delta}^s,\mathbf{0})$, since it verifies the conditions

(a) $V(\boldsymbol{\delta}^s,\mathbf{0}) = 0$;

(b) $V(\boldsymbol{\delta},\boldsymbol{\Omega})$ is positive definite in this domain;

(c) $\dot{V}(\boldsymbol{\delta},\boldsymbol{\Omega})$ is negative semi-definite everywhere for $\lambda \neq 0$ $(\lambda > 0)$, or $\dot{V} \equiv 0$ for $\lambda = 0$.

Hence, the above V function guarantees the system asymptotic stability when $\lambda > 0$, or the simple stability when $\lambda = 0$. On the other hand, the "size" of the stability domain does not depend on $\dot{V}$.

Remarks 3. 1) The above V expression contains two groups of terms : one of the "kinetic energy" type, and one of the "potential energy" type.

2) With the above construction, presence of uniform damping reinforces the system stability (asymptotic instead of simple) but does not allow enlarging the stability domain, as might be physically expected.

3) The V function (3.7) may be considered as "optimal" in the sense that it determines the largest stability domain estimate for the considered system modelling.

4) One of the prices to be paid for the advantage of the

Liapunov's "global" stability assessment, is the inability of the method to account for transfer conductances which may have a significant effect. Their influence comes from the fact that the admittance matrix used here is the one reduced at the generator nodes; this reduction amounts to introducing (some of the) effects of the loads connected to other than generator nodes in the off-diagonal terms of the Y matrix (e.g. see Appendix in Ribbens-Pavella and Evans, 1985a).

5) Neglecting G_{ij} does not modify the form of eqs. (3.2) and (3.3) but implies replacing therein P_{ei} (2.3) by

$$P_{ei} = E_i^2 G_{ii} + \sum_{j=1}^{N} E_i E_j B_{ij} \sin(\delta_i - \delta_j) \ . \tag{3.8}$$

If the terms containing G_{ij} were not neglected, an additional integral term (3.9) would have appeared in (3.7), which is not analytically computable (Evans, 1978) :

$$I(\boldsymbol{\delta}) = \sum_{i=1}^{n} \sum_{j=i+1}^{N} \int_{\delta_{ij}^{s}}^{\delta_{ij}} \{M_j [\sum_{k \neq i}^{N} E_i E_k G_{ik} \cos \delta_{ik}] - M_i [\sum_{k \neq j}^{N} E_k E_j G_{jk} \cos \delta_{jk}]\} \, d\delta_{ij} \ . \tag{3.9}$$

V.3.2.3. Stability domain estimates

V.3.2.3.1. Theoretical determination of $V_{lim} : V_{Th}$. The problem of determining a SDE has been investigated in various ways (e.g. see El-Abiad and Nagappan, 1966; Willems and Willems, 1970, Willems, 1971, 1974). Here we shall follow a geometrical reasoning. With V given by (3.7) and for $\dot{V} = 0$ [1], the curves $V(\boldsymbol{\delta}, \boldsymbol{\Omega}) = C$ (>0) in the state space are, for small values of C, closed hypersurfaces, surrounding the SEP. As C increases, the hypersurfaces "swell" while remaining closed; the surface stops being closed as soon as it meets a multiple point, that is a point for which

$$\frac{\partial V}{\partial \delta_{iN}} = \frac{\partial V}{\partial \Omega_{iN}} = 0 \qquad (i = 1, 2, \dots, n) \ .$$

Now, this multiple point is by construction a UEP, solution of eqs. (3.3).

The practical procedure for computing V_{Th} is summarized as follows :

(a) compute *all* UEPs surrounding the SEP of concern for the system in its final configuration;

[1] For $\dot{V}$ negative semi-definite, the same conclusion is reached.

(b) define V_{Th} as being the value of the V function on the boundary of the closed hypersurface surrounding SEP :

$$V_{Th} \triangleq \min_i V(\delta^{ui}, 0) \qquad (3.10)$$

where $(\delta^{ui}, 0)$ represents the so-called "closest" (in the sense of of (3.10)) UEP to the SEP.

The computations implied by this "theoretical" SDE are obviously extremely heavy, for the number of UEPs to be explored is a priori unknown - and so is their location. And because of the highly nonlinear character of eqs. (3.3b) which have to be solved, any means capable of providing a "first guess" about this information may be extremely useful [1].

The practical procedure suggested below is another means to circumvent these difficulties.

Fig. 6 illustrates the above considerations on the basis of a realistic 3-machine system. (It derives from the reduction of a real power system.) The "equi -V" curves have been plotted in the state plane $(\delta_{13}, \delta_{23})$ for $\Omega_{13} = \Omega_{23} = 0$. Observe that they are organized around δ^s (i.e. the SEP). Incidentally, $V_{Th} = 96$.

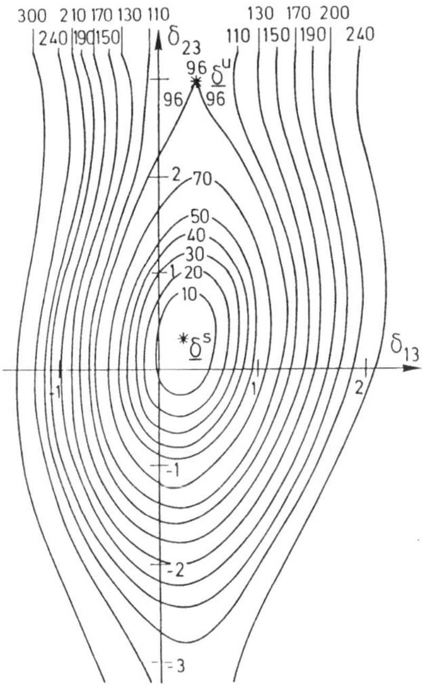

Figure 6

[1] From physical arguments one can propose as many as 2^n *starting points* for solving eqs. (3.3b). This may also be inferred from exploration of the sector conditions (2.10).

V.3.2.3.2. Practical computation of $V_{lim}: V_\ell$. Suggested by Ribbens-Pavella (1975), it is summarized as follows.

1°) Consider all *"type 1"* starting points; these are points corresponding to the physical situation where one machine goes unstable with respect to the remaining system. For instance, the j-th type 1 starting point is expressed by :

$$\boldsymbol{\delta}^{sj} = [\delta^s_{1N}, \delta^s_{2N}, \dots, \delta^s_{j-1,N}, \pi - \delta^s_{jN}, \delta^s_{j+1,N}, \dots, \delta^s_{nN}]^T \ . \tag{3.11}$$

2°) A good approximation of V_{Th} will then be the value

$$V_\ell = \min_{j=1,2,\dots,n} V(\boldsymbol{\delta}^{sj}, \mathbf{0}) \tag{3.12}$$

Justification of the above procedure is based on physical argumentation (Ribbens-Pavella, 1975). Its use contributes to speed up considerably the computation of an approximate value of V_{Th} . Obviously, however, it does not alleviate its conservative character.

V.3.2.4. Alleviating the conservativeness of Liapunov's criterion : PSDEs

Among the many approaches proposed within this context, those recognizing the "relevant" UEP as related to the fault location and to the practical SDE (PSDE) have certainly contributed to make the Liapunov method truly reliable and effective. They came about in the late seventies. In what follows, we describe two of them which are based on sound justifications, while exhibiting interesting practical features.

V.3.2.4.1. The Kyoto approach. For $\lambda = 0$, $\dot{V}(\boldsymbol{\delta}, \boldsymbol{\Omega})$ is identically zero. Conjecturing that the trajectory of a sustained disturbance "almost" coincides with the critically cleared one, Kakimoto et al. (1978,1980, 1981) propose to identify the crossing of the faulted trajectory with the PSDE. Thus, they come up with the following procedure.

(i) At successive clearing times, t_e , compute the value of $V[\boldsymbol{\delta}(t_e), \boldsymbol{\Omega}(t_e)]$ and of its potential part $V_p[\boldsymbol{\delta}(t_e)]$, until reaching a maximum of the latter. This will be considered as the limit value, V_k , taken by V on the boundary of the PSDE :

$$V_k = V_{p\,max}(\boldsymbol{\delta}) \ ; \tag{3.13}$$

(ii) The CCT, t_k , will accordingly be found at the intersection of $V[\boldsymbol{\delta}(t_e), \boldsymbol{\Omega}(t_e)]$ with V_k :

$$V[\boldsymbol{\delta}(t_k), \boldsymbol{\Omega}(t_k)] = V_k \ . \tag{3.14}$$

Fig. 7 illustrates this procedure.

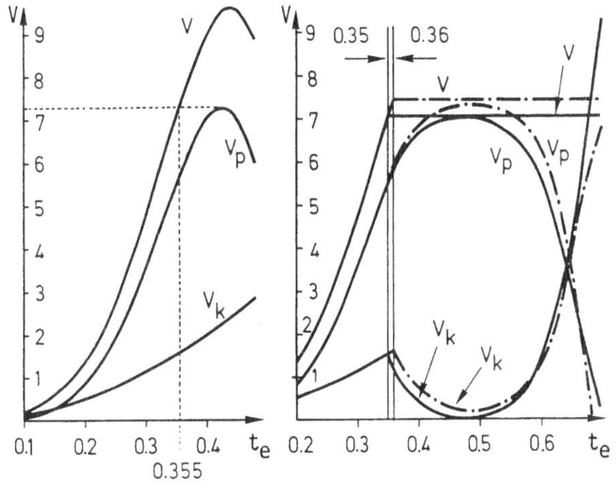

Sustained fault : —— : *fault cleared at 0.35 s.*

CCT = 0.355 s. — · — : *fault cleared at 0.36 s.*

Figure 7

Principle of the Kyoto's method for assessing CCT.
Inspired by Kakimoto and Hayashi, 1981, and applied to a 3ØSC at GB #5 of the 7-machine system.

V.3.2.4.2. The acceleration approach. The underlying conjecture is that the interesting UEP is related to the machine which the first would go out of step, should the considered disturbance be sustained. Moreover, this UEP is assumed to be very close to its corresponding starting point (SP) of form (3.11), so that the value of the V function at SP is a good approximation of its value at UEP. So, this method essentially requires identification of some "relevant" machines. This is described below. For the time being let us assume known what we denote symbolically by "fastest" and "slowest" machines. Then the acceleration approach consists of the following two steps :

(i) At t_{o+} , i.e. immediately after the considered disturbance inception, determine the "fastest" and "slowest" machines. Denoting them respectively by i and N , define the "interesting" UEP to be given by (3.11) and compute accordingly the limit value V_a :

$$V_a = V(\boldsymbol{\delta}^{si}, 0) \tag{3.15}$$

(ii) At successive clearing times, t_e , compute the value of V , $V[\boldsymbol{\delta}(t_e), \boldsymbol{\Omega}(t_e)]$, until reaching V_a ; the CCT, t_a , is given by

$$V[\boldsymbol{\delta}(t_a), \boldsymbol{\Omega}(t_a)] = V_a \tag{3.16}$$

Remarks 4. 1) When a cluster of machines, instead of a single one, are shown to be (almost) equally "fast", the interesting UEP of concern is no longer the SP of form (3.11), of "type 1", but rather a "type higher than 1" appropriately expressed.

2) The identification rules of the "fastest" and "slowest" machines rely on information provided by the values of the accelerations that the system machines acquire at t_{o+} , i.e. immediately after a disturbance inception. The details of the procedure are given in

Toumi et al. (1985). Let us only mention here that, roughly speaking, the "fastest" machine corresponds to that possessing the largest magnitude of initial accelerations, whereas the "slowest" to that with the minimum initial acceleration.

V.3.2.5. *Practical improvements of the energy type function*

V.3.2.5.1. Approximate account of transfer conductances. An approximate way of accounting for transfer conductances consists of adding to the eq. (3.7) the non-analytically computable term (3.9), numerically evaluated along the fault trajectory. The obtained function is no longer a real Liapunov function. From a practical point of view, however, this so-called "Transient energy function" shows very interesting (Athay et al., 1979a, 1979b; Kakimoto et al., 1978, 1980).

V.3.2.5.2. A speeding up procedure. Under certain conditions it is possible to speed up the computation of $\boldsymbol{\delta}(t_e)$, $\boldsymbol{\Omega}(t_e)$ and hence of $V[\boldsymbol{\delta}(t_e),\boldsymbol{\Omega}(t_e)]$ over the fault trajectory, by using a Taylor series expansion instead of the step-by-step method. Indeed, expressing $\delta_i(t_e)$ and $\Omega_i(t_e)$ about t_{o+}, i.e. immediately after a disturbance inception, one gets (Ribbens-Pavella et al., 1976, 1977)

$$\begin{aligned} \delta_i(t_e) &= \delta_i^o + \frac{1}{2}\gamma_i t_e^2 + \frac{1}{24}\ddot{\gamma}_i t_e^4 + \dots \\ \Omega_i(t_e) &= \gamma_i t_e + \frac{1}{6}\ddot{\gamma}_i t_e^3 + \dots \end{aligned} \qquad (3.17)$$

where γ_i denotes the acceleration of machine i at t_{o+}.
This procedure may be shown quite interesting in practice.

V.3.2.6. *The individual energy type function*

Rather than assessing system's stability through global information contained in the V function, the Iowa group proposes to rely on information contained in an individual function. The leading idea is to attribute the system separation which would cause (if ever) instability, to the motion of a certain "critical" machine, say the i-th. Thus, Michel et al. (1983), Fouad et al. (1984) derive the individual energy type function by taking that part of functions (3.7) + (3.9) which corresponds to this machine.

Initially devised with respect to COA, the individual function can also be constructed with N as the reference machine. One gets

$$V_{iN} = \frac{1}{2} M_i M_N \Omega_i^2 - P_{iN}(\delta_{iN} - \delta_{iN}^s) + \\
+ M_N E_i \sum_{\substack{i=1 \\ j \neq i}}^{N} \int_{\delta_{iN}^s}^{\delta_{iN}} E_j [B_{ij} \sin \delta_{ij} + G_{ij} \cos \delta_{ij}] \, d\delta_{iN} \\
- M_i E_N \sum_{j=1}^{N} \int_{\delta_{iN}^s}^{\delta_{iN}} E_j [B_{Nj} \sin \delta_{Nj} + G_{Nj} \cos \delta_{Nj}] \, d\delta_{iN} \tag{3.18}$$

V.3.2.7. *Taylor based energy functions*

Interesting analytic developments of the above Taylor series expansion are obtained when truncating eqs. (3.17) after the second right-hand term for δ_i and consequently after the first one for Ω_i. Indeed, this allows relating linearly rotor angle differences; one gets readily

$$(\delta_{ik} - \delta_{ik}^o) = (\delta_{k\ell} - \delta_{k\ell}^o) \frac{\gamma_{ik}}{\gamma_{k\ell}} \quad ; \quad \Omega_{ik} = \Omega_{k\ell} \frac{\gamma_{ik}}{\gamma_{k\ell}} . \tag{3.19}$$

In turn, this approximation makes analytically integrable the terms (3.9) which account for transfer conductance effects. Moreover, functions (3.7) + (3.9) and (3.18) may now be formulated in terms of any single state variable. Further, evaluation of the PSDEs becomes extremely flexible and straightforward (Toumi et al., 1985).

The above approximation shows extremely interesting within its validity range, i.e. essentially for small t_e's (see also the Closure of the above reference).

V.3.2.8. *Direct criteria for real time operation*

The combination of the above four types of functions (viz. the pure energy type V function (3.7), the transient energy function (3.7) + (3.9), the individual type (3.18) and the Taylor-based function), along with the two PSDEs (viz. the Kyoto and the acceleration approaches), yields a large number of variants. Details of the many contributions along with their pros and cons on the basis of their performances with respect to accuracy, reliability and computational requirements are discussed in Toumi et al. (1985). Some salient conclusions are brought out and summarized in Section V.3.5.

V.3.3. Family of the "energy type" V functions

In this Section we give a very short account of some methodologies used to construct Liapunov functions of the energy type or similar to it.

V.3.3.1. *Calculating analytically first integrals of the state equations*

Two different approaches may be distinguished. The first is not a Liapunov technique stricto sensu but rather intuitive, physically sound one evolving purely energy considerations. Initiated by Magnusson (1947), Aylet (1958) and later on by Gorev and the Soviet School (Union Institute of Scientific and Technological Information and the Academy of Sciences of the U.S.S.R., 1971)[1], this technique consists in applying implicitly the Liapunov direct method. Indeed, these authors "rebuild" the direct criterion through calculation of the system "kinetic" and "potential" energies and consideration of their variations and signs. The lack of the numerical analysis tools and of computational facilities available today, prevented these pioneers from fully succeeding in elaborating effective practical stability tools. Nevertheless, their works remain exemplary for clarity, and give a thorough physical insight.

The second approach concerns the construction of Liapunov functions through definite first integrals of the system state equations, as mentioned in Section V.3.2 : function (3.7), transient energy function (3.7) + (3.9), individual type function (3.18), Taylor based type and their variants.

V.3.3.2. *The generalized Popov criterion*

This criterion is certainly an attractive technique for generating Liapunov functions. Credit for applying it for the first time to the multimachine case goes to J.L. Willems (1970b), and to J.L. Willems and J.C. Willems (1970). These authors used Moore and Anderson's theorem for generalizing the Popov frequency criterion to feedback with multiple nonlinearities.

The Popov criterion leading to Lur'e type "quadratic form plus multi-integral" V functions for studying multi-input - multi-output system stability has also been exploited almost simultaneously in 1974 by Henner (1974) and by Pai and Murthy (1974). The same V function (3.7) is devised once again in the uniform damping case. Inclusion of non-uniform mechanical damping has also been envisaged by the above authors. This has theoretical more than practical appealing, since it does not allow concluding about the size of the SDE; moreover the values of mechanical damping constants are generally difficult to assess.

[1] Western scientists are not very familiar with Russian works in the field : apart from some private translations, these works are generally published in Russian only.

Extension of the Moore and Anderson generalization of Popov's criterion has been used by Kakimoto et al. (1978,1980) to encompass systems multi-argument nonlinearities. Machine model of third (rather than second) order is incorporated so as to consider field flux decay. The resulting Lur'e type V function has an additional term representing the effects of this flux decay. The authors show that a new type of instability arises with the vanishing of the stability region due to flux decay, but this suggests that the inclusion of automatic voltage regulators (AVRs) would give a more realistic picture if flux decay is to be modelled; or perhaps that constant electromotive forces behind transient reactances may be a satisfactory assumption after all - in which case the generated Liapunov function comes back to the form (3.7). Moreover, the authors specify that their method allows accounting for AVRs only when the time constants lie within a limited range of values.

V.3.3.3. *Approximating transfer conductance effects*

Kitamura et al. (1977) use the perturbation method in order to incorporate transfer conductance effects by means of small parameters. The obtained Liapunov function contains a series of terms where the small parameters reflect some of the transfer conductance effects. Its systematic application is however difficult.

Saeki et al. (1983) regard transfer conductances as "external feedback loops". These act as sort of disturbances of the main part of the system which does not contain transfer conductances and for which a Liapunov function is constructed in the usual way. Further investigations could make effective this original contribution.

V.3.3.4. *The structure preserving model*

Another original approach allowing inclusion of the most important part of the system transfer conductances was proposed by Bergen and Hill (1981). The interesting leading idea is the consideration of a "structure preserving model" where the loads are explicitly retained instead of absorbed in the admittance matrix reduced at the generator nodes. In this case, neglecting the transfer conductances amounts to neglecting the resistances of the transmission lines only, which are generally small compared with reactances.

To preserve the unreduced network topology, the authors consider a frequency dependent load (FDL) modelling, where the load variation with frequency is taken to be linear about nominal frequency. The "augmented network" thus resulting comprises $(N+n_o)$ buses, where n_o is the total number of the network buses; it necessitates $(2N-1+n_0)$ state variables for describing its dynamics in the non-uniform damping case.

For the above "structure preserving model", an energy type Liapunov function is once again constructed by Bergen and Hill (1981).

The unreduced network idea has been exploited by Athay and Su (1981) to incorporate static, nonlinear load modelling. A new TE function is derived for this interesting model which, however, is a pseudo-Liapunov function only. In fact, it is a positive definite function but its derivative is not a priori negative (semi-) definite : the sign of the latter has to be examined numerically along the trajectory. This, however, may impose heavy computations.

The same unreduced system's idea has recently been used by Narasimhamurthi and Musavi (1984) who model the active power load as a function of voltage and local frequency and the reactive power load as a function of voltage.

Overall, this nice idea shows potential. Very likely, it is going to receive many developments which sooner or later will arrive at interesting practical achievements (see also the survey by Varaiya et al., 1985).

V.3.3.5. The Hamiltonian approach

Proposed by Kumagai and Wu (1982), this approach uses a nonlinear dynamic circuit model and derives a Hamiltonian formulation for the transient stability analysis of power systems. The method arrives again at the structure preserving model with $(N+n_0)$ buses so as to keep the loads which, however, are represented by a power demand with or without FDL representation and constitutes an interesting contribution.

V.3.4. The Zubov method

This method is a priori appealing because of the possibility it offers to generate Liapunov functions together with their corresponding SDE boundaries. Applied to power systems, the method allows in addition relaxation of current simplifying assumptions, such as accounting for transfer conductances. However, its inherent drawbacks have till now prevailed over the above advantages and made the method unattractive, if not inapplicable to large-scale systems, restricting its interest to the case of "one-machine-infinite-bus" system.

Indeed, in the conventional Zubov method (Yu and Vongasuriya, 1967), a Liapunov function is constructed in a (truncated) series form and its corresponding boundary is estimated. However, this latter does not approach the true boundary monotonically; thus, increasing the number of the series' terms does not guarantee improvement of the stability boundary. This disadvantage is cleared in the generalized Zubov method used by Prabhakara et al. (1974) where the stability boundary is obtained in a closed form. Another difficulty arises in this latter case, however, related to the choice of an appropriate transformation of the variables. And apparently, this difficulty has not yet been overcome in the multimachine case.

V.3.5. Numerical simulations

The simulations reported hereafter are a sample of those performed by the Liège group. Some of them have been published by Toumi et al. (1985), some others by Ribbens-Pavella et al. (1985b). For more "electrical" details the reader may refer to the above publications.

V.3.5.1. Simulation description

The systems. Eight systems have been investigated comprising real (or reduced versions of) power networks along with some test-systems so as to cover a wide range of system characteristics. They are identified as follows

* 3-machine test system
* 6-machine Tunisian EHV simplified power system
* 7-machine CIGRE system
* 9-machine power system
* 14-machine Greek EHV simplified power system (N.B. The configuration used here corresponds to its situation in the early seventies)
* 15-machine test system
* 17-machine Iowa system, reduced version of the network of the State of Iowa
* 40-machine Belgian EHV simplified power system.

Information about data of the above systems may be obtained upon request.

The disturbances. The first group of disturbances is of the standard three-phase short circuit (3ØSC) type. Their location concerns almost always Generators' Bus-bars (GB), apart from a few exceptions relative to Other than generators' Bus-bars (OB).

A second group of disturbances has also been considered; it concerns line trippings and load sheddings.

The results. The results relative to 3ØSC simulations are commented in Section V.3.5.2 and summarized in Tables I to III. Results relative to other types of disturbances are reported in Section V.3.5.3.

The standard step-by-step method is used as the benchmark in our comparisons. Its CCTs have been assessed through numerical integration of the generators' swing curves pursued up to 1.50 sec. All CCTs are expressed in sec. They correspond to the stable (lower bound) case.

A dozen of practical methods or variants have been investigated, seven of which appear in the Tables hereafter.

The simulation results sought to explore two essential methods' qualities : accuracy and computational efficiency. The accuracy of a method is assessed through comparisons of CCTs provided by it and by the step-by-step one, under the same simulation conditions and, whenever possible, with the same system modelling. Evaluating computational efficiency is somewhat more hazardous; an approximate first estimation is only given in what follows.

V.3.5.2. Results obtained with 3ØSC type disturbances

V.3.5.2.1. Organization of Tables I, II and III.

Table 1. It collects information relative to reliability and accuracy aspects. It is organized in the columns numbered 1 to 11 as follows.

1 : disturbance location : unless otherwise specified it concerns Generators' Busbars (GBs);

2 : CCT provided by the step-by-step method : t_c ;

3 : identification of the relevant "fastest" machine, as is provided by the test proposed in Toumi et al. (1985); this is the one used in the individual method (see column 9 below) and, whenever type 1 UEPs are chosen, in the acceleration approach as well. Otherwise, the case is specified in column 6 below and further described in Table II;

4 : identification of the relevant "slowest" machine, as is provided by the test;

5 : CCT provided by the conventional Liapunov criterion (see eq. (3.12)) : $t_{c\ell}$;

6 : CCT provided by the acceleration approach : V function expressed by eq. (3.7) , i.e. without accounting for transfer conductances; V_a expressed by eq. (3.16); the considered relevant machines are those of columns 3 and 4, except for "other than type 1" UEPs indicated (♦).

7 : CCT provided by the "Global" Kyoto approach : V function expressed by eq. (3.7), i.e. without accounting for transfer conductances; V_k expressed by eq. (3.14) : t'_{kG} ;

8 : CCT provided by the "Global" Kyoto approach with transfer conductances included; V function expressed by adding eqs. (3.7) and (3.9) : t_{kG} ;

9 : CCT provided by the individual Kyoto approach; V_{in} is expressed by eq. (3.18) ; t_{kI} ;

#10 : CCT provided by the Taylor based function combined with the acceleration approach : t_T .

Table II. It collects all cases where a cluster of, rather than one machine is to be considered in the acceleration approach. The resulting "other than type 1" UEPs correspond to the machines indicated in this Table.

Table III. It assembles three groups of information.

The first concerns investigated systems' essential topological data :

1 : number of system generators (N);

2 : total number of system buses (load plus generator nodes) $(N+N_L)$;

3 : total number of system branches (B).

The second group summarizes the information of Table I. The compared methods are reported in columns 4 and 5 as follows :

4 : accuracy of the Kyoto "Global" method, listed in # 8 of Table I;

5 : accuracy of the Kyoto "Individual" method, listed in # 9 of Table I.

Moreover, each of the above columns is subdivided in three sub-columns providing the following :

- sub-column I : maximum discrepancy of method's CCT with respect to t_c , in sec.;
- sub-column II : number of simulations yielding CCTs within $t_c \pm 0.02$ limits over total number of simulated 3ØSCs;
- sub-column III : number of simulations yielding CCTs within $t_c \pm 0.01$ limits over total number of simulated 3ØSCs.

The third group of Table II gives a first approximate evaluation of the methods' computing times in % of CPU :

6 : compares the computing time required by the step-by-step method for evaluating t_c ; three trials have been assumed necessary : $t_{e1} < t_c$, $t_{e2} = t_c$, $t_{e3} > t_c$; the t_c considered here are identified in italics in Table I;

7 : compares approximately the ratios in % of the computing times needed respectively by the "Global" (T_G) and by the step-by-step (T_b) methods;

TABLE 1

SYSTEM	1 Fault locat. Bus #	2 CCT s.b.s. t_c	3 "Fastest" gener.	4 Refer. gener. (n)	5 Conven. criterion $t_{c\ell}$	6 Accel. criterion t_a	7 Kyoto global (without) t'_{kG}	8 Kyoto global (with) t_{kG}	9 Kyoto individ. ref. gen. t_{kI}	10 Taylor Accel. indiv. t_T
3-mach.	1	0.54	1	2	0.47	0.47 (♦)	0.72	0.72	0.55	0.48
	2	0.37	2	1	0.40	0.40	0.40	0.37	0.37	0.38
	3	0.32	3	2	0.27	0.33	0.33	0.32	0.33	0.32
6-machine Tunisian	1	*0.30*	1	4	0.36	0.36	0.37	0.30	0.30	0.31
	2	0.86	2	4	0.51	0.98	0.98	0.92	0.90	0.49
	3	0.38	3	4	0.32	0.41	0.42	0.38	0.38	0.39
	4	> 1.50	4	2	1.07	> 1.50	1.42	1.34	1.35	0.88
	5	0.26	5	4	0.17	0.28	0.28	0.27	0.25	0.25
	6	0.39	6	4	0.25	0.41	0.42	0.39	0.39	0.33
7-machine CIGRE	1	0.37	1	2	0.32	0.38	0.37	0.38	0.38	0.36
	2	0.41	2	7	0.41	0.42	0.41	0.41	0.41	0.41
	3	0.39	3	2	0.32	0.39	0.39	0.39	0.39	0.37
	4	0.52	4	2	0.37	0.54 (♦)	0.54	0.51	0.52	0.45
	5	0.35	5	2	0.33	0.38	0.38	0.35	0.35	0.35
	6	0.52	6	2	0.43	0.50	0.50	0.52	0.51	0.48
	7	*0.33*	7	2	0.45	0.34	0.33	0.33	0.33	0.34
9-machine	1	0.45	1	9	0.48	0.48	0.49	0.45	0.45	0.43
	2	0.44	2	8	0.31	0.46	0.46	0.44	0.44	0.43
	3	0.27	3	8	0.22	0.28	0.28	0.27	0.27	0.27
	4	0.60	4	9	0.52	0.64	0.65	0.60	0.61	0.53
	5	0.50	5	9	0.50	0.55	0.55	0.50	0.50	0.53
	6	0.48	6	9	0.34	0.51	0.51	0.48	0.48	0.45
	7	0.60	7	9	0.45	0.65	0.65	0.60	0.60	0.57
	8	0.50	8	5	0.42	0.51	0.53	0.50	0.50	0.49
	9	0.48	9	5	0.48	0.51	0.52	0.49	0.48	0.47
	15 (OB)	0.47	6	8	0.33	0.56 (♦)	0.57	0.54	0.50	0.53
	24 (OB)	0.74	7	9	0.54	0.78	0.80	0.74	0.76	0.57
	26 (OB)	*0.36*	3	8	0.28	0.36	0.40	0.38	0.38	0.36

14-machine Greek	0	0.60	0	77	0.23	0.70 (♦)	0.78	0.77	1.30	0.76
	1	0.44	1	17	0.10	0.54	0.62	0.58	0.60	0.40
	17	0.58	17	21	0.29	0.65 (♦)	0.65	0.62	0.59	0.48
	19	0.52	19	21	0.26	0.53 (♦)	0.55	0.52	0.52	0.45
	21	0.50	21	59	0.17	0.50	0.50	0.49	0.49	0.46
	33	0.64	33	17	0.22	0.67	0.69	0.65	0.65	0.58
	34	0.32	34	17	0.08	0.35	0.36	0.33	0.31	0.31
	48	> 1	48	84	1.02	1.04	1.06	1.00	1.01	1.05
	59	0.48	59	21	0.19	0.49	0.51	0.48	0.48	0.47
	64	0.67	64	17	0.20	0.72	0.75	0.71	0.71	0.60
	76	0.47	76	64	0.17	0.49	0.49	0.48	0.48	0.47
	77	0.47	77	76	0.26	0.47	0.49	0.48	0.50	0.49
	84	0.26	84	0	0.07	0.27	0.27	0.26	0.26	0.25
	89	0.30	89	76	0.09	0.30	0.31	0.30	0.31	0.30
15-machine	M1	*0.36*	M1	M2	0.34	0.38	0.39	0.36	0.36	0.37
	K1	0.36	K1	M2	0.37	0.38	0.39	0.36	0.36	0.37
	Q1	0.43	Q1	G2	0.28	0.45 (♦)	0.47	0.44	0.43	0.40
	O1	0.40	O1	G2	0.28	0.46 (♦)	0.47	0.43	0.43	0.40
	C1	0.42	C1	G2	0.38	0.46	0.47	0.42	0.42	0.43
	G1	0.42	G1	G2	0.38	0.46	0.47	0.42	0.43	0.43
	N1	0.40	N1	K2	0.39	0.39	0.50	0.44	0.42	0.42
	ZLU	0.51	ZLU	M1	0.37	0.55	0.55	0.51	0.52	0.51
17-machine Iowa	G1	0.29	G1	G4	0.11	0.34	0.34	0.32	0.32	0.32
	G2	0.21	G2	G1	0.14	0.24	0.25	0.21	0.23	0.23
	G3	0.27	G3	G14	0.13	0.30	0.30	0.27	0.26	0.27
	G4	0.36	G4	G14	0.12	0.39	0.39	0.38	0.37	0.36
	G5	0.23	G5	G14	0.15	0.25	0.27	0.23	0.23	0.23
	G6	0.20	G6	G4	0.11	0.22	0.23	0.19	0.20	0.21
	G7	0.30	G7	G14	0.11	0.33	0.34	0.30	0.30	0.30
	G8	0.22	G8	G14	0.26	0.27	0.27	0.22	0.22	0.21
	G9	0.25	G9	G14	0.18	0.29	0.29	0.25	0.24	0.25
	G10	0.24	G10	G14	0.23	0.27	0.28	0.24	0.24	0.25
	G11	0.34	G11	G14	0.27	0.39	0.40	0.34	0.34	0.34
	G12	0.22	G12	G14	0.13	0.25	0.26	0.22	0.22	0.22
	G13	0.39	G13	G14	0.14	0.42	0.42	0.41	0.41	0.41
	G14	0.40	G14	G13	0.14	0.41	0.41	0.39	0.41	0.41
	G15	0.26	G15	G14	0.09	0.27	0.27	0.26	0.25	0.26
	G16	0.31	G16	G14	0.20	0.34	0.35	0.31	0.31	0.31
	G17	0.22	G17	G14	0.13	0.23	0.24	0.21	0.21	0.21
	Cooper	0.22	G2	G1	0.11	*0.20*	*0.26*	0.22	0.23	0.22
	RAUN	0.20	G6	G15	0.10	0.22	0.22	0.19	0.21	0.20
	Calhoun	0.33	G16	G14	0.14	*0.26*	*0.44*	0.36	0.34	0.25

TABLE I (continued)

SYSTEM	1	2	3	4	5	6	7	8	9	10
	Fault locat. Bus #	CCT s.b.s. t_c	"Fastest" gener.	Refer. gener. (n)	Conven. criterion $t_{c\ell}$	Accel. criterion t_a	Kyoto global (without) t'_{kG}	Kyoto global (with) t_{kG}	Kyoto individ. ref. gen. t_{ki}	Taylor Accel. indiv. t_T
40-machine Belgian	AWIR3	0.24	AWIR3	MAAS1	0.11	0.24	0.27	0.23	0.24	0.24
	AWIR4	0.23	AWIR4	REVI1	0.16	0.27	0.28	0.23	0.23	0.24
	BAUD3	0.34	BAUD3	ROMK1	0.18	0.33	0.40	0.32	0.34	0.35
	BRES4	0.26	BRES4	REVI1	0.17	0.29	0.30	0.25	0.26	0.27
	CHOO2	0.49	CHOO2	REVI1	0.29	0.52	0.56	0.49	0.50	0.51
	COO1	0.33	COO1	ROMK1	0.17	0.33	0.37	0.34	0.34	0.34
	COUG4	0.36	COUG4	REVI1	0.28	0.31	0.44	0.38	0.36	0.36
	DAMP3	0.30	DAMP3	ROMK1	0.16	0.31	0.37	0.29	0.30	0.31
	DOEL1A	0.25	DOEL1A	MAAS1	0.12	0.25	0.30	0.24	0.24	0.25
	DOEL1B	0.25	DOEL1B	ROMK1	0.11	0.24	0.31	0.25	0.25	0.25
	DROG3	0.38	DROG3	ROMK1	0.13	0.39 (♦)	0.46	0.36	0.37	0.37
	EISD3	0.31	EISD3	ROMK1	0.24	0.24	0.41	0.32	0.32	0.32
	FARC3	0.32	FARC3	ROMK1	0.16	0.31 (♦)	0.64	0.33	0.32	0.33
	FARC4	0.27	FARC4	ROMK1	0.24	0.32	0.33	0.27	0.27	0.26
	GRAM1	0.33	GRAM1	ROMK1	0.10	0.33	0.38	0.33	0.33	0.33
	KALO3A	0.39	KALO3A	ROMK1	0.16	0.37	0.48	0.38	0.39	0.40
	KALO3B	0.39	KALO3B	ROMK1	0.17	0.42 (♦)	0.48	0.38	0.38	0.38
	LANB3	0.28	RODH3A	ROMK1	0.14	0.32 (♦)	0.33	0.27	0.28	0.27
	LANO3A	0.26	EISD3	ROMK1	0.18	0.27 (♦)	0.30	0.25	0.30	0.30
	LANO3B	0.38	LANO3B	REVI1	0.20	0.41	0.46	0.38	0.38	0.38
	MERC1(OB)	0.25	DOEL1A	MAAS1	0.09	0.27 (♦)	0.31	0.24	0.25	0.25
	MERK4	0.30	MERK4	REVI1	0.24	0.37	0.38	0.29	0.30	0.28
	MOL3	0.28	MOL3	ROMK1	0.16	0.31	0.35	0.27	0.28	0.28
	MONC3	0.30	MONC3	ROMK1	0.15	0.33	0.35	0.29	0.29	0.30
	PTBR3	0.32	PTBR3	ROMK1	0.13	0.31	0.40	0.31	0.32	0.33
	PTRE4	0.42	PTRE4	REVI1	0.26	0.63 (♦)	0.76	0.48	0.46	0.33
	RODH3A	0.27	RODH3A	ROMK1	0.14	0.32 (♦)	0.33	0.26	0.26	0.27
	RODH3B	0.31	RODH3B	ROMK1	0.15	0.34	0.36	0.26	0.30	0.31
	RUIE3	0.32	RUIE3	ROMK1	0.13	0.37	0.38	0.31	0.32	0.33
	SCHE3A	0.31	SCHE3A	ROMK1	0.14	0.25	0.37	0.29	0.30	0.31
	SCHE3B	0.28	SCHE3B	ROMK1	0.17	0.26	0.34	0.27	0.28	0.28
	SCLE4	0.33	SCLE4	REVI1	0.26	0.32	0.40	0.34	0.34	0.33
	SERA2	0.33	SERA2	ROMK1	0.17	0.23	0.62	0.37	0.34	0.34
	STAL3	0.26	STAL3	ROMK1	0.18	0.28 (♦)	0.31	0.25	0.28	0.28
	TRIV3	0.32	TRIV3	ROMK1	0.15	0.25	0.65	0.34	0.32	0.33
	VILG4	0.25	VILG4	MAAS1	0.20	0.29	0.30	0.26	0.26	0.26

TABLE II

Exploring the acceleration approach.
Identification of other than Type 1 UEPs
and of their related machines

SYSTEM	Fault location	Involved machines' identification
3-machine	1	(1,3) : Type 2 UEP
6-machine (Tunisian)	4	(5,4,3,6,1) : Type 5 UEP
7-machine (CIGRE)	4	(4,5) : Type 2 UEP
9-machine	15	(3,6) : Type 2 UEP
14-machine (Greek)	0 17 19	(0,21,19,17) : Type 4 UEP (17,19) : Type 2 UEP (19,17) : Type 2 UEP
15-machine	Q1 01	(Q1,01) : Type 2 UEP (01,Q1) : Type 2 UEP
40-machine (Belgian)	DROG3 FARC3 KALO3B LAN3B LANO3A MERC1 PTRE4 RODH3A STAL3	(DORG3 , PTBR3) (FARC3 , FARC4) (SCHE3B , KALO3B) (RODH3A , LANB3) (STAL3 , EISD3) (DOEL1A , DOEL1B) } Type 2 UEP (BRES4 , SCLE4 , PTRE4) : T.3 (RODH3A , LANB3) : T.2 (STAL3 , EISD3) : T.2

TABLE III

1	2	3	4			5			6	7	8	9	10
SYSTEM DATA			METHOD ACCURACY						CPU COMPUTING TIMES (T). RATIOS IN %				
			GLOBAL			INDIVIDUAL							
N	$N + N_L$	B	I	II	III	I	II	III	s.b.s. (T_b)	$\frac{T_G}{T_b}$	$\frac{T_I}{T_G}$	$\frac{T_T}{T_G}$	$\frac{T_r}{T_r+T_I}$
6	20	26	.16	4/5	4/5	.14	4/5	4/5	15	11	60	8	9
7	10	13	.01	7/7	7/7	.01	7/7	7/7	9	14	66	4	3
9	34	36	.07	11/12	10/12	.03	11/12	9/12	18	9	54	5	8
14	92	119	.04	9/10	9/10	.03	9/10	9/10	40	14	41	1.7	8
15	44	99	.04	6/8	6/8	:03	7/8	6/8	32	22	41	1.4	9
17	162	281	.03	9/10	8/10	.02	10/10	9/10	147	8	37	0.2	64
40	163	269	.01	10/10	10/10	.01	10/10	10/10	100	95	19	0.8	37

\# 8,9 : compare the computing times needed respectively by the "Individual approach" (T_I), and by the "Taylor approach" (T_T), all with respect to T_G ; the ratios are expressed in %; note also that these comparisons do not take into account the computing time needed for the computation of the admittance matrix reduced at the generators' nodes;

\#10 : ratio of the CPU time needed to compute Y_r (T_r) with respect to the overall CPU time needed by the "Individual approach".

V.3.5.2.2. Discussion

(i) Accuracy evaluation. * *Conventional direct criterion* (column # 5 of Table I). Notice the erratic results of this column.
* *Influence of transfer conductances* (columns # 6,7,8 of Table I). The results are generally in good agreement with t_c. Including transfer conductances allows further improvements, as it may be seen by comparing columns # 7 and 8 relative to the Kyoto global criteria. The same observation holds for the acceleration criteria according to simulations not reported here : when the loads are localized at GBs, the approximation of transfer conductances neglect is better justified and the accuracy of t_a improves.
* *Global vs. individual functions* (columns # 8,9 of Table I). The two methods behave almost equally well, with, however, a slight advantage of the latter over the former (see also columns # 4,5 of Table III).
* *Taylor's approximation.* For a limited range of CCTs, say for CCTs not exceeding 0.45 sec , it shows accurate.

(ii) Estimating computational efficiency (Table III). N.B. During our simulations, computational efficiency was not sought. Therefore, the Computing Times (CT) can only give a crude indication; this is why we proceed by comparisons. To make them valid, we have used simulations implying similar CCTs in the various systems. Indeed, the rapidity of the step-by-step integration is different in the fault-on and the post-fault periods. On the other hand, we have split the procedures into two parts and estimated their corresponding CCTs separately; the one concerns the reduction of the admittance matrix, the other the rest of the procedure. Note that for 3ØSC at GBs the reduced admittance matrix of the fault-on period may be derived from that of the pre-fault one through the superposition technique suggested in Ribbens-Pavella and Evans (1985a); thus, for this type of disturbances, matrix reduction is computed once for all.

6 : the T_b ratios generally increase with the system size. The 100 % choice is obviously arbitrary;

7 : the gain in CT increases rapidly with the system size;

9 : notice the substantial gain in CT for medium-sized and even more for large systems.

V.3.5.3. Other types of disturbances

The second group of disturbances concerns Line Trippings (LT) and Load Sheddings (LS). Table IV lists some of our investigation results.

TABLE IV

Disturb. identif.	t_c	$t_{c\ell}$	t_a	t'_{kG}	t_{kG}	t_{kI}
1	0.32	0.08	0.35	0.35	0.32	0.33
2	>1	>1	>1	0.38	0.38	0.66
3	>1	>1	>1	0.33	0.33	0.51
4	>1	>1	>1	0.28	0.28	0.35
5	>1	>1	>1	0.44	0.43	0.42

Dist #1 : 14-machine Greek system : LT of the double line 32-72
Dist #2 : 14-machine Greek system : LS of the load 191.7 MW at GB #1
Dist #3 : 14-machine Greek system : LS of the load 344.3 MW at OB #30
Dist #4 : 40-machine Belgian system : LS of the load 267 MW at GB #MOL3
Dist #5 : 40-machine Belgian system : LS of the load 855 MW at OB #AVEL1

Obviously, the Kyoto and the acceleration approaches behave quite differently.

The acceleration methodology. It is bound to the assumptions implied by the V function and its accuracy depends upon which extent it is justified to neglect transfer conductances. Yet, it always lies within reasonable accuracy limits and it never fails to depict the severity of a disturbance.

The Kyoto methodology. It generally proves very accurate with, however, the following two exceptions.

(i) In presence of "mild" disturbances, it may happen that small swing curves oscillations produce local, irrelevant, maxima of V_S : the Kyoto method will "stop" at the first local maximum and declare the system unstable whereas actually it is far from instability (e.g. the actual CCT may be larger than 1.5 sec , whereas the Kyoto's CCT = 0.3 sec). Fig.8a illustrates this case.

(ii) Conversely, in presence of very severe disturbances, it may happen that V_P reaches its maximum for rotor angular differences much larger than 180°, which is in fact very close to system's loss of synchronism (apart from very few peculiar exceptions). In this case, the Kyoto criterion becomes unacceptably overoptimistic. (E.g. the ratio of the actual CCT over that provided by Kyoto may be as low as 30 %.) Fig.8b illustrates such a situation.

To avoid the above serious discrepancies, Van Cutsem et al. (1986) propose to combine the Kyoto approach with the following practical rule. For each t_e compute the maximum rotor angular difference : let $\delta_{max} = \max \delta_{ij} = \max(\delta_i - \delta_j)$ $(i,j = 1,2,\dots,m)$ denote it. Then :

(i) if $V_{P\,max}$ is reached before δ_{max} reaches 90°, declare that CCT is very large (larger than 1 sec);

(ii) conversely, if δ_{max} reaches 180° before V_P reaches its actual maximum, declare $V'_{max} = V_P$ ($\delta = \delta_{max} = 180°$) and apply Kyoto with V'_{max} rather than $V_{P\,max}$;

(iii) otherwise, i.e. if $V_{P\,max}$ is reached for $180° > \delta_{max} > 90°$, use the standard Kyoto procedure.

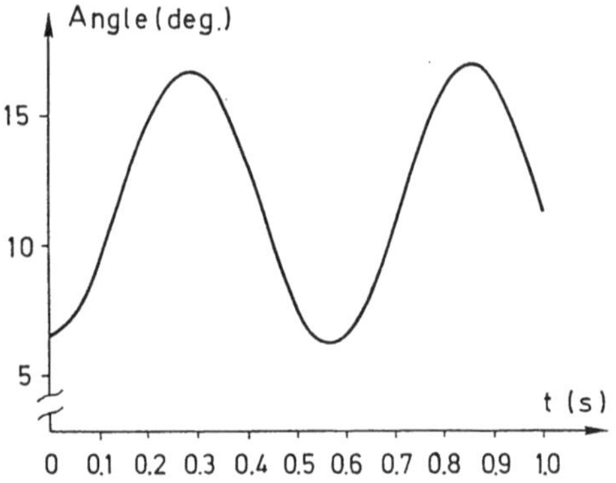

3ØSC at bus #9

Clearing time : t_e = 0.28 sec.
Actual CCT > 1 sec.
(see in Table I)

Figure 8a

Time evolution of $\delta_{max} = \delta_{32}$ of the CIGRE system

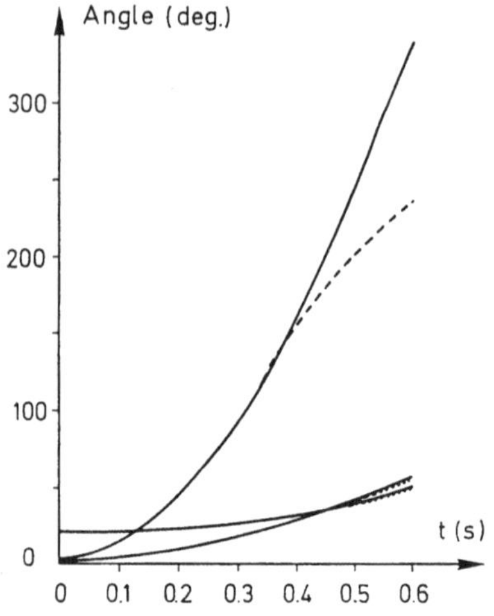

3ØSC at TRIV3

—— : *sustained fault*
--- : *fault (critically) cleared at .36 sec.*

Figure 8b

Swing curve of some rotor angles of the Belgian system

V.3.5.4. Conclusion

Some of the variants devised in Toumi et al. (1985) along with the improvement proposed by Van Cutsem et al. (1986) have been assessed here. Their pros and cons may be summarized as follows.

* With respect to ***computational efficiency*** :

(1) The *Taylor series-based criterion* (combination of the energy-type function - global or individual - with either of the Kyoto and the acceleration PSDE), proves to be the most straightforward;

among the remaining criteria, those which show equally straightforward (i.e. equally - and in fact very - fast) are :

(2) the *Kyoto individual TE function*, i.e. the individual function, in conjunction with the Kyoto approach;

(3) the *acceleration global energy type function*, i.e. the standard energy function in conjunction with the acceleration approach.

* With respect to **accuracy**, criteria (1) and (2) are very accurate : criterion (2) whatever the severity of the disturbance, criterion (1) in the case of very severe disturbances (see below). Their accuracy lies almost always within ± 5 %, whereas criterion (3) is less accurate (± 15 to 20 %).

* With respect to possibilities of devising **stability margins**, criteria (1) and (3) show very interesting, whereas criterion (2) cannot be used. Indeed, in order to assess a priori the "distance to instability" for a given clearing (generally rather short) time, one has to know the limit value of V prior to a step-by-step computation up to CCT; and the Kyoto method necessitates step-by-step computation for even larger time than CCT.

Hence, three direct criteria emerge, called to meet different needs :

- the Taylor series-based criterion : extremely fast, appropriate for stability margins, and very accurate for very large disturbances, it may fail for less severe ones;
- the Kyoto individual criterion : very fast and very accurate, it is not appropriate for use in conjunction with stability margin concept;
- the acceleration criterion : very fast but less accurate than the previous criteria, it is appropriate for stability margin definitions, whatever the disturbance severity.

N.B. In the above discussion, the "very severe disturbances" for which the Taylor criterion is valid correspond to CCTs ranking up to, say, 0.45 sec, and belong to the type "three-phase short circuit" (3ØSC) acting at "important" generators' buses. But then, the question of concern is "how to know for sure whether the considered disturbance belongs indeed to the class of the "very severe" ones ?

A sound answer may be provided by the ranking of the machines' initial (i.e. at the instant of disturbance inception) accelerations. Indeed, for a given power system, the range of these accelerations is typically related to the severity of the disturbance. Table V illustrates the strong dependence of the disturbance severity on the initial accelerations of the Belgian system machines : one can see that the 3ØSC and a line tripping (855 MW) at AVEL1 (which is electrically close to DOEL1) induce machines' accelerations lying in quite different ranges.

TABLE V

1	2	3	4
Bus	3ØSC at AVEL1 γ	Bus	LT at AVEL1 γ
DOEL1A	19.20	DOEL1A	3.44
DOEL1B	17.86	DOEL1B	3.27
SCHE3A	10.95	CHOO2	2.65
AWIR3	10.63	SCHE3A	1.90
SERA2	9.93	SERA2	1.88
PTBR3	9.71	DROG3	1.77
COUG4	9.12	PTBR3	1.72
COO1	9.01	COUG4	1.55
DROG3	8.72	MERK4	1.28
AWIR4	8.71	COO1	1.24
SCHE3B	7.65	TRIV3	1.13
CHOO2	7.49	FARC3	1.04
GRAM1	7.35	MONC3	1.01
VILG4	7.27	PTRE4	0.98
MERK4	7.11	BAUD3	0.95
TRIV3	7.00	FARC4	0.94
FARC3	6.96	AWIR3	0.93
MONC3	6.94	MOL3	0.91
FARC4	6.92	SCHE3B	0.89
DAMP3	6.82	DAMP3	0.83
BRES4	6.47	GRAM1	0.81
MOL3	6.30	VILG4	0.75
BAUD3	5.18	AWIR4	0.74
KALO3A	5.13	BRES4	0.73
PTRE4	5.01	KALO3A	0.67
SCLE4	5.01	RODH3B	0.64
KALO3B	4.87	KALO3B	0.64
EISD3	4.83	SCLE4	0.61
RODH3A	4.50	LANB3	0.59
RODH3B	4.42	RODH3A	0.53
STAL3	4.42	LANO3B	0.50
LANB3B	4.40	REVI1	0.41
LANO3B	3.42	EISD3	0.39
LANO3A	2.57	RUIE3	0.37
RUIE3	2.27	MAAS1	0.29
REVI1	1.08	STAL3	0.27
MAAS1	0.41	ROMK1	0.17
ROMK1	0.38	LANO3A	0.14

V.3.6. Direct criteria for real-time operation. Prospectives and perspectives

1) The above criteria solve efficiently the real-time transient stability problem, provided that the very simplified system modelling that they impose is deemed satisfactory enough.

2) Within the applicability limits determined above, their reliability and accuracy are not affected by the system size.

3) The only lengthy computation they require is the reduction of the admittance matrix at the generators' nodes; there is still room for improvements along this direction.

4) With the exception of the above, the computation time varies quite smoothly with the system size; as regards the step-by-step computation, (whenever) needed during the short period between disturbance inception and clearance, many short-cuts may be used to speed it up (Taylor's expansion, variable step size, etc.).

5) A timely question is whether the conventional stability assessment in terms of disturbance Critical Clearing Time (CCT) is adequate or not. The answer seems to be "yes" and "no". "Yes", for methodological research purposes, since it allows a handy way of explorations and comparisons. "No" for actual effective real-time applications; indeed, there are certainly other, more physically sound, parameters to consider. Thus, it is very likely that the coming years will see the outburst of other, more adequate, alternative stability measures and assessment.

6) Another short-term perspective concerns the practical application and use of the "structure preserving model" proposed by Bergen and Hill (1981). In essence, this model amounts to keeping unreduced the admittance matrix at all the system nodes, as they appear in a standard load flow calculation, rather than reducing it at the generators' nodes. Relaxation of this matrix reduction amounts to avoiding the computationally most demanding step. But much more importantly, it makes the neglect of transfer conductance assumption very realistic and perfectly acceptable; indeed, unlike the reduced admittance matrix, the unreduced one does not include part of the load effects in its non-diagonal terms. Other model refinements may also be obtained more easily with this than with the reduced at the generators' equivalent system model (e.g., see De Marco and Bergen, 1984). These advantages will be obtained at the expense of a greater number of state variables and implied burden. But overall, the advantages will probably predominate over this relatively minor inconvenience.

V.4. VECTOR LIAPUNOV APPROACH

V.4.1. Introduction

Attempts to overcome some of the difficulties met with the scalar functions, have led to a new approach based upon the concept of the decomposition-aggregation method and the vector Liapunov functions. Roughly speaking, this approach consists in decomposing the multimachine system into a number of interconnected subsystems, composed of interconnections and of suitably chosen disconnected (free) parts; for these latter, use is made of standard techniques for constructing scalar V functions and for evaluating SDEs. Then, an aggregate model is constructed which involves a vector Liapunov function based on the subsystems' V functions, and a SDE of the overall system is determined using subsystem's SDEs and certain properties of their interconnections. This methodology appears quite attractive for many reasons : besides allowing consideration of more sophisticated mathematical models for the analysis of disconnected low-order subsystems, it offers possibilities to obtain simple analytical expressions and hence information about the special structural features of a power system; this in turn opens new horizons to the sensitivity analysis problem.

In the studies devoted to this area, various aspects have been examined and developed. They mainly deal with

(i) the power system modelling;
(ii) the system decomposition, i.e. the decomposition of the state equations of the overall system into subsystems and further of each subsystem into its disconnected (free) part and its interconnections;
(iii) the subsystem stability analysis, i.e. the construction of V functions for the free subsystems and the evaluation of their SDEs;
(iv) the system aggregation, i.e. the choice of a scalar Liapunov function for the overall system, resulting from the aggregation rule relating the subsystems' V functions and evolving the definition of an aggregation matrix which accounts for the influence of the interconnections;
(v) the determination of the (asymptotic) SDE of the overall system.

Concerning developments of system modelling (i), let us at once say that all studies take into account - or at least would be able to do so - transfer conductances and mechanical damping - either uniform or non-uniform. This is owing to the fact that by a suitable system decomposition (ii), one can succeed in getting free subsystem's representation of the well-known Lur'e-Postnikov type system, for which suitable V functions and their SDEs are readily available. Note that system decomposition plays an important role not only with respect to the above consideration, but also because the choice of interconnections influences the construction of the aggregation matrix (iv) which in turn influences considerably the degree of conservativeness of the overall method. In fact, the conservativeness, which as we shall see constitutes the weak point of the method, depends essentially upon above steps (iv) and (v). In what follows we develop at first a thorough "classical" analysis based on Bellman's decomposition-aggregation method modified and adapted to the power systems problem (Grujić and Ribbens-Pavella, 1977). Then we briefly outline the main feautres of other aggregation-decomposition forms of power systems. This review merely aims to bring out the essential features of the vector Liapunov function approach at the present stage of progress, with the ultimate goal to pinpoint its specific potential field of practical use.

V.4.2. Stationary large-scale systems decompositions and aggregations in general

V.4.2.1. General comparison of scalar and vector Liapunov functions

In the framework of Liapunov stability considerations there exist different approaches to power systems aggregation, which are not completely unrelated but can lead to different stability requirements imposed on the systems. They were considered in general in the preceding chapters and outlined in the sequel to the extent suitable for power systems.

One is Liapunov's classical aggregation of dynamic systems (Chapter I). If the system is described by an M-th order nonlinear autonomous vector differential equation (2.2) denoted herein as (4.1),

$$\dot{\mathbf{x}} = \mathbf{f}(\mathbf{x}) \tag{4.1}$$

then, by following Liapunov (1892) a scalar positive definite function $v : \mathbb{R}^M \to \mathbb{R}$ is used as an aggregation function for the system. For the sake of clearness, let $v \in C^{(1)}(\mathbb{R}^M)$ be differentiable in $\mathbf{x} \in \mathbb{R}^M$. Its eulerian (total time) derivative $\dot{v}$ along motions $\chi(t;\mathbf{x}_o)$, $\chi(0;\mathbf{x}_o) \equiv \mathbf{x}_o$, of (4.1) is obtained in the form

$$\dot{v}(\mathbf{x}) = [\text{grad } v(\mathbf{x})]^T \mathbf{f}(\mathbf{x}) , \tag{4.2}$$

where

$$\text{grad } v(\mathbf{x}) = \left(\frac{\partial v}{\partial x_1}, \frac{\partial v}{\partial x_2}, \dots, \frac{\partial v}{\partial x_M}\right)^T . \tag{4.3}$$

Here, $\mathbf{x} = (x_1, x_2, \dots, x_M)^T$.

Notice that for the autonomous N-machine power system the dimension M can for instance be 2N-1,

$$M = 2N - 1 , \tag{4.4}$$

and

$$\mathbf{f}(\mathbf{x}) = \mathbf{P}\mathbf{x} + \mathbf{h}(\mathbf{x}) .$$

Use of a scalar positive definite aggregation function v allows the

M-th order system (4.1) to be aggregated into a first order equation (4.2), which can be set in the form

$$\dot{v}(\mathbf{x}) = \psi(\mathbf{x}) \quad , \quad \psi = (\text{grad } v)^T \mathbf{f} \ . \tag{4.5}$$

For Liapunov stability analysis it is necessary and sufficient to prove an appropriate sign property of ψ on a certain connected neighbourhood X of $\mathbf{x}=\mathbf{0}$. For asymptotic stability analysis of $\mathbf{x}=\mathbf{0}$, the set X should be chosen so that $\mathbf{x}=\mathbf{0}$ be the unique equilibrium state of (4.1) in it.

Liapunov's aggregation of (4.1) into (4.2) is called also *scalar Liapunov function aggregation* of (4.1).

In general it is extremely difficult to find a suitable function v for (4.1) such that a required sign property of ψ can be effectively examined on the set X. This problem is fundamental for the Liapunov stability theory and its applications, but it has not yet been solved either in general or for power systems in particular. That is why different majorization approaches were proposed. When they can be applied, they yield exact problem solutions rather than approximate.

One of them is based on the comparison principle (Kamke, 1932; Važewski, 1950; Matrosov, 1968; and Chapter II). It is effective if we can find a scalar function ψ_1 such that

(i) $$\psi(\mathbf{x}) \leq \psi_1[v(\mathbf{x})] \quad (\text{or} \quad \psi(\mathbf{x}) \geq \psi_1[v(\mathbf{x})]\ ) \quad , \quad \forall \mathbf{x} \in X \ , \tag{4.6}$$

and

(ii) the required stability property of (4.1) can be effectively proved for the comparison system

$$\dot{W} = \psi_1(W) \ . \tag{4.7}$$

We can note that a stability property of (4.7) can be in special cases concluded from the motions of (4.7) obtained by solving it, or by repeating the whole above procedure applied now to (4.7) rather than to (4.5).

Another majorization approach is based on the assumption that we can find a scalar function ψ_2 such that :

(1) $$\psi(\mathbf{x}) \leq \psi_2(\mathbf{x}) \quad (\text{or} \quad \psi(\mathbf{x}) \geq \psi_2(\mathbf{x})\ ) \quad , \quad \forall \mathbf{x} \in X \ , \tag{4.8}$$

and

(2) the required sign property of ψ_2 can be effectively verified on X $(\forall \mathbf{x} \in X)$.

From the technical viewpoint, suitable form of ψ_2 is the following :

$$\psi_2(\mathbf{x}) = \mathbf{u}^T(\mathbf{x})\, \mathbf{A}(\mathbf{x})\, \mathbf{w}(\mathbf{x}) \ , \tag{4.9}$$

where $\mathbf{A} : R^M \to R^{s\times s}$, $\mathbf{u},\mathbf{w} : R^M \to R^s$ are dependent on $\mathbf{x}$ in general and s is a positive integer. They are required to possess the adequate sign property on X. The test of these properties is simplified if ψ_2 can be found such that either

(a) $\mathbf{u}(\mathbf{x}) \equiv \mathbf{w}(\mathbf{x})$ and $\mathbf{A}$ is (elementwise) constant,

or

(b) $\mathbf{u}(\mathbf{x}) \equiv \mathbf{b}$ is (elementwise) constant vector and $\mathbf{A}$ is also constant.

The Liapunov aggregation of the system (4.1) does not recognize the system structure. For this reason and in order to find a more effective procedure for both V function construction and the system aggregation, Bellman (1962) and Matrosov (1962) proposed two other approaches. The link between them can be made by following La Salle (1976).

In order to explain their approaches we introduce s scalar functions $V_i = R^{m_i} \to R$, $v_i = R^M \to R$ and vector functions $\mathbf{V} : R^M \to R^s$, $\mathbf{v} = R^M \to R^s$, $m_1, m_2, \ldots, m_s = M$,

$$\mathbf{V} = (V_1, V_2, \ldots, V_s)^T \quad , \quad \mathbf{v} = (v_1, v_2, \ldots, v_s)^T \ . \tag{4.10}$$

The vector functions $\mathbf{V}$ and $\mathbf{v}$ are called *vector Liapunov functions of* (4.1) iff their application allows verification of a required Liapunov stability property of (4.1). In fact, $\mathbf{V}$ and $\mathbf{v}$ are required to possess a certain sign property elementwise (La Salle, 1976, and Chapters II, III).

From the purely mathematical point of view, $\mathbf{V}$, or $\mathbf{v}$, is associated with (4.1) so that

$$\dot{\mathbf{V}}(\mathbf{x}) = \Psi(\mathbf{x}) \quad , \quad \Psi : R^M \to R^s \ . \tag{4.11}$$

or

$$\dot{\mathbf{v}}(\mathbf{x}) = \Psi(\mathbf{x}) \ . \tag{4.12}$$

From the system structural viewpoint, the system (4.1) is decomposed into s interconnected subsystems

$$\dot{\mathbf{x}}_i = \mathbf{f}_i(\mathbf{x}) \ , \ \mathbf{f}_i : R^M \to R^{m_i} \ , \ \mathbf{x}_i \in R^{m_i} \ , \ \forall i = 1,2,\ldots,s \ . \tag{4.13}$$

If the subsystems are linearly interconnected then the functions $\mathbf{h}_i : R^M \to R^{m_i}$ describe their interactions and are called *interactions*, and functions $\mathbf{g}_i : R^{m_i} \to R^{m_i}$ describe *free (disconnected) subsystems*

$$\dot{\mathbf{x}}_i = \mathbf{g}_i(\mathbf{x}_i) \ , \ \forall i = 1,2,\ldots,s \ . \tag{4.14}$$

so that $\mathbf{f}_i = \mathbf{g}_i + \mathbf{h}_i$;
interconnected subsystems are then given by

$$\dot{\mathbf{x}}_i = \mathbf{g}_i(\mathbf{x}_i) + \mathbf{h}_i(\mathbf{x}) \ , \ \forall i = 1,2,\ldots,s \ . \tag{4.15}$$

The functions v_i are associated with the i-th subsystem (4.13), i.e. (4.15), in general. Then, their form is influenced by the form of $\mathbf{f}_i$

(of both $\mathbf{g}_i$ and $\mathbf{h}_i$ respectively). The form of V_i is governed only by the form of $\mathbf{g}_i$, and the form of the dependence of $\mathbf{h}_i$ on $\mathbf{x}$ is not taken into consideration. The aggregation of the system (4.1) proposed by Bellman (1962) and Matrosov (1962) is based on the system decomposition (4.13) in general, or on (4.15) in particular, and the construction of V_i for (4.14) and v_i for (4.13) or (4.15), respectively. Then, (4.11) or (4.12), respectively, is the aggregated operation related to (4.1).

The Bellman-Matrosov aggregation (4.11) or (4.12) of (4.1) is called *the vector Liapunov function aggregation of* (4.1).

Notice that Bellman (1962) proposed the use of $\mathbf{V}$ (4.10) whereas $\mathbf{v}$ (4.10) was introduced by La Salle (1976) and developed in Chapter II.

Further procedure is the same for both (4.11) and (4.12). Hence, it will be explained for (4.11).

The aggregated equation (4.11) can be used in two conceptually different ways which are analogous to those explained for the scalar aggregated equation (4.5).

One possibility is realizable if there exists a vector $\mathbf{w} : R^s \to R^s$ such that (elementwise)

$$\Psi(\mathbf{x}) \le \mathbf{w}[\mathbf{V}(\mathbf{x})] \quad (\text{or} \quad \Psi(\mathbf{x}) \ge \mathbf{w}[\mathbf{V}(\mathbf{x})]\ ) \ , \ \forall \mathbf{x} \in X \ . \tag{4.16}$$

Then, provided the function $\mathbf{w} = (w_1, w_2, \dots, w_s)^T$ satisfies

$$w_i(V_1', V_2', \dots, V_{i-1}', V_i, V_{i+1}', \dots, V_s') \le w_i(V_1'', V_2'', \dots, V_{i-1}'', V_i, V_{i+1}'', \dots, V_s'') \ ,$$
$$V_j' \le V_j'' \ , \ \forall i,j = 1,2,\dots,s \ , \ i \ne j \ , \ \forall \mathbf{V} \in R^s \ , \tag{4.17}$$

the stability property of (4.1) is deduced from that of the comparison system (Kamke, 1932; Ważewski, 1950; Matrosov, 1968, and Chapter II) :

$$\dot{\mathbf{u}} = \mathbf{w}(\mathbf{u}) \quad , \quad \mathbf{u} \in R^s \ . \tag{4.18}$$

Stability test of the comparison system is generally not available when $\mathbf{w}$ is a nonlinear function. Often, the test requires application of the (second) Liapunov method at the last step.

Another possibility for stability analysis based on the use of the aggregated equation is to employ a scalar function v_1,

where $\mathbf{b} \in \mathring{R}_+^s$. Then,

$$v_1(\mathbf{x}) = \mathbf{b}^T \mathbf{V}(\mathbf{x}) \ , \tag{4.19}$$

$$\dot{v}_1(\mathbf{x}) = \mathbf{b}^T \Psi(\mathbf{x}) \ , \tag{4.20}$$

due to (4.11). In fact, we arrive at the problem of testing appropriate sign properties of v_1 (4.19) and $\psi = \mathbf{b}^T \Psi$ (4.20).

The approach of Matrosov (1962) supposes existence of v_i's such that

$v = v_1 + v_2 + \dots + v_s$ and $\psi = \dot{v}$ possess the appropriate sign properties. These considerations show that crucial steps in the stability analysis aimed for practical applications are the following :

A. The system decomposition.
B. The choice of the aggregation functions and of the overall system Liapunov function v.
C. Sign tests of v and $\psi = \dot{v}$.
D. Useful estimation of the asymptotic stability domain.

The first two steps (A and B) considered separately from the last two (C and D) provide a number of different possitibilities. Essentially, *the problem consists in the choice of such an aggregation-decomposition form that sign tests of* v *and* ψ *can be effectively performed together with practically useful estimation of the asymptotic stability domain.* This problem is treated herein in general and in the framework of large-scale power systems in particular.

The vector Liapunov function aggregation of the system is conceptually advantageous over the scalar one for the following :

- use of a more refined mathematical model of the system;
- easier construction of V_i's for subsystems than of v directly for the overall system, and hence, easier construction of v for the whole system via $\mathbf{V}$;
- possible reduction of the order of the aggregation matrix to the number of the subsystems;
- derivation of effective stability criteria;
- condideration of structural properties of the system and influence of variations of its structure on its stability;
- possibility to infer stability property of the overall system from stability of disconnected subsystems and therefore possibility of easier parameter sensitivity analysis;
- easier analytical estimation of the (asymptotic) stability domain.

The drawback of the vector Liapunov function aggregation are majorizations, which are at the up-to-date state of the research almost necessarily involved in it. However, application of the scalar Liapunov function aggregation can also require different majorizations.

Majorizations used in aggregation of the system can be interpreted as follows : interactions which are stabilizing on subspaces of the system state space are considered as destabilizing on the whole state space; hence also on these subspaces.

Majorizations can prevent effective application of the vector Liapunov function aggregation; this is a technical rather than a conceptual drawback of the approach.

V.4.2.2. *Suitable aggregation forms and comparison functions*

The preceding exposition requires deeper elaboration in order to examine effective possibilities of using (4.9), (4.11), (4.12) and (4.20).

The procedure of the vector Liapunov function aggregations starts with a suitable decomposition of the system (4.1) into subsystems (4.15).

With each subsystem (4.15) we associate a positive definite differentiable function V_i and determine $\dot{V}_i$,

$$\dot{V}_i(\mathbf{x}) = [\text{grad } V_i(\mathbf{x}_i)]^T[\mathbf{g}_i(\mathbf{x}_i) + \mathbf{h}_i(\mathbf{x})] \quad , \quad \forall i = 1,2,\dots,s \ . \tag{4.21}$$

For the sake of conciseness let $\psi_i : \mathbf{R}^M \to \mathbf{R}$,

$$\psi_i(\mathbf{x}) = [\text{grad } V_i(\mathbf{x}_i)]^T[\mathbf{g}_i(\mathbf{x}_i) + \mathbf{h}_i(\mathbf{x})] \quad , \quad \forall i = 1,2,\dots,s \ . \tag{4.22}$$

Their exist four different aggregation forms which are based on suitable majorizations of ψ_i's .

One is based on the assumption that there exist numbers α_{ij} , $\alpha_{ij} \geq 0$, $i \neq j$, functions $u_i : \mathbf{R}^{m_i} \to \mathbf{R}$ and open connected neighbourhood $\mathbf{X}_i$ of $\mathbf{x}_i = \mathbf{0}$ such that

$$\psi_i(\mathbf{x}) \leq \sum_{j=1}^{s} \alpha_{ij} u_j(\mathbf{x}_j) \quad , \quad \forall \mathbf{x} \in \mathbf{X} \ , \ \forall i = 1,2,\dots,s \ . \tag{4.23}$$

Here, $\mathbf{X} = \mathbf{X}_1 \times \mathbf{X}_2 \times \dots \times \mathbf{X}_s$ is the cartesian product of all $\mathbf{X}_i$'s . This aggregation-comparison form (in short, aggregation form) generalizes that of Grujić and Šiljak (1973), Chapter III. All u_i's are required to be strictly increasing comparison functions ϕ_{i3} in $\|\mathbf{x}_i\|$, $u_i(\mathbf{x}_i) = \phi_{i3}(\|\mathbf{x}_i\|)$, determined by the disconnected subsystems (4.14) for chosen V_i . This aggregation form appears too stringent for asymptotic stability analysis of power systems when V_i is "quadratic form + integral of nonlinearity".

Another form is of the Lur'e-Postnikov type proposed by Grujić (1974), Grujić et al. (1976), Chapter III. It assumes existence of u_i and f_i functions and real numbers α_{ij} , and β_{ij} such that either (Chapter III) :

$$\psi_i(\mathbf{x}) \leq u_i(\mathbf{x}_i) \sum_{j=1}^{s} \{\alpha_{ij} u_j(\mathbf{x}_j) + \beta_{ij} f_j[u_j(\mathbf{x}_j)]\} \quad , \quad \forall \mathbf{x} \in \mathbf{X} \ , \ \forall i = 1,2,\dots,s \ , \tag{4.24}$$

or (Chapter III) :

$$\psi_i(\mathbf{x}) \leq \sum_{j=1}^{s} \{\alpha_{ij} u_j(\mathbf{x}_j) + \beta_{ij} f_j[u_j(\mathbf{x}_j)]\} \quad , \quad \forall \mathbf{x} \in \mathbf{X} \ , \ \forall i = 1,2,\dots,s \ , \tag{4.25}$$

for $\alpha_{ij} \geq 0$, $\beta_{ij} \geq 0$, $i \neq j$. The non-linearities f_i obey $f_i(u_i) \geq 0$ $\forall u_i > 0$, $f_i(0) = 0$ and $f_i(u_i) \in C(R)$. This aggregation form appears suitable for global asymptotic stability analysis, which is not the case for power systems.

The third possible aggregation of the system is based on the existence of functions $\zeta_{ij} : R^M \to R$ and u_i such that (Chapter III) :

$$\psi_i(\mathbf{x}) \leq u_i(\mathbf{x}_i) \sum_{j=1}^{s} \zeta_{ij}(\mathbf{x})\, u_j(\mathbf{x}_j) \quad , \quad \forall \mathbf{x} \in \mathbf{X} \ , \ \forall i = 1,2,\ldots,s \ . \tag{4.26}$$

In the case of power systems this form appears most suitable for estimating the asymptotic stability domain. Then, $\zeta_{ij}(\mathbf{x}) \equiv \alpha_{ij} = \text{constant}$. In special cases, when $u_i(\mathbf{x}_i) = \phi_{i3}(\|\mathbf{x}_i\|)$, the aggregation form of Michel (1974) results.

Another suitable form for asymptotic stability analysis of power systems is obtained from Grujić (1974) (Chapter III) :

$$\sum_{i=1}^{s} \psi_i(\mathbf{x}_i) \leq \sum_{i,j=1}^{s} \zeta_{ij}(\mathbf{x})\, u_i(\mathbf{x}_i)\, u_j(\mathbf{x}_j) \quad , \quad \forall \mathbf{x} \in \mathbf{X} \ . \tag{4.27}$$

For power systems $\zeta_{ij}(\mathbf{x}) \equiv \alpha_{ij}$ is obtained, where then α_{ij} is ij-th element of the overall power system aggregation matrix $\mathbf{A}$, $\mathbf{A} = (\alpha_{ij})$.

Since this research is aimed to establish both asymptotic stability criteria and asymptotic stability domain estimates, the aggregation form (4.26) will be only used.

Seeking possibilities for the relaxation of the stability conditions and for the best stability domain estimate we considered different decomposition-aggregation forms. Some of them were proposed by Jocić et al. (1977,1978) and by Grujić et al. (1977) and others were new.

The aggregation form (4.26) appeared the most suitable for the power system aggregation. It enables one-shot stability analysis of the whole system without requiring majorizations of the comparison functions u_i in terms of V_i.

The investigations were done with the following comparison functions

$$u_i(\mathbf{x}_i) = \|\mathbf{x}_i\| \quad , \quad u_i(\mathbf{x}_i) = \|\mathbf{w}_i(\mathbf{x}_i)\| \quad \text{or} \quad u_i(\mathbf{x}_i) = (\mathbf{w}_i^T \mathbf{W}_i \mathbf{w}_i)^{1/2}$$

for

$$\mathbf{w}_i(\mathbf{x}_i) = [\phi_i(x_{i1})\ x_{i2}]^T \ ,$$

$$\mathbf{w}_i(\mathbf{x}_i) = \left[\sqrt{(|\phi_i(x_{i1})| + |b_i h_{12}^i \psi_i(x_{i1})|)\,|x_{i1}|}\ \ x_{i2}\right]^T ,$$

or

$$\mathbf{w}_i(\mathbf{x}_i) = \left[\sqrt{\phi_i(x_{i1})\, x_{i1}}\ \ x_{i2}\right]^T .$$

From the viewpoint of the relaxation of the stability conditions

$$\mathbf{w}_i(\mathbf{x}_i) = \left[\sqrt{(|\phi_i(x_{i1})| + |b_i h^i_{12}\psi_i(x_{i1})|)\ |x_i|}\ x_{i2}\right]^T$$

seems to be the best among them. But, its use essentially reduced the estimate of the domain of asymptotic stability. The comparison function $u_i(\mathbf{x}_i) = \|\mathbf{x}_i\|$ seems to enable the best estimate of the domain of asymptotic stability. Its use relaxes the stability conditions more than use of other forms of u_i . For this reason, we accept $u_i(\mathbf{x}_i) = \|\mathbf{x}_i\|$ and the proposed decomposition and aggregation of the power system.

V.4.3. General stability analysis of stationary large-scale systems

V.4.3.1. Stability criterion

Let a large-scale system described by the differential equation (4.1) be decomposed into s interconnected subsystems (4.15). A motion of (4.15) denoted ty $\mathbf{x}_i(t;\mathbf{x}_o)$, $\mathbf{x}_i(0,\mathbf{x}_o) \equiv \mathbf{x}_{i_o}$, is supposed to exist for every $\mathbf{x}_o \in R^M$ and to be continuous in $t \in R_+$ for every $\mathbf{x}_o \in R^M$.

Let X_i be a subset of R^{m_i} and X the cartesian product of all X_i's , $X = X_1 \times X_2 \times \ldots \times X_s$.

In the sequel the following assumption will be used :

Assumption 1. There exist a connected neighbourhood X_i of $\mathbf{x}_i = \mathbf{0}$, functions $V_i : R^{m_i} \to R$ and $u_i : R^{m_i} \to R$, and real numbers α_{ij} , $\alpha_{ij} \geq 0$ $i \neq j$, satisfying (i)-(iv) :

(i) $\mathbf{x} = \mathbf{0}$ is the unique equilibrium state of the system (4.15) in X ,

(ii) V_i is positive definite and differentiable on X_i , $\forall i = 1,2,\ldots,s$,

(iii) u_i is positive definite on X_i , $\forall i = 1,2,\ldots,s$,

(iv) the total time derivative $\dot{V}_i$ (4.21) of V_i taken along motions of (4.15) obeys (4.22), (4.26). ■

Notice that (4.21), (4.22), (4.26) define a system aggregation that links those of Michel (1974) and Grujić (1974). In comparison with Michel (1974) it allows the functions u_i to be any positive definite functions on the set X_i rather than to be those dependent on $\|\mathbf{x}_i\|$ and related to scalar Liapunov functions of disconnected subsystems (4.14) and simultaneously to be strictly increasing. The proposed system aggregation (4.21), (4.22), (4.26) allows also an one-shot both aggregation and stability analysis of the overall system. In comparison with Grujić (1974), it provides a larger estimate set of the asymptotic stability domain.

For stability analysis we need :

Assumption 2. The function V_i of Assumption 1 is *radially increasing* on an $\hat{\epsilon}_i$ -neighbourhood $N_i(X_i,\hat{\epsilon}_i)$ of X_i , that is

$$V_i(\mu_1 x_i) < V_i(\mu_2 x_i) \quad \text{iff} \quad 0 \le \mu_1 < \mu_2 \ ,$$
$$\forall \mu_k x_i \in N_i(X_i,\hat{\epsilon}_i) \ , \ \forall i = 1,2,\dots,s \ , \ \forall k = 1,2 \ ,$$

where

$$N_i(X_i,\hat{\epsilon}_i) = \{x : \inf(\|x-y\| : y \in X_i) \le \hat{\epsilon}_i\} \ , \quad \hat{\epsilon}_i > 0 \ . \ \blacksquare$$

A criterion for asymptotic stability of $x = 0$ of (4.1) is established in this section by referring to Grujić (1974), Michel (1974) and Theorem 23 of Section III.4.2.5 in which $L = A^T B + BA$, $\epsilon = 0$ so that $M(L,\epsilon) = A^T B + BA$.

Theorem 1. *If Assumption 1 holds,* $A = (\alpha_{ij})$ *and there is a diagonal matrix* B *with positive diagonal such that* $A^T B + BA$ *is negative definite; then* $x = 0$ *of the system (4.15) is asymptotically stable.*

This theorem is also valid when $\alpha_{ij} \ge 0$, $i \ne j$, is not satisfied in Assumption 1.

In particular cases determined by negative definiteness of $A^T + A$ it is sufficient to set $B = I$ - the identity $s \times s$ matrix.

La Salle (1976) stated that there is a diagonal matrix B with positive diagonal such that $A^T B + BA$ is negative definite as soon as the Metzler $s \times s$ matrix $A = (\alpha_{ij})$ obeys (4.28) (see Conjecture 1 in Section V.4.3.2 for possible link between this statement and d) of Proposition 13 in Section II.3.4) :

$$(-1)^k \begin{vmatrix} \alpha_{11} & \alpha_{12} & \dots & \alpha_{1k} \\ \alpha_{21} & \alpha_{22} & \dots & \alpha_{2k} \\ \text{---} & \text{---} & \dots & \text{---} \\ \alpha_{k1} & \alpha_{k2} & \dots & \alpha_{kk} \end{vmatrix} > 0 \ , \quad \forall k = 1,2,\dots,s \ . \qquad (4.28)$$

Corollary 1. *If Assumption 1 holds and* A *obeys (4.28) then* $x = 0$ *of the system (4.15) is asymptotically stable.*

Positive result on asymptotic stability of $x = 0$ often is not sufficient for engineering applications, e.g. for power systems stability analysis. Knowledge, or at least an estimate, of the domain of asymptotic stability is needed.

V.4.3.2. *Estimates of the asymptotic stability domain*

In order to define tentative estimates $\mathcal{E}_1$ and $\mathcal{E}_2$ of $\mathcal{D}$ we use $\mathbf{V}$ (4.10), v_1 (4.19) by following Grujić and Šiljak (1973) as well as

$$\gamma_1 = \min \, [b_i V_i^o : i = 1,2,\dots,s] \, . \tag{4.29}$$

Weissenberger (1973) introduced

$$v_2(\mathbf{x}) = \max \left[\frac{V_i(\mathbf{x}_i)}{b_i} : i = 1,2,\dots,s\right] , \tag{4.30}$$

$$\gamma_2 = \min \left[\frac{V_i^o}{b_i} : i = 1,2,\dots,s\right] . \tag{4.31}$$

Here V_i^o is the i-th component of $\mathbf{V}^o$ such that the closure $\bar{\mathcal{V}}_i$ of

$$\mathcal{V}_i = \{\mathbf{x}_i : 0 \le V_i(\mathbf{x}_i) < V_i^o\} \tag{4.32}$$

is connected subset of $\mathcal{X}_i$ and contains $\mathbf{x}_i = \mathbf{0}$. Besides, $\partial\bar{\mathcal{V}}_i$ will denote the boundary of $\bar{\mathcal{V}}_i$ whenever $\bar{\mathcal{V}}_i$ is bounded.

Now we can define $\mathcal{E}_1$ and $\mathcal{E}_2$,

$$\mathcal{E}_i = \{\mathbf{x} : v_i(\mathbf{x}) < \gamma\} \quad , \quad \forall i = 1,2 \, , \tag{4.33}$$

and the closure of $\mathcal{E}_i$ is denoted by $\bar{\mathcal{E}}_i$.

It is to be noted that Weissenberger (1973) presented the excellent analysis of the exponential stability domain estimate of $\mathbf{x} = \mathbf{0}$ of (4.15).

The set $\mathcal{V}$,

$$\mathcal{V} = \mathcal{V}_1 \times \mathcal{V}_2 \times \dots \times \mathcal{V}_s = \{\mathbf{x} : \mathbf{0} \le \mathbf{V}(\mathbf{x}) \le \mathbf{V}^o \, , \, \mathbf{x} \in \mathcal{X}\} \, , \tag{4.34}$$

is the largest possible estimate of $\mathcal{D}$ that we can derive for chosen $\mathbf{V}$ and $\mathbf{V}^o$. In order to establish conditions under which $\mathcal{V}$ is an estimate of $\mathcal{D}$ we introduce the matrix $\mathbf{E} = (e_{ij})$ by

$$e_{ij} = [r_{i1}\delta_{ij} + r_{i2}(1-\delta_{ij})]\,\alpha_{ij} \quad , \quad \forall i,j = 1,2,\dots,s \, . \tag{4.35}$$

Here δ_{ij} is the Kronecker symbol, all $\mathcal{V}_i$'s are assumed bounded and

$$r_{i1} = \frac{\inf \, [u_i(\mathbf{x}_i) : \mathbf{x}_i \in \partial\mathcal{V}_i]}{V_i^o} \quad , \quad r_{i2} = \frac{\sup \, [u_i(\mathbf{x}_i) : \mathbf{x}_i \in \bar{\mathcal{V}}_i]}{V_i^o} \, . \tag{4.36}$$

Now, by referring to Grujić and Ribbens-Pavella (1977), Grujić et al. (1979), one arrives at

Theorem 2. *If Assumptions 1 and 2 hold and if* $\mathbf{E}\mathbf{V}^o < \mathbf{0}$ *then the set* $\mathcal{V}$ *(4.34) is an estimate of* $\mathcal{D}$ *of* $\mathbf{x} = \mathbf{0}$ *of the system (4.15).*

Proof (Grujić and Ribbens-Pavella, 1977; Grujić et al., 1979). Let

$$\mathcal{P}_{1\gamma} = \{\mathbf{V} : 0 \le \mathbf{b}^T\mathbf{V} \le \gamma\} \, , \tag{4.37}$$

$$P_{2\gamma} = \{\mathbf{V} : 0 \leq \frac{V_i}{b_i} \leq \gamma \ , \forall i = 1,2,\dots,s\} \ , \tag{4.38}$$

and

$$E_{i\gamma} = \{\mathbf{x} : 0 \leq V_i(\mathbf{x}) \leq \gamma\} \quad , \ \forall i = 1,2 \ , \tag{4.39}$$

be connected sets containing the origins. Hence,

$$E_{i\gamma} = \{\mathbf{x} : \mathbf{V}(\mathbf{x}) \in P_{i\gamma} \ , \mathbf{x} \in X\} \quad \text{and} \quad E_{i\gamma_i} = \bar{E}_i \quad , \quad i = 1,2 \ , \tag{4.40}$$

as well as

$$E_2 = V \ \text{ iff } \ V_i^o b_i^{-1} = \gamma_2 \quad , \ \forall i = 1,2,\dots,s \ . \tag{4.41}$$

Let $\partial^+ P_{1\gamma}$ and $\partial^+ P_{2\gamma_i}$ be defined by

$$\partial^+ P_{1\gamma} = \{\mathbf{V} : \mathbf{b}^T \mathbf{V} = \gamma \ , \mathbf{V} \in P_{1\gamma}\} \tag{4.42}$$

and

$$\partial^+ P_{2\gamma_i} = \{\mathbf{V} : V_i b_i^{-1} = \gamma \ , \mathbf{V} \in P_{2\gamma}\} \ . \tag{4.43}$$

Let

$$\partial E_{2\gamma_i} = \{\mathbf{x} : \mathbf{V}(\mathbf{x}) \in \partial^+ P_{2\gamma_i} \ , \mathbf{x} \in E_{2\gamma}\} \quad . \tag{4.44}$$

Under Assumption 2 the following is obvious for $\partial E_{1\gamma}$ - the boundary of $E_{1\gamma}$:

Statement I. If $\mathbf{x} \in E_{1\gamma}$ then $\mathbf{V}(\mathbf{x}) \in \partial^+ P_{1\gamma}$ iff $\mathbf{x} \in \partial E_{1\gamma}$.

Let $\partial E_{2\gamma}$ denote the boundary of $E_{2\gamma}$. Now,

Statement II. $\mathbf{x} \in \partial E_{2\gamma}$ iff $\exists i \in \{1,2,\dots,s\}$ such that $\mathbf{x} \in \partial E_{2\gamma_i}$, i.e.

$$\partial E_{2\gamma} = \bigcup_{i=1}^{s} \partial E_{2\gamma_i} \ .$$

Statement III. Under Assumption 2 and for $\epsilon_m = \min \{\hat{\epsilon}_i : i = 1,2,\dots,s\}$,

$$\mathbf{V}(\mathbf{x}) \leq \mathbf{V}(\mathbf{y}) < \mathbf{V}(\mathbf{z}) \ , \ \forall \mathbf{x} \in E_{i\gamma} \ , \ \forall \mathbf{y} \in \partial E_{i\gamma} \ , \ \forall \mathbf{z} \in N(E_{i\gamma}, \epsilon_m) \setminus E_{i\gamma} \ , \ \forall \gamma \in]0,\gamma_i[$$

$$\mathbf{y} = \lambda_1 \mathbf{x} \quad , \quad \mathbf{z} = \lambda_2 \mathbf{x} \ , \ \lambda_k \in R \ , \ \forall k = 1,2 \ , \ \forall i = 1,2 \ .$$

Statements I-III imply Lemma 1.

Lemma 1. If there are : a compact neighbourhood X_i of $\mathbf{x}_i = \mathbf{0}$, functions v_i and positive numbers $\hat{\epsilon}_i$ such that Assumptions 1 and 2 hold and if along motions of the system (4.15) :

$$\dot{v}_i(\mathbf{x}) < 0 \ , \ \forall \mathbf{x} \in \partial E_{i\gamma} \ , \ \gamma \in]0,\gamma_i] \ , \ i \in \{1,2\} \tag{4.45}$$

then $E_{i\gamma}$ is a positively invariant set of the system (4.15).

Statement IV. The function v_2 of equation (4.31) is positive definite and radially increasing on $N(X,\hat{\epsilon})$ for some $\hat{\epsilon} > 0$ due to (ii) of Assumption 1, all $b_i > 0$ and Assumption 2.

Assumption 1 and the definition of E (4.35),(4.36) imply

$$\mathbf{A}\mathbf{u}(\mathbf{x}) \le \mathbf{E}V^{\circ} \quad , \quad \forall \mathbf{x} \in \partial V \tag{4.46}$$

where

$$\mathbf{u} = [u_1, u_2, \dots, u_s]^T \ . \tag{4.47}$$

From equation (4.28) applied to E and (iii) of Assumption 1 it results that

$$\operatorname{sign} \dot{V}_i(\mathbf{x}) = \begin{cases} 0 & \text{if} \quad \mathbf{x}_i = \mathbf{0} \ , \\ \operatorname{sign}\left[\sum_{j=1}^{s} \alpha_{ij} u_j(\mathbf{x}_j) \right] & \text{if} \quad \mathbf{x}_i \neq \mathbf{0} \ , \ \forall i = 1,2,\dots,s \ , \ \forall \mathbf{x} \in X \ , \end{cases} \tag{4.48}$$

where $\operatorname{sign} \xi = \xi |\xi|^{-1}$ for $\xi \neq 0$ and $\operatorname{sign} 0 = 0$. Equations (4.46) and (4.48) and $\mathbf{E}V^{\circ} < \mathbf{0}$ (the condition of Theorem 2) imply

$$\dot{V}(\mathbf{x}) < \mathbf{0} \ , \ \forall \mathbf{x} \in \partial V \quad , \quad \mathbf{x}_i \neq \mathbf{0} \ , \ \forall i = 1,2,\dots,s \ . \tag{4.49}$$

This result, equation (4.48), Assumption 1 and Lemma 1 prove that $\bar{V}$ is a positively invariant set of the system (4.15).

Further,

$$\mathbf{E}V^{\circ} < \mathbf{0} \ , \ V^{\circ} > \mathbf{0} \quad \text{and} \quad e_{ij} \ge 0 \ , \ i \neq j \ , \ \forall i,j = 1,2,\dots,s \ ,$$

imply stability of E, that is, that $\mathbf{E} = (e_{ij})$ obeys relationship (4.28) (see Proposition 13 in Chapter II.3.4).

Let $\psi_i = r_{i1} r_{i2}^{-1}$, $\forall i = 1,2,\dots,s$. Now,

Statement V. $\psi_i \in]0,1[$, $\forall i = 1,2,\dots,s$.

Let $\mathbf{A}^{\psi} = (\alpha_{ij}^{\psi})$, $\alpha_{ij}^{\psi} = \psi_i \alpha_{ii} \delta_{ij} + \alpha_{ij}(1-\delta_{ij})$, $\forall i,j = 1,2,\dots,s$.

Statement VI. $\mathbf{A} \le \mathbf{A}^{\psi}$ as soon as all $\alpha_{ii} \le 0$ and $\alpha_{ij} \ge 0$, $i \neq j$, $\forall i,j = 1,2,\dots,s$.

When E obeys relationship (4.28), $\mathbf{A}^{\psi}$ also obeys relationship (4.28) due to the definitions of $\mathbf{A}^{\psi}$ and E and the property of determinants that multiplication of any (row) column by a positive number (e.g. r_{i2}^{-1}) does not influence the sign of a determinant. Hence :

Statement VII. Stability of E implies stability of $\mathbf{A}^{\psi}$. Because $\mathbf{A}^{\psi}$ is Metzlerian and stable, there is $\mathbf{b} \in \mathring{R}_+^s$, $\mathbf{b} = [b_1, b_2, \dots, b_s]^T$, such that (Proposition 13,c in Section II.3.4) $\mathbf{A}^{\psi}\mathbf{b} < \mathbf{0}$, which implies, together with Statement VI, that $\mathbf{A}\mathbf{b} < \mathbf{0}$ and (Proposition 13,d in Section II.3.4) :

Statement VIII. There is diagonal matrix B with positive diagonal such that $(\mathbf{A}^T\mathbf{B} + \mathbf{B}\mathbf{A})$ is negative definite.

In the references, Grujić and Ribbens-Pavella (1977), Grujić et al. (1979), we accepted $\mathbf{B} = \mathrm{diag}\{b_1,b_2,\dots,b_s\}$ satisfying Statement VIII and used b_i's for $\mathbf{b}$: $\mathbf{b} = [b_1,b_2,\dots,b_s]^T$. Hence,

$$\mathbf{u}^T(\mathbf{x})(\mathbf{A}^T\mathbf{B} + \mathbf{B}\mathbf{A})\,\mathbf{u}(\mathbf{x}) < \mathbf{0} \ , \ \forall(\mathbf{x}\neq\mathbf{0}) \in \bar{\mathcal{V}} \ .$$

Now, the following is obvious :

Statement IX. Both v_1 and $(-\dot{v}_1)$ are positive definite on $\bar{\mathcal{V}}$.

Since $\bar{\mathcal{V}}$ is positively invariant set and Statement IX holds, it results that

$$\lim\,[\,\|\mathbf{x}(t;\mathbf{x}_o)\| : t \to +\infty] = 0 \quad , \quad \forall \mathbf{x}_o \in \bar{\mathcal{V}} \ . \tag{4.50}$$

Hence :

Statement X. $\bar{\mathcal{V}}$ is an estimate of the domain of attraction of $\mathbf{x}=\mathbf{0}$ of the system (4.15).

Let $\epsilon\in\overset{o}{\mathcal{R}}_+$ be arbitrary and $\gamma>0$ be a number such that $\mathcal{E}_{1\gamma} \subset \mathcal{B}_\epsilon$, $\mathcal{B}_\epsilon = \{\mathbf{x} : \|\mathbf{x}\| < \epsilon\}$. Then, $\mathbf{x}_o \in \mathcal{E}_{1\gamma} \cap \bar{\mathcal{V}}$ implies $\mathbf{x}(t;\mathbf{x}_o) \in \mathcal{B}_\epsilon$, $\forall t \in \mathcal{R}_+$, due to the positive invariability of $\bar{\mathcal{V}}$ and Statement IX.

Notice that $\gamma = \gamma(\epsilon)$ is continuous and strictly increasing in $\epsilon \in \mathcal{R}_+$, and that there are positive numbers γ^* and ϵ^* satisfying $\bar{\mathcal{V}} \subset \mathcal{E}_{1\gamma^*} \subset \mathcal{B}_{\epsilon^*}$. Therefore, $\mathbf{x}_o \in \bar{\mathcal{V}}$ guarantees $\mathbf{x}(t;\mathbf{x}_o) \in \mathcal{B}_\epsilon$, $\forall t \in \mathcal{R}_+$, $\forall\epsilon \in [\epsilon^*,+\infty[$. Hence :

Statement XI. $\bar{\mathcal{V}}$ is an estimate of the domain of stability of $\mathbf{x}=\mathbf{0}$ of the system (4.15).

Statements X and XI prove the theorem in view of Definitions 4-7 (Section 2.3 of this Chapter). ∎

Theorem 2 presents quite simple conditions for estimation of $\mathcal{D}$. No stability test of $\mathbf{E}$ is required. However, since $\mathbf{E}$ is a Metzler matrix then $\mathbf{E}\mathbf{V}^\circ < \mathbf{0}$ can be satisfied iff $\mathbf{E}$ is stable (Proposition 13, Section II.3.4). Hence, stability of $\mathbf{E}$ is implicitly required. Under the condition $\mathbf{E}\mathbf{V}^\circ < \mathbf{0}$ we get the best possible estimate of $\mathcal{D}$ for chosen $\mathbf{V}$ and $\mathbf{V}^\circ$. When the conditions of Theorem 2 are not met then we look for other estimates of $\mathcal{D}$ under relaxed conditions. Certainly, these estimates cannot be better than the $\mathcal{V}$ estimate (4.34).

If there exist positive numbers η_{i1} and η_{i2} such that

$$\eta_{i1}u_i^2(\mathbf{x}_i) \le V_i(\mathbf{x}_i) \le \eta_{i2}u_i^2(\mathbf{x}_i) \quad , \ \forall \mathbf{x}_i \in \mathcal{X}_i \ , \ \forall i = 1,2,\dots,s \ , \tag{4.51}$$

then we construct matrix $\mathbf{K} = (k_{ij})$,

$$k_{ij} = [\eta_{i2}^{-1/2}\,\delta_{ij} + \eta_{j1}^{-1/2}(1-\delta_{ij})]\,\alpha_{ij} \quad , \ \forall i,j = 1,2,\dots,s \ , \tag{4.52}$$

and propose the next theorem by Grujić et al. (1979) in which

$$(\mathbf{V})^{1/2} = [V_1^{1/2}, V_2^{1/2}, \dots, V_s^{1/2}]^T .$$

Theorem 3. *If Assumptions 1 and 2 hold and the matrix* $\mathbf{K}$ *fulfils (4.28) then* E_2 *is an estimate of* $\mathcal{D}$ *of* $\mathbf{x}=\mathbf{0}$ *of the system (4.15). If in addition* $\mathbf{K}\mathbf{V}^{\circ 1/2} < \mathbf{0}$ *then* V *(4.34) is an estimate of* $\mathcal{D}$.

Proof (Grujić and Ribbens-Pavella, 1977; Grujić et al., 1979). Inequalities (4.51) and equations (4.52) yield

$$\sum_{j=1}^{s} \alpha_{ij} u_j(x_j) \le \sum_{j=1}^{s} k_{ij} V_j^{1/2}(\mathbf{x}_j) , \quad \forall \mathbf{x} \in X , \ \forall i = 1,2,\dots,s . \tag{4.53}$$

Further, for $\hat{\mathbf{b}} = (\hat{b}_1, \hat{b}_2, \dots, \hat{b}_s)^T$, $b_i = \hat{b}_i^2$, $\forall i = 1,2,\dots,s$,

$$\sum_{j=1}^{s} k_{ij} V_j^{1/2}(\mathbf{x}_j) \le \gamma^{1/2} \sum_{j=1}^{s} k_{ij} \hat{b}_j ,$$
$$\forall \mathbf{x} \in \partial E_{2\gamma_i} , \ \forall \gamma \in]0, \gamma_2] , \ \forall i = 1,2,\dots,s . \tag{4.54}$$

Let $\hat{\mathbf{b}} = -(\mathbf{K})^{-1}\mathbf{c}$ for $\mathbf{c} \in \mathring{R}_+^s$. Now, $\hat{\mathbf{b}} > \mathbf{0}$ because $\mathbf{K}$ obeys relationship (4.28) and $\mathbf{K}$ is a Metzler matrix. This choice of $\hat{\mathbf{b}}$, equations (4.53) and (4.54) and Assumption 1 yield

$$\dot{V}_i(\mathbf{x}) \le \gamma^{1/2} u_i(\mathbf{x}_i) \sum_{j=1}^{s} k_{ij} \hat{b}_j < 0 , \ \forall \mathbf{x} \in \partial E_{2\gamma_i} , \ \forall i = 1,2,\dots,s , \ \forall \gamma \in]0, \gamma_2] . \tag{4.55}$$

The following statements result from equations (4.40), (4.42) and (4.44).

Statement XII. For every $(\mathbf{y} \neq \mathbf{0}) \in R^M$, $\mathbf{y}$ is partitioned as $\mathbf{x}$, there exist $\alpha > 0$, $\gamma = \gamma(\mathbf{y})$, and $i \in \{1,2,\dots,s\}$ such that $\mathbf{y} \in \partial E_{2\gamma(\mathbf{y})i}$. In fact, $\gamma(\mathbf{y}) = \max\{V_i(\mathbf{y}_i)(\hat{b}_i^2)^{-1} : i = 1,2,\dots,s\}$, i.e. $\gamma(\mathbf{y}) = v_2(\mathbf{y})$.

Statement XIII. The function $\gamma : R^M \to R_+$ is continuous : $\gamma(\mathbf{x}) \in C(R^M)$.

Now, statements XII and XIII together with equation (4.55) imply $\forall \mathbf{y} \in E_{2\gamma_2}$, $\exists i \in \{1,2,\dots,s\}$ such that $\mathbf{y} \in \partial E_{2\gamma(\mathbf{y})i}$ and

$$\dot{V}_i(\mathbf{y}) \le \gamma^{1/2}(\mathbf{y}) u_i(\mathbf{y}_i) \sum_{j=1}^{s} k_{ij} \hat{b}_j . \tag{4.56}$$

Let
$$\gamma[\mathbf{x}(t;\mathbf{x}_o)] = \Gamma(t;\mathbf{x}_o) . \tag{4.57}$$

Statement XIV. Statement XIII and continuity of $\mathbf{x}$ in $t \in R_+$ guarantee continuity of Γ in $t \in R_+$, $\forall \mathbf{x}_o \in R^M$.

Let it be supposed that for arbitrary $\mathbf{x}_o \in E_2$, $\mathbf{x}_o \neq \mathbf{0}$, there exists $\gamma_o \in]0, \gamma_2[$ such that $E_{2\gamma_o} \subset E_{2\Gamma(t,\mathbf{x}_o)}$, $\forall t \in R_+$.

Then, there exist $\zeta_o = \zeta(\mathbf{x}_o) \in]0,+\infty[$ and $\xi_o = \xi(\mathbf{x}_o) \in]0,+\infty[$ such that $\zeta_o < \Gamma^{1/2}(t,\mathbf{x}_o)$ and $u_i[\boldsymbol{\chi}_i(t;\mathbf{x}_o)] > \xi_o$, $\forall t \in R_+$, $\forall i = 1,2,\ldots,s$, because of Statement XIV and equation (4.57). Hence,

$$\dot{v}_2[\boldsymbol{\chi}(t;\mathbf{x}_o)] \leq \zeta_o \xi_o \max \left\{ \sum_{j=1}^{s} k_{ij}\hat{b}_j : i = 1,2,\ldots,s \right\} < 0 . \tag{4.58}$$

The right-hand side is negative due to the choice of $\hat{\mathbf{b}}$, $\hat{\mathbf{b}} = -(\mathbf{K}^T)^{-1}\mathbf{c}$, and $\mathbf{c} \in \mathring{R}_+^s$. Inequality (4.58) can be integrated from $t_o = 0$ to $t \in R_+$,

$$v_2[\boldsymbol{\chi}(t;\mathbf{x}_o)] \leq v_{20} + \zeta_o \xi_o t \max \left\{ \sum_{j=1}^{s} k_{ij}\hat{b}_j : i = 1,2,\ldots,s \right\} .$$

This implies the following absurd :

$$v_2[\boldsymbol{\chi}(t;\mathbf{x}_o)] < 0 \ , \ \forall t \in]\tau,+\infty[\ , \ \forall(\mathbf{x}_o \neq \mathbf{0}) \in E_2 \ ,$$

where

$$\tau = v_{20}\left[-\zeta_o \xi_o \max \left\{ \sum_{j=1}^{s} k_{ij}\hat{b}_j : i = 1,2,\ldots,s \right\}\right]^{-1} \in]0,+\infty[\ ,$$

which is a consequence of the assumption that $\gamma_o > 0$ instead of the true value $\gamma_o = 0$. But $\gamma_o = 0$ implies $\lim [E_{2\Gamma(t)} : t \to +\infty] = \{\mathbf{0}\}$. This result proves

$$\lim [\|\boldsymbol{\chi}(t;\mathbf{x}_o)\| : t \to +\infty] = 0 \ , \ \forall \mathbf{x}_o \in E_2 . \tag{4.59}$$

Further, let $\epsilon \in]0,+\infty[$ be arbitrary and $\gamma_\epsilon \in]0,\gamma_2]$ such that

$$E_{2\gamma_\epsilon} \subseteq B_\epsilon \cap E_2 \tag{4.60}$$

Now, we define $\delta = \delta(\epsilon) \in]0,+\infty[$ so that

$$B_\delta \subseteq E_{2\gamma_\epsilon} \tag{4.61}$$

Statement XV. The motion $\boldsymbol{\chi}(t;\mathbf{x}_o)$ can leave $E_{2\gamma}$, $\mathbf{x}_o \in E_{2\gamma}$, iff it passes through $\partial E_{2\gamma}$:

$$\exists \tau \in [0,+\infty[\ , \ \tau = \tau(\mathbf{x}_o,\gamma) \ , \ \boldsymbol{\chi}(\tau;\mathbf{x}_o) \in \partial E_{2\gamma} \ , \ \forall \gamma \in]0,\gamma_2] .$$

Assumption 2, Statements II and XV, and (4.55) prove :

Statement XVI. The set $E_{2\gamma}$ is a positively invariant set of the system (4.15), $\forall \gamma \in]0,\gamma_2]$.

Equations (4.60) and (4.61) and Statement XVI imply (4.62),

$$\boldsymbol{\chi}(t;\mathbf{x}_o) \in B_\epsilon \ , \ \forall t \in R_+ \ , \ \forall \mathbf{x}_o \in B_\delta \ , \ \forall \epsilon \in \mathring{R}_+ . \tag{4.62}$$

Moreover, Statement XVI implies (4.63),

$$\boldsymbol{\chi}(t;\mathbf{x}_o) \in B_\epsilon \ , \ \forall t \in R_+ \ , \ \forall \mathbf{x}_o \in E_2 \ , \ \forall \epsilon \in [\epsilon^*,+\infty[\ , \tag{4.63}$$

where ϵ^* is a positive number such that $E_2 \subset B_{\epsilon^*}$.

Now, the results (4.59)-(4.63) prove that E_2 is an estimate of the domain of asymptotic stability of $x=0$ of the system (4.15). If, in addition, $KV^{o1/2}<0$ then we can set $b_i b_j^{-1} = V_i^o (V_j^o)^{-1}$, which implies $E_2 = V$ because of equation (4.41), and completes the proof by following the proof of Theorem 2. ■

Theorem 1 and Theorem 2 impose conditions on systems through E and K matrix, respectively. In essence, these matrices should obey (4.28), which is sharper than to require A to obey (4.28). Hence, it is interesting to discover under what conditions A matrix can be used for the asymptotic stability domain estimate. One case happens when $u_i = V_i^{1/2}$, $\forall i = 1,2,\dots,s$. Then $\eta_{i1} = \eta_{i2} = 1$, $\forall i = 1,2,\dots,s$, and, therefore, $K = A$ in Theorem 3. In such a case, the conditions of Theorem 3 are more relaxed than those of Weissenberger (1973).

Another case when A can be employed for estimating D is discovered by

Theorem 4. *If Assumptions 1 and 2 are fulfilled and the matrix* $A^T B + BA$ *is negative definite for a diagonal matrix* B *with positive diagonal then the set* E_1 *is an estimate of* D *of* $x=0$ *of the system (4.15).*

Proof. Along motions of equation (4.15) the derivative

$$\dot{v}_1(x) \leq u^T(x)[A^T B + BA]\, u(x) \ , \quad \forall x \in X \ , \tag{4.63}$$

because of Assumption 1. This result, Assumption 2, positive-definiteness of all u_i on X_i, negative definiteness of $A^T B + BA$ and Lemma 1 prove that $E_{1\gamma}$ is a positively invariant set of the system (4.15), $\forall \gamma \in]0,\gamma_1]$. Hence, equation (4.63), positive definiteness of v_1 for $b = [b_1, b_2, \dots, b_s]^T$, b_i being the i-th diagonal element of B, and the last result prove

$$\lim \left[\|\chi(t;x_o)\| : t \to +\infty \right] = 0 \ , \quad \forall x_o \in E_1 \ . \tag{4.64}$$

Let $\epsilon \in]0,+\infty[$ be arbitrary and $\gamma = \gamma(\epsilon)$ such that

$$E_{1\gamma} \subset B_\epsilon \cap E_1 \ . \tag{4.65}$$

Let $\delta = \delta(\epsilon)$ obey

$$B_\delta \subset E_{1\gamma} \ . \tag{4.66}$$

Positive invariability of $E_{1\gamma}$, $\forall \gamma \in]0,\gamma_1]$ and equations (4.65) and (4.66) prove

$$\chi(t;x_o) \in B_\epsilon \ , \quad \forall t \in R_+ \ , \quad \forall x_o \in B_\delta \ , \quad \forall \epsilon \in]0,+\infty[\ . \tag{4.67}$$

Further, for ϵ^* such that $E_1 \subseteq B_{\epsilon^*}$, $\epsilon^* \in]0,+\infty[$, we have :

$$\chi(t;x_o) \in B_\epsilon \ , \quad \forall t \in R_+ \ , \quad \forall x_o \in E_1 \ , \quad \forall \epsilon \in]\epsilon^*,+\infty[\ . \tag{4.68}$$

Equations (4.64), (4.67) and (4.68) prove that $\mathcal{E}_1$ is an estimate of $\mathcal{D}$ in view of Definitions 4-7 (Section 2.3 of this Chapter). ∎

This criterion requires knowledge of $\mathbf{B}$ in order the set $\mathcal{E}_1$ to be completely determined. For the asymptotic stability domain estimate it is not enough to know that there exists a diagonal matrix $\mathbf{B}$ with positive diagonal for which $\mathbf{A}^T\mathbf{B}+\mathbf{BA}$ is negative definite. The first trial for determination of $\mathbf{B}$ can be with $\mathbf{B}=\mathbf{I}$. The second one is with $\mathbf{B}=\text{diag}\{b_1\ b_2\ \dots\ b_s\}$ and b_i - i-th element of $\mathbf{b}$ determined by $\mathbf{b}=-(\mathbf{A}^T)^{-1}\mathbf{c}$ for $\mathbf{c}\in\mathring{R}_+^s$. Then, conditions (4.28) are to be verified for the matrix $\mathbf{A}$ (Proposition 13 in Section II.3.4) for which the following conjecture can be useful.

Conjecture 1. If a Metzler matrix $\mathbf{A}$ obeys (4.28) and $\mathbf{b}=[b_1\ b_2 \dots b_s]^T$ is determined from (4.69),

$$\mathbf{b} = -(\mathbf{A})^T\mathbf{c} \tag{4.69}$$

for $\mathbf{c}\in\mathring{R}_+^s$ then $\mathbf{A}^T\mathbf{B}+\mathbf{BA}$ is negative definite as soon as $\mathbf{B}=\text{diag}\{b_1\ b_2\ \dots\ b_s\}$.

Remark 5. The Conjecture 1 fails for the following *counterexample*. Let

$$\mathbf{A} = \begin{bmatrix} -1 & 10 \\ 20 & -201 \end{bmatrix}, \quad \mathbf{c} = \begin{bmatrix} 1 \\ 1 \end{bmatrix} \in \mathring{R}_+^2$$

$\mathbf{A}$ obeys (4.28) and

$$\mathbf{b} = -(\mathbf{A}^T)^{-1}\mathbf{c} = \begin{bmatrix} 221 \\ 11 \end{bmatrix} \in \mathring{R}_+^2 .$$

Hence,

$$\mathbf{B} = \text{diag}\,\{221 \quad 11\} .$$

However,

$$\mathbf{A}^T\mathbf{B}+\mathbf{BA} = \begin{bmatrix} -442 & 2430 \\ 2430 & -4422 \end{bmatrix}$$

is not negative definite.

Remark 6. If two theorems or all three, among Theorems 2-4, are satisfied then the final estimate of $\mathcal{D}$ is the union of estimates determined by the theorems.

V.4.4. Power systems modelling

V.4.4.1. Uniform damping

Consider an N-machine power system. With the simplifying assumptions adopted in Sections V.2.1 and V.2.5 (constant : E_i , P_{mi} , M_i and system structure), the motion of the i-th machine taken separately is expressed by eqs. (2.1),(2.3).

When moreover we neglect electromagnetic damping, we consider uniform mechanical damping, and we choose N-th machine to act as the reference generator, then we get the state variables (3.1) and the state equations (3.2) (Section V.3.2.1), repeated as (4.70) :

$$\begin{aligned} \dot{\delta}_{iN} &= \Omega_{iN} \\ \dot{\Omega}_{iN} &= -\lambda\Omega_{iN} + M_i^{-1}(P_{mi}-P_{ei}) - M_N^{-1}(P_{mN}-P_{eN}) \quad , \quad \forall i = 1,2,\ldots,n \, . \end{aligned} \tag{4.70}$$

In order that the origin of the state space coincides with one of the equilibrium solutions of (4.70), which is expressed by (3.3) (Section V.3.2), we adopt suitable new state variables and state vector :

$$\mathbf{x}^T = [\sigma_{1N},\Omega_{1N},\sigma_{2N},\Omega_{2N},\ldots,\sigma_{nN},\Omega_{nN}]^T \in R^{2n} \, . \tag{4.71}$$

System (4.28) may now be written as follows :

$$\begin{aligned} \dot{\sigma}_{iN} &= \omega_{iN}, \\ \dot{\omega}_{iN} &= -\lambda\omega_{iN} - M_i^{-1} \sum_{\substack{j=1 \\ j\neq 1}}^{N} f_{ij}(\sigma_{ij}) + M_N^{-1} \sum_{j=1}^{n} f_{Nj}(\sigma_{Nj}) \quad , \quad \forall i = 1,2,\ldots,n \, , \end{aligned} \tag{4.72}$$

where

$$f_{ij}(\sigma_{ij}) = A_{ij} \, [\cos(\sigma_{ij} + \delta_{ij}^{o} - \theta_{ij}) - \cos(\delta_{ij}^{o} - \theta_{ij})] \quad , \quad \forall i,j = 1,2,\ldots,N \, , \tag{4.73}$$

and where we have set $\sigma_{iN} = \delta_{iN} - \delta_{iN}^{o}$ and $\omega_{iN} \equiv \Omega_{iN}$ for the sake of compatibility with the non-uniform damping case.

V.4.4.2. Non-uniform damping

With both mechanical and electromagnetic dampings taken into account, the absolute motion of i-th machine is described by (2.1) (Section V.2.1), where P_{ei} is still expressed by (2.3) (Section V.2.5).

There are 2N-1 state variables in the present case, namely

$$\delta_{1N},\delta_{2N},\ldots,\delta_{nN},\Omega_1,\Omega_2,\ldots,\Omega_N \, .$$

From (2.1) we set the state equations of the system (see Notation) :

$$\begin{aligned} \dot{\delta}_{iN} &= \Omega_i - \Omega_N \quad , \quad i = 1,2,\ldots,n = N-1 \, , \\ \dot{\Omega}_i &= -\lambda_i\Omega_i - \sum_{j=1}^{N} \lambda_{ij}(\Omega_i - \Omega_j) + M_i^{-1}[P_{mi} - P_{ei}(\boldsymbol{\delta})] = f_i(\boldsymbol{\delta},\boldsymbol{\Omega}) \quad , \quad i = 1,2,\ldots,N \, . \end{aligned} \tag{4.74}$$

The equilibrium states of (4.74) are given by :

$$\Omega_1^o = \Omega_2^o = \dots = \Omega_n^o = \Omega_N^o = \Omega^o = \frac{\sum_{i=1}^{N} [P_{mi} - P_{ei}(\boldsymbol{\delta}^o)]}{\sum_{i=1}^{N} D_i} \tag{4.75}$$

$$f_i(\boldsymbol{\delta}^o, \boldsymbol{\Omega}^o) = 0 \quad , \quad i = 1,2,\dots,N \ .$$

Let us make the origin of the state space coincide with one of the equilibrium solutions (4.75) by choosing the new variables ω_i and σ_{iN} ,

$$\omega_i = \Omega_i - \Omega^o \quad , \quad i = 1,2,\dots,N$$

$$\sigma_{iN} = \delta_{iN} - \delta_{iN}^o \quad , \quad i = 1,2,\dots,n \ .$$

The state vector is now :

$$\mathbf{x} = [\sigma_{1N}, \sigma_{2N}, \dots, \sigma_{nN}, \Omega_1, \Omega_2, \dots, \Omega_N]^T \in R^{n+N} \ .$$

The equations of motion (4.74) are transformed into :

$$\begin{aligned} \dot{\sigma}_{iN} &= \omega_i - \omega_N \quad , \quad i = 1,2,\dots,n \ , \\ \dot{\omega}_i &= -\lambda_i \omega_i - \sum_{j=1}^{N} \lambda_{ij}(\omega_i - \omega_j) - M_i^{-1} \sum_{\substack{j=1 \\ j \neq i}}^{N} f_{ij}(\sigma_{ij}) \quad , \quad i = 1,2,\dots,N \ , \end{aligned} \tag{4.76}$$

where f_{ij} are still expressed by (4.73).

V.4.4.3. Non-uniform damping and governors action

With governors modelled as first order porportional speed regulators, the equations of the absolute motion of the i-th machine are given by (2.1), (2.4) (Sections V.2.1 and V.2.5).

The overall system's motion is governed by the (n+2N) = (3N-1) state equations :

$$\dot{\delta}_{iN} = \Omega_i - \Omega_N \quad , \quad i = 1,2,\dots,n = N-1 \ ,$$

$$\dot{\Omega}_i = -\lambda_i \Omega_i - \sum_{j=1}^{N} \lambda_{ij}(\Omega_i - \Omega_j) + M_i^{-1} p_{mi} + M_i^{-1}[P_{mi}^o - P_{ei}(\boldsymbol{\delta})] = f_i(\boldsymbol{\delta}, \boldsymbol{\Omega}, \mathbf{p}) \ ,$$

$$\dot{p}_{mi} = -\mu_i p_{mi} - \alpha_i \Omega_i \quad , \quad i = 1,2,\dots,N \tag{4.77}$$

whose steady state solutions are :

$$\Omega_1^o = \Omega_2^o = \dots = \Omega_N^o = \Omega_o = \frac{\sum_{i=1}^{N} [P_{mi}^o - P_{ei}(\boldsymbol{\delta}^o)]}{\sum_{i=1}^{N} (D_i + \alpha_i \mu_i^{-1})}$$

$$f_i(\boldsymbol{\delta}^o, \boldsymbol{\Omega}^o, \mathbf{p}^o) = 0$$

$$p^{\circ}_{mi} = -\frac{\alpha_i}{\mu_i}\Omega^{\circ} \quad , \quad i = 1,2,\dots,N \ .$$

By adopting the new state variables σ_{iN} , ω_i and p_i ,

$$\sigma_{iN} = \delta_{iN} - \delta^{\circ}_{iN} \quad , \quad i = 1,2,\dots,n \ ,$$
$$\omega_i = \Omega_i - \Omega^{\circ} \ ,$$
$$p_i = p_{mi} - p^{\circ}_{mi} \quad , \quad i = 1,2,\dots,N \ ,$$

we get the state equations

$$\dot{\sigma}_{iN} = \omega_i - \omega_N \quad , \quad i = 1,2,\dots,n$$
$$\dot{\omega}_i = -\lambda_i\omega_i - \sum_{j=1}^{N} \lambda_{ij}(\omega_i - \omega_j) + M_i^{-1}p_i - M_i^{-1}\sum_{\substack{j=1\\ j\neq i}}^{N} f_{ij}(\sigma_{ij}) \tag{4.78}$$
$$\dot{p}_i = -\mu_i p_i - \alpha_i\omega_i \quad , \quad i = 1,2,\dots,N \ .$$

Obviously, the state vector is expressed in this case by

$$\mathbf{x} = [\sigma_{1N},\sigma_{2N},\dots,\sigma_{nN},\omega_1,\omega_2,\dots,\omega_N,p_1,p_2,\dots,p_N] \in \mathbf{R}^{n+2N} \ .$$

V.4.5. Power systems decompositions and aggregations

V.4.5.1. Uniform damping case

V.4.5.1.1. Form 1. This decomposition form was proposed by Grujić, Darwish and Fantin (1977).

For the sake of simplifying further analysis we define functions ρ_i , τ_i , ϕ_{ij} , ψ_{ij} and parameters β_i , γ_i , β_{ij} as follows :

$$\left.\begin{matrix}\rho_i(\sigma_{iN})\\ \tau_i(\sigma_{iN})\end{matrix}\right\} = \cos(\sigma_{iN} + \delta^{\circ}_{iN} \mp \theta_{iN}) - \cos(\delta^{\circ}_{iN} \mp \theta_{iN}) \quad , \quad i = 1,2,\dots,n \tag{4.79}$$

$$\left.\begin{matrix}\phi_{ij}(\sigma_{ij})\\ \psi_{ij}(\sigma_{ij})\end{matrix}\right\} = \cos(\sigma_{ij} + \delta^{\circ}_{ij} \mp \theta_{ij}) - \cos(\delta^{\circ}_{ij} \mp \theta_{ij}) \quad , \quad i,j = 1,2,\dots,N \tag{4.80}$$

$$\beta_i = A_{iN}M_i^{-1} \quad , \quad i = 1,2,\dots,n \ ,$$
$$\beta_{ij} = A_{ij}M_i^{-1} \quad , \quad i,j = 1,2,\dots,n \ ,$$
$$\gamma_i = A_{iN}M_N^{-1} \quad , \quad i = 1,2,\dots,n \ .$$

The mathematical model (4.72) takes the form :

$$\dot{\sigma}_{iN} = \omega_{iN} \ ,$$

$$\dot{\omega}_{iN} = -\lambda\omega_{iN} - \beta_i\rho_i(\sigma_{iN}) + \gamma_i\tau_i(\sigma_{iN}) + \sum_{\substack{j=1\\ j\neq i}}^{n} \{\beta_{Nj}\psi_{Nj}(\sigma_{Nj}) - \beta_{ij}\phi_{ij}(\sigma_{ij})\} ,$$

$$i = 1,2,\dots,n . \qquad (4.81)$$

The above system (4.81) may be readily decomposed into $s=n$ subsystems by choosing the state vector of the i-th subsystem to be

$$\mathbf{x}_i = [\sigma_{iN} \quad \omega_{iN}]^T = [x_{i1} \quad x_{i2}]^T , \quad i = 1,2,\dots,s .$$

It comes

$$\dot{\mathbf{x}}_i = \mathbf{g}_i(\mathbf{x}_i) + \mathbf{h}_i(\mathbf{x}) , \quad i = 1,2,\dots,s , \qquad (4.82)$$

where

$$\mathbf{g}_i(\mathbf{x}_i) = \mathbf{P}_i\mathbf{x}_i + \mathbf{f}_i\rho_i(\sigma_i) , \quad \sigma_i = \mathbf{e}_i^T\mathbf{x}_i , \quad \mathbf{e}_i = [1 \quad 0]^T ,$$

$$\mathbf{P}_i = \begin{bmatrix} 0 & 1 \\ 0 & -\lambda \end{bmatrix} ; \quad f_i = \begin{bmatrix} 0 \\ -\beta_i \end{bmatrix} ; \quad \mathbf{h}_i(\mathbf{x}) = \begin{bmatrix} 0 \\ \chi_i(\mathbf{x}) \end{bmatrix} ;$$

$$\chi_i(\mathbf{x}) = \gamma_i\tau_i(x_{i1}) + \sum_{\substack{j=1\\ j\neq i}}^{n} \{\beta_{Nj}\psi_{Nj}(x_{i1}) - \beta_{ij}\phi_{ij}(x_{i1}-x_{j1})\} . \qquad (4.83)$$

Thus, the subsystem (4.82) is composed of the disconnected subsystems

$$\dot{\mathbf{x}}_i = \mathbf{g}_i(\mathbf{x}_i) = \mathbf{P}_i\mathbf{x}_i + \mathbf{f}_i\rho_i(\sigma_i) \qquad (4.84)$$

and of the interconnections $\mathbf{h}_i(\mathbf{x})$ (4.83).

The state vector of the whole system is

$$\mathbf{x}^T = [\mathbf{x}_1^T, \mathbf{x}_2^T, \dots, \mathbf{x}_s^T] ;$$

we note that $\mathbf{x}_i = \mathbf{0}$ is an isolated equilibrium solution of the i-th subsystem (4.84), $i = 1,2,\dots,s$.

Notice that $\sigma_i = \sigma_{iN}$ holds.

Let the disconnected subsystem (4.84) be aggregated by using V_i function of the Lur'e form :

$$V_i(\mathbf{x}_i) = \mathbf{x}_i^T\mathbf{H}_i\mathbf{x}_i + \zeta_i \int_0^{\sigma_i} \rho_i(\sigma_i)\, d\sigma_i \qquad (4.85)$$

where ζ_i is a non-negative number. In the sector S_i ,

$$S_i = \{\sigma_{iN} : 2(\theta_{in} - \delta_{in}^{o}) \geq \sigma_{iN} \geq -2(\pi - \theta_{in} + \delta_{in}^{o})\} ,$$

V_i is a positive definite function, provided that the (symmetric) matrix $\mathbf{H}_i$ is positive definite, i.e.

$$\mathbf{H}_i = \begin{bmatrix} h_{11}^i & h_{12}^i \\ h_{12}^i & h_{22}^i \end{bmatrix} \quad \text{such that} \quad h_{11}^i > 0 ; \ h_{11}^i h_{22}^i > (h_{12}^i)^2 . \qquad (4.86)$$

However, because of the structure of $\mathbf{P}_i$, matrix $\mathbf{G}_i$ defined by

$$-\mathbf{G}_i = \mathbf{P}_i^T\mathbf{H}_i + \mathbf{H}_i\mathbf{P}_i$$

cannot be positive definite; indeed we find

$$-\mathbf{G}_i = \mathbf{P}_i^T\mathbf{H}_i + \mathbf{H}_i\mathbf{P}_i = \begin{bmatrix} 0 & h_{11}^i - \lambda h_{12}^i \\ h_{11}^i - \lambda h_{12}^i & 2(h_{12}^i - \lambda h_{22}^i) \end{bmatrix}. \tag{4.87}$$

So, for the stability analysis we can only ask for conditions on $\mathbf{H}_i$ under which $\mathbf{G}_i$ is positive semi-definite; therefore the quadratic part of the derivative of V_i along the disconnected subsystem will be at most negative semi-definite. From (4.87) it follows that this is true iff

$$h_{11}^i = \lambda h_{12}^i \quad , \quad 2(h_{12}^i - \lambda h_{22}^i) = -g_{22}^i \quad , \quad g_{22}^i > 0 \; . \tag{4.88}$$

Notice that $h_{12}^i > 0$ is required for $\mathbf{H}_i > 0$ due to $h_{11}^i > 0$ and $\lambda > 0$. Then, along motions of (4.82) the total derivative of V_i (4.85) is

$$\begin{aligned} \dot{V}_i &= (\text{grad } V_i)^T(\mathbf{g}_i + \mathbf{h}_i) \\ &= -g_{22}^i\omega_{iN}^2 - 2\beta_i\rho_i\sigma_{iN}h_{12}^i + \omega_{iN}\rho_i(\zeta_i - 2\beta_i h_{22}^i) + 2\chi_i(h_{12}^i\sigma_{iN} + h_{22}^i\omega_{iN}) \; . \end{aligned} \tag{4.89}$$

We choose ζ_i so that the term $\omega_{iN}\rho_i(\zeta_i - 2\beta_i h_{22}^i)$ is cancelled : $\zeta_i = 2\beta_i h_{22}^i$.

Further, the largest compact interval L_i of σ_{iN}, $L_i = [\ell_{im}, \ell_{iM}]$, is to be chosen so that for certain $\epsilon_i \in]0,+\infty[$ the following is satisfied

$$\sigma_{iN}\rho_i(\sigma_{iN}) \geq \epsilon_i\sigma_{iN}^2 \quad , \quad \forall \sigma_{iN} \in L_i \; .$$

Such $\epsilon_i \in]0,+\infty[$ exists iff $\sin(\theta_{iN} - \delta_{iN}^o) > 0$; this is accepted in what follows. Smaller ϵ_i, larger interval L_i and better estimate of the asymptotic stability domain can be achieved under appropriate conditions, but less possibility for these conditions to be satisfied.

Choosing g_{22}^i, and therefore h_{22}^i, as follows :

$$g_{22}^i = 2\beta_i\epsilon_i h_{12}^i \quad , \quad h_{22}^i = \frac{1 + \beta_i\epsilon_i}{\lambda} h_{12}^i$$

we get from (4.86) that $\mathbf{H}_i$ is positive definite and from (4.89)

$$\dot{V}_i \leq -2\beta_i\epsilon_i h_{12}^i \|\mathbf{x}_i\|^2 + 2\chi_i(h_{12}^i\sigma_{iN} + h_{22}^i\omega_{iN}) \; .$$

It now results that $\mathbf{X}_i$ (Section V.4.3.1) is given by :

$$\mathbf{X}_i = \{\mathbf{x}_i : \sigma_{iN} \in L_i \; , \omega_{iN} \in [-(\lambda h_{12}^i + \tfrac{1}{2}\zeta_i\epsilon_i)\ell_{im}^2 \; , (\lambda h_{12}^i + \tfrac{1}{2}\zeta_i\epsilon_i)\ell_{im}^2]\} \tag{4.90}$$

where $$\xi_{i1} = \sin(\theta_{iN} - \delta^o_{iN}) .$$

Moreover, using majorizations on L_i :

$$|\rho_i| \le \xi_{i1}|\sigma_{iN}| \quad ; \quad |\tau_i| \le \xi_{i2}|\sigma_{iN}|$$

$$\xi_{i2} = \sin(\delta^o_{iN} + \theta_{iN})$$

$$\tau_{ij} = |\sin(\delta^o_{ij} - \theta_{ij})|$$

$$\xi_{i3} = \min\left[\frac{1+\lambda+\beta_i\epsilon_i}{\lambda}, \sqrt{2}\max\left(1, \frac{1+\beta_i\epsilon_i}{\lambda}\right)\right]$$

$$\xi_{i4} = \frac{1+\beta_i\epsilon_i}{\lambda}\gamma_i\xi_{i2}$$

$$\gamma_{ij} = \begin{cases} \dfrac{1+\beta_i\epsilon_i}{\lambda}\tau_{ij} & \text{iff } \sin(\theta_{ij}-\delta^o_{ij}) \ge 0 \\ \xi_{i3}\tau_{ij} & \text{iff } \sin(\theta_{ij}-\delta^o_{ij}) \le 0 \end{cases}$$

and

$$-\phi_{ij}(h^i_{12}\sigma_{iN} + h^i_{22}\omega_{iN}) \le h^i_{12}(\gamma_{ij}\|\mathbf{x}_i\|^2 + \xi_{i3}\tau_{ij}\|\mathbf{x}_i\|\|\mathbf{x}_j\|) .$$

We obtain after trivial calculations and under the assumption that $\xi_{i1} > 0$,

$$\dot{V}_i \le -2h^i_{12}(\beta_i\epsilon_i - \xi_{i4} - \sum_{\substack{j=1\\j\ne i}}^{n}\beta_{ij}\gamma_{ij})\|\mathbf{x}_i\|^2 +$$

$$+2h^i_{12}\xi_{i3}\sum_{\substack{j=1\\j\ne i}}^{n}(\beta_{Nj}\xi_{j2} + \beta_{ij}\tau_{ij})\|\mathbf{x}_j\|\|\mathbf{x}_i\| \quad , \quad \forall \mathbf{x} \in X . \qquad (4.91)$$

Let $h^i_{12} = \frac{1}{2}$. Comparison of (4.91) with (4.21), (4.22) and (4.26) yields

$$u_i(\mathbf{x}_i) = \|\mathbf{x}_i\|$$

and

$$\alpha_{ij} = \begin{cases} -(\beta_i\epsilon_i - \xi_{i4} - \sum\limits_{\substack{j=1\\j\ne i}}^{n}\beta_{ij}\gamma_{ij}) \quad , & i = j \\ \xi_{i3}(\beta_{Nj}\xi_{j2} + \beta_{ij}\tau_{ij}) \quad , & i \ne j \end{cases} \qquad (4.92)$$

Matrix $\mathbf{A} = (\alpha_{ij})$ of Assumption 1 (Section V.4.3.1) is now completely determined by (4.92). Notice that $\epsilon_i \in]0, \xi_{i1}[$ is only meaningful, and $\epsilon_i \in [\xi_{i1}, +\infty[$ is meaningless.

The set X (Assumption 1) is completely determined via (4.90).

V.4.5.1.2. Form 2. Following Grujić and Ribbens-Pavella (1977) we define functions ϕ_i , ψ_i and parameters a_i and b_i as follows :

$$\phi_i(\sigma_{iN}) = \sin(\sigma_{iN}+\delta^o_{iN}) - \sin \delta^o_{iN} \,, \qquad \psi_i(\sigma_{iN}) = \cos(\sigma_{iN}+\delta^o_{iN}) - \cos \delta^o_{iN} \,, \qquad i=1,2,\dots,n \,,$$

$$a_i = (\gamma_i+\beta_i) \sin \theta_{iN} \,, \qquad b_i = (\gamma_i-\beta_i) \cos \theta_{iN} \,. \qquad i=1,2,\dots,n \,.$$

The mathematical model (4.72) takes the form :

$$\dot\sigma_{iN} = \omega_{iN}$$

$$\dot\omega_{iN} = -\lambda\omega_{iN} - a_i\phi_i(\sigma_{iN}) + b_i\psi_i(\sigma_{iN}) + \sum_{\substack{j=1\\ j\neq i}}^{n} \{\beta_{Nj}\psi_{Nj}(\sigma_{Nj}) - \beta_{ij}\phi_{ij}(\sigma_{ij})\} \,, \qquad i=1,2,\dots,n \,. \qquad (4.93)$$

The above system (4.93) may be readily decomposed into $s=n$ subsystems by choosing the state vector of the i-th subsystem to be the same as for the aggregation decomposition form 1,

$$\mathbf{x}_i = [\sigma_{iN} \ \ \omega_{iN}]^T = [x_{i1} \ \ x_{i2}]^T \,, \qquad i=1,2,\dots,s \,.$$

Eq. (4.82) is valid for

$$\mathbf{g}_i(\mathbf{x}_i) = \mathbf{P}_i\mathbf{x}_i + \mathbf{f}_i\phi_i(\sigma_i) \,,$$

$$\mathbf{P}_i = \begin{bmatrix} 0 & 1 \\ 0 & -\lambda \end{bmatrix} ; \quad \mathbf{f}_i = \begin{bmatrix} 0 \\ -\alpha_i \end{bmatrix} ; \quad \mathbf{h}_i(\mathbf{x}) = \begin{bmatrix} 0 \\ b_i\psi_i(\sigma_i) + \chi_i(\mathbf{x}) \end{bmatrix} ;$$

$$\chi_i(\mathbf{x}) = \sum_{\substack{j=1\\ j\neq i}}^{n} \{\beta_{Nj}\psi_{Nj}(\sigma_{Nj}) - \beta_{ij}\phi_{ij}(\sigma_{ij})\} \,. \qquad (4.94)$$

Thus, the interconnected subsystem (4.82) is composed of disconnected subsystems (4.95),

$$\dot{\mathbf{x}}_i = \mathbf{g}_i(\mathbf{x}_i) = \mathbf{P}_i\mathbf{x}_i + \mathbf{f}_i\phi_i(\sigma_i) \,, \qquad i=1,2,\dots,s \,, \qquad (4.95)$$

and interconnections $\mathbf{h}_i(\mathbf{x})$ (4.94).

We accept V_i in the form

$$V_i = \mathbf{x}_i^T\mathbf{H}_i\mathbf{x}_i + \zeta_i \int_0^{\sigma_{iN}} \phi_i(\sigma_{iN})\, d\sigma_{iN} \,.$$

Then, along motions of (4.82) the total time derivative of V_i is found easily along the same lines as for the Form 1.

We choose ζ_i so that term $\phi_i(\sigma_i)$ is cancelled in $\dot V_i$:

$$\zeta_i = 2a_ih^i_{22} \,.$$

Further, we select both a sufficiently small $\epsilon_i > 0$ and a compact interval L_i of σ_{iN} , $L_i = [\ell_{im}, \ell_{iM}]$ on which

$$\sigma_{iN}\phi_i(\sigma_{iN}) \geq \epsilon_i \sigma_{iN}^2 \quad , \quad \epsilon_i \in]0,+\infty[\; .$$

The condition $\epsilon_i \in]0,+\infty[$ is based on $\cos \delta_{iN}^o > 0$. Smaller ϵ_i , larger interval L_i but less possibility for A matrix to be stable. Then, choosing g_{22}^i , and therefore h_{22}^i , as follows :

$$g_{22}^i = 2a_i\epsilon_i h_{12}^i \quad , \quad h_{22}^i = \frac{1 + a_i\epsilon_i}{\lambda} h_{12}^i \; ,$$

we get that $H_i > 0$.

Now, X_i is defined by (4.90), which is the i-th component of X in Assumption 1.

Let $\tilde{\xi}_{i1} = \cos \delta_{iN}^o$. It is assumed that $\tilde{\xi}_{i1} > 0$ so that

$$|\phi_i(\sigma_{iN})| \leq \xi_{i1}|\sigma_{iN}| \quad \text{on} \quad L_i \; .$$

Let

$$\xi_{i2} = \sin(\delta_{iN}^o + \theta_{iN}) \; ,$$

$$\tau_{ij} = |\sin(\delta_{ij}^o - \theta_{ij})| \; ,$$

$$\tilde{\xi}_{i3} = \min\left[\frac{1 + \lambda + a_i\epsilon_i}{\lambda} \, , \, \sqrt{2}\, \max\left(1 \, , \frac{1 + a_i\epsilon_i}{\lambda}\right)\right] \; ,$$

$$\gamma_{ij} = \begin{cases} \dfrac{1 + a_i\epsilon_i}{\lambda} \tau_{ij} & \text{iff} \quad \sin(\theta_{ij} - \delta_{ij}^o) \geq 0 \\ \tilde{\xi}_{i3}\tau_{ij} & \text{iff} \quad \sin(\theta_{ij} - \delta_{ij}^o) < 0 \end{cases} \; ,$$

$$\tilde{\xi}_{i4} = \begin{cases} \dfrac{1 + a_i\epsilon_i}{\lambda} b_i \sin \delta_{iN}^o & \text{iff} \quad b_i \sin \delta_{iN}^o \geq 0 \\ \tilde{\xi}_{i3}|b_i \sin \delta_{iN}^o| & \text{iff} \quad b_i \sin \delta_{iN}^o \leq 0 \end{cases} \; ,$$

$$h_{12}^i = \frac{1}{2} \; .$$

After some trivial calculations we get

$$\dot{V}_i \leq -\|x_i\|^2 \Big(a_i\epsilon_i - \tilde{\xi}_{i4} - \sum_{\substack{j=1 \\ j\neq i}}^{n} \beta_{ij}\gamma_{ij}\Big) + \tilde{\xi}_{i3}\|x_i\| \sum_{\substack{j=1 \\ j\neq i}}^{n} (\beta_{Nj}\xi_{j2} + \beta_{ij}\tau_{ij}) \|x_j\| \; . \tag{4.96}$$

Comparison of (4.96) with (4.21), (4.22) and (4.26) shows that

$$u_i(x_i) = \|x_i\|$$

and

$$\alpha_{ij} = \begin{cases} -(a_i\epsilon_i - \tilde{\xi}_{i4} - \sum\limits_{\substack{j=1\\ j\neq i}}^{n} \beta_{ij}\gamma_{ij}) & , \ i=j \ , \\ \tilde{\xi}_{i3}(\beta_{Nj}\xi_{j2} + \beta_{ij}\tau_{ij}) & , \ i \neq j \ . \end{cases} \tag{4.97}$$

The matrix $\mathbf{A} = (\alpha_{ij})$ is now completely determined by (4.97). Notice that $\epsilon_i \in]0, \tilde{\xi}_{i1}[$ is only meaningful. The choice of $\epsilon_i \in [\tilde{\xi}_{i1}, +\infty[$ is meaningless.

V.4.5.1.3. Comments. In the two decomposition forms above considered, the whole system was decomposed into $s=n$ second order subsystems. The difference between them lies in the definition of subsystems and interactions.

In the first case, free subsystems and interactions are described by

$$\mathbf{g}_i(\mathbf{x}_i) = \begin{bmatrix} 0 & 1 \\ 0 & -\lambda \end{bmatrix} \mathbf{x}_i + \begin{bmatrix} 0 \\ -\beta_i \end{bmatrix} \rho_i \ ,$$

$$\mathbf{h}_i(\mathbf{x}_i) = \begin{bmatrix} 0 \\ \gamma_i\tau_i + \sum\limits_{\substack{j=1\\ j\neq i}}^{n} (\beta_{Nj}\psi_{Nj} - \beta_{ij}\phi_{ij}) \end{bmatrix} ,$$

while in the second case they are described by

$$\mathbf{g}_i(\mathbf{x}_i) = \begin{bmatrix} 0 & 1 \\ 0 & -\lambda \end{bmatrix} \mathbf{x}_i + \begin{bmatrix} 0 \\ -a_i \end{bmatrix} \phi_i \ ,$$

$$\mathbf{h}_i(\mathbf{x}_i) = \begin{bmatrix} 0 \\ b_i\psi_i + \sum\limits_{\substack{j=1\\ j\neq i}}^{n} (\beta_{Nj}\psi_{Nj} - \beta_{ij}\phi_{ij}) \end{bmatrix} .$$

Hence $-\beta_i\rho_i$ and $\gamma_i\tau_i$ are replaced by $-a_i\phi_i$ and $b_i\psi_i$, respectively. If $\gamma_i = \beta_i$ then $b_i = 0$ and the term $b_i\psi_i$ vanishes in the second case, which is important for avoiding its majorization. In the first case, both $-\beta_i\rho_i$ and $\gamma_i\tau_i$ appear for $\gamma_i = \beta_i$ and the latter one is to be majorized. For $\gamma_i \neq \beta_i$ it is not evident a priori which decomposition is more suitable.

V.4.5.2. *Non-uniform damping and governors action*

Various decomposition-aggregation forms were considered by the authors for power systems without and with governors actions (Grujić and Ribbens-Pavella, 1977-1979; Grujić et al., 1979; Shaaban and Grujić, 1984, 1985). Two of them will be only presented to show possibilities and drawbacks rather than to try giving final results.

V.4.5.2.1. Pairwise decomposition-aggregation. Consider an N-machine power system, with mechanical and electromagnetic dampings. The differential equation describing the absolute motion of the i-th machine, represented by a constant voltage behind its transient reactance and with a 1st-order proportional governor, is given by (2.1), (2.3) and (2.4) (Sections V.2.1 and V.2.5), i.e. by (4.98),

$$\begin{aligned} M_i\ddot{\delta}_i &= -D_i\dot{\delta}_i - \sum_{j=1}^{N} D_{ij}(\dot{\delta}_i-\dot{\delta}_j) + P_{mi} + p_{mi} - P_{ei} \\ &= f_i(\boldsymbol{\delta},\dot{\boldsymbol{\delta}},\mathbf{p}_m) \\ \dot{p}_{mi} &= -\mu_i p_{mi} - \alpha_i\dot{\delta}_i \end{aligned} \tag{4.98}$$

where

$$P_{ei}(\delta_{iN},\delta_{jN}) = \sum_{j=1}^{N} E_iE_jY_{ij}\cos(\delta_{ij}-\theta_{ij}) \; . \tag{4.99}$$

The motion of the system taken as a whole is described in terms of (3N-1) state equations derived from equation (4.98) with the following (3N-1) state variables (Ribbens-Pavella, 1971b) :

$$\delta_{1N},\Omega_1,p_{m1},\delta_{2N},\Omega_2,p_{m2},\dots,\delta_{nN},\Omega_n,p_{mn},\Omega_N,p_{mN} \; .$$

Here, the N-th machine acts as the comparison machine. Furthermore, in order to make the origin of the state space coincide with an equilibrium of the system, we adopt the new state variables

$$\sigma_{iN} = \delta_{iN} - \delta^{\circ}_{iN} \qquad \omega_i = \Omega_i - \Omega^{\circ} \qquad p_i = p_{mi} - p^{\circ}_{mi}$$

where (Ribbens-Pavella, 1971b)

$$p^{\circ}_{mi} = -\frac{\alpha_i}{\mu_i}\Omega^{\circ} \quad ; \quad \Omega^{\circ} = \Omega^{\circ} = \frac{P_{mN} - P_{eN}(\boldsymbol{\delta}^{\circ},\boldsymbol{\delta}^{\circ})}{D_N + \dfrac{\alpha_N}{\mu_N}} \quad , \quad \forall i = 1,2,\dots,N \; , \tag{4.100}$$

and where δ°_{iN} , $i = 1,2,\dots,n$, are solutions of

$$f_i(\boldsymbol{\delta}^{\circ},\boldsymbol{\Omega}^{\circ},\mathbf{p}^{\circ}_m) = 0 \quad , \quad \forall i = 1,2,\dots,n \; . \tag{4.101}$$

Hence, the overall system's motion is governed by the (3N-1) state equations :

$$\dot{\sigma}_{iN} = \omega_i - \omega_N \quad , \quad i = 1,2,\dots,n \; ,$$

$$\dot{\omega}_i = -\lambda_i\omega_i - \sum_{j=1}^{N}\lambda_{ij}(\omega_i-\omega_j) + k_i(p_i - \sum_{j=1}^{N}A_{ij}\phi_{ij}) , \qquad (4.102)$$

$$\dot{p}_i = -\mu_i p_i - \alpha_i\omega_i , \quad i = 1,2,\dots,n, N=n+1 .$$

Obviously, the state vector is now

$$\mathbf{x} = [\sigma_{1N},\omega_1,p_1,\sigma_{2N},\omega_2,p_2,\dots,\sigma_{nN},\omega_n,p_n,\omega_N,p_N]^T . \qquad (4.103)$$

One of most interesting decomposition forms appeared to be the following derived from (4.102) (in which the differences $\dot{\omega}_i - \dot{\omega}_N = \dot{\omega}_{iN}$ $\dot{p}_i - \dot{p}_N = \dot{p}_{iN}$, $\forall i = 1,2,\dots,n$, are used) :

$$\dot{\sigma}_{iN} = \omega_{iN} ,$$

$$\dot{\omega}_{iN} = -\Gamma_i\omega_{iN} + M_i^{-1}p_{iN} - (\lambda_i-\lambda_N)\,\omega_N + (M_i^{-1} - M_N^{-1})\,p_N - a_i\phi_i + b_i\psi_i$$
$$+ \sum_{\substack{j=1\\ j\neq i}}^{n} (\Lambda_{ij}\omega_{jN} + \beta_{Nj}\phi_{Nj} - \beta_{ij}\phi_{ij}) , \quad \forall i = 1,2,\dots,n , \qquad (4.104)$$

$$\dot{p}_{iN} = -\alpha_i\omega_{iN} - \mu_i p_{iN} + (\alpha_N-\alpha_i)\,\omega_N + (\mu_N-\mu_i)\,p_N ,$$

where

$$a_i = (\beta_{iN}+\beta_{Ni})\sin\theta_{iN} ,$$

$$b_i = (\beta_{Ni}-\beta_{iN})\cos\theta_{iN} ,$$

and

$$\dot{\omega}_N = -\lambda_N\omega_N + M_N^{-1}p_N + \sum_{j=1}^{n}(\lambda_{Nj}\omega_{jN} - \beta_{Nj}\phi_{Nj}) ,$$

$$\dot{p}_N = -\alpha_N\omega_N - \mu_N p_N . \qquad (4.105)$$

Note that the first n interconnected subsystems are of order 3, the last one is of order 2.

Thus, by defining the subvectors $\mathbf{x}_i$, $\mathbf{x}_N$ as follows :

$$\mathbf{x}_i = [x_{i1},x_{i2},x_{i3}]^T = [\sigma_{iN},\omega_{iN},p_{iN}]^T , \quad \forall i = 1,2,\dots,n ,$$
$$\mathbf{x}_N = [x_{N1},x_{N2}]^T = [\omega_N,p_N]^T , \qquad (4.106)$$

we decompose the overall system of equation (4.102) into N interconnected subsystems of the general form :

$$\dot{\mathbf{x}}_i = \mathbf{g}_i(\mathbf{x}_i) + \mathbf{h}_i(\mathbf{x}) = \mathbf{P}_i\mathbf{x}_i + \boldsymbol{\ell}_i\phi_i(x_{i1}) + \mathbf{h}_i(\mathbf{x}) , \quad \forall i = 1,2,\dots,N . \qquad (4.107)$$

Here, the state vector $\mathbf{x}$ still has the form of equation (4.103); vector $\boldsymbol{\ell}_i$ and $\mathbf{h}_i$ and matrices $\mathbf{P}_i$ are defined as follows :

$$\boldsymbol{\ell}_i = [0 \quad -a_i \quad 0]^T , \qquad \mathbf{P}_i = \begin{bmatrix} 0 & 1 & 0 \\ 0 & -\Gamma_i & M_i^{-1} \\ 0 & -\alpha_i & -\mu_i \end{bmatrix} ,$$

$$\mathbf{h}_i(\mathbf{x}) = \begin{bmatrix} 0 \\ (\lambda_N-\lambda_i)\, x_{N1} + (M_i^{-1}-M_N^{-1})\, x_{N2} + b_i\psi_i + \sum_{\substack{j=1\\ j\neq i}}^{n} (\Lambda_{ij}x_{j2} + \beta_{Nj}\phi_{Nj} - \beta_{ij}\phi_{ij}) \\ (\alpha_N-\alpha_i)\, x_{N1} + (\mu_N-\mu_i)\, x_{N2} \end{bmatrix},$$

$$\forall i = 1,2,\dots,n \ , \qquad (4.108)$$

$$\boldsymbol{\ell}_N = [0\ 0]^T \quad , \quad \mathbf{P}_N = \begin{bmatrix} -\lambda_N & M_N^{-1} \\ -\alpha_N & -\mu_N \end{bmatrix} ,$$

$$\mathbf{h}_N(\mathbf{x}) = \begin{bmatrix} \sum_{j=1}^{n} (\lambda_{Nj}x_{j2} - \beta_{Nj}\phi_{Nj}) \\ 0 \end{bmatrix} . \qquad (4.109)$$

We can now decompose the i-th subsystem of equation (4.107) into a disconnected (free) part of the Lur'e-Postnikov type, and interconnections. The i-th free subsystem has the general form :

$$\begin{aligned} \dot{\mathbf{x}}_i &= \mathbf{P}_i\mathbf{x}_i + \boldsymbol{\ell}_i\phi_i(\sigma_{iN}) \ , \\ \sigma_{iN} &= \mathbf{e}_i^T\mathbf{x}_i \ , \end{aligned} \qquad i = 1,2,\dots,n,N \ , \qquad (4.110)$$

where

$$\mathbf{e}_i = [1\ 0\ 0]^T \ , \ \forall i = 1,2,\dots,n \quad ; \quad \mathbf{e}_N = [0\ 0]^T \ . \qquad (4.111)$$

For the free subsystem of equation (4.110), we accept a Liapunov function of the Lur'e-Postnikov type "quadratic form + integral of the nonlinearity" :

$$\begin{aligned} V_i(\mathbf{x}_i) &= \mathbf{x}_i^T\mathbf{H}_i\mathbf{x}_i + \zeta_i \int_0^{\sigma_{iN}} \phi_i(\sigma_{iN})\, d\sigma_{iN} \ , \quad \forall i = 1,2,\dots,n \ , \\ V_N(\mathbf{x}_N) &= \mathbf{x}_N^T\mathbf{H}_N\mathbf{x}_N \ . \end{aligned} \qquad (4.112)$$

In these expressions, $\mathbf{H}_i$ are symmetric positive-definite matrices :

$$\begin{aligned} \mathbf{H}_i &= (h_{kj}^i) = \mathbf{H}_j^T \in R^{3\times 3} \ , \quad \forall i = 1,2,\dots,n \ , \\ \mathbf{H}_N &= (h_{kj}^N) = \mathbf{H}_N \in R^{2\times 2} \ , \end{aligned} \qquad (4.113)$$

ζ_i are positive numbers, and the nonlinearity $\phi_i(\sigma_{iN})$ obeys the Lur'e sector condition only on the bounded sector S_i , defined by :

$$S_i = \{\sigma_{iN} : \pi - 2\,\delta_{iN}^{\circ} \geq \sigma_{iN} \geq -\pi - 2\,\delta_{iN}^{\circ}\} \ , \qquad (4.114)$$

provided that

$$\frac{\pi}{2} > \delta_{iN}^{\circ} > -\frac{\pi}{2} \ .$$

The functions

$$u_i(\mathbf{x}_i) = \|\mathbf{x}_i\| \quad , \quad i = 1,2,\dots,n,N \; , \tag{4.115}$$

appeared to be the most convenient for the system stability analysis. In what follows, we accept equation (4.115) and proceed further to the determination of matrices $\mathbf{H}_i$, numbers ζ_i , and therefore to the construction of the $\mathbf{A}$ matrix.

In order to determine the quadratic term of $\dot{V}_i$, we calculate the symmetric matrix $\mathbf{G}^i = (g^i_{kj})$ defined by :

$$-\mathbf{G}^i = \mathbf{P}_i^T \mathbf{H}_i + \mathbf{H}_i \mathbf{P}_i \; .$$

Based on expressions (4.108) and (4.113), we find :

$$-\mathbf{G}^i = \begin{bmatrix} 0 & (h^i_{11} - \Gamma_i h^i_{12} - \alpha_i h^i_{13}) & (k_i h^i_{12} - \mu_i h^i_{13}) \\ (h^i_{11} - \Gamma_i h^i_{12} - \alpha_i h^i_{13}) & 2(h^i_{12} - \Gamma_i h^i_{22} - \alpha_i h^i_{23}) & (h^i_{13} - (\Gamma_i + \mu_i) h^i_{23} + k_i h^i_{22} - \alpha_i h^i_{33}) \\ (k_i h^i_{12} - \mu_i h^i_{13}) & (h^i_{13} - (\Gamma_i + \mu_i) h^i_{23} + k_i h^i_{22} - \alpha_i h^i_{23}) & 2(k_i h^i_{23} - \mu_i h^i_{33}) \end{bmatrix} ,$$

$$\forall i = 1,2,\dots,n \; . \tag{4.116}$$

Note that by its very structure $\mathbf{G}^i$ cannot be negative-definite. Let

$$\mathbf{G}^i = \begin{bmatrix} 0 & 0 & 0 \\ 0 & 2 & 0 \\ 0 & 0 & 2b^2 \end{bmatrix} .$$

$$-h^i_{11} + \Gamma_i h^i_{12} + \alpha_i h^i_{13} = 0 \; , \tag{4.117}$$

$$\mu_i h^i_{13} - M_i^{-1} h^i_{12} = 0 \; , \tag{4.118}$$

$$-h^i_{13} + (\Gamma_i + \mu_i)\, h^i_{23} + \alpha_i h^i_{33} - M_i^{-1} h^i_{22} = 0 \; , \tag{4.119}$$

$$-h^i_{12} + \Gamma_i h^i_{22} + \alpha_i h^i_{23} = 1 \; , \tag{4.120}$$

$$\mu_i h^i_{33} - M_i^{-1} h^i_{23} = b^2 \; , \tag{4.121}$$

which yields the expressions of the h^i_{ij} versus h^i_{12} and b^2

$$h^i_{11} = (\Gamma_i + \alpha_i \frac{M_i^{-1}}{\mu_i})\, h^i_{12} \, , \tag{4.122}$$

$$h^i_{12} = h \in \mathring{R}_+ \; , \tag{4.123}$$

$$h^i_{13} = \frac{M_i^{-1}}{\mu_i}\, h^i_{12} \; , \tag{4.124}$$

$$h^i_{22} = \frac{\mu_i(\Gamma_i + \mu_i)\, h^i_{12} + \alpha_i^2 b^2 + \alpha_i M_i^{-1} + \mu_i(\Gamma_i + \mu_i)}{(\Gamma_i + \mu_i)(\alpha_i M_i^{-1} + \mu_i \Gamma_i)} \; , \tag{4.125}$$

$$h_{23}^{i} = \frac{M_i^{-1}(\Gamma_i+\mu_i)\,h_{12}^{i} + \mu_i M_i^{-1} - b^2\alpha_i\Gamma_i}{(\Gamma_i+\mu_i)(\alpha_i M_i^{-1} + \Gamma_i\mu_i)}\,, \tag{4.126}$$

$$h_{33}^{i} = \frac{M_i^{-2}(1 + \Gamma_i\mu_i^{-1})\,h_{12}^{i} + M_i^{-2} + b^2\alpha_i M_i^{-1} + b^2\Gamma_i(\Gamma_i+\mu_i)}{(\Gamma_i+\mu_i)(\alpha_i M_i^{-1} + \Gamma_i\mu_i)}\,. \tag{4.127}$$

Using equations (4.117)-(4.121), the derivative of V_i is obtained along the motion of the whole system :

$$\begin{aligned}\dot V_i = &- 2\omega_{iN}^2 - b^2 p_{iN}^2 - 2h_{12}^i\beta_{iN}\sigma_{iN}\phi_{iN} + \omega_{iN}\phi_{iN}(\zeta_i - 2a_i h_{22}^i) - 2a_i h_{23}^i p_{iN}\phi_i \\ &+ \chi_1^i(2h_{12}^i\sigma_{iN} + 2h_{22}^i\omega_{iN} + 2h_{23}^i p_{iN}) + \chi_2^i(2h_{13}^i\sigma_{iN} + 2h_{23}^i\omega_{iN} + 2h_{33}^i p_{iN})\,,\end{aligned}$$

where χ_1^i and χ_2^i are the following functions :

$$\chi_1^i = -(\lambda_i-\lambda_N)\,\omega_N + (M_i^{-1}-M_N^{-1})\,p_N + b_i\psi_i + \sum_{\substack{j=1\\ j\neq i}}^{n} (\Lambda_{ij}\omega_{jN} + \beta_{Nj}\phi_{Nj} - \beta_{ij}\phi_{ij})$$

$$\chi_2^i = -(\alpha_i-\alpha_N)\,\omega_N - (\mu_i-\mu_N)\,p_N\,.$$

Let us choose $\zeta_i = 2a_i h_{22}^i$, so that the term in $\omega_{iN}\phi_{iN}$ disappears (ζ_i is positive because of the positiveness of h_{22}^i in equation (4.125) whenever we choose $h_{12}^i > 0$).

Let the interval L_i be $L_i = \{\sigma_{iN} : \ell_{iM} \ge \sigma_{iN} \ge \ell_{im}\}$ so that :

$$\sigma_{iN}\phi_i \ge 0 \quad\text{and}\quad |\phi_i| > \epsilon_i|\sigma_{iN}|\,, \quad \forall\sigma_{iN}\in L_i\,. \tag{4.128}$$

By setting $h_{23}^i = 0$ we eliminate the term $-2a_i h_{23}^i p_{iN}\phi_i$. This implies the following relationship between h_{12}^i and b^2 :

$$M_i^{-1}(\Gamma_i+\mu_i)\,h_{12}^i = \alpha_i\Gamma_i b^2 - \mu_i M_i^{-1}\,,$$

which ensures the positive-definiteness of $\mathbf{H}_i$.

Moreover, imposing

$$h_{12}^i = \frac{1}{\epsilon_i a_i}$$

which implies

$$b^2 = \frac{M_i^{-1}(\Gamma_i+\mu_i)\,h_{12}^i + \mu_i M_i^{-1}}{\alpha_i\Gamma_i} \tag{4.129}$$

we get

$$\begin{aligned}\dot V_i \le -2\min(1,b^2)\|\mathbf{x}_i\|^2 &+ |\chi_1^i|(2h_{12}^i\sigma_{iN} + 2h_{22}^i\omega_{iN} + 2h_{23}^i p_{iN}) + \\ &+ |\chi_2^i|(2h_{13}^i\sigma_{iN} + 2h_{23}^i\omega_{iN} + 2h_{33}^i p_{iN})\,.\end{aligned}$$

Using the following :

$$|\alpha\sigma_{iN} + \beta\omega_{iN}| \le Z_2(\alpha,\beta)\|\mathbf{x}_N\|\,,$$

$$|\beta\omega_{iN} + \gamma f_{iN}| \le Z_3(\alpha,\beta,\gamma)\|\mathbf{x}_N\|\,,$$

$$|\psi_i| \le |\sin(\delta^o_{ij})|\,|\sigma_{ij}| = \tilde{\xi}_{ij}|\sigma_{ij}| ,$$
$$\tilde{\xi}_{ij} = |\sin \delta^o_{ij}| ,$$
$$|\sigma_{ij}| \le |\sigma_{iN}| + |\sigma_{jN}| , \tag{4.130}$$

where

$$Z_2(\alpha,\beta) = \min\{\sqrt{2}\max(|\alpha|,|\beta|) , |\alpha|+|\beta|\} ,$$

$$Z_3(\alpha,\beta,\gamma) = \min\{2\max(|\alpha|,|\beta|,|\gamma|) , \sqrt{2}\max(|\alpha|,|\beta|)+|\gamma| , \sqrt{2}\max(|\alpha|,|\gamma|)+|\beta| , \sqrt{2}\max(|\beta|,|\gamma|)+|\alpha| , |\alpha|+|\beta|+|\gamma|\} ,$$

we obtain

$$\dot{V}_i \le -2\min(1,b^2)\|\mathbf{x}_i\|^2 + 2\|\mathbf{x}_i\|\, Z_3(h^i_{12},h^i_{22},h^i_{23}) \times$$
$$\times\Big\{|\lambda_i-\lambda_N|\,|\omega_N| + |M_i^{-1}-M_N^{-1}|\,|p_N| + b_i\xi_{iN}|\sigma_{iN}| + \sum_{\substack{j=1\\ j\neq i}}^{n}\Big[|\Lambda_{ij}|\,|\omega_{jN}| +$$
$$+ (\beta_{Nj}\mu_{Nj} + \beta_{ij}\mu_{ij})|\sigma_{jN}| + \beta_{ij}\mu_{ij}|\sigma_{iN}|\Big]\Big\} +$$
$$+ 2\|\mathbf{x}_i\|\, Z_3(h^i_{13},h^i_{23},h^i_{33})\Big\{|\alpha_i-\alpha_N|\,|\omega_N| + |\mu_i-\mu_N|\,|p_N|\Big\} .$$

This yields

$$\alpha_{ij} = \begin{cases} -2\min(1,b^2) + 2Z_3(h^i_{12},h^i_{22},h^i_{23})(|b_i|\,\xi_{iN} + \sum\limits_{j\neq i}^{n}\beta_{ij}\mu_{ij}) & \text{for } j=i , \\ 2Z_3(h^i_{12},h^i_{22},h^i_{23})\, Z_2(\Lambda_{ij}, \beta_{Nj}\mu_{Nj} + \beta_{ij}\mu_{ij}) & \text{for } i\neq i ,\ j\neq N , \\ 2Z_2\{Z_3(h^i_{12},h^i_{22},h^i_{23})|\lambda_i-\lambda_N| + Z_3(h^i_{13},h^i_{23},h^i_{33})|\alpha_i-\alpha_N| , \\ \quad Z_3(h^i_{12},h^i_{22},h^i_{23})|M_i^{-1}-M_N^{-1}| + Z_3(h^i_{13},h^i_{23},h^i_{33})|\mu_i-\mu_N|\} \\ & \text{for } j=N . \end{cases} \tag{4.131}$$

This determines the first n rows of the aggregation matrix $\mathbf{A}$. Notice that in order to determine the last row of $\mathbf{A}$ we start with V_N expressed by equation (4.112) and calculate $\dot{V}_N$ along motions of equation (4.107); hence

$$\dot{V}_N = \mathbf{x}_N^T(\mathbf{P}_N^T\mathbf{H}_N + \mathbf{H}_N\mathbf{P}_N)\,\mathbf{x}_N + 2\mathbf{x}_N^T\mathbf{H}_N\mathbf{h}_N(\mathbf{x}) .$$

Let $\mathbf{H}_N$ be expressed by

$$\mathbf{H}_N = \begin{bmatrix} 1 & 0 \\ 0 & k \end{bmatrix} .$$

We can calculate $\dot{V}_N$ along the overall system's motion

$$\dot{V}_N = -2\lambda_N\omega_N - 2k\mu_N p_N^2 + 2(M_N^{-1} - k\alpha_N)\,\omega_N p_N + 2\omega_N\,(\sum_{j=1}^{n}\lambda_{Nj}\omega_{jN} - \sum_{j=1}^{n}\beta_{Nj}\phi_{Nj}) .$$

Choosing k so as to cancel the term in $\omega_N p_N$

$$k = M_N^{-1}\alpha_N^{-1}$$

and using the majorizations in equation (4.130) and

$$|\phi_{Nj}| \leq |\sin(\delta_{Nj}^{o} - \theta_{Nj})|\,|\sigma_{Nj}| = \tau_{Nj}|\sigma_{jN}|$$

we get

$$\dot{V}_N \leq -2\min(\lambda_N, \alpha_N^{-1}M_N^{-1}\mu_N)\|\mathbf{x}_N\|^2 + 2\|\mathbf{x}_N\| \sum_{j=1}^{n} [\lambda_{Nj}|\omega_{jN}| + \beta_{Nj}\tau_{Nj}|\sigma_{jN}|] \,.$$

Comparing the latter expression with equations (4.21), (4.22), (4.26) and use of (4.115) yield

$$\alpha_{Nj} = \begin{cases} -2\min(\lambda_N, \alpha_N^{-1}M_N^{-1}\mu_N) & \text{for } j = N \\ 2Z_2(\lambda_{Nj}, \beta_{Nj}\tau_{Nj}) & \text{for } j \neq N\,. \end{cases} \tag{4.132}$$

For a chosen ϵ_i, the limits ℓ_{im} and ℓ_{iM} of L_i are the nonzero solutions of the equation

$$\epsilon_i \ell_{ij} = \sin(\ell_{ij} + \delta_{iN}^{o}) - \sin\delta_{iN}^{o} \qquad j = m, M\,. \tag{4.133}$$

Notice that $\ell_{im} < 0$ and $\ell_{iM} > 0$. Smaller ϵ_i, larger interval L_i and better estimate of the asymptotic stability domain can be achieved under appropriate conditions but with less possibility for these conditions to be satisfied.

Following Walker and McClamrock (1967) and Šiljak (1969) we get :

$$V_i^{o} = \min_{\sigma_{iN}} \left[\sigma_{iN}(\mathbf{e}_i^T \mathbf{H}_i^{-1}\mathbf{e}_i)^{-1} + \zeta_i \int_0^{\sigma_{iN}} \phi_i \, d\sigma_{iN}\right]\Bigg|_{\sigma_{iN} = \{\ell_{im}, \ell_{iM}\}}, \quad \forall i = 1, 2, \ldots, N\,, \tag{4.134}$$

where $\mathbf{e}_i$, $\mathbf{e}_N$ are defined by equation (4.111).

In the present case, and for $u_i = \|\mathbf{x}_i\|$ we get

$$\lambda_{mi} u_i^2 \leq V_i(\mathbf{x}_i) \leq \Lambda_{Mi} u_i^2\,, \tag{4.135}$$

where

$$\lambda_{mi} \equiv \lambda_{mi}(\mathbf{H}_i) \tag{4.136}$$

denotes the minimum eigenvalue and

$$\Lambda_{Mi} \equiv \Lambda_{Mi}(\mathbf{H}_i + \tfrac{1}{2}\zeta_i \xi_{i1}\mathbf{e}_i\mathbf{e}_i^T) \tag{4.137}$$

denotes the maximum eigenvelue of the corresponding matrix. Hence :

$$\eta_{i1} = \lambda_{mi}\,, \quad \eta_{i2} = \Lambda_{Mi}\,. \tag{4.138}$$

Numbers r_{i1}, r_{j2} (4.36) are now found to be expressed by :

$$\begin{aligned} r_{i1} &= [\Lambda_M(\mathbf{H}_i + \tfrac{1}{2}\zeta_i\xi_{i1}\mathbf{e}_i\mathbf{e}_i^T)\, V_i^{o}]^{-1/2}\,, \\ r_{i2} &= [\lambda_m(\mathbf{H}_i)\, V_i^{o}]^{-1/2}\,. \end{aligned} \tag{4.139}$$

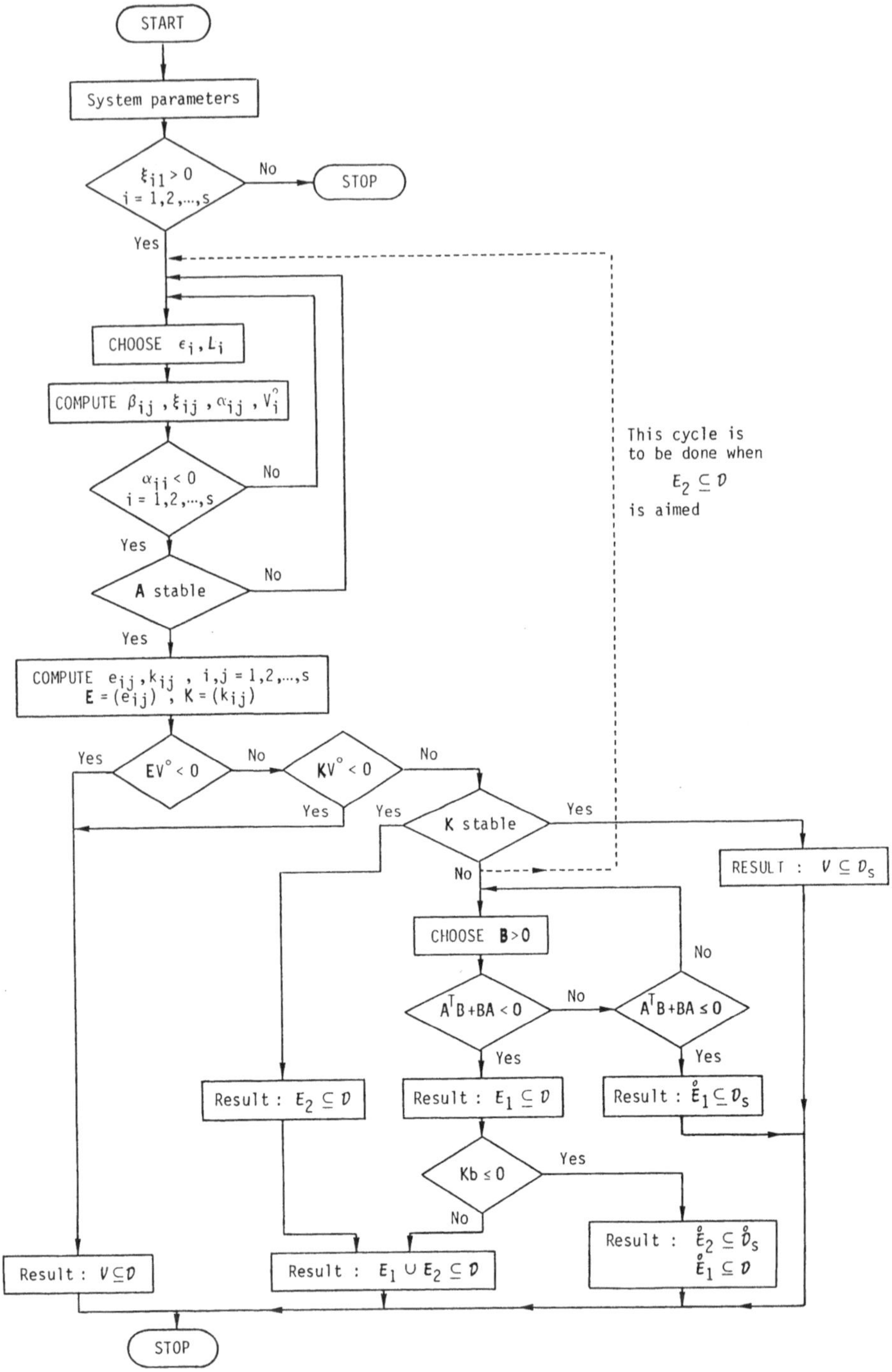

Figure 9. Algorithm for power system stability analysis

We easily find in view of (4.135)-(4.139) that in the present case, matrices **E** and **K** are related by :

$$\mathbf{K} = \mathbf{E}(\mathbf{V}^{\circ})^{1/2}$$

where $(\mathbf{V}^{\circ})^{1/2} = [V_1^{\circ\,1/2}, V_2^{\circ\,1/2}, \ldots, V_s^{\circ\,1/2}]^T$.

We now possess all elements necessary to apply Theorems 1-4.

The power system stability analysis can be performed as indicated by the algorithm in Figure 9.

Numerical example 1. Rather than performing an actual conventional stability analysis, we have preferred to consider a simple 3-machine system, to compute interesting parameters from the vector approach, and to compare their values in different cases. These cases and the derived results are indicated in Table VI where :

- columns I, III, IV and V deal with particular cases treated by Grujić and Ribbens-Pavella (1977) or by Bouffioux (1978);
- column II is relative to the formalism established by Jocić et al. (1978);
- columns VI and VII are relative to the general case treated in this section.

The general data of the 3-machine system considered here are the following (Jocić et al., 1978) :

$E_1 = 1.017$	$E_2 = 1.005$	$E_3 = 1.033$	(p.u.)
$Y_{12} = 0.98 \times 10^{-3}$	$Y_{13} = 0.114$	$Y_{23} = 0.106$	(p.u.)
$\theta_{12} = 86$	$\theta_{13} = 88$	$\theta_{23} = 89$	(degrees)
$\delta_1^{o} = -2$	$\delta_2^{o} = 3$	$\delta_3^{o} = 0$	(degrees)
$M_1 = 0.01$	$M_2 = 0.01$	$M_3 = 0.0101$	(p.u.)

Moreover, in order to compute the aggregation **A** matrix in the different cases, we have adopted

$$\epsilon_1 = \epsilon_2 = 0.1 \; .$$

The particular data corresponding to each case are reported in the first row of Table 1.

The information provided by this table concerns :

- the aggregation matrix, according to expressions (4.131) and (4.132) or other analogue expressions adapted to the corresponding case;
- the minimum and maximum eigenvalues of the corresponding aggregation matrix;

- the limit values of the Liapunov functions of the disconnected subsystems, V_i^o s, according to expression (4.134) or its analogue corresponding to each particular case.

We observe that in all cases - except for that of column II - the **A** matrix is stable. According to the corollary to Theorem 1 of Section V.4.3.1, stability of **A** implies stability of the system's equilibrium point. We also observe that the values of the V_i^o s increase substantially in the case where governors are taken into account; this could imply that the subsystems' stability is reinforced.

TABLE VI

	I	II	III	IV
Data	$\lambda_i = 10 \quad i = 1,\dots,3$ $\lambda_{ij} = 0 \quad i,j = 1,\dots,3 \quad i \neq j$ $\alpha_i = 0 \quad i = 1,\dots,3$ $\mu_i = 0 \quad i = 1,\dots,3$		$\lambda_1 = \lambda_2 = 10 \; ; \; \lambda_3 = 9.9$ $\lambda_{ij} = 0 \quad i,j = 1,\dots,3 \quad i \neq j$ $\alpha_i = 0 \quad i = 1,\dots,3$ $\mu_i = 0 \quad i = 1,\dots,3$	$\lambda_1 = \lambda_2 = 10 \; ; \; \lambda_3 = 9.9$ $\lambda_{ij} = 0.11 \quad i,j=1,\dots,3 \quad i \neq j$ $\alpha_i = 0 \quad i=1,\dots,3$ $\mu_i = 0 \quad i=1,\dots,3$
Aggregation matrix A	$\begin{bmatrix} -1.767 & 0.315 \\ 0.348 & -1.761 \end{bmatrix}$	$\begin{bmatrix} 0.089 & 17.450 \\ 18.950 & 0.089 \end{bmatrix}$	$\begin{bmatrix} -1.770 & 0.315 & 2.028 \\ 0.348 & -1.761 & 2.190 \\ 0.120 & 0.110 & -19.800 \end{bmatrix}$	$\begin{bmatrix} -1.769 & 0.313 & 2.017 \\ 0.346 & -1.763 & 2.178 \\ 0.311 & 0.311 & -19.800 \end{bmatrix}$
$\lambda_m(A)$	-2.095	-18.10	-19.830	-19.870
$\Lambda_M(A)$	-1.434	18.27	-1.409	-1.366
V_1^o	43.43	5.35	43.43	44.33
V_2^o	43.63	5.22	43.63	44.58
V_3^o	-	-	∞	∞

	V	VI	VII
Data	$\lambda_1 = \lambda_2 = 10 \; ; \; \lambda_3 = 9.90$ $\lambda_{12} = 0.10 \; ; \; \lambda_{21} = 0.12$ $\lambda_{31} = 0.11 \; ; \; \lambda_{13} = 0.12$ $\lambda_{32} = 0.11 \; ; \; \lambda_{23} = 0.10$ $\alpha_i = 0 \quad i = 1,\dots,3$ $\mu_i = 0 \quad i = 1,\dots,3$	$\lambda_i = 10 \quad i = 1,\dots,3$ $\lambda_{ij} = 0.11 \quad i,j = 1,\dots,3 \quad i \neq j$ $\alpha_i = 25 \quad i = 1,\dots,3$ $\mu_i = 1 \quad i = 1,\dots,3$	$\lambda_1 = \lambda_2 = 10 \; ; \; \lambda_3 = 9.90$ $\lambda_{12} = 0.10 \; ; \; \lambda_{21} = 0.12$ $\lambda_{31} = 0.11 \; ; \; \lambda_{13} = 0.12$ $\lambda_{32} = 0.11 \; ; \; \lambda_{23} = 0.10$ $\alpha_i = 25 \quad i = 1,\dots,3$ $\mu_i = 1 \quad i = 1,\dots,3$
Aggregation matrix A	$\begin{bmatrix} -1.769 & 0.442 & 2.017 \\ 0.489 & -1.763 & 2.178 \\ 0.311 & 0.311 & -19.800 \end{bmatrix}$	$\begin{bmatrix} -1.788 & 0.286 & 1.846 \\ 0.317 & -1.783 & 1.994 \\ 0.120 & 0.112 & -7.920 \end{bmatrix}$	$\begin{bmatrix} -1.788 & 0.305 & 2.031 \\ 0.448 & -1.783 & 2.194 \\ 0.311 & 0.311 & -7.920 \end{bmatrix}$
$\lambda_m(A)$	-19.870	-7.988	-8.116
$\Lambda_M(A)$	-1.230	-1.417	-1.219
V_1^o	44.33	2.851	2.851
V_2^o	44.58	2.915	2.915
V_3^o	∞	∞	∞

The computation of the **E** matrix derives from expressions (4.35) and (4.36); in the case of column VII we compute

$$E = \begin{bmatrix} -7.32 \times 10^{-4} & 1.88 \times 10^{-2} & 0 \\ 8.32 \times 10^{-3} & -6.92 \times 10^{-4} & 0 \\ 1.94 \times 10^{-2} & 1.92 \times 10^{-2} & 0 \end{bmatrix}$$

Matrix $\mathbf{K}$ is readily computed via expression (4.51) and (4.52); we find

$$\mathbf{K} = \begin{bmatrix} -3.91\times10^{-2} & 1.01 & 2.03 \\ 4.44\times10^{-1} & -3.74\times10^{-2} & 2.19 \\ 1.04 & 1.04 & -3.98 \end{bmatrix} .$$

Neither $\mathbf{E}$ nor $\mathbf{K}$ is stable; therefore, $\mathcal{E}_2$ cannot be used as an estimate of $\mathcal{D}$.

Let us now examine the matrix $\mathbf{A}^T\mathbf{B}+\mathbf{B}\mathbf{A}$, with $\mathbf{B}=\mathbf{B}^T=\mathbf{I}$, the identity matrix. In the case of column VII we get

$$\mathbf{A}^T+\mathbf{A} = \begin{bmatrix} -3.576 & 0.751 & 2.342 \\ 0.751 & -3.564 & 2.505 \\ 2.342 & 2.505 & -15.84 \end{bmatrix} .$$

Obviously, this is a stable matrix; according to Theorem 4, we conclude that

$$\mathbf{E}_1 = \{\mathbf{x} : V_1(\mathbf{x}) \le \gamma_1\}$$

where

$$V_1(\mathbf{x}) = [1 \;\; 1 \;\; 1] \begin{bmatrix} V_1(x_1) \\ V_2(x_2) \\ V_3(x_3) \end{bmatrix} ,$$

$$\gamma_1 = \min[V_1^o, V_2^o, V_3^o] = 2.851 ,$$

is an estimate of $\mathcal{D}$.

The transient stability analysis of the 3-machine system should now be performed by computing $V_1(t_e)$, $V_2(t_e)$, $V_3(t_e)$, where $V_i(t_e)$ denotes the value that the V_i function $(i=1,2,3)$ takes at different clearing times, t_e. The critical clearing time, t_c, is the one for which $V_1(\mathbf{x}_c) = \gamma_1$.

V.4.5.2.2. Triplewise decomposition-aggregation. The present section is mainly contributed by Shaaban[1] (Shaaban, 1983; Shaaban and Grujić, 1985).

An N-machine power system is decomposed into subsystems, each consisting of two machines in addition to the comparison machine. The system is decomposed into $(N-1)/2$ interconnected subsystems for odd number N. Considering transfer conductances, mechanical damping, electro-

[1] H. Shaaban was on leave from the University of Menoufia, El-Kom, Egypt, at the University of Belgrade where his research (Shaaban, 1983) was guided by Lj.T. Grujić.

magnetic damping and speed governor action, the mathematical model of the system is derived, and it is decomposed into $(N-1)/2$ sixth-order and one second-order interconnected subsystems. If N is even then the system is decomposed in $(N-2)/2$ sixth-order, one third order and one second order subsystems.

Each of these systems is decomposed into a free (disconnected) subsystem and interconnections. Each of the sixth-order free subsystems is assumed to include the largest number of nonlinearities, i.e. six nonlinearities. For this subsystem we adopt a scalar Liapunov function in the form "quadratic form + sum of the integrals of the six nonlinearities". A vector Liapunov function, whose components are Liapunov functions of the free subsystems, is constructed, and used for the system aggregation. A square aggregation matrix of the order $(N+1)/2$ is obtained when N is odd, and of the order $(N+2)/2$ when N is even. Stability of this matrix implies asymptotic stability of the system equilibrium.

In this section N is odd, without losing generality. The system (4.102) is decomposed into $(N-1)/2$ interconnected subsystems, each consisting of two machines and the comparison machine, using the triplewise decomposition. It is to be noted that none of the system machines (except the comparison machine) can be included in more than one subsystem.

Now, by introducing the set $J_I = \{i_I, i_I+1\}$ and defining the state vectors $\mathbf{x}_I$ and $\mathbf{x}_N$ as follows :

$$\mathbf{x}_I = [\sigma_{i_I N}, \sigma_{i_I+1,N}, \omega_{i_I N}, \omega_{i_I+1,N}, P_{i_I N}, P_{i_I+1,N}]^T$$
$$= [x_{I_1}\ x_{I_2}\ x_{I_3}\ x_{I_4}\ x_{I_5}\ x_{I_6}]^T ,$$

and

$$\mathbf{x}_N = [\omega_N \quad P_N]^T = [x_{N_1} \quad x_{N_2}]^T , \tag{4.140}$$

we can decompose the system mathematical model (4.103) into $s = (N-1)/2$ sixth-order interconnected subsystems and the second-order interconnected subsystem, which has the general form

$$\mathbf{x}_N = \mathbf{P}_N \mathbf{x}_N + \mathbf{h}_N(\mathbf{x}) , \tag{4.141}$$

where

$$\mathbf{P}_N = \begin{bmatrix} -\lambda_N & M_N^{-1} \\ -\alpha_N & \mu_N \end{bmatrix} ,$$

and

$$\mathbf{h}_N = \begin{bmatrix} \sum_{j=1}^{N-1} \{\lambda_{Nj}\omega_{jN} - M_N^{-1} A_{Nj}\phi_{Nj}(\sigma_{Nj})\} \\ 0 \end{bmatrix} .$$

Each of the sixth-order subsystems may be written in the general form

$$\dot{\mathbf{x}}_I = \mathbf{P}_I \mathbf{x}_I + \mathbf{B}_I \mathbf{f}_I(\boldsymbol{\sigma}_I) + \mathbf{h}_I(\mathbf{x}) \quad \text{for} \quad I = 1,2,\dots,s \,, \tag{4.142}$$

and it can be decomposed into the free (disconnected) subsystems given by

$$\dot{\mathbf{x}}_I = \mathbf{P}_I \mathbf{x}_I + \mathbf{B}_I \mathbf{f}_I(\boldsymbol{\sigma}_I) \,, \tag{4.143}$$

where

$$\boldsymbol{\sigma}_I = \mathbf{C}_I^T \mathbf{x}_I \quad \text{for} \quad I = 1,2,\dots,s \tag{4.144}$$

and the interconnections $\mathbf{h}_I(\mathbf{x})$.

In (4.143), the matrices $\mathbf{P}_I$, $\mathbf{B}_I$ and $\mathbf{C}_I^T$ are constant matrices, and $\mathbf{f}_I(\boldsymbol{\sigma}_I)$ is a nonlinear vector function. Referring to (4.103), we can define the matrix $\mathbf{P}_I$ as

$$\mathbf{P}_i = \begin{bmatrix} 0 & 0 & 1 & 0 & 0 & 0 \\ 0 & 0 & 0 & 1 & 0 & 0 \\ 0 & 0 & -\Gamma_I & \Lambda_I & M_{i_I}^{-1} & 0 \\ 0 & 0 & \bar{\Lambda}_I & -\bar{\Gamma}_I & 0 & M_{i_I+1}^{-1} \\ 0 & 0 & -\alpha_{i_I} & 0 & -\mu_{i_I} & 0 \\ 0 & 0 & 0 & -\alpha_{i_I+1} & 0 & -\mu_{i_I+1} \end{bmatrix} . \tag{4.145}$$

Assuming that the free subsystem (4.143) contains the six nonlinearities given by

$$\begin{aligned}
\phi_{I_1}(\sigma_{I_1}) &= \cos(\sigma_{i_I N} + \delta^{\circ}_{i_I N} - \theta_{i_I N}) - \cos(\delta^{\circ}_{i_I N} - \theta_{i_I N}) \,, \\
\phi_{I_2}(\sigma_{I_2}) &= \cos(\sigma_{i_I+1,N} + \delta^{\circ}_{i_I+1,N} - \theta_{i_I+1,N}) - \cos(\delta^{\circ}_{i_I+1,N} - \theta_{i_I+1,N}) \,, \\
\phi_{I_3}(\sigma_{I_3}) &= \cos(\sigma_{i_I,i_I+1} + \delta^{\circ}_{i_I,i_I+1} - \theta_{i_I,i_I+1}) - \cos(\delta^{\circ}_{i_I,i_I+1} - \theta_{i_I,i_I+1}) \,, \\
\phi_{I_4}(\sigma_{I_4}) &= \cos(\sigma_{i_I+1,i_I} + \delta^{\circ}_{i_I+1,i_I} - \theta_{i_I+1,i_I}) - \cos(\delta^{\circ}_{i_I+1,i_I} - \theta_{i_I+1,i_I}) \,, \\
\phi_{I_5}(\sigma_{I_5}) &= \cos(\sigma_{N i_I} + \delta^{\circ}_{N i_I} - \theta_{i_I N}) - \cos(\delta^{\circ}_{N i_I} - \theta_{i_I N}) \,, \\
\phi_{I_6}(\sigma_{I_6}) &= \cos(\sigma_{N,i_I+1} + \delta^{\circ}_{N,i_I+1} - \theta_{i_I+1,N}) - \cos(\delta^{\circ}_{N,i_I+1} - \theta_{i_I+1,N}) \,,
\end{aligned} \tag{4.146}$$

we can define the following matrices :

$$\mathbf{f}_I(\boldsymbol{\sigma}_I) = [\phi_{I_1}(\sigma_{I_1}), \phi_{I_2}(\sigma_{I_2}), \phi_{I_3}(\sigma_{I_3}), \phi_{I_4}(\sigma_{I_4}), \phi_{I_5}(\sigma_{I_5}), \phi_{I_6}(\sigma_{I_6})]^T \,, \tag{4.147}$$

$$B_I = \begin{bmatrix} 0 & 0 & 0 & 0 & 0 & 0 \\ 0 & 0 & 0 & 0 & 0 & 0 \\ -M_{i_I}^{-1}A_I & 0 & -M_{i_I}^{-1}\tilde{A}_I & 0 & M_N^{-1}A_I & M_N^{-1}\bar{A}_I \\ 0 & -M_{i_I+1}^{-1}\bar{A}_I & 0 & -M_{i_I+1}^{-1}\tilde{A}_I & M_N^{-1}A_I & M_N^{-1}\bar{A}_I \\ 0 & 0 & 0 & 0 & 0 & 0 \\ 0 & 0 & 0 & 0 & 0 & 0 \end{bmatrix},$$
(4.148)

$$C_I^T = \begin{bmatrix} 1 & 0 & 0 & 0 & 0 & 0 \\ 0 & 1 & 0 & 0 & 0 & 0 \\ 1 & -1 & 0 & 0 & 0 & 0 \\ -1 & 1 & 0 & 0 & 0 & 0 \\ -1 & 0 & 0 & 0 & 0 & 0 \\ 0 & -1 & 0 & 0 & 0 & 0 \end{bmatrix}, \qquad (4.149)$$

$$h_I(x) = \begin{bmatrix} 0 \\ 0 \\ \tau_I x_{N_1} + \eta_I x_{N_2} + \sum_{j \in J_I}^{N-1} \{\Lambda_{Ij}\omega_{jN} - M_{i_I}^{-1}A_{i_I j}\phi_{i_I j}(\sigma_{i_I j}) + M_N^{-1}A_{Nj}\phi_{Nj}(\sigma_{Nj})\} \\ \bar{\tau}_I x_{N_1} + \bar{\eta}_I x_{N_2} + \sum_{j \notin J_I}^{N-1} \{\bar{\Lambda}_{Ij}\omega_{jN} - M_{i_I+1}^{-1}A_{i_I+1,j}\phi_{i_I+1,j}(\sigma_{i_I+1,j}) + \\ + M_N^{-1}A_{Nj}\phi_{Nj}(\sigma_{Nj})\} \\ -\alpha_I x_{N_1} - \mu_I x_{N_2} \\ -\bar{\alpha}_I x_{N_1} - \bar{\mu}_I x_{N_2} \end{bmatrix}.$$
(4.150)

In (4.145)-(4.150) the following additional notation is used :

$$A_{ij} = E_i E_j Y_{ij} ,$$

$$A_I = E_{i_I} E_N Y_{i_I N} ; \quad \bar{A}_I = E_{i_I+1} E_N Y_{i_I+1,N} ; \quad \tilde{A}_I = E_{i_I} E_{i_I+1} Y_{i_I,i_I+1} ,$$

$$\sigma_{i_I N} = \delta_{i_I N} - \delta^{\circ}_{i_I N} ; \quad \sigma_{i_I+1,N} = \delta_{i_I+1,N} - \delta^{\circ}_{i_I+1,N} ,$$

$$\sigma_{i_I,i_I+1} = \sigma_{i_I N} - \sigma_{i_I+1,N} = \delta_{i_I,i_I+1} - \delta^{\circ}_{i_I,i_I+1} ,$$

$$\omega_{i_I N} = \omega_{i_I} - \omega_N ; \quad \omega_{i_I+1,N} = \omega_{i_I+1} - \omega_N ,$$

$$\tau_I = \lambda_N - \lambda_{i_I} ; \quad \bar{\tau}_I = \lambda_N - \lambda_{i_I+1} ,$$

$$\Lambda_I = \lambda_{i_I,i_I+1} - \lambda_{N,i_I+1} ; \quad \bar{\Lambda}_I = \lambda_{i_I+1,i_I} - \lambda_{N i_I}$$

$$\Lambda_{Ij} = \lambda_{i_I j} - \lambda_{Nj} ; \quad \bar{\Lambda}_{Ij} = \lambda_{i_I+1,j} - \lambda_{Nj} ,$$

$$\Gamma_I = \lambda_{i_I} + \lambda_{Ni_I} + \sum_{j \neq i_I}^{N} \lambda_{i_I j} \quad ; \quad \bar{\Gamma}_I = \lambda_{i_I+1} + \lambda_{N,i_I+1} + \sum_{j \neq i_I+1}^{N} \lambda_{i_I+1,j} \ ,$$

$$\mu_I = \mu_{i_I} - \mu_N \quad ; \quad \bar{\mu}_I = \mu_{i_I+1} - \mu_N \ ,$$

$$\alpha_I = \alpha_{i_I} - \alpha_N \quad ; \quad \bar{\alpha}_I = \alpha_{i_I+1} - \alpha_N \ ,$$

$$\eta_I = M_{i_I}^{-1} - M_N^{-1} \quad ; \quad \bar{\eta}_I = M_{i_I+1}^{-1} - M_N^{-1} \ .$$

It is obvious that the state vector of the whole system is given now by

$$\mathbf{x} = [x_1^T, x_2^T, \ldots, x_s^T, x_N^T]^T \ . \tag{4.151}$$

For each of the sixth-order free subsystems, we accept a V_I function of the form

$$V_I(\mathbf{x}_I) = \mathbf{x}_I^T \mathbf{H}_I \mathbf{x}_I + \sum_{\ell=1}^{6} \psi_{I\ell} \int_0^{\sigma_{I\ell}} \phi_{I\ell}(\sigma_{I\ell}) \, d\sigma_{I\ell} \quad \text{for } I = 1,2,\ldots,s \ , \tag{4.152}$$

where $\mathbf{H}_I$ is a sixth-order symmetric positive-definite matrix, $\psi_{I\ell}$ are arbitrary positive numbers, and the nonlinearities $\phi_{I\ell}(\sigma_{I\ell})$ are given by (4.146).

For the last $(s+1)$-th second-order free subsystem, we construct the Liapunov function

$$V(\mathbf{x}_N) = \mathbf{x}_N^T \begin{bmatrix} C & 0 \\ 0 & 1 \end{bmatrix} \mathbf{x}_N \tag{4.153}$$

where C is an arbitrary positive number.

Along the motion of the free subsystem (4.143) we compute $\dot{V}_{If}$,

$$\dot{V}_I(\mathbf{x}_I)_f = \mathbf{x}_I^T(-\mathbf{G}_I)\,\mathbf{x}_I + 2\,\mathbf{f}_I^T(\boldsymbol{\sigma}_I)\,\mathbf{B}_I^T \mathbf{H}_I \mathbf{x}_I + \sum_{I=1}^{6} \psi_{I\ell} f_{I\ell}(\sigma_{I\ell})\,\dot{\sigma}_{I\ell}$$

$$\forall I = 1,2,\ldots,s \ , \tag{4.154}$$

where $\mathbf{G}_I$ is a symmetric matrix, defined as

$$-\mathbf{G}_I = \mathbf{P}_I^T \mathbf{H}_I + \mathbf{H}_I \mathbf{P}_I \ . \tag{4.155}$$

Substituting the matrix $\mathbf{P}_I$ from (4.145) in (4.155), the matrix $\mathbf{G}_I$ is computed, and it is found that this matrix cannot be positive definite (the first two diagonal elements are zeros).

Now, under the condition $\Lambda_I = \bar{\Lambda}_I = 0$, and by choosing the matrix $\mathbf{H}_I$ in the form

$$\mathbf{H}_I = \begin{bmatrix} h_{11}^I & 0 & h_{13}^I & 0 & h_{15}^I & 0 \\ 0 & h_{22}^I & 0 & h_{24}^I & 0 & h_{26}^I \\ h_{13}^I & 0 & h_{33}^I & 0 & 0 & 0 \\ 0 & h_{24}^I & 0 & h_{44}^I & 0 & 0 \\ h_{15}^I & 0 & 0 & 0 & h_{55}^I & 0 \\ 0 & h_{26}^I & 0 & 0 & 0 & h_{66}^I \end{bmatrix}, \qquad (4.156)$$

where

$$h_{11}^I = (\Gamma_I + \frac{M_{i_I}^{-1}\alpha_{i_I}}{\mu_{i_I}})\, h_{13}^I ,$$

$$h_{15}^I = M_{i_I}^{-1} h_{13}^I / \mu_{i_I} ,$$

$$h_{22}^I = (\bar{\Gamma}_I + \frac{M_{i_I+1}^{-1}\alpha_{i_I+1}}{\mu_{i_I+1}})\, h_{24}^I ,$$

$$h_{26}^I = M_{i_I+1}^{-1} h_{24}^I / \mu_{i_I+1} ,$$

$$h_{33}^I = \frac{1 + K_I}{\Gamma_I}\, h_{13}^I ,$$

$$h_{44}^I = \frac{1 + K_I}{\bar{\Gamma}_I}\, h_{24}^I ,$$

$$h_{55}^I = \frac{1}{\alpha_{i_I}} (M_{i_I}^{-1} h_{33}^I + \frac{M_{i_I}^{-1}}{\mu_{i_I}}\, h_{13}^I) ,$$

$$h_{66}^I = \frac{1}{\alpha_{i_I+1}} (M_{i_I+1}^{-1} h_{44}^I + \frac{M_{i_I+1}^{-1}}{\mu_{i_I+1}}\, h_{24}^I) ,$$

we obtain the matrix $\mathbf{G}_I$ in the form

$$\mathbf{G}_I = \begin{bmatrix} 0 & 0 & 0 & 0 & 0 & 0 \\ 0 & 0 & 0 & 0 & 0 & 0 \\ 0 & 0 & 2K_I h_{13}^I & 0 & 0 & 0 \\ 0 & 0 & 0 & 2K_I h_{24}^I & 0 & 0 \\ 0 & 0 & 0 & 0 & 2\mu_{i_I} h_{55}^I & 0 \\ 0 & 0 & 0 & 0 & 0 & 2\mu_{i_I+1} h_{66}^I \end{bmatrix}. \qquad (4.157)$$

It is to be noted that positive definiteness of the matrix $\mathbf{H}_I$ of equation (4.156) can be guaranteed (K_I is an arbitrary positive number) only under the two conditions $h_{13} > 0$ and $h_{24} > 0$.

Now, substituting from equations (4.147)-(4.149), (4.156) and (4.157) into equation (4.154), and selecting the positive numbers

$$\psi_{I_1} = 2M_{i_I}^{-1}A_I h_{33}^I \qquad \psi_{I_2} = 2M_{i_I+1}^{-1}\bar{A}_I h_{44}^I$$
$$\psi_{I_3} = 2M_{i_I}^{-1}\tilde{A}_I h_{33}^I \qquad \psi_{I_4} = 2M_{i_I+1}^{-1}\tilde{A}_I h_{44}^I$$
$$\psi_{I_5} = 2M_N^{-1}A_I h_{33}^I \qquad \psi_{I_6} = 2M_N^{-1}\bar{A}_I h_{44}^I \quad ,$$

we obtain

$$\begin{aligned}\dot{V}_I(x_I)_f = & -2K_I h_{13}^I x_{I_3}^2 - 2K_I h_{24}^I x_{I_4}^2 - 2\mu_{i_I} h_{55}^I x_{I_5}^2 - 2\mu_{i_I+1} h_{66}^I x_{I_6}^2 - \\ & -2M_{i_I}^{-1}A_I h_{13}^I x_{I_1}\phi_{I_1}(\sigma_{I_1}) - 2M_{i_I+1}^{-1}\bar{A}_I h_{24}^I x_{i_2}\phi_{I_2}(\sigma_{I_2}) - \\ & -2M_{i_I}^{-1}\tilde{A}_I \phi_{I_3}(\sigma_{I_3})(h_{13}^I x_{I_1} + h_{33}^I x_{I_4}) - \\ & -2M_{i_I+1}^{-1}\tilde{A}_I \phi_{I_4}(\sigma_{I_4})(h_{24}^I x_{I_2} + h_{44}^I x_{I_3}) + \\ & +2M_N^{-1}A_I\phi_{I_5}(\sigma_{I_5})(h_{13}^I x_{I_1} + h_{24}^I x_{I_2} + h_{44}^I x_{I_4}) + \\ & +2M_N^{-1}\bar{A}_I\phi_{I_6}(\sigma_{I_6})(h_{13}^I x_{I_1} + h_{24}^I x_{I_2} + h_{33}^I x_{I_3}) \quad . \end{aligned} \tag{4.159}$$

Let us now introduce the positive constants $\epsilon_{I\ell} \in]0,\xi_{I\ell}[$, for $\ell = 1,2,5,6$, $\xi_{I\ell}$ are determined by $\xi_{i\ell} = \sin(\theta_{i\ell}-\delta_{i\ell}^{\circ})$ so that

$$\sigma_{I\ell}\phi_{I\ell}(\sigma_{I\ell}) \geq \epsilon_{I\ell}\sigma_{I\ell}^2 \quad \text{for} \quad \ell = 1,2,5,6 \tag{4.160}$$

is satisfied on a compact interval $L_{I\ell}$ of $\sigma_{I\ell}$, i.e.

$$L_{I\ell} = [\underline{U}_{I\ell} \quad \bar{U}_{I\ell}] \ , \tag{4.161}$$

where $\underline{U}_{I\ell}$ and $\bar{U}_{I\ell}$ are the negative and positive solutions, respectively, of

$$\phi_{I\ell}(\sigma_{I\ell}) = \epsilon_{I\ell}\sigma_{I\ell} \quad \text{for} \quad \ell = 1,2,5,6 \ . \tag{4.162}$$

It is important to note that if the value of ϵ_I is taken smaller, the interval L_I given by (4.161) becomes larger, and so we obtain a larger estimate for the system stability domain.

Now, by adding and subtracting from the right-hand side of (4.159) the non-negative expression

$$\begin{aligned} & 2M_{i_I}^{-1}\tilde{A}_I h_{13}^I[\sigma_{I_3}\phi_{I_3}(\sigma_{I_3}) - \xi_{I_3}^{-1}\phi_{I_3}^2(\sigma_{I_3})] \\ & \qquad + 2M_{i_I+1}^{-1}\tilde{A}_I h_{24}^I[\sigma_{I_4}\phi_{I_4}(\sigma_{I_4}) - \xi_{I_4}^{-1}\phi_{I_4}^2(\sigma_{I_4})] \ , \end{aligned} \tag{4.163}$$

where ξ_{I_3} and ξ_{I_4} are determined as

$$\xi_{I_3} = \sin(\theta_{i_I,i_I+1} - \delta_{i_I,i_I+1}^{\circ}) \quad \text{and} \quad \xi_{I_4} = \sin(\theta_{i_I,i_I+1} + \delta_{i_I,i_I+1}^{\circ}) \tag{4.164}$$

we can "majorize" the right-hand side of (4.159) after trivial calculations, as

$$\dot{V}_I(\mathbf{x}_I)_f \leq -\lambda_I^* \|\mathbf{x}_I\|^2 \quad , \quad \forall I = 1,2,\ldots,s \ , \tag{4.165}$$

where λ_I^* is the minimal (positive) eigenvalue of the symmetric matrix $\mathbf{G}_I^*$, which is given as

$$\mathbf{G}_I^* = \begin{bmatrix} g_{11}^I & g_{12}^I & 0 & g_{14}^I & 0 & 0 & 0 & g_{18}^I \\ g_{12}^I & g_{22}^I & g_{23}^I & 0 & 0 & 0 & g_{27}^I & 0 \\ 0 & g_{23}^I & g_{33}^I & 0 & 0 & 0 & 0 & g_{38}^I \\ g_{14}^I & 0 & 0 & g_{44}^I & 0 & 0 & g_{47}^I & 0 \\ 0 & 0 & 0 & 0 & g_{55}^I & 0 & 0 & 0 \\ 0 & 0 & 0 & 0 & 0 & g_{66}^I & 0 & 0 \\ 0 & g_{27}^I & 0 & g_{47}^I & 0 & 0 & g_{77}^I & 0 \\ g_{18}^I & 0 & g_{38}^I & 0 & 0 & 0 & 0 & g_{98}^I \end{bmatrix}, \quad (4.166)$$

where

$$g_{11}^I = 2A_I h_{13}^I (M_i^{-1}\epsilon_{I_1} + M_N^{-1}\epsilon_{I_5}), \qquad g_{22}^I = 2\bar{A}_I h_{24}^I (M_{i_I+1}^{-1}\epsilon_{I_2} + M_N^{-1}\epsilon_{I_6}),$$

$$g_{33}^I = 2K_I h_{13}^I, \qquad g_{44}^I = 2K_I h_{24}^I,$$

$$g_{55}^I = 2\mu_{i_I} h_{55}^I, \qquad g_{66}^I = 2\mu_{i_I+1} h_{66}^I,$$

$$g_{12}^I = -M_N^{-1}(A_I h_{24}^I \xi_{Ni_I} + \bar{A}_I h_{13}^I \xi_{N,i_I+1}), \quad (4.167)$$

$$g_{14}^I = -M_N^{-1} A_I h_{44}^I \xi_{N,i_I}, \qquad g_{18}^I = -M_{i_I+1}^{-1}\tilde{A}_I h_{24}^I,$$

$$g_{23}^I = -M_N^{-1}\bar{A}_I h_{33}\xi_{N,i_I+1}, \qquad g_{27}^I = -M_{i_I}^{-1}\tilde{A}_I h_{13}^I,$$

$$g_{38}^I = -M_{i_I+1}^{-1}\tilde{A}_I h_{44}^I, \qquad g_{47}^I = -M_{i_I}^{-1}\tilde{A}_I h_{33}^I,$$

$$g_{77}^I = 2M_{i_I}^{-1}\tilde{A}_I h_{13}^I / \xi_{I_3}, \qquad g_{88}^I = 2M_{i_I+1}^{-1}\tilde{A}_I h_{24}^I / \xi_{I_4},$$

$$\xi_{Ni_I} = \sin(\theta_{i_I N} + \delta_{i_I N}^{\circ}) \quad \text{and} \quad \xi_{N,i_I+1} = \sin(\theta_{i_I+1,N} + \delta_{i_I+1,N}^{\circ}).$$

It is important to note that the numbers K_I, h_{13}^I, h_{24}^I, ϵ_{I_1} and ϵ_{I_2} ($\epsilon_{I_5} = \epsilon_{I_6} = 0.001$ are chosen to be slightly larger than zero) should be chosen so that we can guarantee positive definiteness of the matrix $\mathbf{G}_I^*$.

Now, we proceed to majorize in the right-hand side of what follows :

$$[\text{grad } V_I(\mathbf{x}_I)]^T h_I(\mathbf{x}) = 2(h_{13}^I x_{I_1} + h_{33}^I x_{I_3}) \times$$

$$\times \Big[\tau_I x_{N_1} + \eta_I x_{N_2} + \sum_{j \notin J_I}^{N-1} \{\Lambda_{Ij}\omega_{jN} - M_{i_I}^{-1} A_{i_I j}\phi_{i_I j}(\sigma_{i_I j}) + M_N^{-1} A_{Nj}\phi_{Nj}(\sigma_{Nj})\}\Big] +$$

$$+ 2(h_{24}^I x_{I_2} + h_{44}^I x_{I_4}) \times$$

$$\times\left[\bar{\tau}_I x_{N_1} + \bar{\eta}_I x_{N_2} + \sum_{j\notin J_I}^{N-1}\{\Lambda_{Ij}\omega_{jN} - M^{-1}_{i_I+1}A_{i_I+1,j}\phi_{i_I+1,j}(\sigma_{i_I+1,j}) + M_N^{-1}A_{Nj}\phi_{Nj}(\sigma_{Nj})\}\right]$$

$$+\ 2(h^I_{15}x_{I_1} + h^I_{55}x_{I_5})(-\alpha_I x_{N_1} - \mu_I x_{N_2}) + 2(h^I_{26}x_{I_2} + h^I_{66}x_{I_6})(-\bar{\alpha}_I x_{N_1} - \bar{\mu}_I x_{N_2}) . \tag{4.168}$$

Noting that $\sum\limits_{j\notin J_I}^{N-1}$ is equivalent to $\sum\limits_{K\neq I}^{s}\ \sum\limits_{j\in J_K}$, and introducing the following majorizations :

$$|f_{i_Ij}(\sigma_{i_Ij})| \le \xi_{i_Ij}(|x_{I_1}| + |x_{J_1}|) \quad , \quad \xi_{i_Ij} = \sin(\theta_{i_Ij} - \delta^\circ_{i_Ij}) ,$$

$$|\phi_{i_I+1,j}(\sigma_{i_I+1,j})| \le \xi_{i_I+1,j}(|x_{I_2}| + |x_{J_1}|) \quad , \quad \xi_{i_I+1,j} = \sin(\theta_{i_I+1,j} - \delta^\circ_{i_I+1,j}) , \tag{4.169}$$

we can majorize the right-hand side of (4.168) in the form

$$[\mathrm{grad}\ V_I(\mathbf{x}_I)]^T h_I(\mathbf{x}) \le 2Z_{IN}\|\mathbf{x}_I\|\|\mathbf{x}_N\| + \tilde{\lambda}_I\|\mathbf{x}_I\|^2 + 2\sum_{K\neq I}^{s} Z_{IK}\|\mathbf{x}_I\|\|\mathbf{x}_K\| ,$$

$$\forall I = 1,2,\dots,s \tag{4.170}$$

where

$$Z_{IN} = Z_2(Z_{IN_a}, Z_{IN_b}) , \tag{4.171}$$

$$Z_{IN_a} = Z_2\left[Z_3\{(\tau_I h^I_{13} - \alpha_I h^I_{15}), \tau_I h^I_{33}, \alpha_I h^I_{55}\} ; Z_3\{(\bar{\tau}_I h^I_{24} - \bar{\alpha}_I h^I_{26}), \bar{\tau}_I h^I_{44}, \bar{\alpha}_I h^I_{66}\}\right] ,$$

$$Z_{IN_b} = Z_2\left[Z_3\{(\eta_I h^I_{13} - \mu_I h^I_{15}), \eta_I h^I_{33}, \mu_I h^I_{55}\} ; Z_3\{(\bar{\eta}_I h^I_{24} - \bar{\mu}_I h^I_{26}), \bar{\eta}_I h^I_{44}, \bar{\mu}_I h^I_{66}\}\right] ,$$

and $\tilde{\lambda}_I$ is the maximal eigenvalue of the fourth-order symmetric matrix $\mathbf{Q}_I$, whose elements are given as

$$q^I_{11} = q^I_{22} = q^I_{33} = q^I_{44} = q^I_{12} = q^I_{14} = q^I_{23} = q^I_{34} = 0 ,$$

$$q^I_{13} = M^{-1}_{i_I} h^I_{33} \sum_{K\neq I}^{s}(A_{i_Ii_K}\xi_{i_Ii_K} + A_{i_I,i_K+1}\xi_{i_I,i_K+1}) , \tag{4.172}$$

$$q^I_{24} = M^{-1}_{i_I+1} h^I_{44} \sum_{K\neq I}^{s}(A_{i_i+1,i_K}\xi_{i_I+1,i_K} + A_{i_I+1,i_K+1}\xi_{i_I+1,i_K+1}) .$$

In (4.170), Z_{IK} is defined as

$$Z_{IK} = Z_2[Z_2(\bar{Z}_{IK}, \bar{\bar{Z}}_{IK}) ; Z_2(\tilde{Z}_{IK}, \tilde{\tilde{Z}}_{IK})] , \tag{4.173}$$

where

$$\bar{Z}_{IK} = Z_2\Big[(M_N^{-1}A_{Ni_K}\xi_{Ni_K} + M^{-1}_{i_I}A_{i_Ii_K}\xi_{i_Ii_K})\, Z_2(h^I_{13}, h^I_{33}) ;$$

$$(M_N^{-1}A_{Ni_K}\xi_{Ni_K} + M^{-1}_{i_I+1}A_{i_I+1,i_K}\xi_{i_I+1,i_K})\, Z_2(h^I_{24}, h^I_{44})\Big] ,$$

$$\bar{\bar{Z}}_{IK} = Z_2\Big[(M_N^{-1}A_{N,i_K+1}\xi_{N,i_K+1} + M^{-1}_{i_I}A_{i_I,i_K+1}\xi_{i_I,i_K+1})\, Z_2(h^I_{13}, h^I_{33}) ;$$

$$(M_N^{-1}A_{N,i_K+1}\xi_{N,i_K+1} + M^{-1}_{i_I+1}A_{i_I+1,i_K+1}\xi_{i_I+1,i_K+1})\, Z_2(h^I_{24}, h^I_{44})\Big] ,$$

$$\tilde{Z}_{IK} = Z_2\left[|\lambda_{i_I i_K} - \lambda_{Ni_K}|\ Z_2(h_{13}^I, h_{33}^I)\ ;\ |\lambda_{i_I+1,i_K} - \lambda_{Ni_K}|\ Z_2(h_{24}^I, h_{44}^I)\right] ,$$

$$\tilde{\tilde{Z}}_{IK} = Z_2\left[|\lambda_{i_I,i_K+1} - \lambda_{N,i_K+1}|\ Z_2(h_{13}^I, h_{33}^I)\ ;\ |\lambda_{i_I+1,i_K+1} - \lambda_{N,i_K+1}|\ Z_2(h_{23}^I, h_{44}^I)\right] .$$

We can define the first $(N-1)/2$ rows of the aggregation matrix $\mathbf{A} = [\alpha_{IK}]$ as follows :

$$\alpha_{IK} = \begin{cases} -(\dot{\lambda}_I^* - \tilde{\lambda}_I) & K = I \\ 2Z_{IK} & K \neq I \\ 2Z_{IN} & K = \frac{N-1}{2} \\ & \text{for} \quad I = 1,2,\dots,s = \frac{N-1}{2} . \end{cases} \tag{4.174}$$

To define the last $((N+1)/2\text{th})$ row of the matrix $\mathbf{A}$, the total time derivative of V_N, along the motion of the subsystem (4.141) is computed as

$$\begin{aligned} \dot{V}_N(\mathbf{x}) &= -2C\lambda_N x_{N_1}^2 - 2\mu_N x_{N_2}^2 + 2x_{N_1}x_{N_2}(CM_N^{-1} - \alpha_N) \\ &\quad - 2x_{N_1}\sum_{j=1}^{N-1}(\lambda_{Nj}\omega_{Nj} + M_N^{-1}A_{Nn}\phi_{Nj}(\sigma_{Nj})) . \end{aligned} \tag{4.175}$$

Choosing the constant $C = M_N\alpha_N$, we can eliminate the term in $x_{N_1}x_{N_2}$, and then equation (4.175) is rewritten as

$$\begin{aligned} \dot{V}_N(\mathbf{x}) &= -2(M_N\alpha_N\lambda_N x_{N_1}^2 + \mu_N x_{N_2}^2) - 2x_{N_1}\sum_{K=1}^{s}\{\lambda_{Ni_K}\omega_{Ni_K} + M_N^{-1}A_{Ni_K}\phi_{Ni_K}(\sigma_{Ni_K}) \\ &\quad + \lambda_{N,i_K+1}\omega_{N,i_K+1} + M_N^{-1}A_{N,i_K+1}\phi_{N,i_K+1}(\sigma_{N,i_K+1})\} . \end{aligned} \tag{4.176}$$

The left-hand side of equation (4.176) may be directly majorized as

$$\begin{aligned} \dot{V}_N(\mathbf{x}) &= -2\min(M_N\alpha_N\lambda_N , \mu_N)\|\mathbf{x}_N\|^2 + \\ &\quad + 2\sum_{K=1}^{s} Z_2\left[M_N^{-1}Z_2(A_{Ni_K}\xi_{Ni_K} , A_{N,i_K+1}\xi_{N,i_K+1})\ ;\ Z_2(\lambda_{Ni_K} , \lambda_{N,i_K+1})\right]\|\mathbf{x}_N\|\,\|\mathbf{x}_K\| . \end{aligned}$$

This implies

$$\alpha_{s+1,K} = \begin{cases} -2\min(M_N\alpha_N\lambda_N , \mu_N) & K = s+1 \\ 2Z_2\{M_N^{-1}Z_2(A_{Ni_K}\xi_{Ni_K} , A_{N,i_K+1}\xi_{N,i_K+1})\ ;Z_2(\lambda_{Ni_K} , \lambda_{N,i_K+1})\} , \\ & K = I \quad \text{for} \quad I = 1,2,\dots,s = \frac{N-1}{2} . \end{cases} \tag{4.177}$$

The system aggregation matrix is completely defined now by (4.174) and (4.177).

Numerical example 2. In this example, the seven-machine system shown in Fig. 10 is considered as a simple power system. Choosing machine 7 as the comparison machine, the system is decomposed into three interconnected subsystems, as indicated in Fig. 11.

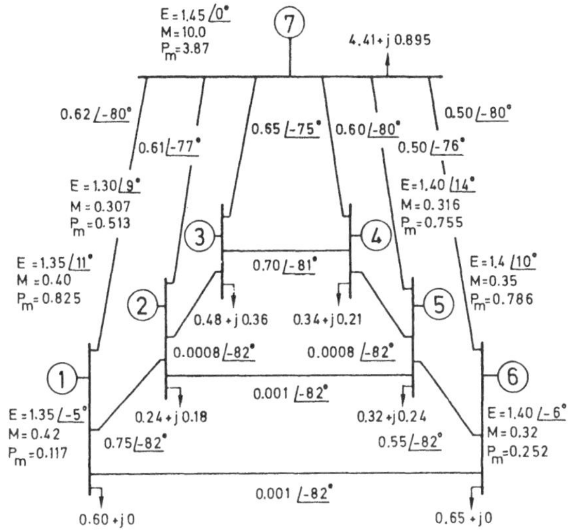

Figure 10

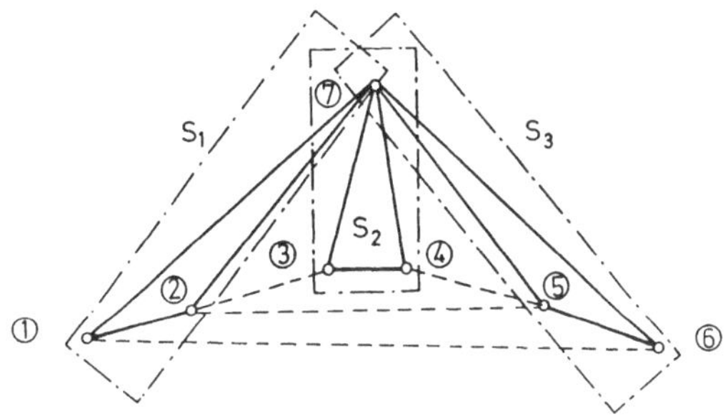

Figure 11

The following parameters are selected :

$$\lambda_i = 8.0\ ,\ i = 1,2,\dots,6\ ;\ \lambda_7 = 8.1\ ;$$

$$\lambda_{ij} = 0.10\ ,\ i \neq j\ ,\ i,j = 1,2,\dots,7\ ;$$

$$\mu_i = 20\ ;\ \alpha_i = 25\ ,\ i = 1,2,\dots,7\ ;$$

$$h^j_{13} = h^j_{24} = 1.0\ ,\ j = 1,2,3\ ;$$

$$K_1 = 2.0\ ;\ K_2 = 2.6\ ;\ K_3 = 2.1\ ;$$

$$\epsilon_{j_1} = \epsilon_{j_2} = 0.70\ ,\ j = 1,2\ ;\ \epsilon_{3_1} = \epsilon_{3_2} = 0.65\ .$$

Then the matrices $\mathbf{G}^*$ (4.166) and $\mathbf{Q}$ (4.172) are constructed for each of the three subsystems, and the eigenvalues λ^* and $\tilde{\lambda}$ for these matrices are determined. Finally, the aggregation matrix of (4.174) and (4.177) is computed as

$$\mathbf{A} = \begin{bmatrix} -1.5013 & 0.6213 & 0.5688 & 9.5094 \\ 0.6841 & -2.0747 & 0.5727 & 12.9961 \\ 0.6880 & 0.6278 & -1.5126 & 11.9691 \\ 0.4839 & 0.4480 & 0.4051 & -40.0000 \end{bmatrix} .$$

This matrix satisfies the conditions (4.28) and thus it is a stable matrix. This implies asymptotic stability of the system equilibrium due to Corollary 1 to Theorem 1.

We can now go further, to determine a stability domain estimate for the whole system. Therefore the matrix $\mathbf{A}^T\mathbf{B} + \mathbf{B}^T\mathbf{A}$, with the matrix $\mathbf{B} = \mathbf{B}^T$ in the form $\mathbf{B} = \mathrm{diag}\{1 \;\; 1 \;\; 1 \;\; 20\}$, is computed and found to be negative definite.

According to Theorem 4 we conclude that

$$E_1 = \{\mathbf{x} : V(\mathbf{x}) \le \gamma_1\} , \tag{4.178}$$

where

$$V(\mathbf{x}) = V_1(\mathbf{x}_1) + V_2(\mathbf{x}_2) + V_3(\mathbf{x}_3) + 20\, V_4(\mathbf{x}_4) ,$$

and

$$\gamma_1 = \min\,(V_1^o, V_2^o, V_3^o, 20\, V_4^o) ,$$

is an estimate of the system asymptotic stability domain.

Using the following equation (see Appendix of Shaaban, 1983) :

$$V_I^o = \min_{m=1,2,\dots,6} \; \min_{x_{I_m} \in \{\underline{x}_{I_m}, \bar{x}_{I_m}\}} \left\{ (\mathbf{x}_I)^T H_I \mathbf{x}_I + \sum_{\ell=1}^{6} \psi_{I\ell} \int_0^{\sigma_{I\ell}} \phi_{I\ell}(\sigma_{I\ell})\, d\sigma_{I\ell} \right\}$$

where $\underline{x}_{I_m}$ and $\bar{x}_{I_m}$ are lower and upper bounds of x_{I_m}, we get

$$V_1^o = 10.37 \;,\quad V_2^o = 21.12 \quad \text{and} \quad V_3^o = 13.82 \;.$$

Using these values $(V_4^o = \infty)$, the constant $\gamma_1 = 10.37$ is determined.

In terms of the original physical variables $\boldsymbol{\delta}$, $\boldsymbol{\omega}$ and $\mathbf{P}$ of the system, the estimate E_1 is written as

$$E_1 = \{(\boldsymbol{\delta},\boldsymbol{\omega},\mathbf{P}) : (V_1(\boldsymbol{\delta},\boldsymbol{\omega},\mathbf{P}) + V_2(\boldsymbol{\delta},\boldsymbol{\omega},\mathbf{P}) + V_3(\boldsymbol{\delta},\boldsymbol{\omega},\mathbf{P}) + 20V_4(\omega_7, P_7)) \le 10.37\}$$

where

$$V_1(\boldsymbol{\delta},\boldsymbol{\omega},\mathbf{P}) =$$

$$= 11.7\,(\delta_{17} + 0.09)^2 + 11.8\,(\delta_{27} - 0.19)^2 + 0.35\,\{(\omega_1 - \omega_7)^2 + (\omega_2 - \omega_7)^2\} +$$

$$+ 0.04\{(P_1-P_7)^2 + (P_2-P_7)^2\} + 2(\delta_{17} + 0.09)\{(\omega_1-\omega_7) + 0.12(P_1-P_7)\} +$$
$$+ 2(\delta_{27} - 0.19)\{(\omega_2-\omega_7) + 0.13(P_2-P_7)\} + 2.0 \sin(\delta_{17} - 1.74) +$$
$$+ 2.03 \sin(\delta_{27} - 1.8) + 2.25 \sin\{(\delta_{17}-\delta_{27}) - 1.71\} +$$
$$+ 2.36 \sin\{(\delta_{27}-\delta_{17}) - 1.71\} - 0.08\{\sin(\delta_{17} + 1.74) + \sin(\delta_{27} + 1.8)\} +$$
$$+ 1.08\,\delta_{17} - 1.205\,\delta_{27} + 8.85 ,$$

$$V_2(\boldsymbol{\delta},\boldsymbol{\omega},\mathbf{P}) =$$
$$= 12.8(\delta_{37} - 0.16)^2 + 12.7(\delta_{47} - 0.24)^2 + 0.41\{(\omega_3-\omega_7)^2 + (\omega_4-\omega_7)^2\} +$$
$$+ 0.06\{(P_3-P_7)^2 + (P_4-P_7)^2\} + 2(\delta_{37} - 0.16)\{(\omega_3-\omega_7) + 0.16(P_3-P_7)\} +$$
$$+ 2(\delta_{47} - 0.24)\{(\omega_4-\omega_7) + 0.16(P_4-P_7)\} + 3.31 \sin(\delta_{37} - 1.83) +$$
$$+ 3.19 \sin(\delta_{47} - 1.75) + 3.44 \sin\{(\delta_{37}-\delta_{47}) - 1.73\} +$$
$$+ 3.34 \sin\{(\delta_{47}-\delta_{37}) - 1.73\} - 0.10\{\sin(\delta_{37} + 1.83) + \sin(\delta_{47} + 1.75)\} +$$
$$+ 0.904\,\delta_{37} - 0.863\,\delta_{47} + 13.395 ,$$

$$V_3(\boldsymbol{\delta},\boldsymbol{\omega},\mathbf{P}) =$$
$$= 12.3(\delta_{57} - 0.17)^2 + 12.6(\delta_{67} + 0.11)^2 + 0.36\{(\omega_5-\omega_7)^2 + (\omega_6-\omega_7)^2\} +$$
$$+ 0.05\{(P_5-P_7)^2 + (P_6-P_7)^2\} + 2(\delta_{57} - 0.17)\{(\omega_5-\omega_7) + 0.14(P_5-P_7)\} +$$
$$+ 2(\delta_{67} + 0.11)\{(\omega_6-\omega_7) + 0.16(P_6-P_7)\} + 2.07 \sin(\delta_{57} - 1.82) +$$
$$+ 2.26 \sin(\delta_{67} - 1.75) + 2.2 \sin(\delta_{57}-\delta_{67}) - 1.71 +$$
$$+ 2.4 \sin\{(\delta_{67}-\delta_{57}) - 1.71\} - 0.072\{\sin(\delta_{57} + 1.82) + \sin(\delta_{67} + 1.75)\} +$$
$$- 1.167\,\delta_{57} + 1.90\,\delta_{67} + 9.144$$

and $$V_4(\omega_7,P_7) = 250\,\omega_7^2 + P_7^2 .$$

V.5. CONCLUSION

Two direct methods for transient stability analysis of power systems have been developed and examined : the scalar and the vector Liapunov approaches.

The scalar approach seems to have by now reached maturity, owing to the efforts produced by many researchers during the last twenty-five years : the numerous studies carried out in this field have succeeded in removing the major handicap i.e. its conservative character. Concerning the necessity of adopting simplified system representation, it may not be - and in our opinion is not - a drawback of the method, when used for first contingency dynamic security monitoring. In fact, the Liapunov or Liapunov-like criteria should be considered as complementary, rather than competitive with the standard numerical integration procedure, each method possessing its own field of application.

For the vector Liapunov approach it is still too early to tell how interesting it could become for practical applications. A priori, its essential characteristics make this approach likely to be more appropriate for decentralized control purposes, rather than for global stability investigations. Once again, by their very nature, the scalar and the vector approaches appear to be complementary rather than competitive.

Whatever the evolution in this field, one by now may claim that power system transient stability has been a very difficult and extremely challenging domain of research; yet one of the most efficient and fruitful applications of the Liapunov stability theory.

Araki, M. (1975), Application of M-matrices to the stability problems of composite dynamical systems. *J. Math. Anal. and Appl.*, *52*, No.3, 309-321.

Araki, M., M. Saeki, and B. Kondo (1980), Application of a new stability criterion of composite systems to multimachine power systems. *IEEE Trans. Autom. Control, AC-25*, 480-483.

Athay, T., R. Podmore, and S. Virmani (1979) , A practical method for the direct analysis of transient stability. *IEEE Trans PAS, PAS-98*, 573.

Athay, T., and I. Sun (1981), A structure preserving model for power system stability analysis. *IEEE Trans. on PAS, PAS-100*, No.1, 25.

Aylett, P.D. (1958), The energy-integral criterion of transient stability limits of power systems. *IEE Proc., 105-C*, No.8, 527.

Barbier, C., L. Carpentier, and F. Saccomanno (1978), Tentative classification and terminologies relating to stability problems of power systems. *Electra*, No.56.

Bellman, R. (1962), Vector Liapunov functions. *J. SIAM Control, 1*, No.1, 32-34.

Bergen, A.R., and D.J. Hill (1981), A structure preserving model for power system stability analysis. *IEEE Trans. on PAS, PAS-100*, No.1, 25.

Bouffioux, A. (1978), *Fonctions vectorielles de Liapunov pour systèmes énergétiques*. Travail de fin d'études, University of Liège, 1-95.

Chen, Y.K., and R. Schinzinger (1980), Lyapunov stability of multimachine power systems using decomposition-aggregation method. *IEEE PES Winter Meeting*, Paper A 80 036-4.

Chorlton, A., and G. Shackshaft (1972), Comparison of accuracy of methods for studying stability. Northfleet exercise. *Electra, 23*, 9.

Dandeno, P.L. (1977)., Synchronous machine stability constants; requirements and realization (prepared by IEEE Joint Working Group on this subject). *IEEE PES Winter Meeting*, New York, Paper No. 177, 210.

Di Caprio, U., and F. Saccomano (1969), Application of the Liapunov's direct method to analysis of multi-machine power system stability. *3rd. Power Systems Computation Conf.*, Rome (Italy).

Di Caprio, U., and F. Saccomano (1970), Non-linear stability analysis of multi-machine electric power systems. *Ricerche di Automatica, I*, No.1.

El-Abiad, A.H., and K. Nagappan (1966), Transient stability regions of multi-machine power systems. *IEEE Trans. on PAS, PAS-85*, No.2, 169.

Evans, F.J. (1978), Prospect for dynamic security monitoring in large scale electric power systems. *7th World IFAC Congress*, Helsinki, 1.

Fiedler, M., and V. Ptak (1962), On matrices with non-positive off-diagonal elements and positive principal minors. *Czech. math. J., 12*, 382-400.

Fouad, A.A. (1975), Stability theory - Criteria for transient stability. *Proc. Eng. Foundation Conf. on Systems Eng. for Power : Status and Prospects*, Henniker (New Hampshire), 421.

Fouad, A.A., V. Vittal, and T. Oh (1984), Critical energy for direct transient stability using individual machine energy functions. *IEEE Trans. PAS, PAS-103*, 2199-2206.

Gantmacher, F.R. (1974), *The Theory of Matrices*, Chelsea Publ., New York.

Gelopoulos, D.P., and J.W. Lamont (1980), Stability program and output analysis survey. *IEEE Winter Power Meeting*, Paper No. A80083-3.

Grujić, Lj.T. (1974a), On multi-level absolute stability analysis of large-scale systems. Part 1. *Automatika* (Zagreb), Nos. 1-2, 67-72.

Grujić, Lj.T. (1974b), Stability analysis of large-scale systems with stable and unstable subsystems. *Int. J. Control, 20*, No.3, 453-463.

Grujić, Lj.T. (1975), Novel development of Lyapunov stability of motion. *Int. J. Control, 22*, No.4, 529-549.

Grujić, Lj.T. (1976), General stability analysis of large-scale systems. *Proc. IFAC Symp. on Large-Scale Systems*, Udine, 203-213.

Grujić, Lj.T. (1981a), Liapunov-like solutions for stability problems of the most general stationary Lur'e-Postnikov systems". *Int. J. Systems Sci., 12*, No.7, 813-833.

Grujić, Lj.T. (1981b), On absolute stability and the Aizerman conjecture. *Automatica, 17*, No.2, 335-349.

Grujić, Lj.T., and D.D. Šiljak (1973), Asymptotic stability and instability of large-scale systems. *IEEE Trans. Aut. Cont., AC-18*, No.6, 636-645.

Grujić, Lj.T., J.C. Gentina, and P. Borne (1976), General aggregation of large-scale systems by vector Lyapunov functions and vector norms. *Int. J. Control, 24*, No.4, 529-550.

Grujić, Lj.T., M. Darwish, and J. Fantin (1977), Coherence, vector Lyapunov functions and large-scale power systems. *Preprint IFAC Workshop on Contr. and Manag. of Integrat. Ind. Comp.*, Pergamon Press, London, 145-159.

Grujić, Lj.T., and M. Ribbens-Pavella (1977), *Large-Scale Power Systems : Decomposition, Aggregation and Stability*. University of Liège, Internal report.

Grujić, Lj.T., and M. Ribbens-Pavella (1977), New approach to stability domain estimate of large-scale power systems. *Revue E, VIII*, No.10, 241-249.

Grujić, Lj.T., and M. Ribbens-Pavella (1978), Relaxed large-scale systems stability analysis applied to power systems. *7th World IFAC Congress*, Helsinki, 27-34.

Grujić, Lj.T., M. Ribbens-Pavella, and J. Sabatel (1978), Scalar vs. vector Liapunov functions for transient stability analysis of large-scale power systems. *MECO*, Athens, 700-707.

Grujić, Lj.T., M. Ribbens-Pavella, and A. Bouffioux (1979a), Asymptotic stability of large-scale systems with application to power systems. Part I : Domain estimation. *Electrical Power and Energy Systems, 1*, No.3, 151-157.

Grujić, Lj.T., M. Ribbens-Pavella, and A. Bouffioux (1979b), Asymptotic stability of large-scale systems with application to power systems. Part II : Transient analysis. *Electrical Power and Energy Systems, 1*, No.3, 158-165.

Gorev, A.A. (1960), *Selected Works in Problems of Stability of Electric Systems*. M.L. GEI (in Russian).

Hahn, W. (1967), *Stability of Motion*. Springer Verlag, Berlin.

Henner, V.E. (1974), Multi-machine power system Liapunov function using the generalized Popov criterion. *Int. J. of Control, 19*, No.5, 969.

IEEE Committee Report (1982), Proposed terms and definitions for power system stability. *IEEE Trans. PAS, PAS-101*, No.7, 1894.

Jocić, Lj.B., M. Ribbens-Pavella, and D.D. Šiljak (1977), On transient stability of multimachine power systems. *JACC*, 627-632.

Jocić, Lj.B., M. Ribbens-Pavella, and D.D. Šiljak (1978), Multimachine power systems : stability decomposition and aggregation. *IEEE Trans. Aut. Control, AC-23*, No.2, 325-332.

Jocić, Lj.B., and D.D. Šiljak (1978), Decomposition and stability of multimachine power systems. *7th World IFAC Congress*, Helsinki, 21-25.

Kakimoto, N., Y. Ohsawa, and M. Hayashi (1978), Transient stability analysis of electric power systems via Lur'e type Lyapunov function with effect of transfer conductances. *Trans. IEE of Japan, 98*, Nos. 5/6, 63.

Kakimoto, N., Y. Ohsawa, and M. Hayashi (1980), Transient stability analysis of multimachine power systems with field flux decays via Lyapunov's direct method. *IEEE Trans. PAS, PAS-99*, No.5, 1819.

Kakimoto, N., and M. Hayashi (1981), Transient stability analysis of multimachine power system by Lyapunov's direct method. *Proc. 20th IEEE Conf. on Dec. and Control*, 464.

Kamke, E. (1932), Zur Theorie der Systeme Gewöhnlicher Differencial-Gleichungen, II. *Acta Mathematica, 58*, 57-85.

Kitamura, S., T. Dohomoto, and Y. Hurematsu (1977), Construction of a Lyapunov function by the perturbation method and its application to the transient stability problem of power systems with non-negligible transfer conductances. *Int. J. of Control, 26*, No.3, 405.

Krasovskii,N.N. (1963), *Certain Problems of the Theory of Stability of Motion*. FIZMATGIZ, Moscow (in Russian).

La Salle, J.P. (1976), The stability of dynamical systems. *SIAM*.

La Salle, J.P., and S. Lefschetz (1961), *Stability by Lyapunov's Direct Method, with Applications*. Academic Press, New York.

Liapunov, A.M. (1892), *General Problem of Stability of Motion*. The Math. Soc. of Kharkov, Kharkov (in Russian).

Mahalanabis, A.K., and R. Singh (1979), Frequency domain criteria for transient stability of multimachine power systems. *IFAC Symp. on Computer Applications in Large-Scale Power Systems*, New Delhi, 177-181.

Mahalanabis, A.K., and R. Singh (1980), On the analysis and improvements of the transient stability of multimachine power systems. *IEEE PES Winter Meeting*, Feb., Paper A 80 039-8, 1-10.

Magnusson, P.C. (1947), Transient energy method of calculating stability. *AIEE Trans.*, *66*, 747.

Matrosov, V.M. (1962), On the theory of stability of motion. *Prikl. Math. Mekh.*, *26*, 992-1002 (in Russian).

Matrosov, V.M. (1968a), Comparison principle and vector Lyapunov functions. I. *Diff. Urawn*, *4*, No.8, 1374-1386 (in Russian).

Matrosov, V.M. (1968b), Comparison principle and vector Lyapunov functions. II. *Diff. Urawn*, *4*, No.10, 1740-1752 (in Russian).

Matrosov, V.M. (1969a), Comparison principle and vector Lyapunov functions. III. *Diff. Urawn*, *5*, No.7, 1171-1185 (in Russian).

Matrosov, V.M. (1969b), Comparison principle and vector Lyapunov functions. IV. *Diff. Urawn*, *5*, No.12, 2128-2143 (in Russian).

Michel, A.N. (1974), Stability analysis of interconnected systems. *SIAM J. Control*, *12*, No.3, 554-579.

Michel, A.N., A.A. Fouad, and V. Vittal (1983), Power system transient stability using individual machine energy functions. *IEEE Trans. on Circuits and Systems*, *CAS-30*, 266.

Moore, J.B., and B.D.O. Anderson (1968), A generalization of the Popov criterion. *J. Franklin Inst.*, *285*, 488.

Pai, M.A. (1981), *Power System Stability*. North Holland, Control Series.

Pai, M.A., and P.G. Murthy (1974), New Liapunov functions for power systems based on minimal realizations. *Int. J. Control*, *19*, No.2, 401.

Pai, M.A., and C.I. Narayana (1975), Stability of large scale power systems. *Proc. Sixth IFAC Congress*, Boston, Mass., 1-10.

Persidskii, S.K. (1969), Concerning problem of absolute stability. *Aut. i Telemekh.*, *12*, 5-11 (in Russian).

Popov, V.M. (1962), Absolute stability of non-linear systems of automatic control. *Automation Remote Control*, *22*, 857.

Prabhakara, F.S., A.H. El-Abiad, and A.J. Kovio (1974), Applications of generalized Zubov's method to power system stability. *Int. J. of Control*, *20*, No.2, 203.

Quazza, G. (1976), Large-scale control problems in electric power systems - A survey. *IFAC Symp. on Large Scale Systems*, Udine (Italy).

Ribbens-Pavella, M. (1969), *Théorie Générale de la Stabilité Transitoire de n Machines Synchrones*. Thèse de Doctorat. Coll. des publ. de la Fac. des Sc. Appl., Université de Liège.

Ribbens-Pavella, M. (1971a), Transient stability of multi-machine power systems by Lyapounov's direct method. *IEEE Winter Power Meeting*, Paper No. 71CP17-PWR.

Ribbens-Pavella, M. (1971b), Critical survey of transient stability studies of multimachine power systems by Liapunov's direct method. *Proc. of 9th Allerton Conf. on Circuit and System Theory*, Univ. of Illinois, 151-167.

Ribbens-Pavella, M. (1975), On-line measurements of transient stability power system index. *Computerized Operation of Power Systems (COPOS)*, Elsevier, Savulsescu (Ed.), 176.

Ribbens-Pavella, M., and B. Lemal (1976), Fast determination of stability regions for on-line transient power system studies. *Proc. IEE, 123*, No.7, 689.

Ribbens-Pavella, M., B. Lemal, and W. Pirard (1977), On-line operation of Lyapunov criterion for transient stability studies. *IFAC Symp.*, Melbourne, 292.

Ribbens-Pavella, M., and F.J. Evans (1981), Direct methods in the study of the dynamics of large-scale electric power systems - An overview. *8th World IFAC Congress*, Kyoto, Japan, August, 2931-2938.

Ribbens-Pavella, M., and F.J. Evans (1985a), Direct methods for studying dynamics of large-scale electric power systems - A survey. *Automatica, 21*, 1-21.

Ribbens-Pavella, M., Th. Van Cutsem, R. Dhifaoui, and B. Toumi (1985b), Energy-type Liapunov-like direct criteria for rapid transient stability analysis. *Proc. of the Int. Symp. on Power System Stability*, Iowa, May.

Saeki, M., M. Araki, and B. Kondo (1983), A Lur'e type Lyapunov function for multimachine power systems with transfer conductances. *Int. J. Control, 42*, No.3, 607-619.

Santalo, L.A. (1976), *Encyclopedia of Mathematics and its Applications, 1*, Reading, Mass; Addison-Wesley.

Sastry, V.R., and P.G. Murthy (1972a), Discussion of J.L. Willems "Direct methods for transient stability studies in power system analysis" and reply by author. *IEEE Trans. Aut. Control, AC-17*, No.3, 415.

Sastry, and P.G. Murthy (1972b), Derivation of completely controllable and completely observable state models for multi-machine power system stability studies. *Int. J. of Control, 16*, No.4, 777.

Shaaban, H. (1983), *Transient Stability Analysis of Electric Power Systems Under Structural Perturbations Via Vector Liapunov Functions*. Ph.D. thesis, University of Belgrade.

Shaaban, H., and Lj.T. Grujić (1985), Transient stability analysis of large-scale power systems with speed governor via vector Liapunov functions. *IEE Proc., 132*, No.2, 45-52.

Šiljak, D.D. (1969), *Nonlinear Systems. The Parameter Analysis and Design*, Wiley.

Šiljak, D.D. (1978), *Large-Scale Dynamic Systems : Stability and Structure*. North-Holland, New York.

Tavora, C.J., and O.J.M. Smith (1972), Characterization of equilibrium and stability in power systems. *IEEE Trans. PAS, PAS-91*, No.3, 1127.

Union Institute of Scientific and Technological Information and the Academy of Sciences of the U.S.S.R. (1971), *Criteria of Stability of Electric Power Systems*. Report, Electric technology and electric power series, Moscow (in Russian). (Contains 132 references).

Van Cutsem, Th., B. Toumi, Y. Xue, and M. Ribbens-Pavella (1986), Direct criteria for structure preserving models of electric power systems. *Proc. of the IFAC Symp. on Power Systems and Power Plant Control*, Beijing, China, August.

Varaiya, P., F.F. Wu, and R.L. Chen (1985), Direct methods for tran-ie sient stability analysis of power systems : recent results. *Proc. of the IEEE.*

Walker, J.A., and N.H. McClamrock (1967), Finite regions of attraction for the problem of Lur'e. *Int. J. Control, 6*, 331-336.

Ważewski, J. (1950), Systèmes des équations et des inégalités différentielles ordinaires aux deuxièmes membres monotones et leurs applications. *Ann. Soc. Pol. Math., 23*, 112-166.

Weissenberger, S. (1973), Stability regions of large-scale systems. *Automatica, 9*, 653-663.

Willems, J.L. (1970a), *Stability Theory of Dynamical Systems*. Nelson, London.

Willems, J.L. (1970b), Optimum Lyapunov functions and stability regions for multi-machine power systems. *Proc. IEE, 117*, No.3, 573.

Willems, J.L. (1971), Direct methods for transient stability studies in power system analysis. *IEEE Trans. Aut. Control, AC-16*, No.4, 332.

Willems, J.L. (1974), Partial stability approach to the problem of transient power system stability. *Int. J. of Control, 19*, No.1, 1.

Willems, J.L., and J.C. Willems (1970), The application of Lyapunov methods to the computation of transient stability regions for multi-machine power systems. *IEEE Trans. PAS, PAS-89*, No.5, 795.

Yu, Y.N., and K. Vongasuriya (1967), Nonlinear power system stability study by Lyapunov function and Zubov's method. *IEEE Trans. PAS, PAS-86*, No.12, 1480.

Zubov, V.I. (1961), Methods of Lyapunov and their application. *U.S. Atomic Energy Commission, Translation AEC-tr-4439*, 83.

POSTFACE

System non-stationariness can be originated by its time-varying physical nature. An adequate system mathematical model is then also time-varying. However, the system description via a state deviation is non-stationary despite the system is physically time-invariant as soon as it is non-linear and its reference motion is time varying. For such a physical and/or mathematical reason it is important to study non-stationary non-linear systems. Their greater complexity than that of stationary non-linear systems results in a great variety of stability properties. They depend in general on the initial moment t_o. The classical stability theory tackles mainly the stability problems whose solutions are independent of t_o, which has been a basic assumption at the very beginning of Liapunov's dissertation (Liapunov, 1892) and for the development of the stability theory for decades.

The stability theory for non-stationary non-linear systems in general, taking into account the influence of t_o, is presented in Chapter I. It shows that stability properties should be tested on time-varying sets rather than on time-invariant sets due to non-stationariness of the system and/or its mathematical model expressed in terms of the state deviation. The general absolute stability results presenting necessary and sufficient Liapunov like conditions are used as a basis for new linear systems like absolute stability criteria related to non-stationary Lur'e systems with multiple nonlinearities. Besides, new relaxed general asymptotic stability conditions are established for time-varying non-linear singularly perturbed systems.

Liapunov like stability conditions are expressed in terms of *existence* of a v function (or, functional family) with specified features called a system Liapunov function (or, Liapunov functional family), respectively. Hence, a fundamental scientific problem and the greatest obstacle for wider effective practical application of Liapunov's (direct, second) method is the problem of a Liapunov function construction for a given system (or for a class of systems). Higher the system dimension, more difficult the problem. More complex the system structure, more complex the problem. Greater the system dynamical complexity (number and/or form of nonlinearities), more severe the problem. In order to simplify the problem, Bellman (1962) proposed the vector Liapunov function concept for a Liapunov aggregation of a (complex, interconnected) large-scale system decomposed into interconnected subsystems.

In Chapter II the concept of matrix Liapunov functions is proposed and developed. This new mathematical tool together with a novel development of the comparison principle presented therein constitute a powerful basis for a new stability theory direction. This is shown via new general stability criteria.

One-shot Liapunov function construction appears more real for subsystems than for the overall system. A number of different aggregation forms are developed in Chapter III. They are used to derive simple overall system algebraic stability conditions via the Liapunov functions of the subsystems. They lead to new forms of Liapunov functions for the whole system even if it is subject to arbitrary structural and/or parameter variations. Hence, both structural and/or parameter robustness is assured for various stability properties of non-stationary non-linear large-scale systems. Moreover, their robustness is guaranteed for cone variations of nonlinearities. An intrinsic advantage of the results is a qualitative simplification of the stability conditions reduced to a stability test of an aggregation matrix with constant entries and with the dimension inherently reduced to the number of the subsystems.

The stability theory of non-stationary non-linear large-scale systems with multiple time-scales and subject to singular perturbations as well as to arbitrary structural variations is presented in Chapter IV. Both non-uniform and uniform time scaling are treated. The aggregation forms established therein are accompanied by simple algebraic conditions expressed via Liapunov functions of the subsystems. Hence, all preceding types of stability robustness are also assured for non-stationary nonlinear large-scale singularly perturbed systems.

Chapter V presents an analysis of stability achievements via both scalar and vector Liapunov functions.

Within the vector Liapunov concept framework, three different estimates of the asymptotic stability domain of the equilibrium state can be guaranteed under simple and relaxed algebraic conditions based on knowledge of subsystems Liapunov functions. The vector Liapunov function approach appears also attractive for possible use of a more refined mathematical model of a power system, simple stability test, and easy asymptotic stability domain estimation. This is shown in general by developing various aggregation forms in terms of subsystems Liapunov functions. However, a number of majorizations used for construction of the overall system aggregation matrix can impose too sharp conditions on the system.

Effective applications of the large-scale systems stability theory is crucially dependent on solving the next three qualitative problems :

Problem 1. How to construct a vector Liapunov function for a given large-scale system ?

This problem essentially means how to select Liapunov functions for subsystems, i.e. for lower order (non-stationary non-linear) systems.

Several important contributions to solving this problem have appeared since the time this book was completed (1982) (Grujić et al., 1984). Michel et al. (1982) developed a methodology for computer construction of Liapunov functions. Vanelli and Vidyasagar (1985) proved necessary and sufficient conditions for a set to be the domain of attraction of the origin. The conditions are expressed in terms of existence of a suitable Liapunov function. They also presented a new iterative procedure for the construction of a system Liapunov function. Other interesting new procedures for the Liapunov function construction are due to Genesio et al. (1985) and Chin (1986). Grujić (1986b) proved necessary and sufficient conditions for asymptotic stability of the origin, for its domain and for a Liapunov function construction. The preceding results are promising for both effective determination of Liapunov functions of the subsystems and effective estimation of the asymptotic stability domains of their equilibria. Hence, they constitute a strong new basis for effective solutions of Problem 1 in general and advanced application of results presented in Chapters II-V.

These results can be directly combined with those of Chapter V for their wider applications.

Problem 2. Which comparison functions mostly relax majorizations of the aggregation procedure, requirements on interactions and stability conditions on the overall system by assuring simultaneously the order reduction of its aggregation matrix at least to the number of its subsystems ?

Michel and Miller (1983) proposed use of stability preserving mappings for stability analysis. It is interesting to explore deeper whether stability preserving mappings can provide a tool for solving Problem 2. Another new promising mathematical tool for solving Problem 2 is the concept of the matrix Liapunov functions by Martynyuk (Chapter II), which has been further developed by himself (1984-1986), and independently in another original way by Djordjević (1983,1986). A link among the concepts of scalar, vector and matrix Liapunov functions is shown by Grujić (1986a).

All these and other approaches should reduce the majorizations stiffness so that stabilizing actions of interactions over subspaces are adequately preserved in the aggregation matrix.

Problem 3. What is the best possible effective estimation of the attraction and/or asymptotic stability domain of an equilibrium state ?

Bondi and Gambardella (1986) contributed to solving the attraction domain estimation in the framework of non-stationary non-linear systems without requiring positive definiteness of Liapunov functions of subsystems. Grujić (1986b) established conditions for exact determination of the asymptotic stability domain of the origin in the framework of stationary non-linear systems. These results linked with those solving Problems 1 and 2 open new horizons for the asymptotic stability domain estimations specified in Chapter V.

Shaaban and Grujić (1984-1986) discovered new relaxed decomposition-aggregation forms for power systems, which provide a basis for more effective application of the vector Liapunov function concept to more refined mathematical models of power systems. Arbitrary structural variations of power systems are also incorporated in their analysis, which is a further development of the methodology presented in Chapter V.

With regards to the power system stability problem, our final qualitative conclusion is the following : vector or matrix Liapunov approach has potential : it simplifies the overall system Liapunov function construction and the asymptotic stability domain estimation; it could cope with the problem of testing asymptotic stability under arbitrary structural perturbations; it enables accounting for more refined mathematical models than those amenable to one-shot application of a scalar Liapunov function. However, the latter imposes less (if not much less) stringent conditions on the system. Called to meet urgent practical needs, it has been continually developing. No doubt, by the time this book appears, the situation will have further changed. Things change faster in engineering sciences than in the book writing process !

Bellman, R. (1962), Vector Lyapunov functions. *J. SIAM Control*, Ser.A, *1*, No.1, 32-34.

Bondi, P., and L. Gambardella (1986), On the asymptotic behaviour of the solutions of large-scale systems. *Proc. IMACS-IFAC Symposium*, June 3-6, IDN, Villeneuve d'Ascq, France, 235-240.

Chin, P.S.M. (1986), A general method to derive Lyapunov functions for non-linear systems. *Int. J. Control, 44*, No.2, 381-393.

Djordjević, M.Z. (1983), Stability analysis of interconnected systems with possible unstable systems. *Systems & Control Letters, 3*, 165-169.

Djordjević, M.Z. (1983), Stability analysis of large-scale systems whose subsystems may be unstable. *Large Scale Systems, 5*, 252-262.

Djordjević, M.Z. (1986), Stability analysis of nonlinear systems by the matrix Lyapunov method. *Proc. IMACS-IFAC Symposium*, June 3-6, IDN, Villeneuve d'Ascq, France, 209-212.

Genesio, R., M. Tartaglia, and A. Vicino (1985), On the estimation of asymptotic stability region : state of the art and new proposals. *IEEE Trans. on Automatic Control, AC-30*, No.8, 747-755.

Grujić, Lj.T. (1986a), Large-scale systems stability. *IMACS-IFAC Symp.* June 3-6, IDN, Villeneuve d'Ascq, France, tutorial.

Grujić, Lj.T. (1986b), Stability domains of general and large-scale systems. *Proc. IMACS-IFAC Symp.*, June 3-6, IDN, Villeneuve d'Ascq, France, 267-272.

Grujić, Lj.T., A.A. Martynyuk, and M. Ribbens-Pavella (1984), *Large-Scale Systems Stability Under Structural and Singular Perturbations.* Naukova Dumka, Kiev (in Russian) (it was completed for print 1982).

Liapunov, A.M. (1892), *The General Problem of Stability of Motion.* Publ. Harkov's Mathematical Society, Harkov (in Russian).

Martynyuk, A.A. (1984), The Lyapunov matrix-function. *Nonlinear Analysis, Theory, Methods & Applications, 8*, No.10, 1223-1226.

Martynyuk, A.A. (1985a), On Liapunov matrix functions and stability of motion. *Proc. of the Academy of Science of USSR, 280*, No.5, 1062-1066 (in Russian).

Martynyuk, A.A. (1985b), On application of the Lyapunov matrix-functions in the theory of stability. *Nonlinear Analysis, Theory, Methods, and Applications, 9*, No.12, 1495-1501.

Martynyuk, A.A. (1986), Liapunov matrix-function and stability theory. *Proc. IMACS-IFAC Symp.*, June 3-6, IDN, Villeneuve d'Ascq, France, 261-265.

Michel, A.N., N.R. Sarabudla, and R.K. Miller (1982), Stability analysis of complex dynamical systems : some computational methods. *Circuits, Systems and Signal Processing, 1*, No.2, 171-202.

Michel, A.N., R.K. Miller, and B.H. Nam (1982), Stability analysis of interconnected systems using computer generated Lyapunov functions. *IEEE Trans. on Circuits and Systems, CAS-29*, No.7, 431-440.

Michel, A.N., and R.K. Miller (1983), On stability preserving mappings. *IEEE Trans. on Circuits and Systems, CAS-30*, No.9, 671-680.

Michel, A.N., B.H. Nam, and V. Vittal (1984), Computer generated Lyapunov functions for interconnected systems : improved results with application to power systems. *IEEE Trans. on Circuits and Systems, CAS-31*, No.2, 189-199.

Shaaban, H., and Lj.T. Grujić (1984), An improved pair-wise decomposition aggregation approach to stability analysis of multimachine power systems. *R.A.I.R.O. Automatique / Systems Analysis and Control, 18*, No.1, 63-77.

Shaaban, H., and Lj.T. Grujić (1985), Transient stability analysis of large-scale power systems with speed governor via vector Lyapunov functions. *IEE Proc., 132*, Pt.D, No.2, 45-52.

Shaaban, H., and Lj.T. Grujić (1986a), Liapunov stability of large-scale power systems under structural perturbations using decomposition aggregation method. *Proc. IMACS-IFAC Symp.*, June 3-6, IDN, Villeneuve d'Ascq, France, 195-200.

Shaaban, H., and Lj.T. Grujić (1986b), Improvement of large-scale power systems decomposition-aggregation approach. *Int. J. of Electrical Power and Energy Systems*, in print.

Shaaban, H., and Lj.T. Grujić (1986c), The decomposition-aggregation method applied to a multimachine power system. *Large Scale Systems, 10*, 115-132.

Shaaban, H., and Lj.T. Grujić (1986d), Transient stability analysis of multimachine power systems using two decomposition-aggregation approaches. *Control - Theory and Advanced Technology, 2*, No.3, in print.

Vanelli, A., and M. Vidyasagar (1985), Maximal Lyapunov functions and domains of attraction for autonomous non-linear systems. *Automatica, 21*, No.1, 69-80.

Lecture Notes in Control and Information Sciences

Edited by M. Thoma

Vol. 43: Stochastic Differential Systems
Proceedings of the 2nd Bad Honnef Conference of the SFB 72 of the DFG at the University of Bonn June 28 – July 2, 1982
Edited by M. Kohlmann and N. Christopeit
XII, 377 pages. 1982.

Vol. 44: Analysis and Optimization of Systems
Proceedings of the Fifth International Conference on Analysis and Optimization of Systems Versailles. December 14–17, 1982
Edited by A. Bensoussan and J. L. Lions
XV, 987 pages, 1982

Vol. 45: M. Arató
Linear Stochastic Systems with Constant Coefficients
A Statistical Approach
IX, 309 pages. 1982

Vol. 46: Time-Scale Modeling of Dynamic Networks with Applications to Power Systems
Edited by J. H. Chow
X, 218 pages. 1982

Vol. 47: P. A. Ioannou, P. V. Kokotovic
Adaptive Systems with Reduced Models
V, 162 pages. 1983

Vol. 48: Yaakov Yavin
Feedback Strategies for Partially Observable Stochastic Systems
VI, 233 pages, 1983

Vol. 49: Theory and Application of Random Fields
Proceedings of the IFIP-WG 7/1 Working Conference held under the joint auspices of the Indian Statistical Institute
Bangalore, India, January 1982
Edited by G. Kallianpur
VI. 290 pages. 1983

Vol. 50: M. Papageorgiou
Applications of Automatic Control Concepts to Traffic Flow Modeling and Control
IX, 186 pages. 1983

Vol. 51: Z. Nahorski, H.F. Ravn, R.V.V. Vidal
Optimization of Discrete Time Systems
The Upper Boundary Approach
V, 137 pages 1983

Vol. 52: A. L. Dontchev
Perturbations, Approximations and Sensitivity Analysis of Optimal Control Systems
IV, 158 pages. 1983

Vol. 53: Liu Chen Hui
General Decoupling Theory of Multivariable Process Control Systems
XI, 474 pages. 1983

Vol. 54: Control Theory for Distributed Parameter Systems and Applications
Edited by F. Kappel, K. Kunisch, W. Schappacher
VII, 245 pages. 1983.

Vol. 55: Ganti Prasada Rao
Piecewise Constant Orthogonal Functions and Their Application to Systems and Control
VII, 254 pages. 1983.

Vol. 56: Dines Chandra Saha, Ganti Prasada Rao
Identification of Continuous Dynamical Systems
The Poisson Moment Functional (PMF) Approach
IX, 158 pages. 1983.

Vol. 57: T. Söderström, P. G. Stoica
Instrumental Variable Methods for System Identification
VII, 243 pages. 1983.

Vol. 58: Mathematical Theory of Networks and Systems
Proceedings of the MTNS-83 International Symposium
Beer Sheva, Israel, June 20–24, 1983
Edited by P. A. Fuhrmann
X, 906 pages. 1984

Vol. 59: System Modelling and Optimization
Proceedings of the 11th IFIP Conference
Copenhagen, Denmark, July 25-29, 1983
Edited by P. Thoft-Christensen
IX, 892 pages. 1984

Vol. 60: Modelling and Performance Evaluation Methodology
Proceedings of the International Seminar
Paris, France, January 24–26, 1983
Edited by F. Bacelli and G. Fayolle
VII, 655 pages. 1984

Vol. 61: Filtering and Control of Random Processes
Proceedings of the E.N.S.T.-C.N.E.T. Colloquium
Paris, France, February 23–24, 1983
Edited by H. Korezlioglu, G. Mazziotto, and J. Szpirglas
V, 325 pages. 1984

For information about Vols. 1–42 please contact your bookseller or Springer-Verlag.

Lecture Notes in Control and Information Sciences

Edited by M. Thoma and A. Wyner

Vol. 62: Analysis and Optimization
of Systems
Proceedings of the Sixth International
Conference on Analysis and Optimization
of Systems
Nice, June 19–22, 1984
Edited by A. Bensoussan, J. L. Lions
XIX, 591 pages. 1984.

Vol. 63: Analysis and Optimization
of Systems
Proceedings of the Sixth International
Conference on Analysis and Optimization
of Systems
Nice, June 19–22, 1984
Edited by A. Bensoussan, J. L. Lions
XIX, 700 pages. 1984.

Vol. 64: Arunabha Bagchi
Stackelberg Differential Games
in Economic Models
VIII, 203 pages, 1984

Vol. 65: Yaakov Yavin
Numerical Studies
in Nonlinear Filtering
VIII, 273 pages, 1985.

Vol. 66: Systems and Optimization
Proceedings of the Twente Workshop
Enschede, The Netherlands, April 16–18, 1984
Edited by A. Bagchi, H. Th. Jongen
X, 206 pages, 1985.

Vol. 67: Real Time Control of Large Scale Systems
Proceedings of the First European Workshop
University of Patras, Greece, Juli 9–12, 1984
Edited by G. Schmidt, M. Singh, A. Titli,
S. Tzafestas
XI, 650 pages, 1985.

Vol. 68: T. Kaczorek
Two-Dimensional Linear Systems
IX, 397 pages, 1985.

Vol. 69: Stochastic Differential Systems –
Filtering and Control
Proceedings of the IFIP-WG 7/1 Working Conference
Marseille-Luminy, France, March 12-17, 1984
Edited by M. Metivier, E. Pardoux
X, 310 pages, 1985.

Vol. 70: Uncertainty and Control
Proceedings of a DFVLR International Colloquium
Bonn, Germany, March, 1985
Edited by J. Ackermann
IV, 236 pages, 1985.

Vol. 71: N. Baba
New Topics in Learning Automata
Theory and Applications
VII, 231 pages, 1985.

Vol. 72: A. Isidori
Nonlinear Control Systems:
An Introduction
VI, 297 pages, 1985.

Vol. 73: J. Zarzycki
Nonlinear Prediction
Ladder-Filters for Higher-Order
Stochastic Sequences
V, 132 pages, 1985.

Vol. 74: K. Ichikawa
Control System Design based on
Exact Model Matching Techniques
VII, 129 pages, 1985.

Vol. 75: Distributed Parameter
Systems
Proceedings of the 2nd International
Conference, Vorau, Austria 1984
Edited by F. Kappel, K. Kunisch,
W. Schappacher
VIII, 460 pages, 1985.

Vol. 76: Stochastic Programming
Edited by F. Archetti, G. Di Pillo,
M. Lucertini
V, 285 pages, 1986.

Vol. 77: Detection of
Abrupt Changes in Signals
and Dynamical Systems
Edited by M. Basseville,
A. Benveniste
X, 373 pages, 1986.

Vol. 78: Stochastic
Differential Systems
Proceedings of the 3rd Bad Honnef
Conference, June 3–7, 1985
Edited by N. Christopeit, K. Helmes,
M. Kohlmann
V, 372 pages, 1986.

Vol. 79: Signal
Processing for Control
Edited by K. Godfrey, P. Jones
XVIII, 413 pages, 1986.

Vol. 80: Artificial Intelligence
and Man-Machine Systems
Edited by H. Winter
IV, 211 pages, 1986.

Lecture Notes in Control and Information Sciences

Edited by M. Thoma and A. Wyner

Vol. 81: Stochastic Optimization
Proceedings of the International Conference, Kiew, 1984
Edited by I. Arkin, A. Shiraev, R. Wets
X, 754 pages, 1986.

Vol. 82: Analysis and Algorithms of Optimization Problems
Edited by K. Malanowski, K. Mizukami
VIII, 240 pages, 1986.

Vol. 83: Analysis and Optimization of Systems
Proceedings of the Seventh International Conference of Analysis and Optimization of Systems
Antiba, June 26-27, 1986
Edited by A. Benoussan, J. L. Lions
XVI, 901 pages, 1986.

Vol. 84: System Modelling and Optimization
Proceedings of the 12th IFIP Conference
Budapest, Hungary, September 2–6, 1985
Edited by A. Prékopa, J. Szelezsán, B. Strazicky
XII, 1046 pages, 1986.

Vol. 85: Stochastic Processes in Underwater Acoustics
Edited by Charles R. Baker
V, 205 pages, 1986.

Vol. 86: Time Series and Linear Systems
Edited by Sergio Bittanti
XVII, 243 pages, 1986.

Vol. 87: Recent Advances in System Modelling and Optimization
Proceedings of the IFIP-WG 7/1 Working Conference
Santiago, Chile, August 27-31, 1984
Edited by L. Contesse, R. Correa, A. Weintraub
IV, 199 pages, 1987.

Vol. 88: Bruce A. Francis
A Course in H_∞ Control Theory
XI, 156 pages, 1987.

Vol. 89: G. K. H. Pang, A. G. J. MacFarlane
An Expert Systems Approach to Computer-Aided Design of Multivariable Systems
XII, 325 pages, 1987.

Vol. 90: Singular Perturbations and Asymptotic Analysis in Control Systems
Edited by P. Kokotovic, A. Bensoussan, G. Blankenship
VI, 419 pages, 1987.

Vol. 91 Stochastic Modelling and Filtering
Proceedings of the IFIP-WG 7/1 Working Conference
Rome, Italy, Decembre 10-14, 1984
Edited by A. Germani
IV, 209 pages, 1987.

Vol. 92: L. T. Grujić, A. A. Martynyuk, M. Ribbens-Pavella
Large-Scale Systems Stability Under Structural and Singular Perturbations
XV, 366 pages, 1987.